MEASUREMENTS, MECHANISMS, AND MODELS OF HEAT TRANSPORT

MEASUREMENTS, MECHANISMS, AND MODELS OF HEAT TRANSPORT

ANNE M. HOFMEISTER
Washington University, St. Louis, MO, United States

Elsevier
Radarweg 29, PO Box 211, 1000 AE Amsterdam, Netherlands
The Boulevard, Langford Lane, Kidlington, Oxford OX5 1GB, United Kingdom
50 Hampshire Street, 5th Floor, Cambridge, MA 02139, United States

British Library Cataloguing-in-Publication Data
A catalogue record for this book is available from the British Library

Library of Congress Cataloging-in-Publication Data
A catalog record for this book is available from the Library of Congress

ISBN: 978-0-12-809981-0

Publisher: Candice Janco
Acquisition Editor: Marisa LaFleur
Editorial Project Manager: Katerina Zaliva
Production Project Manager: Bharatwaj Varatharajan
Cover Designer: Greg Harris

Typeset by MPS Limited, Chennai, India

Contents

6. Transport Behavior of Common, Pourable Liquids: Evidence for Mechanisms Other Than Collisions

7. Thermal Diffusivity Data on Nonmetallic Crystalline Solids from Laser-Flash Analysis

8. A Macroscopic Model of Blackbody Emissions With Implications

ANNE M. HOFMEISTER, EVERETT M. CRISS AND ROBERT E. CRISS

9. Transport Properties of Metals, Alloys and Their Melts From LFA Measurements

E.M. CRISS AND A.M. HOFMEISTER

10. Heat and Mass Transfer in Glassy and Molten Silicates

ALAN G. WHITTINGTON

11. Modeling Diffusion of Heat in Solids

12. Conclusions and Future Work

Preface

Heat transfer is a nonequilibrium process that is important on scales from interatomic to intergalactic, and therefore is relevant to a myriad of scientific problems. For example, heat transfer must be understood to design electronic devices that do not overheat, to decipher the thermal state and evolution of the Earth and its sister rocky planets, and to probe the beginning of the universe. Behavior of large objects, such as the Earth, are complex because many length scales, diverse materials, and very short to very long times are involved. Misunderstandings exist because conventional contact methods of thermal conductivity have been the lab mainstay for ~200 years, but incorporate systematic errors of opposing signs, and are collected under conditions resembling equilibrium, yet are distinct from equilibrium. Interestingly, the pioneering heat transfer studies focused on common building blocks of the Earth, since minerals were then used as technological materials. Even today, many synthetic materials are related to common rock-forming minerals, for example, silica-rich glasses. Thus, faulty heat transfer measurements underlie many studies in engineering and materials science. Above all, disequilibrium conditions are required for heat transfer, yet the equilibrium approximation is invoked, which greatly limits the applicability of existing theoretical models. For these reasons, this book concentrates on extracting the basic principles of heat transfer from reliable, time-dependent measurements under disequilibrium conditions.

This book begins with a nontraditional approach that combines thermodynamics and spectroscopy, because the well-known connection between heat and light in space has been greatly underutilized in probing heat transfer. Fundamental issues remain unresolved even though quantitative investigations of heat transfer began with Fourier long ago. A key problem is that all microscopic models are based on the kinetic theory of gas, which describes heat transfer in terms of elastic collisions of hard spheres. Atoms are not hard spheres and collisions must be inelastic for temperature to evolve: thus, the conventional depiction of microscopic behavior cannot possibly be accurate.

How can something so basic have been missed? This is largely due to out-of-sequence developments in thermodynamics, heat transfer, spectroscopy, and atomic models. An additional problem is that temperature is actually measured, whereas heat energy is modeled. Furthermore, many important developments occurred long ago, and a lack of reliable data at critical times prevented coherence between the developing theories of heat transfer and spectroscopy. The historical timeline also led to radiative transfer being greatly misunderstood, and improperly implemented on diverse scales, impacting models on objects ranging in size from tiny electronic circuitry to models of enormous nebulae undergoing collapse into planets, stars, and spiral galaxies.

Although thermal conductivity has been measured since ~1800, two significant advances have occurred relatively recently. One was the 1961 development of the flash technique by Parker and his colleagues, which permitted accurate measurement of thermal diffusivity as a function of time. Thermal diffusivity governs thermal evolution under non-equilibrium and near-equilibrium conditions. Second, the problematic circumstance where high-frequency light (heat) crosses a sample without diffusing was remediated in the late 1990s by use of coatings and improved models. Authors include Blumm, Degiovanni, Fricke, Hoffman and Mehling, see Chapter 4. The findings in this book are derived from data recently collected with these superior techniques and improved data collection and analysis on a wide variety of minerals, ceramics, rocks, metals, glasses, and melts.

The present book investigates the physical process of heat transfer in the lab (i.e., over $\sim$mm scales), which must be understood before more complicated entities, such as the Earth, can be fruitfully investigated. A variety of behaviors exist in the measurements because heat transport accompanies motions of many different entities. The character of these motions can vary considerably even on a lab scale, and can involve light of various frequencies, vibrating atomic bonds, tiny electrons, molecules of any size, stretching of polymers, and/or the flow of polymerized silicate melts. Many of these processes also involve simultaneous transport of mass. Due to the multitude of possible processes and the virtually limitless range in length and time scales, heat transport is important to most processes in the universe. Because heat transfer is so ubiquitous, it is studied in many disciplines, predominantly pure physics, applied physics, and engineering. Hence, much information exists. But at the same time, many key phenomena remain unexplained or incorrectly explained.

The goals of this book are to present a unified approach to heat transfer in condensed matter that is congruent with all available data and consistent with fundamental physical principles. Understanding the behavior on the microscopic scale from lab measurements is the subject in this volume. This improved understanding will then be applied to the much larger scales associated with interiors of the planets in a companion volume that focuses on the Earth.

Several chapters were contributed by colleagues or were greatly improved by their collaborations or criticisms, making coverage of a broad range of topics possible. In particular, the quality and clarity of the book were substantially improved by the critical reviews of Bob Criss and Everett Criss, who contributed many ideas, crucial discussion, and participated in the key chapters. Bob did far more than his fair share of editing. Graduate students Derick Roy and Jesse Merriman also provided helpful reviews. Juergen Blumm and Alan Whittington commented on Chapter 4. Undergraduate student Jesse Shi formatted references, compiled data on gases and liquids, and helped assemble the large tables in Chapter 7 and Chapter 10. Help from librarians Clara McLeod, Ryan Wallace, and Alison Verbeck was essential in locating historic references. Many websites are referenced per the suggestion of Genevieve Criss, a high school chemistry teacher.

Preparation of this book was supported by my active NSF grants EAR-1321857 and 1524495. However, the content of the are those of the authors, but not necessarily of NSF. This work represents the culmination of almost two decades of work in heat transfer, and a longer career in spectroscopy, both of which were supported by the Packard Foundation, NSF, and to a lesser extent NASA, as well as U.C. Davis and Washington U.

Much appreciation is also due to the staff at Elsevier for their guidance, suggestions, and patience. I thank Louisa Hutchins, the acquisition editor, for inviting me to contribute this book and for obtaining helpful reviews. Marisa LaFleur played a key role in the early and late stages of book development. The high quality of the printed book rests on the concerted efforts of Bharatwaj Varatharajan and his staff in production. Special thanks are due to the developmental editor, Katerina Zaliva, for her encouragement and substantial efforts in bringing the book to the finish line.

Anne M. Hofmeister

St. Louis, MO, United States

August 28, 2018

CHAPTER

1

The Macroscopic Picture of Heat Retained and Heat Emitted: Thermodynamics and its Historical Development

Anne M. Hofmeister and Robert E. Criss

OUTLINE

Measurements, Mechanisms, and Models of Heat Transport
DOI: https://doi.org/10.1016/B978-0-12-809981-0.00001-2

The progress of science ... is not wholly unlike a pack of hounds, which in the long run perhaps catches its game, but ... the louder voiced bring many to follow them nearly as often in a wrong path as in a right one; where the entire pack even has been known to move off bodily on a false scent. ***Samuel Pierpont Langley (1889)***

You can't unfry an egg, but there is no law against thinking about it. ***Don Herold (1953)***

The macroscopic perspective is always the starting point in scientific endeavors because this consists of a simple description of things that we can measure or sense. The macroscopic picture is inseparable from laboratory measurements. Importantly, no special assumptions need to be made about the nature of matter and only a relatively small number of quantities need to be considered. Much of the material on macroscopic, classical thermodynamics presented in this chapter will be familiar to the reader, but our approach diverges from most presentations by drawing attention to flaws and inconsistencies in historic analyses. Although modern studies have remediated many of the incorrect historic analyses, misconceptions remain and some current paradigms in heat transfer, geophysics, planetary science, and astronomy are underlain by historic errors.

Heat is energy, the nature of which is long debated. Idealized, *time-independent* transfer of heat (and its tie to work) is the basis of thermodynamic laws. The equilibrium, conservative, static, and idealized processes covered in this chapter greatly differ from the disequilibrium, diffusive, and dissipative conditions causing real heat flow, which is *time-dependent*: the macroscopic picture involving time is covered in Chapter 3. We also need to understand how energy in its purest form (light) moves and how it interacts with material: basic material is given in Section 1.3 and is developed further in Chapter 2. These three chapters provide a self-contained statement of the classical physics underpinnings, which is predominantly macroscopic.

The field of physics developed through explaining experimental data: this approach is used to understand heat transfer in this book. Much of current understanding of thermodynamics and heat is based on (1) analogs between thermal and mechanical behavior and (2) experiments and models for gas behavior, which are covered in many textbooks on thermodynamics (e.g., Zemansky and Dittman, 1981; Richet, 2001). The present book covers gas behavior where it differs from solid behavior: the difference can be crucial. We consider behavior of mass, because mass diffusion is similar to heat diffusion, but easier to understand in many ways. Although this chapter does not quantify time-evolution, the fact that time is involved in any process needs to be discussed. Variables and abbreviations are defined when introduced, and are compiled in Appendix A for convenience of the reader.

Section 1.1 explores our understanding of heat and conservation laws from classical thermodynamics. Section 1.2 concerns entropy and its connection to heat and space occupied. The role of space has been short-changed in classical discussions, because steam engines were the key technological problem of the 1800s. Section 1.3 covers emissions from hot bodies, emphasizing that acceptance of visible light and heat emissions being the same phenomenon postdates classical thermodynamics. The out-of-synch developments explain why radiative transfer was not incorporated in historical and seminal works on thermodynamics. Inadequately considering volume, overemphasizing reversibility, and omitting radiative loss have consequences for problems on diverse scales.

1.1 ENERGY AND THE FIRST LAW OF THERMODYNAMICS

Physics rests on universal principles. Among the most important are conservation laws, which were developed over many centuries, and are essential to understand heat and its transport. To discuss conservation, it is necessary to differentiate between a *system* (what we are interested in) and its *surroundings* (which we tend to ignore, but should not). We are primarily interested in *closed* systems, which exchange energy but not matter with the surroundings. *Isolated* systems, which exchange neither matter nor energy, and *open* systems, which exchange both, are relevant in some circumstances. These classifications exist because the behavior of the surroundings is important. These various exchanges, as noted above, are considered in terms of the initial and final states, and thus are commonly treated as being more-or-less instantaneous (Fig. 1.1). Great care must be taken in considering thought experiments, as there is always the danger of "unfrying the egg" in our imagination.

Conservation of mass was recognized in the ancient world, but how this was maintained during combustion (i.e., heat production during burning) was not established scientifically until 1786–1789 by Antoine Lavoisier, who proved that heat was weightless. Conservation of linear momentum, and also of angular momentum, originated in Isaac Newton's third law of 1687. Newton and many others did not use the term *energy*, but rather force balance was applied (Todhunter, 1873). An exception was Gottfried Leibnitz, who considered the energy of motion, which he defined as twice the kinetic energy and named *vis viva* (living force) in ~1695. Because the forces considered by Newton do not produce heat and do not consume mass, mechanical energy was implicitly conserved, and so Liebnitz's approach did not gain favor at that time. Such forces are now labeled *conservative*. Under this restriction, conservative potentials were explored prior to 1800, by Newton, Colin Maclaurin, and many other notables.

The concept of energy and its manifestation in various forms and conservation thereof became important and accepted through the development of thermodynamics. For details on the history of this field and related developments, see for example, Partington (1960), Segrè (1984), Purrington (1997), Müller and Müller (2009), or Fegley (2015). Thermodynamics stems from efforts after ~1800 to unify the phenomenon of heat transfer with conservative descriptions of motion. This combination is actually an oxymoron, but was required at that time to advance understanding. This contradiction underlies microscopic models of heat transport, but the consequences have mostly gone unnoticed (Chapter 5). That heat is produced during mechanical work was recognized by Benjamin Thompson (Count Rumford) near 1798. The connection between heat and work was quantified and extended by J.R. Mayer, Hermann Helmholtz, and James Joule from 1840 to 1860. This effort culminated in the descriptions of energy conservation by Helmholtz and Joule in 1847 and 1850, respectively, known as the first law of thermodynamics:

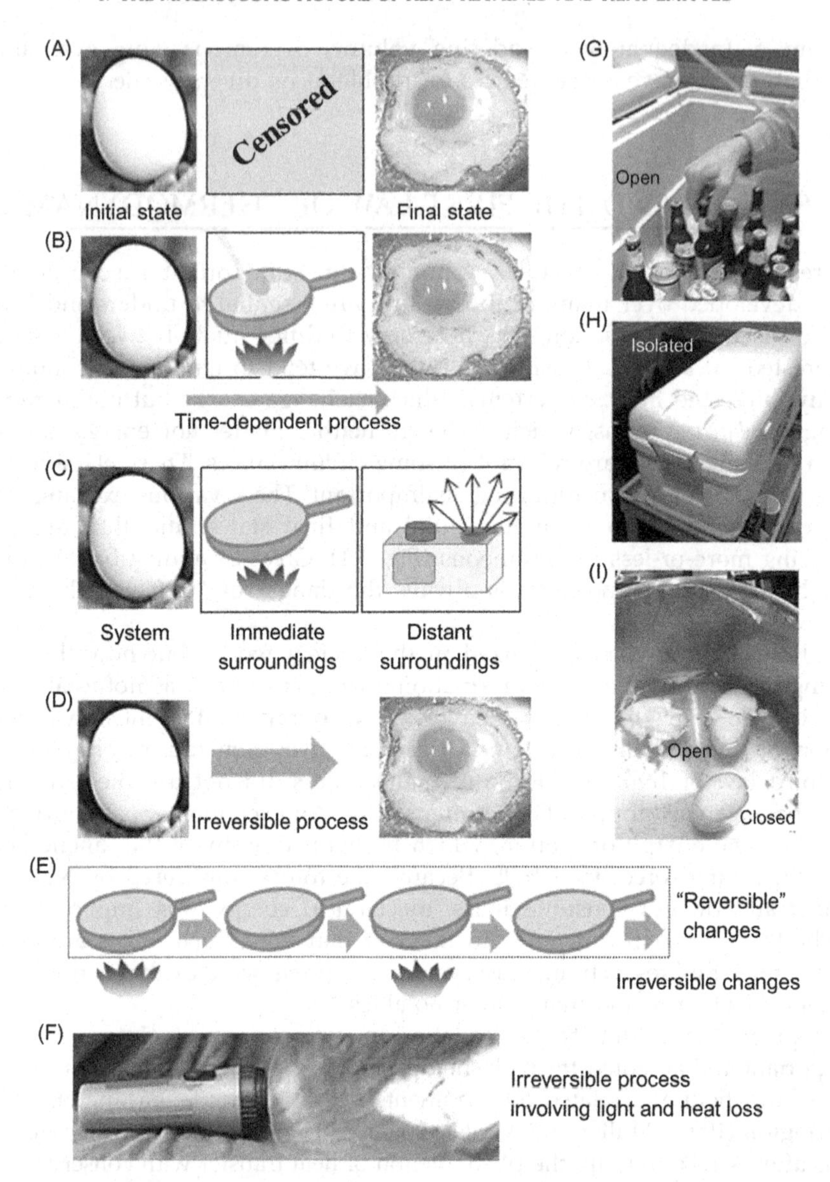

FIGURE 1.1 Schematics illustrating thermodynamic analysis of a heat-transfer problem. (A) Thermodynamic idealization of frying an egg, where only the initial and final states are considered. (B) Frying an egg, illustrating the initial and final states, and the process. (C) Comparison of the system (the egg) and its surroundings, showing the importance of the distance heat can travel to the analysis. Arrows indicate rays of radiant energy. (D) An irreversible process. Because time does not run backwards and heat is always dissipated, the real-world is irreversible. (E) Idealized, reversible process where a pan is considered as a system. If the pan is not damaged by excessive heat or by reaction with the food being cooked, it can be used for cooking many times. However, the changes in the surroundings (the burner providing heat, and the kitchen receiving heat) are irreversible. (F) Emissions from a flashlight to the surroundings, illustrating irreversible loss of both heat and light. (G) Open system, where matter and energy are freely exchanged. (H) Isolated systems are idealizations, as schematized by the contents of a closed cooler. (I) Closed systems are reasonable approximations to many processes, for example, hard-boiling an egg. However, if the eggshell breaks, the system is open.

$$dE = \delta q - \delta w = dq - dw \tag{1.1}$$

where E is the internal energy of the system, q is heat added, and w is work done. This book uses d for a total differential, which in thermodynamics represents an infinitesimal increment and thus leads to the total derivative, ∂ for a partial derivative, δ for a small change and Δ for a large change.

Rejection of their manuscripts for publication in scientific journals lead Joule to utilize a relatives' connection with a newspaper and Helmholtz to circulate a private pamphlet. Mayer's more specific statements were published in a scientific journal via begging the editor. Nonetheless, the first law was accepted shortly thereafter, leading to W. G. McQ. Rankine's proposal of potential energy (see Clausius, 1879, who preferred the name "ergal").

The thermodynamic statement of energy conservation initially met with opposition because prior to the development of Eq. (1.1), the prevailing view was that heat is a substance, and so by analogy to mass conservation, heat should be conserved independent of work (w = force times distance). Specifically, heat was considered a weightless, invisible fluid (the *caloric*, a term introduced by Guyton de Morveau in 1787 to signify heat as a measurable quantity), as discussed further below and in Chapter 3. Importantly, thermodynamics concerns the initial and final states, and thus deals with net transfers of heat, without considering time. Thus, thermodynamic analyses usually ignore heat conduction, radiative transfer, shear stress, temperature and/or pressure gradients, and gravitation. These omissions foretell problems not only in applying classical concepts to processes on planetary or astronomical scales, but moreover suggest flaws exist in the classical concepts themselves. Heat transfer is a dissipative phenomenon and therefore can neither be conservative nor reversible.

Finally, we emphasize that thermodynamics was developed in the 1800s largely by considering gases. These are the simplest state of matter, because interactions between the well-separated molecules of gases are miniscule, making them amenable to theoretical analysis. Applying the results for gases to condensed matter contains pitfalls, particularly as regards heat and heat flow (Section 1.2, Chapter 5). Here, we mention that flow of heat always involves the translation of gas molecules: these processes are inseparable and identical in a gas, whereas in a solid, the flow of heat occurs independently of the flow of matter.

For a detailed expose on the effect of time in thermodynamic studies through considering diverse applications and examples, see Müller and Müller (2009). Regarding the literature on nonequilibrium thermodynamics, a starting point is the seminal work of Prigogine (1967). Chapter 3 concerns a macroscopic approach to time in thermodynamics.

1.1.1 Conversion of Mass to Energy

Understanding how mass (M) converts to energy in nuclear reactions and relativistic behaviors in the 1900s ushered in modern physics. This book concerns nonrelativistic situations where Newton's laws hold. Regarding nuclear reactions, which are an important source of heat inside planets, we can reasonably neglect the associated mass loss, as follows. In planetary materials, significant heat is generated by the radioactive decay of U, Th, and K^{40}. Estimated proportions in the bulk Earth (van Schmus, 1995) are only

0.02, 0.07, and 0.02 ppm, respectively. Only a tiny fraction of the atomic mass is lost during any one nuclear disintegration, and a very small fraction of these rare species decay over long time intervals. Far below a billionth of Earth's mass has been lost over geologic history. Hence, this book does not consider nuclear reactions. Neither are relativistic phenomena germane, and hence mass and energy are treated below as being separately conserved.

1.1.2 Temperature Differs from Heat

In the caloric model (heat as an imponderable fluid), heat and temperature (T) are considered as the same phenomenon. Near 1800, Joseph Black distinguished sensible heat, which raised temperature, from latent heat, which did not. A well-known example of the latter is the heat of fusion, where the energy applied goes into melting the material, rather than raising its temperature (Section 1.1.4). Temperature is one of the four state variables because it greatly affects physical properties. Its importance to heat flow is huge because temperature differences govern the direction and commonly the rate of heat transfer (Section 1.2; Chapter 3).

In typical laboratory settings, temperature is externally controlled, in which case it is an *independent* variable. In addition, T is an *intensive variable,* because it does not depend on the size of the system in question. In great contrast, heat is an extensive variable because its amount depends directly on system size. *Extensive* variables sum, but intensive variables do not. Appendix A indicates whether a variable is intensive or extensive. Some intensive variables are ratios of two extensive variables: a familiar example is density in laboratory measurements: $\rho = \text{mass/volume}$, which is a physical property of the material. Other examples are hardness and the index of refraction. Notably, mass tends to be ignored in thermodynamic problems because it is conserved. However, mass is important because it takes up space, which requires energy. Additionally, whether a variable is intrinsic or extrinsic is determined by its relationship to mass. As occurs for the temperature–heat pair, mass is distinct from, but related to, space. However, as discussed below, these concepts have led to confusion and should be replaced by the more robust categories of independent vs dependent variables, which was developed later.

Laboratory experiments are the basis for thermodynamic laws. Thermodynamics of planets are more complicated because interior temperatures are affected by internally generated heat and self-gravitation. Understanding heat and heat flow in the laboratory and over microscopic scales is a prerequisite to addressing such problems which involve large length and time scales.

1.1.3 Volume, Pressure, and the Equation of State

Volume (V), like temperature, is another state variable. Its importance is evidenced in its connection with system size, which controls the quantity of heat and other extensive variables. Volume is quantitatively linked to temperature, pressure (P) which is the third state variable, and mass, with the latter being independent of T and P. For problems addressed in this book, mass is conserved, and thus is used to ascertain whether a variable is intensive or extensive.

The interrelationships of the state variables V, P, and T for constant mass are known as the *equation of state* (EOS). In the laboratory, P and T are typically controlled, and so the response of V to changes in P and T is monitored. Measurements are commonly reported as the material properties of thermal expansivity (α) and compressibility (β), although the latter is commonly represented as a reciprocal called the bulk modulus (B). These important material properties are:

$$\alpha = \frac{1}{V}\frac{\partial V}{\partial T}\bigg|_P; \quad \beta = -\frac{1}{V}\frac{\partial V}{\partial P}\bigg|_T = \frac{1}{B_T}, \tag{1.2}$$

where the subscripts indicate the variable being held constant. The connection of bulk moduli with acoustic velocities is extremely useful in geophysics. Cross-derivatives also exist, and experiments may be performed under other conditions, such as at constant V in a sealed container, or under the controlled condition of no heat flow. To be succinct, general, and correct, the EOS is represented by some function f:

$$f(V,P,T) = 0 \tag{1.3}$$

In most laboratory situations, pressure is externally imposed and does not depend on the size (mass) of the system in question. For this reason, much literature considers P to be an intensive quantity (Richet, 2001). However, this is not generally true, as is evident from three examples.

1. For a container filled with two or more different gases, each gas provides a partial pressure, which sum to provide P of the mixture. The partial pressures are related to the fraction of each gas. In contrast, each gas in the container has the same temperature (presuming equilibrium) which is equal to that of the mixture. In this example, partial pressure is extensive and sums, while temperature is intensive and does not sum. Regarding volume, the relevant volume is the whole container. Volume is the independent and key variable.
2. Section 1.2.6 discusses the important free-expansion thought experiments. As in the mixture, volume is not defined by the mass, but by the walls of the container, and is thus independent and fundmental.
3. For a large, self-gravitating object, pressure at the center depends on the object mass. Inside the body, P varies from this central value to zero at the surface (for an airless body). Thus, P can be considered an extensive quantity, but the mathematical role is that of a dependent variable. The behavior of real bodies is complicated by phase transitions and stratification of multiple types of materials, in accord with density. For an ideal body made of one material with no phase transitions, the volume depends on Newton's law of gravitation and the compressibility of the material. One can argue that the exterior volume depends also on the mass and is thus extensive. However, to describe the interior, the distance from the center is the independent quantity, and therefore, volume is the independent state variable, and is the key variable, along with temperature.

Free expansion of the ideal gas has been incorrectly analyzed (Section 1.2.6). The error stems from viewing pressure as intensive and from omission of volume considerations in early thermodynamic analyses. Consequently, the stability criterion for a self-gravitating nebula, where internal pressure is governed by the superjacent mass under the force of gravity, has been misunderstood (e.g., Bonner, 1956; see discussion by Hofmeister and Criss, 2012). Crucially, because gravity depends on radius through Newton's laws, V is one of the independent variables. The other is T, which provides thermal agitation and competes against the crushing force of gravity. For this case, P should be cast as a function of the independent variables V and T (e.g., Müller, 2001).

Extensive and intensive variables were useful concepts historically and when considering many laboratory experiments, but not free expansion, and are particularly muddied when considering large, complex systems. Independent vs. dependent is instead the crucial distinction for thermodynamic analysis: see Hofmeister and Criss (in review) for further discussion.

1.1.4 Heat Capacity, Enthalpy, and Latent Heat

How much heat is held by a system limits how much heat can possibly flow. Holdings are quantified by the *heat capacity* (C), which is defined in terms of the amount of heat needed to provide a temperature change of a system, that is, $C = \delta q/\delta T$ (where we use a capital letter to indicate a per mole or per volume basis). We use lower case c to designate heat capacity on a per mass basis. The latter case is sufficiently important to denote this quantity as *specific heat*. For any representation, heat capacity values obtained in the laboratory depend on the conditions of measurement. Constant pressure describes most laboratory experiments, whereas constant volume is more useful in theoretical studies, and particularly for microscopic descriptions (see below and Chapter 5). The difference exists because materials expand upon heating, and whereas expansion is small for a solid, reaching ~3% as T increases to 1000K, it is significant for a gas. A fundamental property is the heat capacity at constant volume:

$$C_V = \left(\frac{\partial E}{\partial T}\right)\bigg|_V \tag{1.4a}$$

But, the most commonly measured property is heat capacity at constant pressure:

$$C_P = \left(\frac{\partial H}{\partial T}\right)\bigg|_P \tag{1.4b}$$

where H is enthalpy and is defined as $H = E + PV$. From Eq. (1.1), the change in enthalpy equals the change in the heat content under constant pressure (∂q_P).

Major changes in the material or system (such as reactions, phase transitions, or combustion) are manifest in the heat capacity and in the enthalpy, simply because energy is exchanged in making the transformation. For example, a specific amount of heat-energy is required to melt a solid, which is then stored in the liquid (denoted latent heat). When the liquid cools, and refreezes, this stored heat is released. The description for heat evolved (*exothermic*) or absorbed (*endothermic*) during a structural transition or reaction is the *heat of reaction*, which we designate as ΔH_{react}. Note that Δ refers to the difference between the final and initial state, and so the sign of this transformational heat is defined by the direction of reaction (e.g., the heat needed to melt a material is called the heat of fusion and is positive). Whether P or V was held constant during the transformation should be specified, but this is often omitted or implied. An illustrative example is a combustion reaction in a sealed container.

1.1.5 Types of Work

Many types of work exist. Because the goal is subsequent application of our findings to planets, we consider pressure–volume work, mechanical work (including when heat-producing friction is involved), and gravitational work.

Work always involves at least one extensive variable (e.g., $P\partial V$ terms which describe externally applied pressure) and is therefore extensive, along with internal energy, and enthalpy, as required by Eq. (1.1). If P is indeed externally applied, then it is the independent variable. Because the tie of work to heat is of central importance to this book, we next consider an important, nonconservative example.

1.1.6 Understanding Frictional Heating Requires the First Law

Effects of frictional heating have been misunderstood in many textbook discussions of blocks sliding downhill and in planetary science applications. The importance of this problem motivated Sherwood and Bernard (1984) to discuss the generation of frictional heat in mechanical systems in detail. By applying the first law to the classic example of a block sliding downhill, they demonstrated that friction heats both the block and the hill, in equal amounts. A plethora of evidence exists for this partitioning, including Rumford's cannon-boring experiments (Fig. 1.2).

In most geophysical analyses of the heat of formation of Earth's core, frictional heat is presumed to be associated with segregated iron particles moving downwards, after Birch's (1965) analysis. The erroneous but oft-stated view that all of the gravitational energy change was sequestered in Earth's core (e.g., Stacey and Stacey, 1999) is based on the familiar, purely mechanical integral of Newton's second law that is presented in many textbooks. From Sherwood and Bernard (1984), thermodynamics must be satisfied in addition to conservation of energy and momentum.

From first law considerations, the sinking and rising particles that bring about core formation, can each receive only half of the frictional heat generated during their journey. At most, half of the heat possibly evolved in core formation can be cast into the core region. This is a stringent limit, based on the first law and also presumes that sliding, not rolling occurs (Fig. 1.2) and that all heat is retained. Instead, the Earth loses heat to space over time: that is, the system is not isolated. For further discussion, see Hofmeister and Criss (2015).

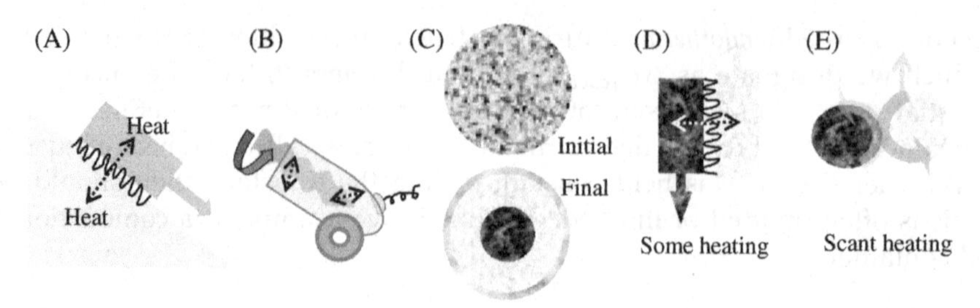

FIGURE 1.2 Schematics illustrating that heat-transfer associated with friction is a complicated process. (A) A block sliding down a hill. Heat is generated at the interface and enters both the block and the hill, by diffusion. The amount generated is controlled by the first law, but the process involves a rate. (B) Count Rumford's cannon drilling experiments. More conveniently, one could simply drill a hole in a metal block: heating of both the drill bit and the metal block can be easily confirmed. (C) Core formation in the Earth. The initial and final states are incompletely characterized regarding temperature, and a homogeneous initial Earth may never have existed, but the entire change in gravitational potential, has been considered to heat the Earth, after Birch (1965). Both radiative transfer and the negative change in potential energy prohibit retention of all energy (Hofmeister and Criss, 2015). (D) Grain sliding mechanism. Heat is generated at the interface, but the coefficient of friction is always below unity, so the process cannot change all kinetic energy into heat, unless the friction halts the motion after some distance traveled. Were friction truly important, formation of a core would have been impeded. (E) Grain rolling mechanism, with negligible heat production. This concept underlies the operational principle of bearings.

1.2 ENTROPY AND MORE LAWS OF THERMODYNAMICS

The second law, which preceded the discovery of the fourth and final state variable *entropy* (S), was founded on Nicolas Carnot's 1824 book discussing the operation of idealized, reversible steam engines. Because entropy is conceptually difficult and easily misunderstood, we first focus on this law.

The significance of Carnot's work was recognized by Rudolf Clausius and William Thomson (Lord Kelvin), and led to their statements in 1850 and 1853, respectively, that:

- No process is possible whose sole result is the transfer of heat from a colder to a hotter body.
- No cyclic process is possible whose sole result is the absorption of heat from a reservoir and the conversion of this heat into work.

The latter verbalization is a paraphrase by Zemansky and Dittman (1981) of Kelvin's statement of the second law, combined with that of Max Planck, which was provided in 1879. A similar statement was made by James Clerk Maxwell who considered the mechanical work performed in order to forcibly move heat in the "wrong" direction (refrigeration). Each of the above presentations prohibit perpetual motion machines, as does the first law. These statements of the second law require that machines known as refrigerators shed heat to an external reservoir (the kitchen), in order to cool a different reservoir (a box with perishable food) through work. The aptness of this description of heat transfer is easily confirmed.

Carnot's discussion also led Kelvin to derive his *absolute temperature scale* in 1848 and 1852 papers, based on measurements of M. Regnault in 1847. However, this attribution overlooks the unpublished work of Jacques Charles in the 1780's and further investigations of gas by Louis Gay-Lussac near 1800 which pointed to a zero-point volume. Absolute zero can be straightforwardly determined from the ideal gas law, which rests on these efforts, and those of Boyle.

Clausius developed the concept of entropy over 1854 to 1865 by examining the contrasting behavior of irreversible and reversible processes. The subsections to follow delve into this contrasting behavior which is linked to heat transfer.

1.2.1 Reversibility, Thermal Equilibrium, and the Zeroth Law

All natural, spontaneous processes are irreversible. Irreversible processes either do not satisfy thermal equilibrium or involve dissipation. The latter phenomenon can involve either heat or concentrations of matter, or some combination thereof. Diffusion of matter is exemplified by isothermal expansion of a gas or by gas mixing (Section 1.2.6). Thermally dissipative processes include friction, viscous drag, inelastic deformation or collisions, electrical resistance, simple cooling, and importantly, radiative transfer (Fig. 1.1F). The latter is often overlooked, because the irreversible changes to the surroundings are typically ignored, following the historical approach. Simple cooling is explored in detail, because it concerns heat alone (Section 1.2.3). In a thermally dissipative process, heat is lost to the surroundings while motions in the system are degraded (such as Brownian motions of molecules). Because heat flow has a specific direction, dissipative processes can only operate in one direction. Importantly, dissipative processes are time-dependent and thus, can only be treated approximately in the classical thermodynamic formulation of Fig. 1.1. In actuality, heat flow must be incorporated in realistic thermodynamic models, not just heat transfer in an impossible static system.

Communicability of thermal equilibrium between systems, which was demonstrated experimentally long ago, and traces to one of Euclid's axioms, is known as the zeroth law of thermodynamics:

- If body A is in equilibrium with body B, and body B is in equilibrium with body C, then body A is also in equilibrium with body C.

This statement was set forth as a postulate by Fowler and Guggenheim (1939). If the bodies in thermal equilibrium are touching, they must have equal temperatures. Situations without physical contact that involve radiative transfer need some additional discussion (Section 1.3).

Reversible processes cannot be dissipative and must occur in thermal equilibrium, and thus are unattainable idealizations. Processes that can be represented as a series of steps in equilibrium are traditionally treated as reversible. This viewpoint serves as a starting point for discussing the second law, following Clausius, but is problematic. Restoring a system actually involves irreversible changes in the surroundings. The traditional focus on the system, while ignoring the surroundings, is terribly flawed. Yet, this idealization permeates thermodynamics, as is evident below.

1.2.2 Isothermal vs Adiabatic

Idealized cases describing time-independent heat transfer (Fig. 1.3) serve as starting points for discussion of heat. Two different processes can be studied experimentally: these are tied to two different time behaviors. In an *isothermal* process, all parts are in thermal equilibrium. Processes such as perfect "isothermal compression" are idealizations that would require infinite time to complete. Traditionally, it is argued that very slow experiments approach this ideal (Olander, 2000). Slow experiments permit the system to equilibrate with the surroundings, and thus require a buffer which is generally ignored. A much different example is the ice bucket in Fig. 1.3A, where the contents are isothermal as long as some ice and some water remain in the container. From these examples, the behavior of the surroundings determines whether isothermal conditions are met, while the rate of change with time is of secondary importance.

Adiabatic systems are thermally insulated from their surroundings, prohibiting heat transfer between the system and its surroundings, even for the case of unequal temperatures. Avoiding radiative exchange with the surroundings (Section 1.3) is problematic, but this discussion will be postponed to later in the book. Adiabatic conditions are approached by using insulating materials and by conducting experiments over short times. Heat transfer occurs over finite times, at some rate that depends on how heat moves and what it moves through (examples are in Section 1.3). Heat transfer can be made negligibly small if the change in the system occurs instantaneously or nearly so, as in a rapidly rising helium balloon (Fig. 1.3B).

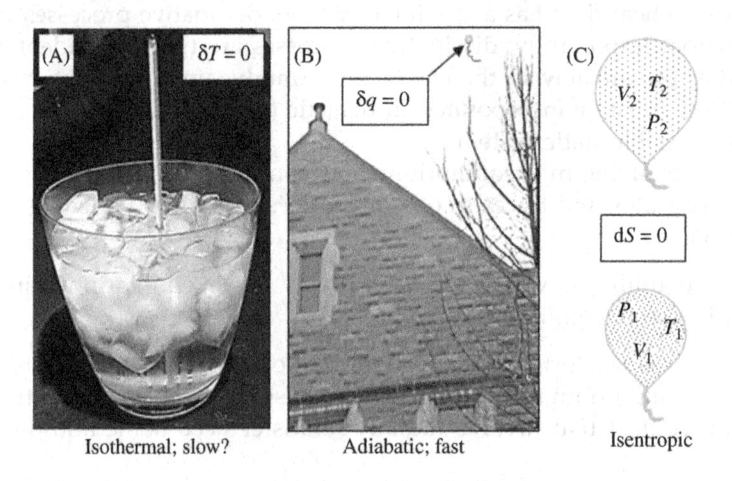

FIGURE 1.3 Schematics illustrating idealizations applied to thermodynamic analysis of heat-transfer problems. (A) Isothermal conditions are exemplified by an ice bath. As long as some ice remains, the temperature of the system is unchanged. (B) Adiabatic conditions are exemplified by the rise of a balloon. The arrow points to the balloon. If its travel is fast, no heat is exchanged with the surrounding atmosphere, even if the balloon is inadequately insulated. The balloon expands as the externally imposed, atmospheric pressure lessens with height. This process is irreversible: to retrieve the balloon requires a long string and expenditure of energy by a machine or person (the surroundings). (C) Conditions for an isentropic change for gas in a balloon. To maintain constant S during the change, we need to control changes in each of T, P, and V, which are constrained by the EOS and the internal energy. How isentropic and adiabatic processes are related is discussed below.

The adiabatic case for an ideal gas is important to thermodynamics and heat transfer. The internal energy depends only on temperature because the molecules are modeled as point masses (e.g., Pippard, 1974). Under this unrealizable idealization, setting $\delta q = 0$ in Eq. (1.1) relates pressure–volume work to a simple function of temperature: $\delta w = dE = f(T)$. The EOS of an ideal gas also relates P, V, and T in a simple way ($PV = NR_{gc}T$, where N is the number of moles and R_{gc} is the gas constant). Combining these relationships (e.g., Fegley, 2015) gives the conditions for an adiabatic change:

$$\frac{T_2}{T_1} = \left(\frac{P_2}{P_1}\right)^{\frac{\gamma-1}{\gamma}} = \left(\frac{V_1}{V_2}\right)^{\gamma-1} \tag{1.5}$$

where $\gamma = C_P/C_V$ is a dimensionless numerical constant that is equal to 5/3 for ideal monatomic gas and 7/5 for ideal diatomic gas.

1.2.3 Entropy Defined

By considering an idealized, reversible processes, Clausius defined the quantity *entropy* as:

$$dS = \frac{dq_{rev}}{T}. \tag{1.6}$$

It is now known that heat production is path dependent, so the quantity is rather an inexact small change, δq. However, Eq. (1.6) is the common starting point in textbook discussions of entropy and the second law, which has led to misunderstandings. The crux of the problem is as follows: In an idealized reversible cycle, heat is passed back and forth without loss. One way for no net losses to occur in a cycle, is for the work to be conservative, in which case, no heat transfer occurs. Obviously this is a problem, although $\delta q = 0$ means $dS = 0$ and conditions are indeed reversible (nothing happened). Alternatively, if the process is reversible, the system must either return to its initial temperature or its temperature did not change in any part of the cycle. For the former case, heat must in one part of the cycle move from a colder to a hotter body, which is impossible. That heat transfer is one-way underlies the second law, as presented in 1850. In the aforementioned case of constant T, no transfer of heat can occur. The situation is "reversible" only because nothing happened.

The picture presented of reversibility associated with Eq. (1.6) is that of Fig. 1.1E, where the system, but not the surroundings, are considered. The incremental changes invoked in textbook discussions do not fix the above problems, because any heat transfer is irreversible (one cannot put the light back into the flashlight, Fig. 1.1F). Steam engines, which are central to historical discussions of irreversibility, are extremely inefficient and involve copious transfer of heat to the surroundings.

Clausius originally considered the name "content of transformation." Interestingly, Rankine introduced a state function in 1854 that corresponded to the entropy function. He is not considered the inventor, because his discussion was limited to reversible cases (Kragh and Weininger, 1996).

Clausius coined the name entropy in 1865 when describing changes during an irreversible process. In comparing a realistic irreversible process to an idealized reversible process, he recognized that entropy change is incompletely described by Eq. (1.6). Clausius realized that some changes in a real system reduce its ability to do work on the surroundings. To address this loss of ability, which differs from adiabatic changes, Clausius arrived at the concept of "uncompensated heat:"

$$dS_{sys} = \frac{\delta q_{rev}}{T} + \frac{\delta q_{uncomp}}{T} \tag{1.7}$$

However, Eq. (1.7) is incomplete. This was derived considering the energy associated with heat, without understanding that occupying space also requires energy. Considering entropy without its volume dependence has created some misunderstandings, especially as regards self-gravitating bodies. How entropy depends on volume is covered in Section 1.2.6.

In completely general terms, any substance can be described in terms of three variables (P, V, T). How these interrelate is described by the EOS. The EOS (Eq. (1.3)) can always be cast as $P = f(V,T)$, an example being the popular Birch-Murnaghan equation. Thus, the energy of a substance will also be some function of V and T, and anything related to the energy, will be a function of V and T. Entropy, being tied to heat-energy, necessarily depends on V as well as T. Because two variables describe any characteristic of the system, the change in entropy is represented by:

$$dS = \left.\frac{\partial S}{\partial T}\right|_V dT + \left.\frac{\partial S}{\partial V}\right|_T dV \tag{1.8}$$

Eq. (1.8) is general, and can describe a substance or a system: reversibility is immaterial. It is also immaterial that the substance or system might be more conveniently represented as V depending on P and T, because in this case energy and entropy will also depend on P and T, and then dS still has two parts.

From Eq. (1.4a) and the first law, the heat change at constant volume (dq_V) equals $C_V dT$. From thermodynamic identities (the Jacobian relations, see Section 1.2.5.1; Bridgman, 1914; Fegley, 2015), $\partial S/\partial T|_V = C_V/T$. The terms to the right of the equal sign in Eqs. 1.7 and 1.8 are identical, and this equality is unconnected with reversibility. Therefore, the rightmost terms in Eqs. 1.7 and 1.8 must also be identical, and this equality is also independent of reversibility. Thus, the uncompensated heat describes changes in volume that were omitted by considering "reversible" heat transfer.

Because this "mortgaged energy" and the second law are notoriously difficult to understand, further discussion is warranted. We next explore changes involving heat transfer and temperature, which are obviously important in development of the concept of entropy. We then cover the relationship of the adiabat to the isentrope (Section 1.2.5.2) and the importance of volume changes to entropy (Section 1.2.6), both of which have been misunderstood.

1.2.3.1 Heat Transfer Between Two Bodies and the Second Law

The alternative verbalization of the second law provided by Clausius is:

- Die Entropie der Welt strebt einem maxiumu zu (The entropy of the universe is increasing).

The above scientific concept was hotly contested during Clausius' life (Kragh and Weininger, 1996). Moreover, extrapolation of his statement into a Weltanschauung caused a philosophical ruckus in the Victorian era (Lewis and Randall, 1923). In his 1879 book, Clausius devoted his Chapter 13 to defending his heat transfer verbalization of the second law given at the beginning of Section 1.2.3, but does not even mention the above statement. The need to minimize hostile reception should be evident.

To evaluate Clausius' contested statement, we now discuss the transfer of heat between two bodies (Fig. 1.4) where volume changes are negligible. If the two bodies constituting the system are perfectly isolated from the surroundings (Fig. 1.4A), then the entropy of the system increases, while no change occurs in the surroundings, and so the total entropy increases, which is the entropy of the universe, as indicated by the arrows below panel A. The stipulation of isolation is very important. If radiative transfer exists, then the 2-body system is no longer isolated and entropy is affected by communication of radiant heat to the surroundings. Panels B and C in Fig. 1.4 show end-member choices for the initial temperature of the surroundings, and how the heat would move in these two cases. Different formulations for the entropy transfer of the system will result for different temperatures of the surroundings. In all cases that describe heat transfer between the two blocks (Fig. 1.4A–C), the total S increases, but S of the surroundings can either increase or decrease depending on its starting temperature.

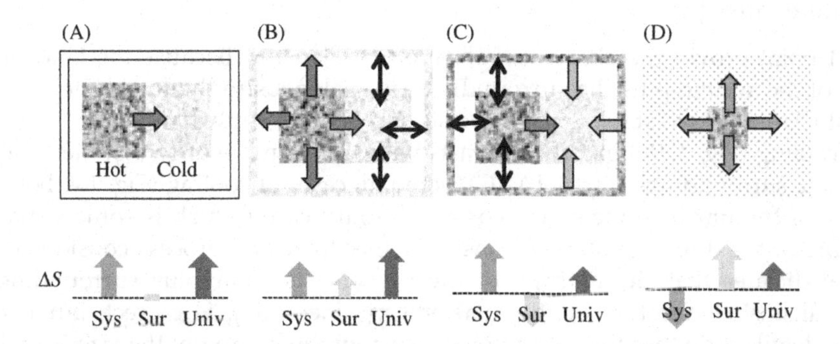

FIGURE 1.4 Schematics of cooling. Top row are the experiments whereas the bottom row shows the entropy changes for the system, surroundings, and their sum, which is the universe. (A) The thought experiment of Maxwell, where one hot plus one cold body (stipples) constitutes the system. Heat will move entirely from the hot body to only the cold body, if and only if no interaction exists with the surroundings (white box). Such an isolated system could be realized if the box is a perfect insulator and furthermore is a perfect reflector of heat and light, which is impossible (see Section 1.3; Chapter 2). (B) End-member case where the surroundings is initially in thermal equilibrium with the colder of the two bodies. In this case, heat is transferred from the hot body to both the cold body and the surroundings, so the entropy of the hot body actually decreases, but everything else increases by a greater collective amount. (C) End-member case where the surroundings is initially in equilibrium with the hot body. In this case, heat is transferred from both the surroundings and the hot body to the cold body. Here, the entropy of the surroundings and hot body decrease, but the total entropy still increases. (D) Simple cooling of a body to its surroundings. The total entropy increases, but that of the body (the system) decreases.

1.2.3.2 *Simple Cooling and the Second Law*

We now consider an even simpler example of irreversible cooling of a single body (the system) placed in an encapsulating medium (Fig. 1.4D), which constitutes the surroundings. For the body to cool, heat is transferred to the medium, which must be colder than the body even if ever so slightly. From conservation of energy, the amount of heat-energy lost by the body is received by the surroundings. For tiny differences, we can calculate entropy as an increment, with reference to the starting temperatures of $T_{sys} > T_{sur}$. Simple cooling provides:

$$dS_{total} = dS_{sys} + dS_{sur} = -\frac{\delta q}{T_{sys}} + \frac{\delta q}{T_{sur}} > 0 \tag{1.9}$$

The key result, $dS_{total} > 0$, holds for any spontaneous process.

Simple cooling (Fig. 1.4D) illustrates key aspects of the second law.

- The change in *total S* is positive, because the temperatures must be different for heat transfer to occur. In contrast, the quantity of heat exchanged (dq) is identical for both components, with the colder part gaining what the hotter part loses; hence, the difference in sign.
- Because the surroundings become warmer, its entropy increases, that is, the change dS_{sur} is positive. However, the surroundings may be huge and may not warm very much, because heat is an extensive quantity. Regardless of the magnitude, the surroundings cannot become warmer than the system. Heat transfer stops when the temperatures are equal.

The first point proves the above verbalization of Clausius, because the Universe, being composed of the system plus the surroundings, is analogous to the total. The second point proves that Clausius' two statements of the second law are equivalent.

Simple cooling (Fig. 1.4D) greatly contrasts with the example of two bodies that are isolated from the surroundings (Fig. 1.4A–C), yet in all cases of cooling (Fig. 1.4, bottom row), the entropy of the universe increases. For an arbitrary case (which is some variant of the cases explored), the direction of heat transfer is specific to the process considered, but the second law dictates that the entropy of the universe (system plus surroundings) must increase. Clausius' declaration of the total entropy increasing is correct, but it does not specify any details for either the system or the surroundings. Again, the sizes of the system and surroundings are important because heat and entropy are extensive quantities. Because of this fact, and due to the first law, and the zeroth law in view of radiative transfer (below), the surroundings must be considered, even though the changes are infinitesimal. By analogy, consider a ball thrown against a wall: momentum is conserved only if one considers the tiny motion of the massive wall. The same holds for energy in this example.

Confusion has existed about the possible behavior of entropy, due to a statement in Maxwell's book (1908, p. 163), that the "entropy of the *system* always increases," which he misattributed to Clausius. In contrast, Clausius did not mention the system, but rather explained that the entropy of the *universe* must increase. Maxwell considered the situation of Fig. 1.4A, but did not stipulate that the 2-body system was thermally isolated. During Maxwell's life, light and heat transfer were considered to be different things (ibid., p. 16).

In modern sources, thermal isolation of the system is associated with its entropy increasing. However, some important percepts such as how heat behaves during planetary accretion or core formation are based on the above historic misunderstanding of the second law.

As regards heat, it is important to recognize that Clausius' statement of entropy of the universe increasing presumes adiabatic conditions exist for the system plus surroundings, that is, energy cannot leave the universe. The equivalence of light and heat emissions (Section 1.3) means that when radiative transfer exists, which can cover immense distances within short time intervals, the size of the system needed to meet adiabatic conditions must be considered when employing the second law in its once hotly contested form.

1.2.4 Entropy, Heat, and the Third Law

The third law began as Walter Nernst's heat theorem of 1905–1907:

- As the temperature tends to zero, the magnitude of the entropy change in any reversible process tends to zero.

Support was provided in the experiments and theory of F.E. Simon over 1927–1930, but acceptance came much later, following W.F. Giauque's magnetic cooling experiments in 1947.

Alternative versions of the third law also originated with Nernst:

- By no finite series of processes is the absolute zero attainable.
- A body cannot be completely deprived of its heat.

Although behavior near 0K may not seem relevant to the planets, omission of this requirement from historic models of gravitational contraction has led to the thermodynamically unsupportable view that gravitational potential energy, which is negative, is converted to heat, which is a positive quantity, during contraction (Hofmeister and Criss, 2012).

1.2.5 State Functions, Variables, and Interrelationships

The important role of heat-energy in thermodynamics is evident in the four (extensive) state functions, as follows. Considering conservation laws leads to the internal energy (E) which reflects the energy contained in the atomic arrangements and motions, the latter of which contains a substantial thermal component. Summing $E + PV$ yields H, the enthalpy, which represents the total heat content of a system in a constant pressure environment, because some energy is needed for the system to occupy space. In more detail, E describes the energy associated with elevating the temperature of the system above 0K and the energy needed to occupy space, that is, by increasing V from the null value. For an ideal gas, E is solely thermal due to the kinetic energy of moving point masses describing this material. Because point masses do not actually occupy space, E of the ideal gas does not depend on P or V. As a consequence, the average molecular translation (i.e., the mean free path, discussed in later chapters) defines the space that this subset occupies. This key difference with solids foretells of problems in basing heat transport on the kinetic theory of gas (Chapter 5).

Recognition of entropy as a state variable leads to the last two state functions, which are important and commonly used. The Helmholtz free energy ($F = E\text{-}TS$) represents the combination of internal energy and configurational energy in terms of the most fundamental independent variables, volume (i.e., space) and temperature. Helmholtz developed his free energy in 1847, naming it after the German word for work (Arbeit) by considering heat, since the concept of entropy was developed later. J.W. Gibbs developed the last state function ($G = H\text{-}TS$) in 1875, referring to it as available energy. The common name "free energy" for G is unfortunate. Because ΔG excludes pressure–volume (PV) work, *free enthalpy* is the proper descriptor (e.g., Nordstrom and Munoz, 1986); but *Gibbs function* (Pippard, 1974) has precedence and is more recognizable. Notably, Gibb's recognized that the most important function is that of Helmholtz, whereby V and T are the independent variables. Gibbs considered the number of particles as being the third independent variable in his statistical analysis (see Purrington, 1997), whereas the related quantity of mass is considered in the present book, which is geared for application to planets.

Gibbs also developed the state function of enthalpy while referring to it as a heat function: the now popular name for H stems from the 1909 work of Heike Kamerlingh-Onnes. Internal energy was defined during the development of the first law. The state functions are sometimes referred to as thermopotentials, but this label is inconsistent with internal energy for an ideal gas being entirely kinetic energy, and with the internal energy of real gases being largely kinetic.

1.2.5.1 Interrelationships of Thermodynamic Properties and Data Needed to Address Problems

The four state functions (E, F, G, H) and four state variables (P, S, T, V) are interrelated. Details are provided in most thermodynamics textbooks (e.g., Anderson and Crerar, 1993).

In brief, ties are encapsulated in Maxwell's four relationships. The first is

$$\left.\frac{\partial T}{\partial V}\right|_S = -\left.\frac{\partial P}{\partial S}\right|_V \tag{1.10}$$

The remaining three relationships of Maxwell can be determined from the chain rule of derivatives

$$\left(\frac{\partial z}{\partial x}\right)_y \left(\frac{\partial x}{\partial y}\right)_z \left(\frac{\partial y}{\partial z}\right)_x = -1 \tag{1.11}$$

From Maxwell's relations and the definitions given above, each state function is associated with two *independent* variables. The appropriate pairings are given in Table 1.1.

These pairings are evident from the differentials, such as

$$dG = VdP - SdT; \quad dF = -SdT - PdV \tag{1.12}$$

Linking each state function with two independent variables means that the two remaining state variables must be *dependent* variables. If F is the appropriate function for the situation, then the dependent variables are P and S, as indicated in Table 1.1. These dependent variables depend on the independent variables, and hence when F is appropriate:

$$P = f(V, T) \quad \text{and} \quad S = f^{\dagger}(V, T) \quad \text{for} \quad F(V, T) \tag{1.13}$$

where the unspecified functions f and $f^{\dagger}$ depend on the system or substance of interest.

TABLE 1.1 State Functions

Symbol[a]	Popular Name	Preferred Name	Key Derivatives[b]	Functionality
$E(S,V)$	Internal energy	Internal energy	$\left.\frac{\partial E}{\partial V}\right\|_S = -P$; $\left.\frac{\partial E}{\partial S}\right\|_V = T$	$P(S,V)$; $T(S,V)$
$H(S,P)$	Enthalpy	Enthalpy	$\left.\frac{\partial H}{\partial P}\right\|_S = V$; $\left.\frac{\partial H}{\partial S}\right\|_P = T$	$V(S,P)$; $T(S,P)$
$F(T,V)$	Helmholtz free energy	Free energy	$\left.\frac{\partial F}{\partial V}\right\|_T = -P$; $\left.\frac{\partial F}{\partial T}\right\|_V = -S$	$P(T,V)$; $S(T,V)$
$G(T,P)$	Gibbs free energy	Gibbs function	$\left.\frac{\partial G}{\partial P}\right\|_T = V$; $\left.\frac{\partial G}{\partial T}\right\|_p = -S$	$V(T,P)$; $S(T,P)$

[a]*Optimal independent variables are indicated.*
[b]*These relations provide the dependent variables, as listed in the right most column.*

Note that whether a variable is intensive or not is unrelated to whether a variable is independent or not. In typical laboratory experiments where T and P are controlled, the optimal state function is G, whereas inside a planet, the optimal state function is F. This requirement stems from the interior pressure depending on planetary mass and therefore, P is an extensive quantity in self-gravitating systems. For statistical calculations, F is also optimal, as recognized by Gibbs, and as obviated in the well-known models of heat capacity by Debye and Einstein. The proper pairings of state functions with variables is connected with division of energy into quantities originating from the heat inventory and that needed to occupy space.

Importantly, using pairings other than those listed in Table 1.1 provide an incomplete description that is insufficient to address a thermodynamic problem (Pippard, 1974). For example, knowing $E(T,V)$ does provide a heat capacity (Eq. (1.4)), yet this information is insufficient to derive the EOS. Consequently, detailed independent information on $f(P,T,V)$ is essential.

An important consequence of the interrelationships between the state functions and variables is that the thermodynamics of a system is completely specified by two sets of data (Pippard, 1974):

- The EOS, which relates V, T, and P.
- Either of the specific heats along any line on the relevant (P,T) or (V,T) diagram, assuming that the line traverses all conditions of interest.

Item 1 specifies the space occupied, which requires energy, whereas item 2 quantifies the effect of heating, see Eqs. (1.4a), (1.4b) and below Eq. (1.8).

Finally, the relationships of state functions and state variables, which are governed by the chain rule, and are encapsulated in Maxwell's relationships, lead to a large number of links between the physical properties of materials (see tables in Bridgman, 1914; Shaw, 1935; Fegley, 2015). One property that appears in diverse thermodynamic problems, and has been linked to heat transfer (e.g., Liebfried and Schlömann, 1954), is the thermal Grüneisen parameter:

$$\gamma_{th} = \frac{\alpha B_T}{\rho c_V} = \frac{\alpha B_S}{\rho c_P} \tag{1.14}$$

This physical property also connects the two specific heats (or the two heat capacities) and the two bulk moduli:

$$B_S = B_T(1 + \alpha\gamma_{th}T); \quad c_P = c_V(1 + \alpha\gamma_{th}T) \tag{1.15}$$

where $B_S = -V\frac{\partial P}{\partial V}\Big|_S$.

Other state functions are possible. Considering the importance of entropy, Planck proposed that $-G/T$ should be a state function, although this never took hold. In contrast, physical chemist Jacobus van't Hoff completely avoids mention of entropy and free energy in his 1884 textbook, while providing a pragmatic approach. Likewise, the chemist Nernst replaced entropy with free energy, focusing on heat (Section 1.2.3), and stressed that entropy was a matter of differences. Persistent conflicts, such as that between Nernst and Planck, have arisen because modern atomic theory was developed subsequent to classical thermodynamics, and because different problems were important to chemists and physicists in the early 1900s.

1.2.5.2 The Relationship of Isentropic to Adiabatic Processes

Many books state that reversible adiabatic processes are isentropic (S is constant). Typically, this statement is supported by an analysis of the ideal gas (e.g., see Fegley, 2015). However, as discussed above, reversibility was a useful idealization important to the development of entropy, but does not exist. If something happened, it did, and the universe is forever different. We therefore take a different approach.

An adiabat is defined by the absence of heat transfer, that is, $\delta q = 0$. From an identity for E and the first law, considering a per mole basis:

$$dE = \frac{\partial E}{\partial T}\Big|_V dT + \frac{\partial E}{\partial V}\Big|_T dV = C_V dT + \frac{\partial E}{\partial V}\Big|_T dV = \delta q + \delta w = 0 - PdV \tag{1.16}$$

The RHS of Eq. (1.16) is not perfectly general because we have limited work to PdV terms, that is, we assumed external application of pressure. Under this limitation, the adiabatic gradient is described by:

$$\frac{\partial V}{\partial T}\Big|_q = -\frac{C_V}{P + (\partial E/\partial V)_T} \tag{1.17}$$

The isentropic (constant S) gradient is evaluated using one of Maxwell's relationships:

$$\frac{\partial V}{\partial T}\Big|_S = -\frac{\partial S}{\partial P}\Big|_V \tag{1.18}$$

Heat capacity is related to S, giving:

$$\frac{C_V}{T} = \frac{\partial S}{\partial T}\Big|_V = \frac{\partial S}{\partial P}\Big|_V \frac{\partial P}{\partial T}\Big|_V \quad \text{or} \quad \frac{\partial S}{\partial P}\Big|_V = \frac{C_V}{T(\partial P/\partial T)_V} \tag{1.19}$$

For an adiabat to equal an isentrope requires Eq. (1.17) to match Eq. (1.18). Combining the above:

$$T\left(\frac{\partial P}{\partial T}\right)_V = P + \left(\frac{\partial E}{\partial V}\right)_T \tag{1.20}$$

From tables of the Jacobian relationships (e.g., Anderson and Crerar, 1993):

$$\left(\frac{\partial E}{\partial V}\right)_T = \frac{T\alpha V - P\beta V}{\beta V} = \alpha T B_T - P \quad \text{and} \quad \left(\frac{\partial P}{\partial T}\right)_V = \alpha B_T \tag{1.21}$$

which proves that Eq. (1.20) is a thermodynamic identity when work is limited to PdV terms. Hence, adiabats are isentropes if $dw = -P_{ext}dV$.

We did not invoke reversibility. However, the analysis considers internal energy as a function of T and V, whereas the independent variables for E are actually V and S. Consequently, the problem is underspecified, and additional constraints apply. The un stated assumption is that pressure is *externally* imposed.

As an illustration, consider a helium balloon rising rapidly (Fig. 1.3). Ascent is effectively adiabatic, if it greatly outpaces heat loss. As the balloon ascends through the atmosphere, the pressure on the balloon decreases. Because the dependence of P on height is determined by the surroundings, this variable is intensive and independent. For an incremental change, the work is $P_{ext}dV$. Hence, the adiabat for gas in the balloon is identical to its isentrope, which can be confirmed from the EOS and heat capacities of an ideal gas (Fegley, 2015) or of a van der Waals gas (Berberan-Santos et al., 2008). These equivalences exist because temperature in both models is regulated entirely by collisions. The process of a rising balloon is clearly irreversible. But, Eq. (1.16) is valid because P is externally imposed, and, concomitantly, P is an intensive, and more importantly, independent variable.

Due to reliance on the ideal gas during the development of thermodynamics, these two different situations (adiabatic vs isentropic) are considered equivalent in much literature, which has been justified by invoking reversibility. The above shows that the equivalence of adiabats and isentropes is unconnected with reversibility. Regarding laboratory studies, adiabats will be isentropes for many experiments. For example, elasticity studies provide a quantity described as the adiabatic bulk modulus, while designating this material property as B_S (below Eq. (1.15)). The experiments indeed provide $B_S = B_q$, because two conditions are met: not only is heat flow avoided in these measurements, but pressure is externally imposed by the operator's use of an apparatus.

Describing processes in self-gravitating bodies is not so simple. Inside the planet, pressure depends on the total mass (and material properties) and so Eq. (1.16) is not correct and adiabats and isentropes are not identical. To emphasize that externally applied pressure is but one circumstance, we next discuss a familiar, but misunderstood, case involving VdP terms and varying S.

1.2.6 Compression and Expansion of an Ideal Gas

Many textbooks on thermodynamics discuss the simple processes of either expanding or compressing an ideal gas in order to illustrate differences between reversible and irreversible processes as well as entropy changes. This section describes the flaws in these various analyses.

Olander (2000) describes thought experiments on "reversible" compression, slow and adiabatic, where a stack of tiny weights are incrementally added to the top of a piston in

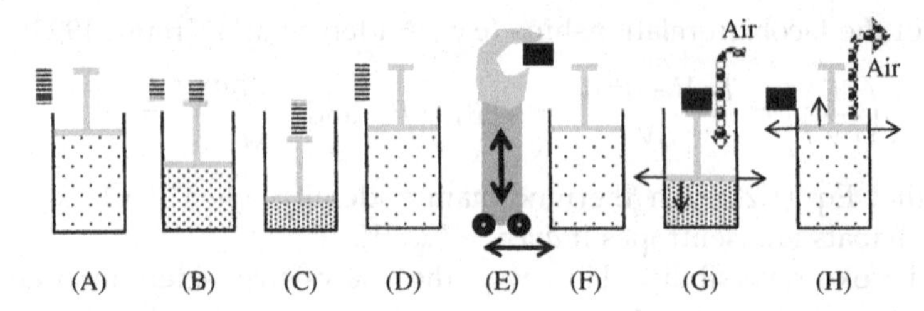

FIGURE 1.5 Schematics of air spring experiments. (A–D) An idealized "reversible" cycle involving infinetesimal displacements, but neglecting interactions with the surroundings. (E) Adjustable height robotic arm on wheels needed to slide small or large weights horizontally. (F–H) Real air spring undergoing a finite displacement. Ornate arrows depict air flow. Thin arrows show motion of heat from the moving piston to gas inside the piston and the room. For the large weights, the system can also be restored to its initial condition, but it takes longer to dissipate the heat generated and work needs to be done to lift the single weight to its original position.

order to compress a gas, then removed slowly to allow the gas to expand (Fig. 1.5). If the compression/expansion is performed slowly, heating of the piston is minimized, thereby maintaining isothermal conditions. Although the experiment is designed to return to the original configuration vertically, the experimenter must accelerate and decelerate the weights horizontally, performing work. The surroundings absorb the small amount of heat generated during compression and supplies heat to offset cooling during expansion. Gas in the room also flows in response to piston motions. "Reversibility" of this system (and any other system) requires changes in the surroundings.

In Joule's recognizably irreversible experiment, gas is allowed to expand from a restricted volume into an adjacent evacuated chamber. In actuality, the stopcock used by Joule was heated, as was the surrounding bath of water, which was used to maintain isothermal and adiabatic conditions (Fegley, 2015). But, water has a huge heat capacity plus the temperature increase of the bath was too small to be resolved with equipment at that time. Joule and his contemporaries were aware of this, and ensuing discussions considered free expansion in terms of the ideal gas. Behavior of an ideal gas reveals some important aspects of entropy, but as in the case of Maxwell's confusing discussion of entropy changes during simple cooling (Fig. 1.4A), the system and its communication with the surroundings must be clearly specified.

We certainly can imagine a configuration where negligible heat is produced and isothermal conditions are maintained. The "box" in Fig. 1.6A is thermally insulating and a low-friction piston is held by an electromagnet. If the magnet is turned off, the piston will advance (rapidly) to the right as the gas expands. We can ignore the fact that the gas is doing work on the piston, by envisioning this piston to offer no resistance, but an ideal vacuum which is devoid of matter is experimentally unknown. "Good" laboratory vacuums contain water and other molecules, providing a density of $\sim 5 \times 10^3$ atoms cm^{-3}. For comparison, densities in molecular clouds (considered to be very dense astrophysical environments where stars are formed) are inferred to range from 100 to 10^3 atoms cm^{-3}, whereas densities between galaxies (considered to be extremely rarified environments) are estimated as 10^{-4}–10^{-6} atoms cm^{-3}. Vacuums are discussed further in Section 1.2.7. Moreover, because the gas molecules carry heat and heat is not considered to be

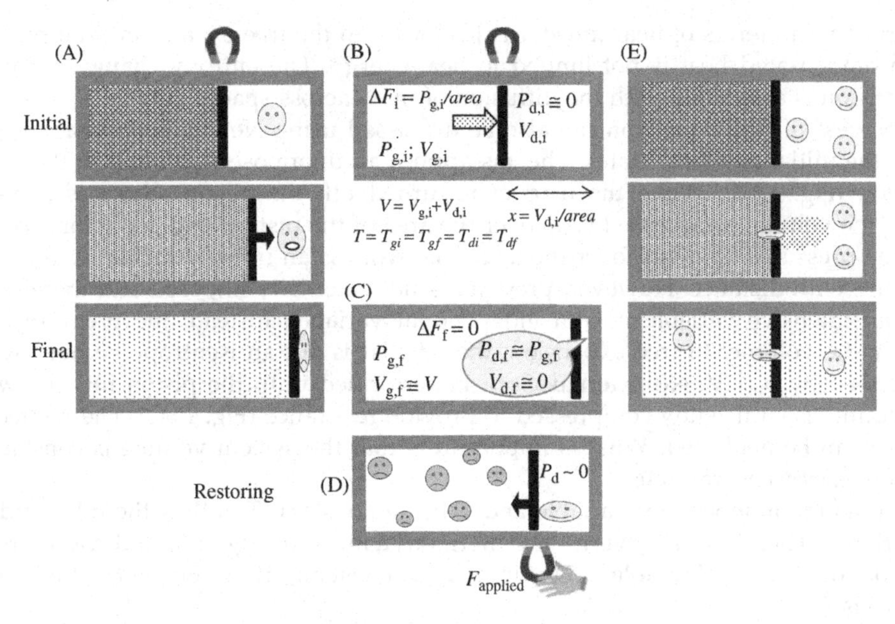

FIGURE 1.6 Schematic of the irreversible and isothermal expansion of an ideal gas in an adiabatic enclosure. Because mass moving in this thought experiment carries heat, the system is defined by the final volume of the gas, and therefore the gas and the vacuum constitute two subsystems. (A) Idealization of a gas on one side of a partition, and very few molecules on the other side, which describes a laboratory vacuum. Upon removal of the magnet, the denser gas expands, but a single molecule prevents compression to zero volume. The denser gas does work to compress the dilute gas, yet no net heat is generated, because the dilute gas does opposing work on the dense gas, and because heat cannot escape the enclosure. (B) Initial conditions of the isothermal experiment of part (A), showing that the gas exerts a pressure on the barrier separating the gas from the rarified medium. (C) Final conditions of the experiment of part (A). (D) The reverse experiment, in which case external work is done to move the magnet, causing the gas to heat. (E) Alternative thought experiment involving a leaky partition between dense gas and a near vacuum. The few molecules in the "vacuum" will mix with the dense gas as the latter diffuses through the pinholes. Entropy should increase due to the mixing, but temperature does not change.

transferred in this experiment, the gas and "vacuum" together necessarily constitute the system and together define V. These must be at the same temperature initially and finally (Fig. 1.6B).

In free expansion of an ideal gas, there is no temperature change and no heat loss, so there is no entropy change arising solely from the heat-energy of the system. Instead, changes occur in component volume and component pressure, simultaneously. In Fig. 1.6A, the system is the gas plus the "vacuum," which is a very dilute gas. The surroundings are outside the gray box and there is no communication. The total volume of the system $= V$ is the sum of the subsystem volumes, which is unchanged during the expansion. Instead, the pressures of the subsystems change in this irreversible experiment: gas pressure decreases while "vacuum" pressure increases. Thus, free expansion experiment involves $V\delta P$ terms where V describes the system. The final pressures of the subsystems are equal, but are lower than the initial pressure of the gas. The gas has lost its ability to do work on the "vacuum" part of the system.

Describing S in terms of heat transfer is irrelevant to the free expansion example: therefore, uncompensated heat is not limited to heat-energy. The entropy change in the ideal gas expansion is associated with the diffusion of mass across space.

The process of free expansion can also be discussed using Newtonian physics and considering the initial and final states. The gas applies an unopposed pressure to the frictionless piston (Fig. 1.6B). When the magnet is turned off, the piston advances a distance equaling the volume compressed (V_{di}) over the area of the piston, with a net force depending on the pressure difference over the area. The work is an integral involving the change in pressure with distance. However, pressure is not externally imposed and therefore P is neither an intensive variable nor an independent variable. Instead, the pressure is controlled by the volume. In free expansion, the adiabat is not an isentrope because work is *not* described by $P_{ext}dV$. No external work is performed until the piston hits the wall or the "vacuum" is sufficiently compressed to provide resistance (Fig. 1.6C). These effects are small and can be neglected. What is important is that the system volume is constant and that P is the *dependent* variable.

With V as the independent variable and T being constant, T is thus the other independent variable. Therefore, F governs the thermodynamics of Fig. 1.5, and the entropy is a function of T and V (Table 1.1). Since T is constant, from Eq. (1.8), the relevant derivative is:

$$\left.\frac{\partial S}{\partial V}\right|_T = \frac{\alpha}{\beta} = \frac{NR_{gc}}{V} \quad \text{(for ideal gas)} \tag{1.22}$$

where we evaluated the derivative using the Jacobians and then applied the EOS for ideal gas. Integration provides the well-known relation:

$$\Delta S = S_f - S_i = NR_{gc} \ln\frac{V_f}{V_i} \quad \text{(for ideal gas)} \tag{1.23}$$

Eq. (1.23) is a consequence of the thermodynamic functions and their interrelationships (Section 1.2.5) plus the EOS of ideal gas, and does *not* require irreversible conditions. This equation is reasonable for a real gas, until compression is sufficiently high that the finite volume of the molecules limits compression. This situation describes the physics underlying pressure-driven phase transitions.

To emphasize that reversibility is unrealizable, we note that our thought experiment could be reversed by using the magnets to pull the frictionless piston to the left (Fig. 1.6D). Without friction, there should be no heating in the insulated enclosure. However, $P_{ext}dV$ work is now being done on the gas, because the magnet–piston arrangement provides an externally derived force. Thus, the temperature of the gas rises. The rate of compression is immaterial. This imagined step is no different than the experimenter loading weights during slow compression in the experiment of Fig. 1.5. Changes in the surroundings must be considered in a thought experiment to avoid unfrying the egg!

To emphasize that S depends on V, we consider an alternative experiment (Fig. 1.6E), where two different gases with the same T exist on either side of a partition. If the partition has "holes" through which the gas molecules can pass, then over time, the double chamber will evolve to having equal proportions of the two kinds of gas on each side. No heat is evolved, and the isothermal conditions hold, but the entropy of the final state is clearly much larger. The thought experiment of Fig. 1.6E involves diffusion of mass over space, but not heat transfer. This parallelism between flow of mass and heat will prove useful in understanding the latter phenomenon later in the book.

The free-expansion example underscores that heat and mass transfer are both linked with entropy. Clausius focused on the T component of entropy and its connection with heat because steam engines were important during his life. Clausius' work on entropy was not completed, because important developments continued well after Clausius' death in 1888 (e.g., the Jacobian relationships of Bridgman, 1914). Based on classical attempts to understand heat through mechanics (the kinetic theory of gas, Chapter 5), and that potential energy describes energy that is stored (energy of position) and is inferred whereas mechanical energy describes what can be directly measured (energy of motions), we further explore the meaning of entropy and its relationships to heat, mass, and configurations later in this book.

1.2.7 Is the Third Law Complete?

If some volume could exist without any molecules, there are no motions, no heat, and no temperature. Regarding space and its vastness, over time, even the smallest volume is crossed by some particle or atom due to the random motions of matter that is not bound in orbits around stars or in and around galaxies: these motions are required because the matter in space, which is a dilute gas, cannot attain absolute zero. Thus, a perfect vacuum is unattainable. As a corollary, it is also impossible for mass to occupy no volume. This conclusion can be reached by recognizing that a perfect vacuum over some finite volume could be produced by confining whatever mass was inside the finite volume to zero space. Likewise, if infinite pressure were possibly attainable, mass could be forced to occupy zero volume. But, even the single-proton nucleus of the hydrogen atom or a single electron occupies space.

Nernst's arguments for the unattainability of absolute zero apply, leading Hofmeister and Criss (in review) to suggest an expansion of his third law statements:

- By no finite series of processes can either absolute zero, or perfect vacuum, or zero volume, or infinite pressure be attained.
- A body cannot be completely deprived of its heat, nor of the space that it occupies.

Our proposed expansion of the third law rests first on the discussion in Section 1.2.4 that the energy of material occurs in two forms: one is thermal and the other is the energy needed to occupy space. This division underlies the concept of the Helmholtz free energy. Second, changes in entropy arise when temperature, or volume, or both are changed. These changes are independent for the case of the ideal gas. Therefore, S depends on both T and V. To uphold the second law requires considering the effect of changing both T and V on the systems and on the surroundings. Assessments can be very tricky when a container does not exist, as for astronomical nebulae.

The above analysis points to Helmholtz's free energy, *F*(*V*,*T*), being of key importance to many thermodynamic problems. From this analysis, relying on the ideal gas will cause problems in high-pressure analyses because infinite compression is associated with the concept of point masses. The ideal gas does not represent a real substance.

1.3 THE RELATIONSHIP OF LIGHT AND HEAT

Substances emit energy, especially if stimulated. For example, applying a high-frequency alternating current to an electric conductor produces radio waves; exposing certain minerals to sunlight causes them to glow visibly; whereas heating a metal filament in a vacuum (the light bulb) produces both emissions that we can see with our eyes and feel with our skin. The latter output, more precisely described as thermal emissions, do not require a filament or a vacuum, and are produced by heating any material. Although we now recognize that all such emissions are the same phenomenon as *electromagnetic* (EM) waves, recent thermodynamics books still describe lightbulbs as emitting ~90% heat and ~10% light, when what is actually meant is that the output is ~90% infrared radiation and ~10% visible radiation. The terminology (heat vs. light) is consistent with our physical perceptions, but misrepresents the underlying physics! This confusion accompanied the development of classical thermodynamics. For example, Maxwell stated in his 1871 book that all heat inside the body is of the same kind, whereas radiation between bodies is of many different kinds. The 1908 version contains this error on p. 16, but on p. 230–240, a long essay explains why radiation of heat or of light are the same phenomenon. This update was provided by Rayleigh after Maxwell's death.

Under the incorrect and ingrained distinction that heat and light are different phenomena, classical thermodynamics and an understanding of heat as energy developed. This section focuses on the history of the infrared region, which is presented in detail by Barr (1960). Note that macroscopic behavior of the visible region was extensively explored much earlier, using glass prisms and lenses, which established refraction and reflection as the key properties of visible light well before 1700 (e.g., Halliday and Resnick, 1966). Separating the microscopic from macroscopic as regards light sensu lato is difficult, as exemplified by the centuries long debate on whether light consisted of corpuscles or was a wave. Debates in the 1800s on the nature of molecules and atoms are also relevant, and are discussed in Chapter 5 regarding the kinetic theory of gas. Resolution of these debates and an understanding of the connection of thermal and visible emissions was achieved as the field of spectroscopy developed over the 1800s and into the 1900s. For a thorough account, see McGucken (1969).

Importantly, light of all kinds is attenuated (i.e., gradually absorbed with distance) as it traverses through matter, and thus this phenomenon is dissipative, just like heat flow. Attenuation of light is covered in Chapter 2. The present chapter explores light in terms of an instantaneous, energy-conserving exchange (after Fig. 1.1A) to the extent that this approach is possible.

1.3.1 Studies Beyond the Rainbow in the 1800s

Investigations circa 1776 of the rainbows created by refracting sunlight with a glass prism showed that thermometer output varied with color, but nothing was made of this observation. In 1800, Sir Frederick William Herschel reported that the Sun was outputting significant energy adjacent to the visually perceived rainbows. Herschel initially considered that this invisible radiation beyond the red end of the rainbow to be of the same nature as visible light, on the basis of refraction and reflection, but upon further measurements changed his mind. About 80 years of controversy followed, with many investigators disputing Herschel's claim that this invisible radiation even existed. Discordant results of various investigations were indeed bona fide, due to the use of prisms made from different types of glass as demonstrated by Thomas Seebeck in 1819. Seeback's study was ignored because a slightly earlier discovery led the scientific community down the wrong path: in 1801 Johann Ritter discovered that the Sun was also outputting energy beyond the violet end of the spectrum. Because these ultraviolet emissions were detected neither by human vision nor by thermometers but instead by chemical effects, the consensus view in the early 1800s was that three kinds of radiation existed: calorific (or heating), luminous (or visible), and actinic (or chemical).

Seeback's 1821 discovery of the thermoelectric effect did not change majority opinion, but contributed to Andre Ampère supporting that calorific and luminous emissions are identical. A breakthrough came in 1831 when Leopoldo Nobili invented the thermopile. Nobili died in 1835, but his junior colleague Macedonio Melloni diligently explored these invisible radiations with the thermopile, demonstrating in 1843 that invisible calorific emissions are identical to visible light. Melloni's many thorough studies, up to his death in 1854, lay the foundations of spectroscopic measurements of thermal emissions. By 1873, the term *infrared* was applied to this region of the EM spectrum, which we can sense with our skin. However, acceptance by the community came about a decade later, following invention of the bolometer (Langley, 1881), substitution of gratings for prisms (Desains and Curie, 1880), and many other supporting measurements (Barr, 1960). The experience motivating Langley's quote provided at the beginning of this chapter should be evident. Unfortunately, Langley is remembered more for his failed airplane design, rather than for his important achievements in the study of light.

Two pieces of corroborating evidence for a single phenomenon are extremely important to understanding heat and its flow. The first development involved ascertaining the relationship between the length of the waves and energy, which is inverse. The visible region was explored in 1803 with Thomas Young's pinhole experiments, whereas progress in the infrared was delayed until ~1880 upon developments of accurate gratings (Barr, 1960). The other development, which is of utmost importance to understanding heat, concerns the macroscopic connection of emissions to temperature, as follows:

1.3.2 Characterization of Emitted Light

Color, which is perceived by our organs of sight, arises in materials preferentially absorbing certain wavelengths in the visible region. The same preferential absorption occurs in the infrared, which was first discovered by Melloni (1850). Quantitative measurement of how materials interact with light as a function of wavelength is called spectroscopy, usually with some qualifier indicating the EM region being explored. Optical spectroscopy (which includes the infrared to the ultraviolet) reached its high point in physics in the 1970s, and was central to the author's PhD dissertation on minerals in 1984. This research area continues to be of great interest in astronomy and planetary science, due to remote sensing applications (e.g., King et al., 2004). Interactions of light with materials are strongly affected by temperature (e.g., Kachare et al., 1972): From whence comes the tie to thermodynamics and the great investment of effort.

Although absorption of light by a material is the most common type of measurement, the emanations or emissions of a heated material are central to thermodynamics and heat transfer. Most materials preferentially interact with light of some frequency (or wavelength). An ideal material, known as a *blackbody* interacts with light of any wavelength in the same manner: This behavior is described by constant emissivity (ε) of unity and the absence of reflections. With such behavior, only temperature affects infrared emissions. Objects with lower, but constant, ε are termed graybodies, whereas objects with idealized (but unachievable) complete reflectivity accompanied by $\varepsilon = 0$ are denoted as whitebodies. The physics underlying blackbody behavior has received much attention, because understanding this behavior is a key to understanding not only macroscopic but also microscopic behavior. The history is described by Schirrmacher (2003) and Robitaille (2009).

A good experimental approximation to a blackbody is provided by sampling output from a tiny hole in a hollow sphere, where the interior walls are kept at a constant temperature. Because a requirement is that light can only enter into the cavity via the small hole, opaque materials such as metals have been used for the hollow sphere, which was typically lined with graphite or graphitic paint. Using this experimental configuration led to the conclusion that the specific material is of little consequence to producing blackbody radiation. However, this is not true. As discovered by Balflour Stewart in 1858, emissivity is related to 1—Reflectivity for opaque materials, and the material indeed has an effect. In addition, as discovered a century later, whether radiation emerging from a cavity is black depends strongly on whether the radiation being injected into the hole is roughly independent of frequency (see Robitaille, 2009). Sunlight, which was historically used, provides a fairly uniform radiation field because its peak is broad. Plus, the historical technology did not well separate different infrared frequencies emerging from the cavity through the small hole.

For opaque materials like metals and particularly graphite, the interior emissions are repeatedly reflected and absorbed from the walls, resulting in essentially isotropic radiation. Hence, the experimental particulars to first order provide what is denoted as cavity radiation. Due to emissions in the standard cavity experiments being little affected by wavelength, the radiation that escapes the tiny hole is only affected by the interior temperature, and thus closely approximates the ideal of blackbody radiation. From energy conservation, over all wavelengths, the light absorbed should equal the light emitted, as deduced by P. Prevost in 1791.

Through experimentation on sodium vapor, which permitted monitoring individual wavelengths, Kirchhoff concluded that the spectral properties of emissivity (ε) and absorptivity (a) are equal at each and every wavelength. This law of Kirchhoff is only true when there is no net heat transfer to or from the surface. Gas, which Kirchhoff studied, lacks a definable surface. Thus, consideration must be given to conditions when the light is partially absorbed and partially transmitted. Exact and quantitative statements which consider the combined reflection and attenuation of light were not made until a century later (Bates, 1978, provided the final word: see Chapter 2).

Kirchhoff's overgeneralization was useful in explaining the shape of blackbody curve by Planck (Schirrmacher, 2003; see below). Unfortunately, its limitations have been overlooked and Kirchhoff's law is misapplied in modern studies both in the microwave region (discussed by Robitaille, 2009) and in the near-infrared to visible (discussed by Hofmeister, 2014).

1.3.3 Connection of Blackbody Radiation to Temperature

A different type of experiment was used to quantify the relationship of total blackbody emissions to temperature. In the 1860s, John Tyndall measured the amount of heat leaving a glowing platinum wire. In examining these experiments, Josef Stefan concluded in 1879 that the total radiant energy emitted per area and per unit time ($\Re$) goes as the fourth power of temperature:

$$\Re = \sigma_{SB}\left(T_{\text{body}}^4 - T_{\text{surroundings}}^4\right) \tag{1.24}$$

The proportionality constant ($\sigma_{SB} = 56.703$ nW m^2 K^4) is known as the Stefan-Boltzmann constant, because shortly thereafter (1884) Ludwig Boltzmann derived the T^4 law from thermodynamic principles. This exercise did not yield a value for σ, but was important because Boltzmann's derivation provides strong support for heat and visible light being the same phenomenon.

The temperature dependence of the Stefan-Boltzmann law, which when extended into the visible, describes the surface emissions of the variously colored stars (Mutlaq, 2016), is strong. Eq. (1.24) together with the thermodynamic connection of energy with temperature (see e.g., the heat capacities of Section 1.1), and the connection of energy and wavelength mandate that blackbody radiation at any given temperature must have a specific wavelength dependence. The curve of energy vs wavelength is known as a *spectrum*.

Early measurements showed that the blackbody spectrum has a peak. From thermodynamic considerations, Wilhelm Wien (1893) derived his displacement law:

$$\lambda_{\max} = \frac{b}{T} \quad \text{where } b = 2897.8\ \mu\text{m K}^{-1} \tag{1.25}$$

Wien considered the effect of slow, adiabatic expansion of the cavity on the interior waves of light. Under such circumstances, thermal equilibrium persists, and so the energy of the waves grows proportionately with temperature. This deduction, coupled with energy inversely depending with wavelength, provides Wien's displacement law. Improved measurements of blackbody curves (Lummer and Pringsheim, 1899) confirmed Eq. (1.25) and provided experimentally determined values for Wein's displacement constant, b. These works ended the debate regarding whether heat emissions are the same phenomenon as visible emissions.

The analyses of Rayleigh (1900) and Jeans (1905) did not suggest that a spectral maximum exists, but rather provided a continuous increase of intensity with energy, known as the ultraviolet catastrophe. That the complicated derivation of Rayleigh-Jeans yielded a result that was inconsistent with the simply derived and fairly accurate spectrum of Wien (see Halliday and Resnick, 1966, and figures in Chapter 2) was apparently not considered in detail. One explanation is that another discovery diverted attention from the macroscopic nature of light to its microscopic origins. Specifically, Planck (1901) provided an alternative formula that nicely matched blackbody spectra, and furthermore Planck's assumptions were compatible with Einstein's (1905) description of the photoelectric effect in terms of energy packets (quantization). We revisit the ultraviolet catastrophe in Chapter 8.

Irrespective of details as to the microscopic nature of light and its interactions with materials, by 1899 it was irrefutably established that visible light and heat-energy traveling in air or vacuum are the same phenomenon. The logical conclusion is then that visible light and heat-energy inside a solid are also the same phenomenon, because a solid can be obtained by condensing gas and because both visible light and invisible heat flow through solids. That heat transfer is tied to transmittal of IR light in the solid was proposed in 1999 by the author, and is evaluated in this book based on subsequent theoretical developments and newer experimental data.

1.3.4 Radiation and Thermodynamic Laws

Radiative transfer is an important process. For example, Earth's surface is maintained near room temperature by receipt of sunlight during the ~12-hour-day, while the energy received is reradiated to space, for all 24 hours and over the whole globe, although this process is most evident on the night side. Over the course of the 24-hour spin period, the energy fluxes to and from the Earth balance out, providing thermal equilibrium and a stable temperature. The internal heat lost by the Earth is tiny in comparison, about 0.1% of the Sun's intercepted flux, and can be ignored in ascertaining Earth's surface temperature. Earth's surface temperature also receives light from the Moon, other planets, and stars, but the latter fluxes are miniscule compared to that from the bright, relatively close Sun, powered by its intense nuclear fires (Fig. 1.7A,B). Hence, solar flux sets Earth's surface T to near 293K. This result can be confirmed quantitatively, using Eq. (1.25), considering areas appropriate to the received and emitted fluxes.

Conservation of energy also describes radiative transfer, but heat from essentially point sources such as the Earth and Sun fans out in all directions (Fig. 1.7A) at a very rapid speed. The process is not adiabatic unless the radiation itself is considered as a system. If one envisions a universe of matter and an interwoven universe of light (Criss and Hofmeister, 2001), then total energy is obviously conserved. But, when considering smaller scales involving radiative transfer (e.g., Fig. 1.1C,F), energy is irreversibly lost to the surroundings. The loss may not seem noticeble or important, such as for the ice bucket (Fig. 1.3), but it still exists, and causes the process to be irreversible. One cannot put the light back into a flashlight!

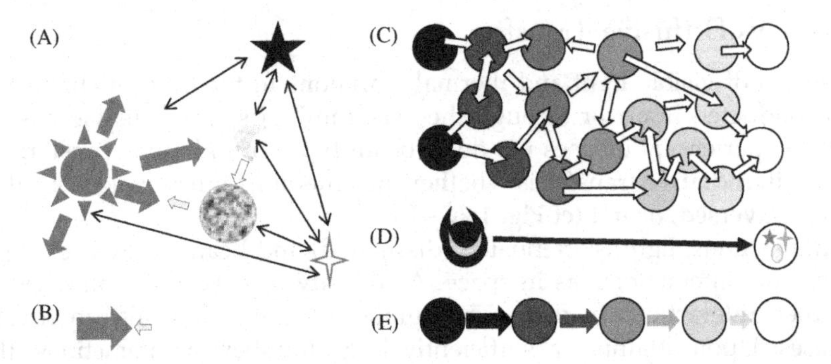

FIGURE 1.7 Aspects of ballistic and diffusive radiative transfer. (A) Familiar example of ballistic transfer, where the light does not interact with the medium it is crossing. This case dominates our view of radiation and thermal emissions. The Earth receives light from the Sun and many other bodies, while heat from the Earth travels the backwards paths. The dominant flux in the Solar system arises from the Sun (dark gray arrows) which is shed radially. (B) Comparison of fluxes from the Sun to the Earth, illustrating that the larger flux from the hotter body dominates. (C) Example of diffusive radiative transfer in a smoke cloud, via light (arrows) transferring energy from the hot end (dark particles) to the cold end (white particle). Radiation goes in all directions, such that the net flux is from hot-to-cold, which constitutes diffusion. (D) Example of a ballistic transfer inside the smoke, where on rare occasion light moves a large distance, without interacting with any particles. The proportions of ballistic and diffusive transfers will depend on the density of the particles in the smoke. (E) Net result of diffusion of radiant energy down the temperature gradient. Here, the random (Brownian) components of the motion of the heat-energy are ignored.

The second law is upheld in these exchanges. This law does not prevent heat from the cold Earth from reaching the Sun. However, this "backwards" transfer is not a sole result: The net flux is from the hot Sun to the cold Earth (Fig. 1.7A,B). In Clausius' time, focusing of rays from a cold to a hot body was offered as counter evidence of the second law. Eq. (1.25), which shows the hotter body emits more intense radiation, was not included in Clausius' book. This result and the two-way transfer of radiation along any path, negates any such argument.

Whenever there is a change in temperature, there is a loss of heat to the surroundings, which can be proximal (Fig. 1.1E,G; 1.3A), remote (Fig. 1.1C,F) or at astronomically far (Fig. 1.7). Unless all heat is back reflected (which is impossible), there is a loss of energy. In all cases, the entropy of the universe increases (Fig. 1.4) and the process is irreversible. Radiation needs to be considered at both high and low temperatures, with due regard to flux balance and attenuation. Many different situations are explored as the book unfolds.

Radiative behavior elucidates Nernst's third law. If a body could be "forced" to reach absolute zero, then it could not emit thermal radiation. However, the surroundings would be emitting some form of radiation. Maintaining absolute zero would require that no radiation of any wavelength reach the body. This is impossible because no perfect reflector exists. To attain absolute zero, the frigid body must be in an enclosure that isolates it from contact and radiative transfer, but the enclosure will be at some finite temperature, albeit quiet cold. A series of enclosures is possible, and very cold temperatures are attainable in modern experiments, but unless all material in the universe is at absolute zero, some radiant energy will be transferred from enclosure to enclosure and eventually to the body (e.g., Fig. 1.7E).

1.3.5 Ballistic vs Diffusive Transfer

Our discussion of visible light and thermal emissions in the present chapter centers on experiments conducted in air or through the "vacuum" of space. This view is compatible with everyday experience, but does not describe all types of radiative transfer. Of overriding importance in radiative transfer is whether the emissions interact significantly with the medium being traversed, or not (cf Fig. 1.7C–E).

Through interactions, light is attenuated (dissipated) and heat-energy is exchanged. Low density means few interactions, as in space. As density increases, the number of particles and thus of their interactions increase. The number of interactions also increases as path-length increases. Upon attaining a sufficiently large number of interactions, the process becomes diffusive and time-dependent (Fig. 1.7C), although this still has a direction, down the temperature gradient (Fig. 1.7E). When interactions are insignificant, the process is known as direct or *ballistic* (Fig. 1.7D). We use the latter term after Blumm et al. (1997) to provide more contrast with the other end-member process of repeated interactions over a small distance, *diffusive* radiative transfer. It should be recognized that whether radiative transfer is diffusive or ballistic depends not only on density and path-length (Fig. 1.7C–E) but also on the wavelength of light (Chapter 2), and that both processes occur simultaneously and gradationally to varying degrees. This situation makes radiative transfer notoriously difficult to understand. Chapters 2, 8, and 11 provide quantitative discussions.

1.3.6 Momentum and Pressure of Light

EM waves transport not only energy, but also momentum. The latter transfer provides radiation pressure. This behavior was predicted by Maxwell by relating the momentum transfer to the energy transferred and the speed of light, and was confirmed about 30 years later by the experiments of Nicolls and Hull in 1903 (see Halliday and Resnick, 1966).

Thermodynamic assessments focus on energy (and entropy) while ignoring momentum transfer. This aspect of light, which is shared by heat, was not known when classical thermodynamics was developed, but clearly is important when radiative transfer is involved. This key result will be utilized later in this book.

1.4 SUMMARY

Classical thermodynamics was developed before atomic structure and the equivalence of heat and light were understood, resulting in serious shortcomings. Some are recognized: for example, "thermodynamic reversibility" is used to semantically address the impossibility of reversibility in the real world. Other shortcomings exist because the dynamic process of heat flow was not incorporated in classical thermodynamics, more properly called thermostatics, even though Fourier's laws were known earlier, and because "heat" and "light" remain distinct concepts in everyday language.

This introductory chapter reveals inconsistencies associated with overemphasis of reversibility, and offers some solutions. We show that temperature and volume are the key state variables and that Helmholtz' free energy is the fundamental state function; this result has been masked by the focus on extensive vs intensive variables, rather than on independent vs dependent variables whose distinction is far more important. Hence, we assign the "reversible heat" of Clausius to entropy contributions at constant volume, and his "uncompensated heat" to entropy changes at constant temperature. The importance of volume to entropy leads us to extend the third law. We show that equating adiabats to isentropes is not connected with reversibility, but rather is solely a consequence of imposed external pressure. Irreversibility is manifest in ubiquitous blackbody emissions. After quantifying light and emissions in Chapter 2, the remainder of the book addresses the dynamics of heat transport.

References

Anderson, G.M., Crerar, D.A., 1993. Thermodynamics in Geochemistry. Oxford University Press, New York, NY.

Barr, E.S., 1960. Historical survey of the early development of the infrared spectral region. Am. J. Phys. 28, 42–54.

Bates, J.B., 1978. Infrared emission spectroscopy. Fourier Transform IR Spect. 1, 99–142.

Berberan-Santos, M.N., Bodunov, E.N., Polliani, L., 2008. The van der Waals equation: analytical and approximate solutions. J. Math. Chem. 43, 1437–1457.

Birch, F., 1965. Energetics of core formation. J. Geophys. Res. 70, 6217–6221.

Blumm, J., Henderson, J.B., Nilsson, O., Fricke, J., 1997. Laser flash measurement of the phononic thermal diffusivity of glasses in the presence of ballistic radiative transfer. High Temp. High Press. 29, 555–560.

Bonner, W.B., 1956. Boyle's Law and gravitational instability. Mon. Not. R. Astr. Soc. 116, 351–359.

Bridgman, P.W., 1914. A complete collection of thermodynamic formulas. Phys. Rev. 3, 273–281.

Clausius, R., 1879. Mechanical Theory of Heat (W. R. Browne, Translator). Macmillan and Co., London, UK.

Criss, R.E., Hofmeister, A.M., 2001. Thermodynamic cosmology. Geochim. Cosmochim. Acta 65, 4077–4085.

Desains, P., Curie, P., 1880. Rechershes sur la détermination des longueurs d'onde des rayons calorifiques à basse température. Paris Acad. Sci. Compt. Rend. 90, 1506–1509.

Einstein, A., 1905. Über einen die Erzuegang and Verwandlung des Lichtes betreffenden heuristischen Gesichtspunkt. Annalen der Physic 17, 132–148.

Fegley Jr., B., 2015. Practical Chemical Thermodynamics for Geoscientists. Academic Press\Elsevier, Waltham, MA.

Fowler, R., Guggenheim, E.A., 1939. Statistical Thermodynamics. Cambridge University Press, Cambridge, UK.

Halliday, D., Resnick, R., 1966. Physics. John Wiley & Sons Inc, New York, NY.

Hofmeister, A.M., 2014. Carryover of sampling errors and other problems in far-infrared to far-ultraviolet spectra to associated applications. Rev. Mineral. Geochem. 78, 481–508.

Hofmeister, A.M., Criss, R.E., 2012. A thermodynamic model for formation of the Solar System via 3-dimensional collapse of the dusty nebula. Planet. Space Sci. 62, 111–131.

Hofmeister, A.M., Criss, R.E., 2015. Evaluation of the heat, entropy, and rotational changes produced by gravitational segregation during core formation. J. Earth. Sci. 26, 124–133.

Hofmeister, A.M., Criss R.E., in review. Reconciling classical thermodynamics with heat and mass transfer: implications for the kinetic theory of gas. Can. J. Phys.

Jeans, J.H., 1905. On the partition of energy between matter and ether. Phil. Mag. London (Ser. 6) 10, 91–98.

Kachare, A., Andermann, G., Brantley, L.R., 1972. Reliability of classical dispersion analysis of LiF and MgO reflectance data. J. Phys. Chem. Solids 33, 467–475.

King, P.L., Ramsey, M.S., Swayze, G.A., 2004. Infrared spectroscopy in geochemistry exploration geochemistry, and remote sensing, Short Course Series, vol. 33. Mineralogical Association of Canada, London, Ontario.

Kragh, H., Weininger, S.J., 1996. Sooner silence than confusion: the tortuous entry of entropy into chemistry. Hist. Stud. Phys. Biol. Sci. 27, 91–130.

Langley, S.P., 1881. The actinic balance. Am. J. Sci. (Ser. 3) 21, 187–198.

Langley, S.P., 1889. The history of a doctrine. Am. J. Sci. 37, 1–23.

Liebfried, G., Schlömann, E., 1954. Warmleitund in elektrische isolierenden Kristallen. Nach Ges Wissenschaften Goettingen Mathematik und Physik K1, 71–93.

Lewis, G.N., Randall, M., 1923. Thermodynamics and the Free energy of Chemical Substances. McGraw-Hill, New York, NY.

Lummer, O., Pringsheim, E., 1899. Die Vertheilung der Energie im Spectrum des schwartzen Körpers und des blacken Platins. Verhandl. Deut. Physik. Ges. 1, 215–235.

Maxwell, J.C., 1908. The Theory of Heat. Longmans, Green, and Co., London (This edition is posthumus, with corrections and additions to the 10th edition in 1891 by Lord Rayleigh. The 1st edition was published in1871.).

McGucken, W., 1969. Nineteenth-Century Spectroscopy. The Johns Hopkins Press, Baltimore, MD and London, UK.

Melloni, M., 1850. La Thermochrôse ou la Coloration Calorifique. Joseph Baron, Naples, Italy, https://play.google.com/books/reader?id = 5NgPAAAAQAAJ&printsec = frontcover&output = reader&hl = en&pg = GBS.PR3 (accessed 21.02.17).

Müller, I., 2001. Grundzüge der Thermodynamik—mit historischen Anmerkungen, third ed. Springer, Berlin, Germany.

Müller, I., Müller, W.H., 2009. Fundamentals of Thermodynamics and Applications. Springer Verlag, Berlin, Germany.

Mutlaq, J., 2016. Star Colors and Temperatures: The AstroInfo Project. https://docs.kde.org/trunk5/en/kdeedu/kstars/ai-colorandtemp.html (accessed 21.02.17).

Nordstrom, D.K., Munoz, J.L., 1986. Geochemical Thermodynamics. Blackwell Scientific, Palo Alto, CA.

Olander, D.R., 2000. Compression of an ideal gas as a classroom example of reversible and irreversible processes. Int. J. Eng. Ed. 16, 524–528.

Partington, J.R., 1960. first ed. A Short History of Chemistry, 1937. MacMillan and Co., London, UK.

Pippard, A.B., 1974. The Elements of Classical Thermodynamics. Cambridge University Press, London, UK.

Planck, M., 1901. Über das Gesetz der Energievertilung im Normalspektrum. Ann. Physik (Ser. 4) 4, 553–563.

Prigogine, I., 1967. Introduction to Thermodynamic of Irreversible Processes, third ed. Interscience\John Wiley and Sons, New York. NY.

Purrington, R.D., 1997. Physics in the Nineteenth Century. Rutgers University Press, New Brunswick, NJ.

Rayleigh, L., 1900. Remarks on the law of complete radiation. Phil. Mag. London (Ser. 5) 49, 539–540.

Richet, P., 2001. The Physical Basis of Thermodynamics with Applications to Chemistry. Kluwer Academic/Plenum Publishers, New York, NY.

Robitaille, P.M., 2009. Kirchhoff's law of thermal emission: 150 years. Prog. Phys. 4, 3–13.

Schirrmacher, A., 2003. Experimenting theory: The proofs of Kirchhoff's radiation law before and after Planck. Hist. Stud. Phys. Biol. Sci. 22, 299–355.

Segrè, E., 1984. From Falling Bodies to Radio Waves. W.H. Freeman and Company, New York, NY.

Shaw, A.N., 1935. The derivation of thermodynamical relationships for a simple system. Phil. Trans. Roy. Soc. London A 234, 299–328.

Sherwood, B.A., Bernard, W.H., 1984. Work and heat transfer in the presence of sliding friction. Amer. J. Phys. 52, 1001–1007.

Stacey, F.D., Stacey, C.H.B., 1999. Gravitational energy of core evolution: implications for thermal history and geodynamo power. Phys. Earth Planet. Inter. 110, 83–93.

Todhunter, I., 1873. A History of the Mathematical Theories of Attraction and Figure of the Earth. MacMillan and Co., London, UK, Reprinted by Dover Publications, New York 1962.

van Schmus, W.R., 1995. Natural radioactivity of the crust and mantle. In: Ahrens, T.J. (Ed.), Global Earth Physics: A Handbook of Physical Constants. American Geophysical Union, Washington, DC., pp. 283–291.

Wien, W. 1893. Eine neue Beziehung den Strahlen schwarze Körper zum zweiten Haupsatz den Wärmetheorie. In: Sitzungsbereichte der Koniglich Preussicchen Akademie der Wissenschaften, Berlin, pp. 55–62.

Zemansky, M.W., Dittman, R.H., 1981. Heat and Thermodynamics, sixth ed. McGraw-Hill, New York, NY.

CHAPTER

2

Macroscopic Analysis of the Flow of Energy Into and Through Matter From Spectroscopic Measurements and Electromagnetic Theory

OUTLINE

Measurements, Mechanisms, and Models of Heat Transport
DOI: https://doi.org/10.1016/B978-0-12-809981-0.00002-4

We wish to raise our feeble voice against innovations that can have no other effect than to check the progress of science, and renew all those wide phantoms of the imagination which Bacon and Newton put to flight from her temple. ***Anonymous attack of Henry Brougham, 1803, on Thomas Young's experimental proof of the wave-like nature of light.***

That light (what we see) and heat (what we feel) traveling through air or vacuum are the same phenomena is now well-known. However, heat transfer within a solid is treated differently: most researchers consider elastic waves as the main carrier of interior heat, after Debye (1914). This situation in part stems from the vibrations of molecules or ions in condensed matter, which are connected with the heat capacity of a solid, not being probed by spectroscopy until ~1930 due to instrumental limitations (Sawyer, 1963).

Thus, when models of heat transfer were being developed, information on vibrations was limited to gases with light molecules. Only recently has it been demonstrated that infrared (IR) light participates in diffusing heat inside solids (Hofmeister et al., 2014), although unwanted perturbation of heat transfer measurements by higher frequency light was recognized much earlier (e.g., Kellett, 1952; Tan et al., 1991). The latter process is the ballistic end-member of radiative transfer (Chapter 1, Section 1.3.5). Radiative transfer also occurs as a diffusive process, both in gas and condensed matter. To understand both radiative processes, we need to explore spectroscopy, which describes how light interacts with a medium or material. For an introduction to IR spectroscopy in geoscience, see Hirschmugl (2004). Vibrational spectroscopy is covered by Chalmers and Griffiths (2006). For a general text covering many techniques, see Hollas (1996). For comprehensive coverage of spectroscopy in a broad sense, see Lindon et al. (2016). Useful websites are listed after the references.

Introducing spectroscopy early on in this heat transfer book differs from traditional approaches, which begin with Fourier's equation and steady-state conditions. Because temporal and spatial dissipation of energy describes heat transfer, steady-state is an approximation, just as equilibrium is an approximation in thermodynamics. In both cases, the final states may not be achieved, even after a long time. Furthermore, Fourier's pioneering studies on the flow of heat (Chapter 3) were developed under the view that heat is a fluid (the caloric). As discussed in (Chapter 1), principles prior to understanding the equivalence of heat and light has led to some misunderstandings. Although the time-dependent process of diffusion can be confusing, it is related to a simpler phenomenon: namely, attenuation of a signal. Attenuation of light is well-understood, and is a central topic of the present chapter. Lastly, IR spectroscopy provides information on interatomic vibrations, which contain the thermal energy in a solid and so this chapter provides background needed to understand the microscopic picture of heat.

The properties of light at visible wavelengths have been studied since ancient times. The nature of light was hotly debated over several centuries (e.g., Weiss, 1996), and even now, its manifestation over enormous scales (the red shift) involves controversy (Arp, 2003). Many efforts have been devoted to the field of spectroscopy (which quantifies how light interacts with matter), because this field is important to chemists, physicists, and astronomers, who consider different aspects of light, and because spectroscopy is useful to both pure and applied science. Spectroscopy was developed in the 1800s, by chemists in efforts to understand the constituency of matter, and by astronomers in efforts to decipher the chemical elements present in the Sun, whereas physicists focused more on the nature of light (see e.g., Barr, 1963; McGucken, 1969). Spectroscopy now encompasses a much broader range of experimental techniques: this chapter (and book) concerns methods where some part of the incident energy is directly absorbed by the sample, therefore heating the sample, and excludes methods involving (inelastic) scattering, such as Raman and Brillouin spectroscopies.

We do not need to understand behavior across the entirety of the EM spectrum, because temperatures in the laboratory and the planets are fairly low. The relevant blackbody emissions peak in the IR and trail into visible and ultraviolet (UV), which both involve the same processes, even though IR and UV wavelengths are invisible.

We begin with what is most familiar: travel of light in a "vacuum," which is in actuality a very dilute gas (Chapter 1, Section 1.2.6). We then move to spectroscopy as practiced by chemists, which centers on the absorption technique. As a prelude to discussing microscopic behavior, this chapter introduces how light traveling inside a material is affected by its density, chemical bond types, wavelength, path length, and strength of the absorptions. Examples of interactions with gas, liquid, and various types of condensed matter are provided to make the discussion concrete. We explore the physics of reflection, so that the effects of interfaces can be understood, and to quantify the less familiar process of emission. Light interacts with matter in a myriad of ways because different microscopic interactions are stimulated by different energies, and because both particle-like and wave behaviors exist.

2.1 PROPERTIES AND MOVEMENT OF LIGHT IN SPACE OR VERY DILUTE GAS

Our eyes being built in spectrometers permitted investigation of visible light long ago. The idea of light as rays originated with Empedodes, circa 450 BC. Ptolemy investigated reflection and refraction, circa 130 AD. Because light is used to explore the universe, advances involve astronomical data as well as laboratory studies. The history of light is covered in many books (e.g, Weiss, 1996). Key discoveries which led to recognizing that radiant heat and light are the same phenomenon are covered in Chapter 1, Section 1.3.3.

As in all branches of science, advancements follow inventions. Newton propelled his career and science forward upon improving telescope design: he used mirrors to remove chromatic aberrations associated with glass lenses. His rival Robert Hooke popularized the study of light through entertaining descriptions of diverse observations he made using the microscope. Many investigations during the 1600s lead to the description of light as an invisible, weightless fluid, which was the consensus view at that time This viewpoint provides the conceptual basis for the discredited hypotheses of the caloric (Chapter 1) and of the luminiferous ether (below).

Wave behavior, being inherent to a fluid, was the basis of Christiaan Huygen's theory and the focus of the many studies that followed from ~1700 to the early 1800s (Barr, 1963). In great contrast was Newton's weighty statement, "Light consists of a vast number of exceedingly small particles shaken off in all directions from a luminous body." This corpuscle hypothesis was shared by Laplace, Descartes, and many others. Hence, the conclusive evidence of light being a wave provided by Young's slit experiments of ~1801 were met with hostility (see the introductory quote). Young further showed that these waves were transverse, which was confirmed by Fresnel. Corpuscular theory was ruled out in 1850 by Foucault's measurement of the speed of light being slower in water than air.

A macroscopic theory of EM waves was developed by Maxwell (1865). The original formulation did not actually concern light, but rather linked the laws of electricity and magnetism established by Gauss and Faraday, while extending the formulation of Ampère (Halliday and Resnick, 1966). Proof of light being an EM wave came much later, with the 1888 experiments of Hertz. Additional confirmation of Maxwell's equations was established by measuring radiation pressure in 1901 by Lebedev and in 1903 by Nichols and Hull.

Further advances in understanding light were based on its interaction with matter. The present chapter is limited to the macroscopic viewpoint that light is some disturbance that propagates at a certain speed in and across a medium. The immediate need is to quantify properties of light (this section) and attenuation of the signal (Sections 2.2 and 2.4), neither of which requires knowledge of the microscopic process.

The properties of the light traveling in a medium whereby interactions are negligible or unimportant are described in the following subsections. To summarize, travel of light has a direction and a speed. Light is energy, which in the particle view is described in terms of Planck's constant (h) times *frequency* (ν). The latter is related to a *wavelength* (λ) via the speed, presumed to be constant in low-density media. The total energy of a light beam is described by its *intensity* (I). If light is described as a wave, then its intensity is related to *amplitude* of the wave. The basics summarized in this section are covered in many reviews and books.

2.1.1 Speed of Light in Highly Dilute Media

Light is extremely fast in air or the "vacuum" of space. In 1675, using astronomical observations of the Galilean moons, Ole Roemer estimated the speed of light: his results were within 0.7 times modern determinations. Laboratory measurements by H.L. Fizeau provided the modern value of $c = 3.0 \times 10^8 \text{ m s}^{-1}$ in 1849. Efforts by A.A. Michaelson spanning 50 years, and of many others, provided six significant figures. Michaelson's most celebrated efforts were with E.W. Morley, in interferometric experiments to test for the existence of the *luminiferous ether*, which was the name for the medium that was hypothesized to carry travel of light waves across space. Michaelson and Morley theorized that presence of the *luminiferous ether* would produce interference fringes when their interferometer was rotated. No shift in fringes was detected in 1881, and this perplexing result was taken as the absence of the ether. Resolution of this puzzle awaited further developments in physics.

2.1.2 Direction in Highly Dilute Media

Unless interrupted, light travels in one direction: thus the phenomenon is one-dimensional. This directionality is evident in geometrical optics, where lenses are designed by ray tracing, presuming that the optical components have dimensions much larger that the wavelength of the light (see e.g., Halliday and Resnick, 1966).

Two different end-member geometries are germane (Fig. 2.1). Light can emanate radially from a point source, in which case s is the sole variable; this describes radiation from stars at some modest distance. Alternatively, light can travel in parallel rays along a Cartesian direction, which generally describes the laboratory conditions, for example, laser experiments or spectrometers. Light rising from a hot-plate is also represented by parallel rays, as long as edge effects can be ignored (discussed further in Chapter 3). Earth is so far from our Sun that the arriving rays arriving at any given location can be considered as plane parallel.

Cylindrical geometries certainly exist, but are far less applicable, and tend to be "converted" to parallel rays. An example is that the source of light in IR spectrometers is a rod of SiC (the globar), but various optical elements and apertures downstream from its emissions produce nearly parallel rays that are then focused on the sample.

2.1.3 Spectral Regions and Wave Characteristics

Wavelength is important to physical phenomena, whereas frequency is connected with energy and thus is of great relevance to radiative transfer. For any wave in a medium:

$$\lambda = \frac{\text{speed}}{\nu} \tag{2.1}$$

For EM waves, the speed of light is provided by:

$$c^2 = \epsilon_0 \mu_0 \tag{2.2}$$

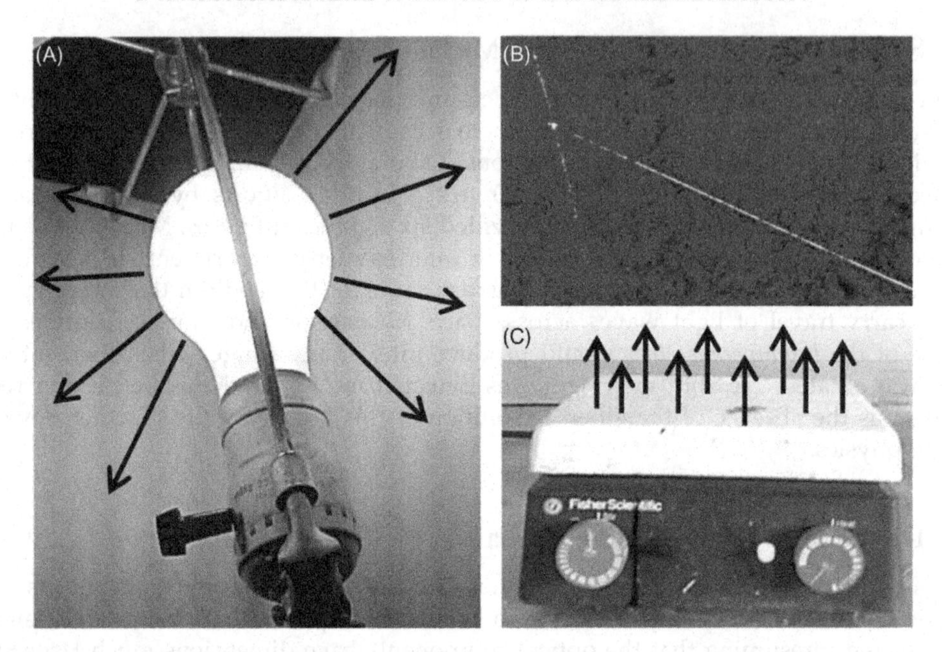

FIGURE 2.1 Examples of the travel of light. (A) An ordinary light bulb sheds rays of light radially in many directions, fairly evenly illuminating the lampshade. This example approximates the radial trajectories from a point source. (B) A low-power, He–Ne laser, with a thin, coherent beam. The arrow is above the beam. The beam exemplifies parallel rays. (C) Picture of a hot-plate. Although we cannot "see" the IR light emitted, the ray paths are vertical, because the many tilted rays average to a forward direction, ignoring edge effects (quantified in Chapter 3).

where ϵ is the permittivity and μ is the permeability. Eq. (2.2) was a prediction of Maxwell. This lead to the concept of the EM spectrum and to the discovery of radio waves by Hertz c. 1890. As discussed below, the fundamental laws of optics can be derived from Maxwell's equations.

In air, and the "vacuums" in the laboratory and space, the speed of light is a constant since μ_0 and ϵ_0 do not vary much between these two mediums. Dense materials slow the travel of light (Section 2.2). For our macroscopic analysis of light, then, we can use either wavelength or frequency, as convenient. Which is convenient is partially dictated by the equipment used to measure light because different energy light is measured in different ways with different materials. The other factor is numerology: because light wavelengths cover a huge range, different units give easily recognizable values. For the IR regions associated with heat at moderate temperatures (Fig. 2.2), units of μm are appropriate for wavelength. Commonly, wavenumbers ($=2\pi/\lambda$, units of cm^{-1}) for ν are used, where $1\ \text{cm}^{-1} = 3 \times 10^{10}$ Hz (inverse seconds).

2.1.4 Intensity

The tie of blackbody radiation to temperature is important to understanding heat (Chapter 1, Section 1.3.3). Eq. (1.24) gives the total flux. Spectra of ideal emitters are most

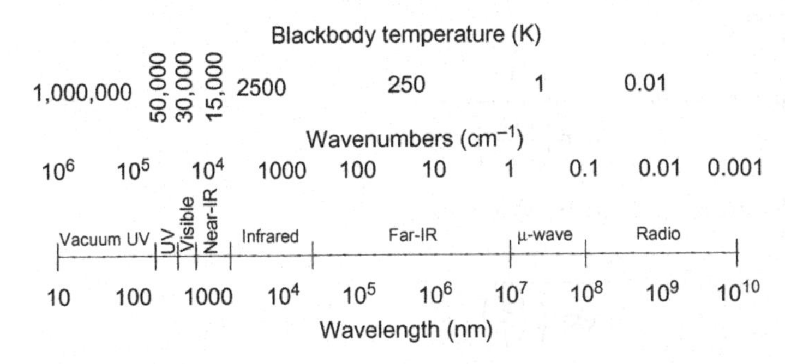

FIGURE 2.2 Spectral regions. The axes are logarithmic for wavelength and wavenumber, which is used in IR studies and is proportional to frequency in Hz (=s^{-1}). Tics define the limits of the spectral regions named in this figure. Blackbody temperatures as listed are appropriate for roughly the middle of each spectral region.

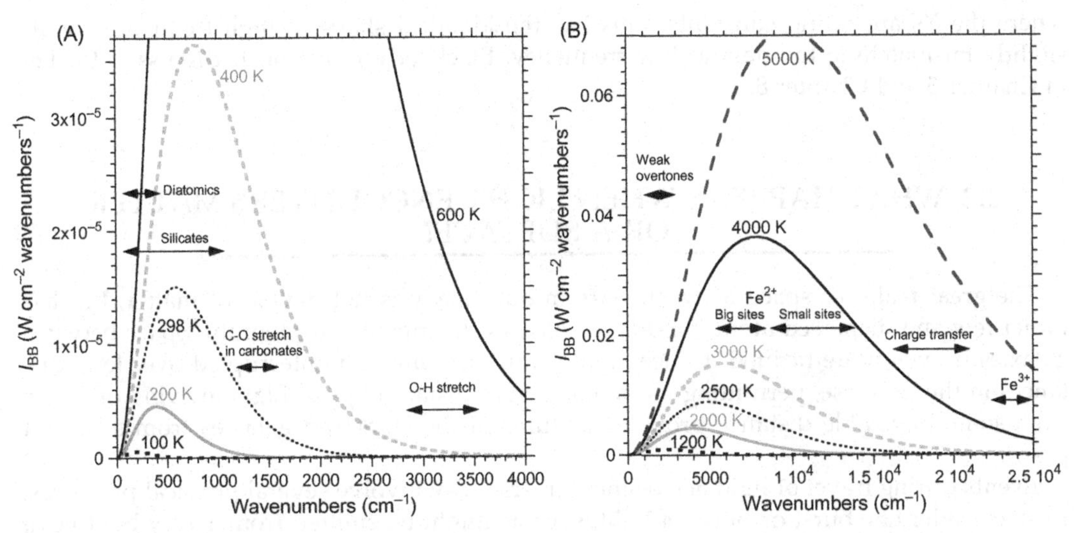

FIGURE 2.3 Planck curves, with temperatures labeled. Horizontal double arrows indicated ranges of absorption for common minerals. (A) Low temperature, where light interacts with lattice vibrations. (B) High temperature, where overtones and electronic d–d transitions are important.

intense at a certain wavelength that is characteristic of the temperature of the body (Eq. (1.25); Fig. 2.2). Fig. 2.3 shows blackbody spectra for temperatures encountered in the laboratory and likely inside the Earth. The shape of the curve was reproduced by Planck (1901) as:

$$I_{BB}(\lambda, T) = \frac{2hc^2}{\lambda^5}\frac{1}{\exp\left(\frac{hc}{\lambda k_B T}\right) - 1} \tag{2.3a}$$

where h is Planck's constant and k_B is Boltzmann's constant. The dependence on frequency ($\underset{\sim}{\nu}$ in Hertz) differs, due to the inverse relationship of Eq. (2.1):

$$I_{BB}(\nu, T) = \frac{2h\underset{\sim}{\nu}^3}{c^2}\frac{1}{\exp\left(\frac{h\underset{\sim}{\nu}}{k_B T}\right) - 1} \tag{2.3b}$$

If wavenumbers are used (Fig. 2.3):

$$I_{BB}(\nu, T) = 2hc^2\nu^3 \frac{1}{\exp\left(\frac{hc\nu}{k_B T}\right) - 1} \tag{2.3c}$$

For comparison, the classical formula derived by Wien is:

$$I_{BB}(\lambda, T) = \frac{b_1}{\lambda^5} \frac{1}{\exp\left(\frac{b_2}{\lambda T}\right)} \tag{2.4}$$

where the b's are fitting constants. This has the identical shape at high frequency, while slightly mismatching the data at low frequency. Blackbody radiation is discussed further in Chapter 5 and Chapter 8.

2.2 WHAT HAPPENS WHEN LIGHT ENCOUNTERS MATTER OR A SURFACE?

The great realm of space between stars in our galaxy is not devoid of matter, but has rather few and dispersed atoms, and is denoted as the interstellar medium. Space between galaxies is even more rarified. Hence, light can travel almost uninterrupted over long distances in the universe, permitting us to image very distant objects. Light arriving on Earth from some incredible distance provides pictures of the stars and galaxies from a distant past.

Events during travel of light across the universe may involve several physical processes. First, consider one burst or pulse of light, such as might be emitted from a very brief nova (star explosion) in distant galaxy. Detecting the nova depends on insignificant *absorption* (irreversible removal of intensity or flux) of this burst during its long journey. Rather than considering a small negative effect, a large positive quantity may be easier to understand. That is, travel of the nova pulse across space and time involves very high *transmission* (forwards conveyance of unaffected light). Concomitantly, *reflection* (backwards redirection) is also negligible, or the light would be diverted towards some other observer on another planet. Although we have discussed light traveling in space, the same process exist for light traversing shorter distances in the atmosphere, or in a laboratory spectrometer or on a laboratory benchtop. These processes are sketched in Fig. 2.4, along with *refraction* (bending of the rays during forward travel).

We are keenly interested in processes that transfer energy. Exchange of energy occurs during absorption when the flux is diminished, but not during reflection or transmission. Note that multiple reflections are described in terms of *scattering*, which does not transfer energy (absorption is required). Commonly, the process of redirection is described as being *elastic*, or involving no loss of transmitted energy. Conversely, the term *inelastic* is used to denote a scattering event wherein the scattered particle loses some portion of its

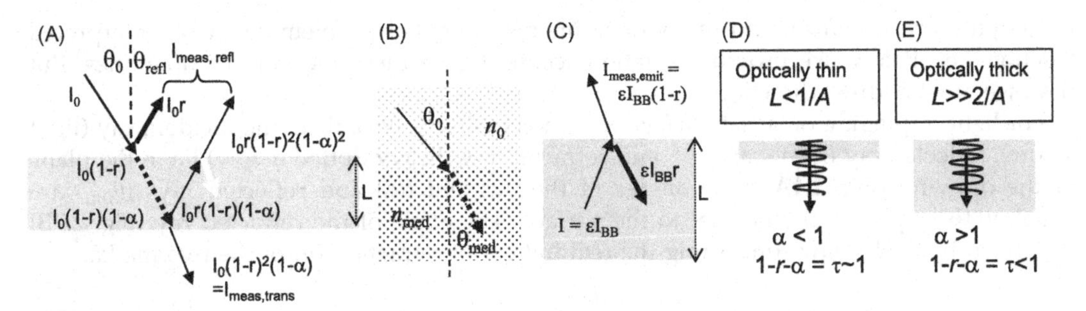

FIGURE 2.4 Schematics relevant to light received versus spectral functions. One back reflection is shown per surface at near-normal incidence (angles are exaggerated and the white arrow showing a secondary reflection is not labeled). (A) Transmission (lower face) and reflection (upper face) experiments. The light received by a detector (I_{meas}) relative to incident intensity (I_0) is a combination of reflection and absorption, depending on thickness, L. (B) Detail of refraction angles. (C) Emission experiments, where I_{BB} is blackbody intensity. (D) Effect of optically thin versus optically thick conditions on spectral functions. These divisions pertain to emission as well as transmission and reflection experiments. Source: *Part (A) reprinted with permissions of Oxford University Press Journals from Figure 1 in Hofmeister, A.M., Keppel, E., Speck, A.K., 2003. Absorption and reflection spectra of MgO and other diatomic compounds. Mon. Not. R. Astron. Soc. 345, 16–38; Parts (C–E) reprinted with permissions of Mineralogical Society of America from Fig. 1 in Hofmeister, A.M., 2014a. Carryover of sampling errors and other problems in far-infrared to far-ultraviolet spectra to associated applications. Rev. Mineral. Geochem. 78, 481–508.*

energy. The macroscopic description of light does address inelastic collisions. These are essential to Raman spectroscopy.

We now wish to quantify the losses incurred by our nova pulse from its start at the distant galaxy until its receipt by the observer here on Earth. To describe these interactions, which are sketched in Fig. 2.4, engineering definitions are used: that is, we quantify how much light is actually absorbed by the sample (which is the interstellar medium in our present example), and/or how much is actually reflected, and/or how much is actually emitted (Brewster, 1992; Siegel and Howell, 1972). In spectroscopic measurements, it is convenient to consider *intensity* (I) which is the total energy, rather than *flux* ($\Im$), which is energy per unit area. Intensity is a more useful quantity because of the way in which spectrometers operate. Usually the amount of light transmitted with and without the sample in the beam are measured (sample versus reference) and then are compared. The intensity is related to the amplitude of an EM wave, which can also be described in terms of the *electric field* (∃). In the spectroscopy, chemistry, and mineralogical literatures, common logarithms are mostly used to describe light loss: this convention is incorporated in commercial software. However, the physics of attenuation is mathematically described by the exponential function and thus the natural logarithm pertains, and is used here, unless noted otherwise.

2.2.1 Macroscopic Description for Light Crossing Matter

Absorption and reflection of light are commonly measured because these can be directly and easily determined quantitatively. Emissions can also be directly measured, but require greater effort, and many experiments are either problematic or misinterpreted (Section 2.6). Below, we provide equations centering on these measurable quantities. But first, we must address direction:

For light impinging on a flat surface, which is easiest to visualize, the incident ray (light arrow), reflected ray (heavy arrow), and refracted ray (heavy dotted arrow) lie in the plane of the drawing (Fig. 2.4A). The angles of the incident (θ_0) and reflected rays (θ_{refl}) are equal, with respect to the normal to the surface. The angle of the refracted ray (Fig. 2.2B) is defined by Snell's law concerning the real indices of refraction (n) of the two media:

$$\frac{\sin\theta_0}{\sin\theta_{med}} = \frac{n_{med}}{n_0} \tag{2.5}$$

where n_0 is very close to unity for space and dilute gas. Since n depends on wavelength (see below), so does the bending of the rays. One example is a rainbow (Fig. 2.5A and B) which is produced by refraction and internal reflection, which leads to the rays fanning (dispersion). The above-described angular dependencies of reflection and refraction can be derived from Maxwell's equations with considerable mathematical complexity, but is obtained rather simply from the 1678 wave theory of Huygens, or from the 1650 principle of the minimum path by Fermat.

Equations describing light traversing a medium presume conservation of energy. If the only heat supplied to a material arrives by radiation, then the general equation is:

$$I = \frac{I_{trns}}{I_0} + \frac{I_{abs}}{I_0} + \frac{I_{refl}}{I_0} + \frac{I_{emit}}{I_0} = \Upsilon + \alpha + R + \varepsilon \tag{2.6}$$

where I_0 is beam intensity impinging on the sample, I_{trns} is the intensity of light transmitted by the sample, I_{abs} is the intensity of light actually absorbed, I_{refl} is the intensity of light reflected on the first bounce from the front face (Fig. 2.4A), I_{emit} is the light emitted,

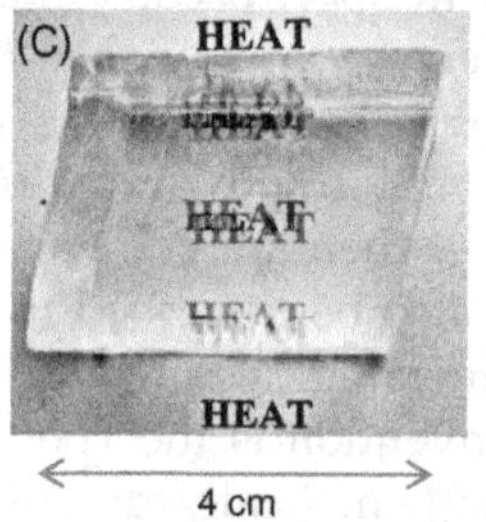

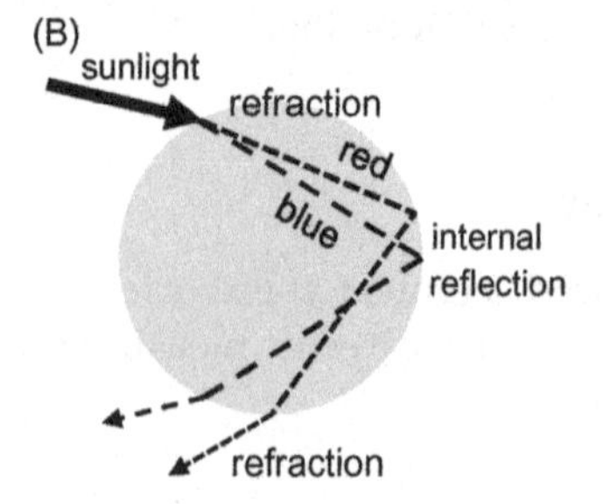

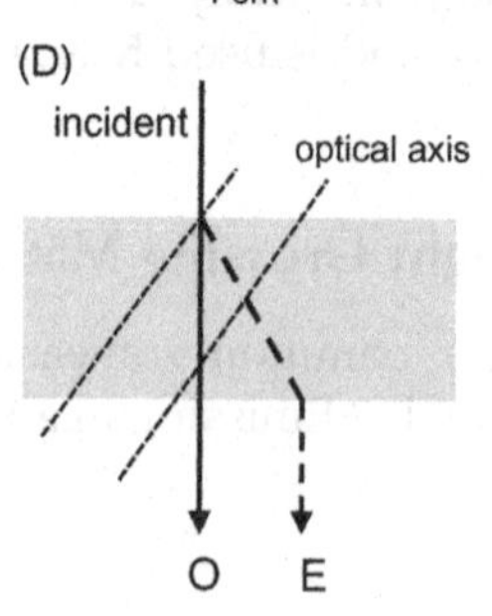

FIGURE 2.5 Refraction phenomena. (A) A rainbow, which results from refraction and internal reflection of visible light inside raindrops, which disperses the light. (B) Schematic of light paths inside a droplet. (C) Double refraction seen for a large calcite crystal. The text under the crystal is "HEAT." A reflection is seen on the bottom side of the crystal. (D) Schematic of light paths in the calcite crystal.

Υ is *transmittivity*, α is *absorptivity*, R is *reflectivity*, and ε is *emissivity*. This problem would be indeed difficult to analyze, especially as Eq. (2.6) is a simplified version assuming parallel rays. Fortunately, the emissivity term is not needed for laboratory measurements around at room temperature because sample emissions that are much less intense than the incident beam (sources are typically a ~1000K SiC electrically heated bar or a bright filament in a light bulb). Thus, conditions common in experiments are amenable to using a simpler form:

$$I = \frac{I_{trns}}{I_0} + \frac{I_{abs}}{I_0} + \frac{I_{refl}}{I_0} = \Upsilon + \alpha + R \tag{2.7}$$

2.2.1.1 Absorption versus Transmission

Because experiments record the amount of light exiting the sample (Fig. 2.4A), we actually measure transmission (I_{tran}/I_0), which is <100%, due to the combined losses due to reflection and absorption. To compute, the true absorption coefficient (A) requires at least accounting for the two first-order back reflections:

$$A(\lambda)L = -\ln\left[\frac{I_{trns}(\lambda)}{I_0(\lambda)}\right] - 2\ln[1 - R(\lambda)] \tag{2.8}$$

where normal incidence and parallel light rays crossing a flat sample with parallel faces of thickness L were assumed. One could also compute the transmission, but this, too, is affected by surface reflections (Fig. 2.4A). It must be emphasized that Eq. (2.8) is an approximation: we have assumed that that second order and higher reflections are not important in our particular experiment.

Eq. (2.8) reveals several important aspects the interaction of light with a material, and how we characterize this interaction. First, both absorption and reflection data are needed for a complete description of the interactions. Yet, when the absorptions are either very strong or very weak, we can provide approximate values (Section 2.4). This feature makes absorption spectra useful. Second, how light interacts with the solid depends on wavelength. Thus, implicit in Eq. (2.8) is that the interaction of light with a material will be characterized by another parameter, the interval of wavelength over which the interaction occurs.

Absorption spectra are discussed before reflection, in Section 2.5, because these measurements are easier to understand than reflection or emission spectra. Moreover, absorption describes the attenuation of light and thus the conversion of light to internal energy.

2.2.1.2 Absorption and Momentum

If the energy of the light (E) is totally absorbed by the material, then momentum (p) is also delivered to the material. Maxwell predicted that $p = E/c$. The consequence is radiation pressure, which was experimentally confirmed in 1903 by Nicolls and Hull. The effect is complicated by light being partially reflected and partially absorbed at a surface (see Halliday and Resnick, 1966 and Sections 2.4 and 2.5).

2.2.2 Other Variables Representing Interaction of Light With a Material

Although the absorption coefficient and reflectivity are directly and most commonly measured, two other pairs of variables can be used to describe how light interacts with matter. We mention these pairings because these are used in quantitative analysis of spectra, such as the damped harmonic oscillator model which is used to model heat transfer (Chapter 11).

The *complex index of refraction* ($\tilde{n} = n + ik$) is used to analyze reflectivity data and pertains to many radiative transfer models, where n is the *real index of refraction*. Also, direct measurements of n provide an independent constraint on R in transmitting regions (e.g., Hofmeister et al., 2009). Several names are used for the imaginary part, k, some of which are confusing. We use *absorption index* (e.g., Brewster, 1992) as opposed to "extinction coefficient" (e.g., Bell, 1967) because the latter is commonly used to describe the combined effects of physical scattering and absorption in radiative transfer models (e.g., Shankland et al., 1979). Together, n and k are referred to as *optical functions*, which is preferred over historic use of "optical constants," because these spectral properties depend on wavelength. The absorption coefficient (A) and the absorption index (k) are closely related:

$$A = 4\pi\nu k \tag{2.9}$$

(Wooten, 1972). If k is obtained from reflectivity data (Section 2.5), then Eq. (2.9) gives absolute values of A.

The complex dielectric function ($\ni = \ni_1 + i\ni_2$) links optical properties with electrical properties of a material. Although this representation is completely equivalent, the connection with light measurements is less direct, and thus dielectric functions are less frequently used to represent spectroscopic data. However, values of this property from reflectivity measurements (Section 2.5) can be cross-checked against direct measurements of dielectric properties at very high and very low frequency (Subramanian et al., 1989). Such comparison provides information on the accuracy of the other derived functions, and is important at low frequency, where n is not independently measured.

2.2.3 Wave Optics

2.2.3.1 *Diffraction and Interference*

When conditions of geometrical optics are not met, that is, when component dimensions are comparatively small, $L << \lambda$, diffraction and interference occur. The experimental and theoretical importance of these two properties of EM waves cannot be overstated. Interference effects underlie Michaelson's experimental setup for measuring the speed of light and the operation of modern IR spectrometers. Diffraction in three-dimensions of EM radiation with very short wavelengths (X-rays) by the atoms periodically arranged inside a material allows its crystalline structure to be ascertained. Halliday and Resnick (1966) and other physics textbooks provide a detailed discussion, the history, quantification, and illustrations of these two phenomena. A short summary is provided here.

When a wave impinges on a single slit with small width (or pinhole, as well as a thin obstacle, like a wire) it is *diffracted*, where the light wave fans out beyond the slit, making an arc. This behavior is known for all waves (in water, or sound) and was observed by many researchers for light in the 1600s. It was not until the early 1800s, when Fresnel used the wave theory of Huygens to describe Young's double slit experiment that the wave theory of light was mathematically established.

Slightly different equations are needed depending on whether the rays are radial or parallel. The latter describes sunlight, which was used in historical experiments.

When a wave is forced through two slits (or two pinholes or wires), the fanned out rays both constructively and destructively interfere. Demonstration of *interference* effects when sunlight passed through double pinholes by Young in 1801 proved the wave-like nature of light and provided numerical values for its wavelength in the visible. Although we tend to think in terms of the variability of the visible (color), this is not really the case: variation is perceived because our eyes detect only a very narrow range of wavelengths. Were it not for sunlight having a broad peak in the yellow, close to where our eyes are sensitive, interference would have been difficult to establish. This work was summarized by Young (1804). However, no one believed Young until after Fresnel's experiments and publications over 1816–1824. Young died in 1829.

Interference effects are important in studies involving light today. Gratings, which proved useful in demonstrating the equivalence of the calorific with the luminous in air (Chapter 1, Section 1.3.1), consist of parallel, multiple slits or wires. Gratings are now commonly used as polarizers in the IR region. Thin films also provide interference effects. Constructive interference occurs when

$$2L = j\frac{\lambda}{n(\lambda)} \tag{2.10}$$

where L is film thickness, and j is an integer. Fringe spacing is useful in spectroscopy (Griffiths and de Haseth, 1986), when sample thickness is too thin to reliable measure in a microscope, because $n \sim 1$ in nonabsorbing regions, and conversely, if L is known, n can be determined.

2.2.3.2 Polarization and the Transverse Nature of Light

Diffraction and interference effects occur for both longitudinal and transverse waves, and so cannot be used to distinguish between these types. Sound waves in air and compressional waves in elastic media are *longitudinal acoustic* (LA) waves, where the displacement is parallel to the direction of propagation. For *transverse acoustic* (TA) waves, displacement is perpendicular to the travel. These shear waves travel inside solid elastic media, but cannot propagate inside a gas or a liquid because there is no mechanism for driving motion perpendicular to the propagation of the wave. Hence, in the 1700s, light waves were thought to be longitudinal. Note that ripples on a pond are transverse, but these are a surface disturbance powered by wind.

In the early 1800s, Augustin Fresnel and D.F. Arago demonstrated that unpolarized visible light traveling through calcite ($CaCO_3$) crystals splits into two beams, with no fringes whatsoever (Fig. 2.5C). Furthermore, the two emerging beams are polarized such that the polarized directions are orthogonal (Fig. 2.5D) as discovered by Malus (1811). This behavior, known as double refraction, does not occur when the waves are longitudinal. Malus did not correctly explain his experiments because he believed in corpuscles. Young provided the explanation in an 1817 letter to Arago, in terms of polarization of transverse waves of light. Calcite having a very large directional dependence of its index of refraction and its occurrence as clear, large rhombohedrons provided this important experimental test.

Conclusive evidence was actually provided earlier by experiments of Malus (1809) on light polarized by reflection. Malus showed that light intensity varies with the angle as $\cos^2(\theta)$, which behavior requires transverse waves, although this was not accepted until double refraction was explained.

Additional behaviors associated with transverse EM waves are illustrated in Fig. 2.6. For example, maintaining the desired polarization in an experiment involving reflection (Fig. 2.6F) requires what is denoted as the S-polarization (Jackson, 1975).

Understanding light propagation should help us to decipher mechanisms of heat diffusion, independent of a microscopic model. In air, EM waves are *transverse optic* (TO). In denser media, EM waves measured are predominantly TO, but *longitudinal optic* (LO) modes also exist. LO modes are best probed through their strong effect on reflectivity measurements (Section 2.5), but LO modes exist to some degree in absorption experiments due to non-normal incidence or the sample being thick or wedged (Berreman, 1963; also, Fig. 2.6). Visible and IR light can cross gas, liquids, and solids, and anisotropic responses are observed, as described in the next two sections. The same should hold for heat, but experiments analogous to Young's have yet to be performed. Acoustic waves measured in elasticity experiment also have two polarizations, LA and TA, which are important during propagation through a solid medium, although TA modes cannot travel through liquids or gases. Young, who also contributed to the field of elasticity, was initially troubled by his inference of TO behavior for light, because this meant that the ether was a solid (Barr, 1963).

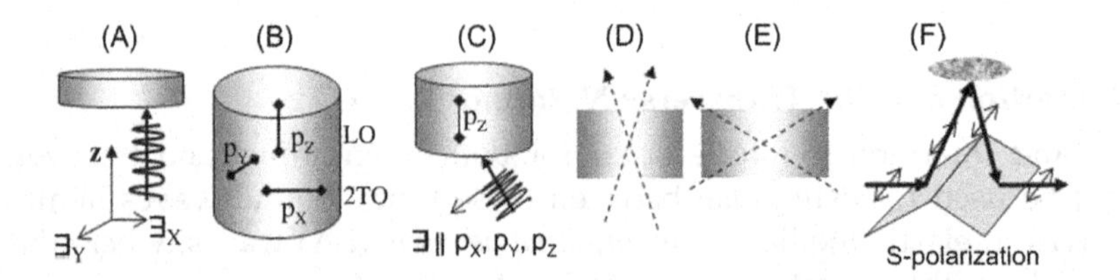

FIGURE 2.6 Polarization effects associated with direction and path length of electromagnetic (EM) waves. (A) An EM wave traveling along the *z*-direction has its amplitude along the *x*- and *y*-directions. (B) Perspective view of atom pairs vibrating in three directions in the solid. Those with dipoles along the *x*- and *y*-directions are directly stimulated by light, because the electric field of the dipole is proportional to the polarization and in the same direction. These are the *transverse optic* (TO) modes of the orientation shown. Indirect coupling stimulates the *longitudinal optic* (LO) modes, which are equivalent to TO modes for the perpendicular orientation of the same sample. For longer paths, more LO modes are excited. (C) During non-normal incidence, the EM wave has amplitude in the direction of all vibrating dipoles, so all are stimulated. As the angle increases from normal, more LO component is present, thus mixing polarizations. (D) Edge view of a moderately focused beam. (E) Edge view of a highly focused beam which has a longer path length, and greater mixing of polarizations. (F) Schematic of the S-polarization for reflectivity measurements. Source: *Modified after Fig 4 in Hofmeister, A.M., 2014a. Carryover of sampling errors and other problems in far-infrared to far-ultraviolet spectra to associated applications. Rev. Mineral. Geochem. 78, 481–508, with permissions to reprint from Mineralogical Society of America.*

2.3 SPECTROSCOPIC INSTRUMENTATION

Spectroscopic instruments and data acquisition are covered in brief, with a focus on the IR region. For an in-depth presentation of optical methods and assessment of data of many materials over a wide wavelength range, see Palik (1998) or Lindon et al. (2016).

That the interaction of light with matter depends on frequency (Figs. 2.2 and 2.3) means that different spectral regions require different equipment. Essential components are a source which generates EM waves, a means to isolate or separate the frequencies, and a detecting system to record the light received. In optical spectroscopy, data are collected on a reference, giving I_0, and a sample, giving I_{meas}. For the visible and adjacent regions, a prism or grating on a rotating plate is used to disperse the rays of light and then a slit is used to limit the light striking the detector to a narrow range of frequencies (Sommer, 1989). This approach works well in the near-IR and at higher ν. Today, data in the near-IR through the UV ($\sim$9000–52,000 cm^{-1}) are routinely acquired with inexpensive single- and double-beam commercially available spectrometers, including miniatures. This literature is huge and includes many applications. Higher frequencies involving electronic transitions of the elements were a focus $\sim$1890 upon development of a vacuum-UV instrument by Victor Schumann. However, continuous monitoring of I versus frequency is difficult above $\sim$80,000 cm^{-1}, due to characteristics of the light sources and materials used for windows and prisms (e.g., Hunter, 1982). Such instrumentation (French et al., 2007) is expensive and rare. Such data are important to astronomical studies, due to the high blackbody temperatures of large stars, whereas the tapering of the blackbody curve in the UV at 5000K (Fig. 2.3) indicates that the visible region is important for planetary interiors. Higher frequencies (soft X-rays) are explored using synchrotron radiation with other techniques (e.g., ellipsometry: Dardono and Jaworowski, 2010).

Due to material property limitations, mid-IR spectrometers were developed later than visible, c. 1930 (Sawyer, 1963). Advances in grating manufacture circa 1934 and in detectors during World War II permitted chemists to study fundamental vibrations of organic and inorganic materials. However, grating instruments have many limitations in the IR region, which led chemists of the 1950s to use other means for studying compounds. Infrared spectroscopy was revitalized around 1960 by the development of Fourier Transform spectrometers (FTIR). At this point in time, spectroscopy was of great importance to physics. Subsequent advances in computer technology made these instruments widespread by $\sim$1980. Today, commercially available FTIR spectrometers typically cover 100–14,000 cm^{-1}, but can extend down to 1 cm^{-1} and up to $\sim$50,000 cm^{-1}, although the latter extension is not particularly advantageous over prism instruments. See Griffiths and de Haseth (1986) for further information.

For wavelengths longer than those considered as optical, different techniques and instrumentation are used. For microwave spectroscopy, see Robitaille (2009) or Townes and Schawlow (2012). The latter book also addresses millimeter spectroscopy. These regions pertain to cryogenic conditions (Figs. 2.2 and 2.3) and to the rotational transitions of molecules. As in IR, a dipole moment must exist. Heat transfer in this region is exemplified by microwave ovens exciting the rotations of the water molecule. Regarding heat transfer studies of solids, extrapolating data in the far-IR to lower frequency generally suffices, due to the combination of low absorption strengths and low intensity of I_{BB} at long wavelengths.

2.4 PROBING MATTER THROUGH ABSORPTION SPECTROSCOPY

Although reflection measurements are needed to completely describe the interaction of light with condensed matter, many modern and historical studies measure only absorption. This approach provides a complete picture for gases, which lack a definable reflecting surface. On this basis, Kirchhoff's law:

$$\alpha(\lambda) = \varepsilon(\lambda) \tag{2.11}$$

(absorptivity equals emissivity) is valid for his experiments on sodium vapor. For condensed matter, neglecting reflection gives a maximum value for the absorption coefficient A (the logarithmic version of a, Eq. (2.8)). This approximation is reasonable for spectral regions with low reflectivity, mainly the visible and near-IR. Lipson (1960) describes how to obtain limiting values in transmitting regions. An extension of Lipson's approach provides A and reflectivity R from spectral data involving multiple thicknesses of a sample (Hofmeister et al., 2009). Single-crystals are required to avoid internal reflections. Other situations can be analyzed semi-quantitatively by baseline subtraction, which removes much of the effects of reflection from measurements that actually record transmission (Fig. 2.4). Thus, measuring absorption spectra can reveal the essentials of how light interacts with matter.

Absorption is directly related to heat transfer. This technique characterizes what wavelengths (or frequencies) interact with the material, and hence describes the amount of energy taken up. Absorption also provides a window into the microscopic behavior of the atoms. In particular, light in the IR to UV regions has energies that match those of vibrational and electronic transitions. We return to the nature of these interactions in Chapter 5, but mention here that the absorption peaks are specific to the chemistry and structure of a material, and thus provide a means of characterizing or identifying a material. Hence, this type of spectroscopic measurement is largely the domain of chemists, and played a key role in revealing that atoms are divisible in the 1800s (McGucken, 1969).

One concern in this section is how the *path length* (thickness for normal incidence) and various other factors affect the absorption of light by various types of material. The factors considered here pertain to heat transfer. The focus in this section is on electrical insulators (e.g., silicate and oxide minerals) which are partially transparent because ionic or covalent bonding involves localized electrons. The vibrations in non-metallic materials involve motions of ions without significant electronic screening and thereby produce distinct vibrational peaks. The connection of radiant emissions at low temperature with the IR region (Figs. 2.2 and 2.3) motivates exploring vibrational spectra in some detail: our spectral examples mainly concern this region. Although use of focusing optics is common, normal incidence is generally presumed in data analysis. This approximation only becomes a problem with extremely converging optics (e.g., Hofmeister, 2010).

In pursuing absorption spectroscopy, it is necessary to recognize how reflections qualitatively affect these measurements. To illustrate the concepts, we provide real data (i.e., concrete examples). One important goal is to understand the conditions under which the diffusive and ballistic types of radiative transfer occur (Fig. 1.6). Another important goal is to understand the energy transfer from a macroscopic perspective. Regarding the physics, absorption of light converts the energy received into the internal energy of the material, and thus spectroscopy involves heat transfer. Due to the speed of light being enormous, the reactions are nearly instantaneous. However, finite time exists between an absorption event and re-emission, which slows the speed of the light ray across the sample. In weakly absorption regions, the speed is c/n, and thus light is still somewhat slower than in space. From Fig. 1.3, spectroscopy cannot be isothermal; but the temperature change is small in the weakly absorbing regions. Rapidity is consistent with meeting adiabatic conditions as regards interactions with the surroundings, except that a small parcel of light-energy is added to the material. The incremental T and energy changes associated with a spectroscopic experiment through a largely transparent region make it highly amenable to thermodynamic analysis. In contrast, for strongly absorbing regions heat is slowly diffused: this behavior is covered in Chapter 3.

2.4.1 Spectra of Gas, Absorbance, and Ballistic Conditions

The field of absorption spectroscopy began with the study of gases. In 1802, William Wollaston noticed dark lines in the visible region of the solar spectrum. Joseph Fraunhofer considerably advanced these efforts. Laboratory studies of 1833 by William Miller showed that different gases absorb at a specific wavelengths. Identification and the principles of spectral analysis were developed in 1850–1860 largely through the collaborative efforts of Robert Bunsen and Gustav Kirchhoff. Bunsen used his eponymous burner to volatilize chemical elements, while Kirchhoff measured spectral emission lines of the vapor, and deduced Eq. (2.11).

Shortly thereafter, spectra of gas and volatilized metals were also obtained in the laboratory using electricity by Anders Ångstrom, while volatilized salts (NaCl) were probed in the visible by A. Mitsherlich. Spectra of metal vapors just below the red were studied ~1873 by E. and H. Becqueral. The astronomical importance of H motivated measurements ~1890 in the UV and visible by Lyman and Balmers, and at lower frequencies by Paschen, Rubens, and others.

Many early spectra consist of lines arising from atomic transitions. This is not true for molecules. A spectrum of the atmosphere in the laboratory is shown in Fig. 2.7. Although the air in a room may seem rarified, it is full of many different molecules that intersect light rays. Infrared spectrometers are evacuated to remove these strong absorptions. The species shown are CO_2, which has two strong absorptions with finite but narrow widths (~650–690 and 2310–2370 cm^{-1}) and H_2O, which has broad and strong envelopes of many narrower bands (~0–400, 1300–2000, and 3540–3950 cm^{-1}) and similar, but weaker, envelopes that reach 41,082 cm^{-1} (not shown: see Grechko et al., 2009). Observing these very weak bands, which amazingly reach the first dissociation limit of the water molecule, require long paths or high concentrations.

Vapor peaks are often complex and broad because molecules not only have vibrations due to thermal energy, but also can rotate (e.g., Bernath, 2016). Rotational transitions were discovered by Eva von Bahr (1913), who pioneered high-resolution near-IR studies. Because light stimulates both types of motion, the peaks of water vapor Fig. 2.7 are the sum of hundreds of narrow absorption bands. For high-resolution spectra which

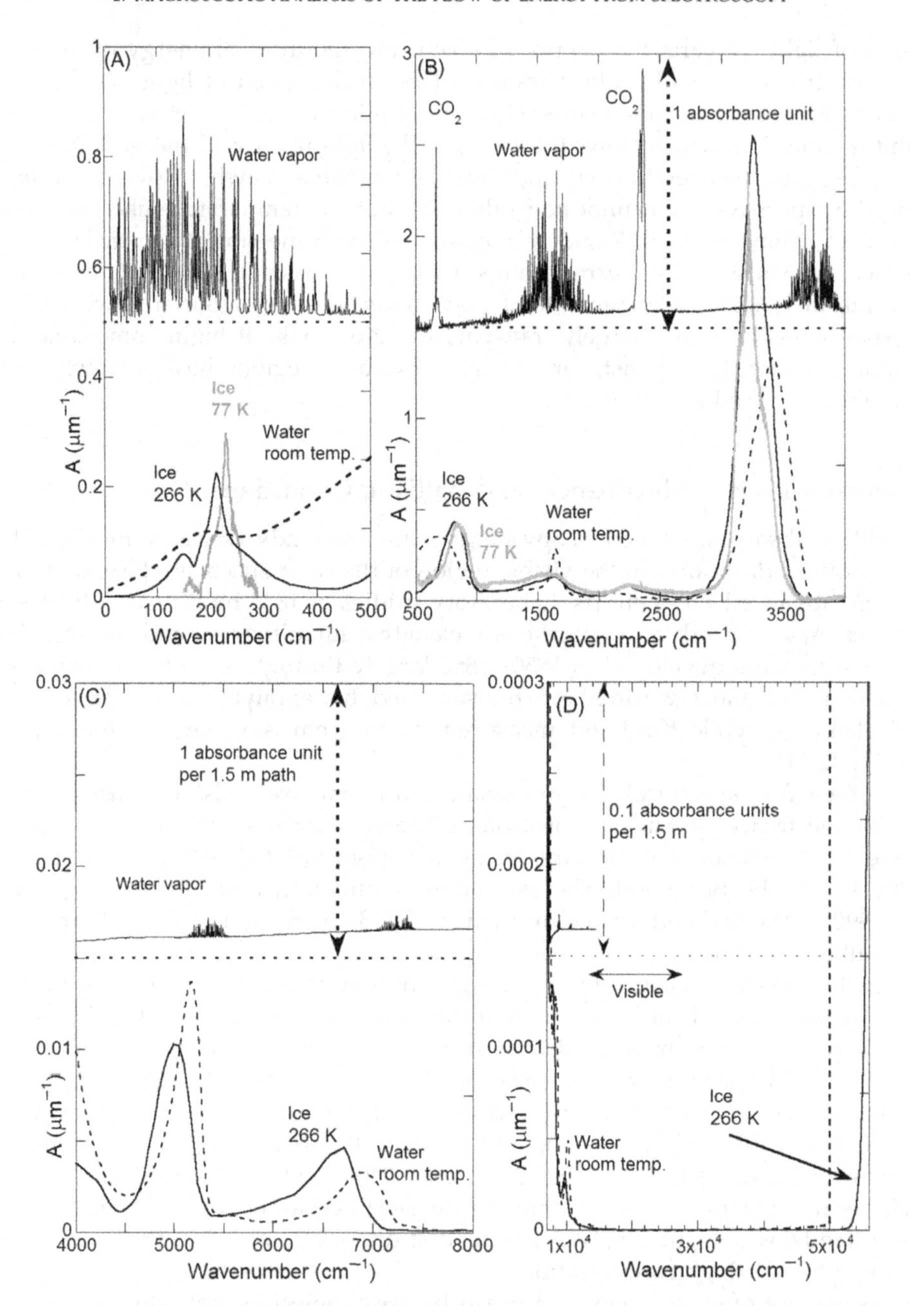

FIGURE 2.7 Spectra of water in its various phases. Each panel shows a different frequency range, mostly connected with instrumentation, with a *y*-axis scale chosen to best show the peaks. (A) Far-IR region. (B) Mid- to near-IR. (C) Near-IR. (D) Near-IR to UV regions. Upper part of all panels: data collected on atmospheric vapor in the spectrometer bench on different days, which were scaled to the 30% humidity during the data collection of (B). However, the far-IR measurements of vapor (A) did not overlap sufficiently to permit reliable scaling: values

distinguish these many bands, see the seminal study at low frequencies by Benedict and Plyler (1951). The copious high frequency data on water are reviewed by Bernath (2002) and Tennyson et al. (2014). Considerable data exist on gases and molecules frozen in inert gas matrices, as summarized by Nakamoto (1978). For diverse gases, see NASA Astrobiology Institute (2017). Pure rotations exist at lower frequencies. For water, rotational bands occur in the millimeter range (Matsushima et al., 1995). Spectral data on atmospheric gases are of great interest, due to different roles played by various species in global warming.

Importantly, a molecule or an atom does not have a definable surface. Therefore, correcting for reflection is not needed, so absorption coefficients are obtained from the intensity loss and path length only, see Eq. (2.8).

From Fig. 2.7, each peak is characterized by the frequency of its maximum (peak position), its absorption strength (i.e., maximum intensity), and its width. The latter is quantified as *full width at half maximum* (fwhm). However, due to complexity of shapes and/or the presence of poorly resolved peaks, peak area is often reported rather than fwhm. This is computed by integration and the term *integrated intensity* (*I.I.*) is used. The absorption of water vapor in this spectrum is much stronger than those of CO_2, considering the area under the envelopes. To quantify the strength of the absorption of any species, we would need to know the number of molecules in the path, which can be done accurately with a tube of gas.

Spectra are also characterized by the number of peaks. The number of peaks is related to the transitions that are allowed. Regarding vibrations, the number of peaks can be ascertained from the symmetry of the molecules (e.g. Cotton, 1971) and from how the symmetry is reduced in assembling a crystal structure (Fateley et al., 1972).

Even if we cannot accurately measure the thickness or optical path, transmission (absorption) measurements can still be made, because we can still determine the product AL. This important quantity is denoted *absorbance*, $\forall = AL$.

If the gas is a weak absorber or the path length is relatively short, it is very difficult to resolve the peak above the noise. This condition is *ballistic*: the light crosses the sample with negligible interactions (Fig. 1.6). The absorptions may still exist, but are below the

shown in (A) are similar to those depicted in (B). For the bands of water vapor in the mid-IR, the absorbance is about 0.67 for the 1.50 m path length in the spectrometer, that is, $A_{max} \sim 1\ \mu m^{-1}$ for water vapor at about 30% relative humidity. Reference spectra were collected from the evacuated bench. Above the "jog" at 9333 cm^{-1}, data on vapor from Schermaul et al. (2001) are shown, which were scaled to match our measurements in the near-IR. Schermaul et al.'s (2001) data at higher concentration had a flat baseline, whereas our baseline increased towards the visible (scattering from airborne dust), providing the break in slope. At 29,000 cm^{-1}, Wilson et al. (2016) provide an upper limit on vapor of 0.0004 μm^{-1}, which is much stronger than that of water. Lower part of all panels: spectra from condensed phases for the y-axes scales as shown. Gray: data on a thin film of crystalline ice condensed on a disk held at liquid nitrogen temperature (~77K), where A is approximate in all panels because the effect of reflections were not removed. The strongest peak is affected the most because A and R are correlated (see text). Black solid curve: ice at mostly 266K computed from compiled k values of Warren and Brandt (2008): this approach accounts for surface reflections. Our absorption peaks are slightly shifted and narrower due to the lower temperature. For our uncorrected data, reflection only has a noticeable effect for the most intense peak near 3300 cm^{-1}. Dashed curve: water at room temperature from compiled k values of Bertie and Lan (1996) below the visible while the visible and above are A-values compiled by Wozniak and Dera (2007). The sharp increase at 50,000 cm^{-1} is attributed to impurities in their sample (sea water). Our direct absorption measurements of water between BaF_2 disks are quite similar to compiled data on k in the mid- and near-IR.

level of detection in the experiment (e.g., bands of vapor in the visible segment do not show even in the expanded scale of Fig. 2.7D).

2.4.2 Spectra of Liquids, Attenuation, Mean Free Path, and Optically Thin Conditions

Limitation of the early period of spectroscopy to transparent substances and near-IR wavelengths (Kayser, 1908; Kayser and Koner, 1934) delayed collection of data on vibrations of condensed matter. Considerable data on liquids now exist. Our example is liquid H_2O, which has been studied in great detail. Water peaks are broad and nearly featureless (Fig. 2.7). Some peaks occur in roughly the region as those of vapor, but seem to be shifted to lower frequencies. Roughly, the absorption coefficients are the same for ice, water, and vapor (accounting for the relative humidity). The shapes of the bands differ greatly between gas and the condensed phases, plus the number of overtones is few in condensed matter (more examples below): the microscopic basis is covered in Chapter 5.

A crucial difference between obtaining spectra from a gas and from a liquid is that the latter has a well-defined surface, which reflects light. Typically, liquid is held in a container for measurement, in which case the reflections involve the surfaces both of the container and the liquid. For this case, collecting a reference spectrum of the container, and/or performing a baseline subtraction, removes most, but not all, of the loss of light due to reflections. Obtaining absolute intensities from liquids is difficult (e.g., French et al., 2007; Keefe et al., 2012), as in solids, see below.

A second important difference is that liquid is much denser than gas of the same chemical composition. In obtaining our spectrum of water vapor, the path length of 1.50 m intersected relatively few H_2O molecules at 30% relative humidity. But to provide a spectrum of liquid water which is quantitative (i.e., the sample transmits some light at all frequencies), an appropriate path length is $\sim 5\ \mu m$. Measured absorption strengths for water range from $\sim 10^{-5}$ to $1.5\ \mu m^{-1}$ (Fig. 2.7). Strong bands for any substance are difficult to quantify through absorption measurements because very thin samples are needed. In this case, reflectivity data are collected (Section 2.5) and Eq. (2.9) is used to provide A.

Quantifying the interaction of light with condensed matter via transmission measurements requires some light to cross the sample at the relevant frequencies. If the sample is so thick as to be opaque, then the spectra cannot be recorded. Conditions where a significant amount of light crosses a sample are known as *optically thin*. But what is significant? Rearranging Eq. (2.8) and neglecting reflections:

$$I_{trns}(\lambda) = I_0(\lambda)\exp^{-LA(\lambda)} = I_0(\lambda)\exp^{-L/\mathrm{mfp}(\lambda)} \qquad (2.12)$$

where the mean free path (mfp) is defined here as $1/A$. Essentially, the signal is exponentially attenuated as it travels through the sample. About an order of magnitude reduction in signal strength occurs when the absorbance $\forall = AL$ exceeds unity. Generally, conditions are optically thin for wavelengths where $AL < 1$ (Siegel and Howell, 1972). This value is not exact, because the effect of reflections was neglected. If the material has a high reflectivity in the region of interest (i.e., at the peak), then the back reflection (Fig. 2.4) reduces the throughput and so the sample must be "thinner" to prevent the reflections and the absorptions together from blocking the light. This situation describes the mid-IR, where vibrations interact strongly with light, and thicknesses below ~μm are needed to quantify the absorption spectra. Conversely, in a spectral region were the sample is a poor reflector, the product *AL* can be higher, ~2 or larger.

At frequencies far from the peaks (either at lower or higher frequencies), the sample negligibly absorbs and the conditions are ballistic. To characterize the sample, then, one needs to measure many different thicknesses for a given material to see if a peak is masked by the noise, and to differentiate the range of lengths appropriate to optically thin and ballistic conditions. For highly transparent materials, it is difficult to determine the absorption coefficient *A* in the transparent near-IR. Many of the data sets in Palik (1998) do not report values when *A* drops below 10^{-8} cm^{-1} because accurate values are not measurable in the laboratory, due to limitations on the sizes of samples and sample chambers.

From considering gas and liquid spectra, thickness is first-order parameter in characterizing spectra. Whether useful data can be collected at any given wavelength depend on the thickness. Given the character of any given peak (its maximum *A*) and Eq. (2.12), there is a limited range of thicknesses where absorption spectroscopy yields useful and reliable information for the peak of interest.

2.4.3 Spectra of Electrically Insulating Solids and Optically Thick Conditions

Melloni made the first IR measurement of solids about 1840. Even in the 1890s, studies were limited to the transparent near-IR due to the strength of the fundamental modes in condensed matter. For example, Rubens and collaborators measured glass, calcite, and mica over ~1–~5 μm (1000–4000 cm^{-1}) (Kayser, 1900, 1908). Measurements of salts were made to ~20 μm (down to 500 cm^{-1}) because their strong vibrations exist at even lower frequencies (e.g., Ferraro, 1984). Therefore, early investigations of solids probed overtones and impurity bands.

The ice spectrum in Fig. 2.7 was collected at a lower temperature than of water, which sharpens the peaks, as shown by comparing ice absorptions at 77 and 266K. Nonetheless, the signatures of crystalline ice and amorphous water are remarkably similar. The similarity exists because the vibrational peaks are derived from motions of nearest neighbor atoms, and because ice is a molecular solid. Molecular bonding is not changed much between the liquid and solid phases. Spectra of vapor and condensed phases differ significantly (Fig. 2.7) because gas molecules are well-separated and do not interact except during collisions.

The effect of order is evident for spectra collected at the same temperature from condensed matter with strong ionic–covalent bonds. Fig. 2.8 compares IR spectra of glass and crystalline silicates with fairly simple chemical compositions. Disorder broadens the peaks at the expense of the maximum *A*. Hence, disordered samples are overall more transparent, and optically thin or ballistic conditions exist over a wider range of thickness.

As thickness increases, less light is transmitted (Eqs. (2.8) and (2.12)). The effect on absorbance is nonlinear (Fig. 2.9). Increasing thickness of materials with several peaks of different strengths in the mid-IR region changes the spectral profile, and also the band

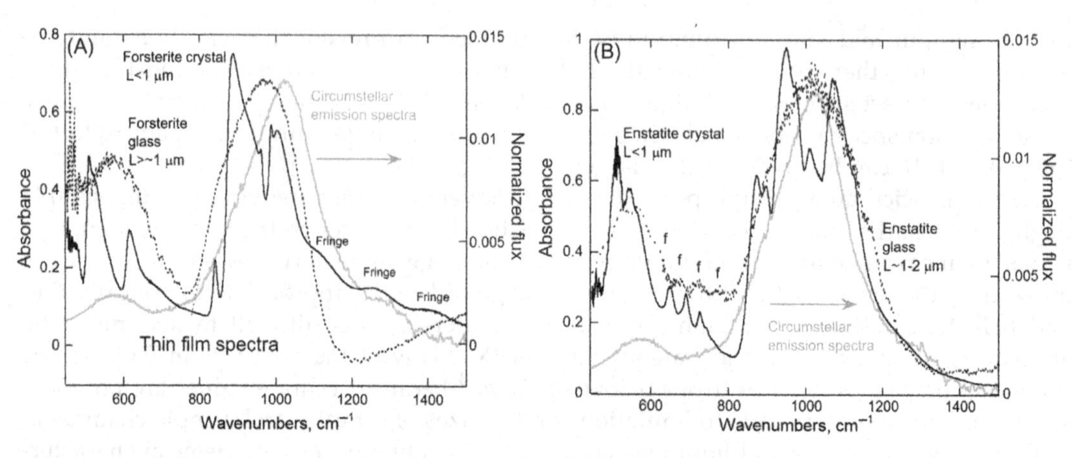

FIGURE 2.8 Comparison of mid-IR spectra from thin films to emission spectra from dust in space. (A) Forsterite (Mg_2SiO_4), which a very hard material and not conducive to making even films. (B) Orthoenstatite ($MgSiO_3$), which is softer and makes better films. In both parts: the thin films were made by crushing grains in a diamond anvil cell, and represent an average of polarizations for the crystals. Glass is less absorbent, so thicker films were used. The rise towards low frequency is connected with broad interference fringes, and is more pronounced from glass due to its higher transparency. The empty cell is used for the reference, for details see Hofmeister et al. (2003). Interference fringes are indicated. The emission spectrum of dust in space (right *y*-axis) is an average of different environments, after Sloan and Price (1995) and Sloan et al. (2003).

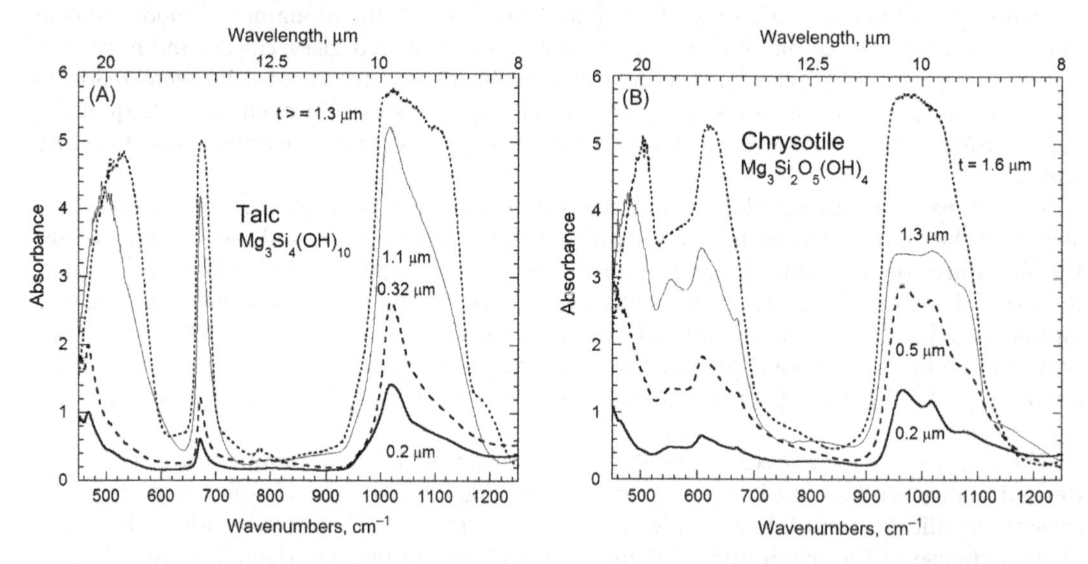

FIGURE 2.9 Spectra of thin films of chrysotile (A) and talc (B), showing the effect of thickness. Thickness was determined from comparing the absorbance of the O–H stretching bands in the thin film to that obtained from a thicker film of 6 μm. This comparison was made possible by the softness of these minerals, which permitted high quality films to be made, and by hydroxyl being stoichiometric. See Hofmeister and Bowey (2006) for details and additional measurements.

width, because as thickness increases, more LO modes participate (Fig. 2.6). Similar results are obtained for quartz (Hofmeister et al., 2000). Strong peaks are affected more than weak in the thickness range explored. Overlapping peaks are blurred, and eventually, a flat topped, almost square peak is produced for the most intense bands ($\sim$1000 cm^{-1} for chrysotile). This sample is actually opaque at 1000 cm^{-1}, but because the film is imperfect, a tiny amount of light reaches the detector and so the absorbance is not infinite. Due to this known problem in distortion, most software limits the computed value of absorbance $\forall$ to about 4, because higher values are not trustworthy due to either neglecting reflectivity (Eq. (2.8)) or light leakage. When $\forall = AL > 4$, the conditions are essentially opaque. However, this situation (complete absorption of heat) is not very interesting. Of greater interest are *optically thick* conditions, when $AL > 2$, which are related to radiative diffusion. As in ascertaining optically thin conditions, the cutoff is approximate, because the value of reflectivity matters.

Most of the peaks in Fig. 2.9 describe experimental conditions that are neither optically thin, nor optically thick, but are somewhere in between. The author is not aware of a name for this situation, which can be encountered during radiative transfer (Fig. 1.6). For shorthand, we suggest *optically distorted*, by analogy to sounds produced from inexpensive stereo equipment at full volume. Distortion is evident in the nonlinear dependence of absorbance on thickness (Fig. 2.9). Obviously, values of A for any peak cannot be accurately determined unless conditions are optically thin at the peak maximum.

Spectra gathered under optically thick conditions over some frequency range would have a constant absorbance over that range, equal to the maximum dictated by the software. However, commercial spectrometers generally will not collect the data if the entire spectral region is optically thick. If part of the region is optically thick (i.e., where the sample strongly absorbs), then flat topped peaks are seen, as in Fig. 2.9. If dispersions of powered in a matrix are used to produce a sample (the KBr pressed pellet technique), then the spectra are a mix of light going through the matrix and through the edges of the sample, with transmission at the peaks being blocked by the large grains. The resulting peaks are optically distorted, and poorly represent intrinsic behavior of ordered, crystalline material (summarized by Hofmeister et al., 2000). The problem also pertains to radiative transfer models (Chapter 8).

2.4.4 Effects of Frequency, Polarization, Temperature, and Pressure on Spectra

Spectra of a material depend greatly on the frequency region that is being studied, as indicated by Fig. 2.7 for H_2O in its three states. Because different types of transitions are associated with some range in energy, different spectral regions probe different processes (Fig. 2.2). The various processes are likewise associated with a temperature (Fig. 2.3). Elevated temperature can expand the frequency range where certain types of transitions occur because higher energy states are excited: water vapor exemplifies this behavior (Bernath, 2002). When the samples are not isotropic (cubic symmetry), the transitions depend on crystallographic orientation, and so measurements with polarized light provide information about structure (discussed further below).

Drastic changes exist in spectra of any given material with frequency (Figs. 2.7–2.10). In particular, absorbance in the IR is about 100 to 1000 times stronger than in the

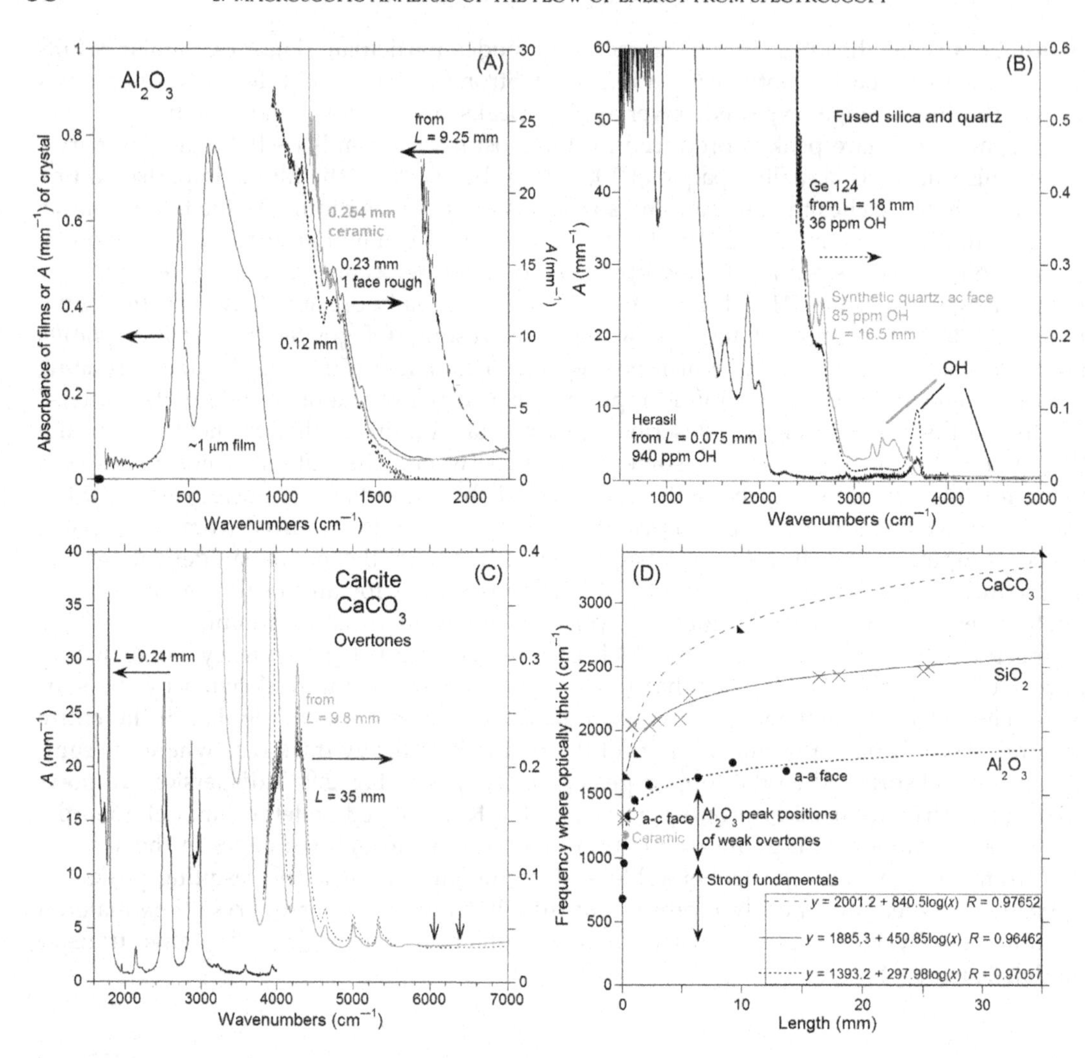

FIGURE 2.10 Absorption spectra, illustrating the weakness of bands in the transparent near-IR. (A) Corundum, showing the much greater strength of the mid-IR bands than those in the near-IR. Thickness is as indicated, but A is reported. Dots: dielectric measurements which are not affected by reflection (Palik, 1998). Fine line: thin film was created by crushing powder in a diamond anvil cell. This is underlain by broad interference fringes, artificially enhancing A in the far-IR. Gray: high purity ceramic. Other samples are single crystals oriented with faces perpendicular to the c-axis. No peaks were discerned above 1500 cm^{-1}, even for a 13.7 mm sample. Dashed line: expanded view of A for a thick sample. (B) Slabs of silica (glass and crystal) with varying amounts of OH impurities, as labeled. Thin herasil glass (solid curve) is optically distorted in the mid-IR, where A is typically 1000 mm^{-1}. Overtones occur at the same positions for crystal (gray), but are broad in glass (dotted). All peaks above 3000 cm^{-1} are O–H impurities. Small peaks near 2900 and 2950 cm^{-1} are organic contamination. (C) Calcium carbonate, which has narrower peaks than silicates. Due to low symmetry of the structure, many peaks exist, up to 6500 cm^{-1} (arrows) as observed for the thick crystals (right axis). (D) Dependence of the frequency below which spectra are optically thick on the path length. Least squares fits are shown in the box. Various circles = Al_2O_3. Only the c-face is fit. X: both forms of silica; triangles: calcite, where polarizations are mixed.

surrounding microwave or near-IR regions (Fig. 2.10). Typically, the near-IR region contains overtone-combinations of the fundamentals and is affected by vibrations of O–H impurities and some electronic transitions of Fe^{2+} when in particularly large sites; see reviews by Libowitzky and Beran (2006) and Rossman (2014). Absorbance in the visible-UV is connected with electronic transitions. For minerals, Fe^{2+} is very important, due to the breadth and strength of its transitions (Fig. 2.11). Considerable strength exists in the

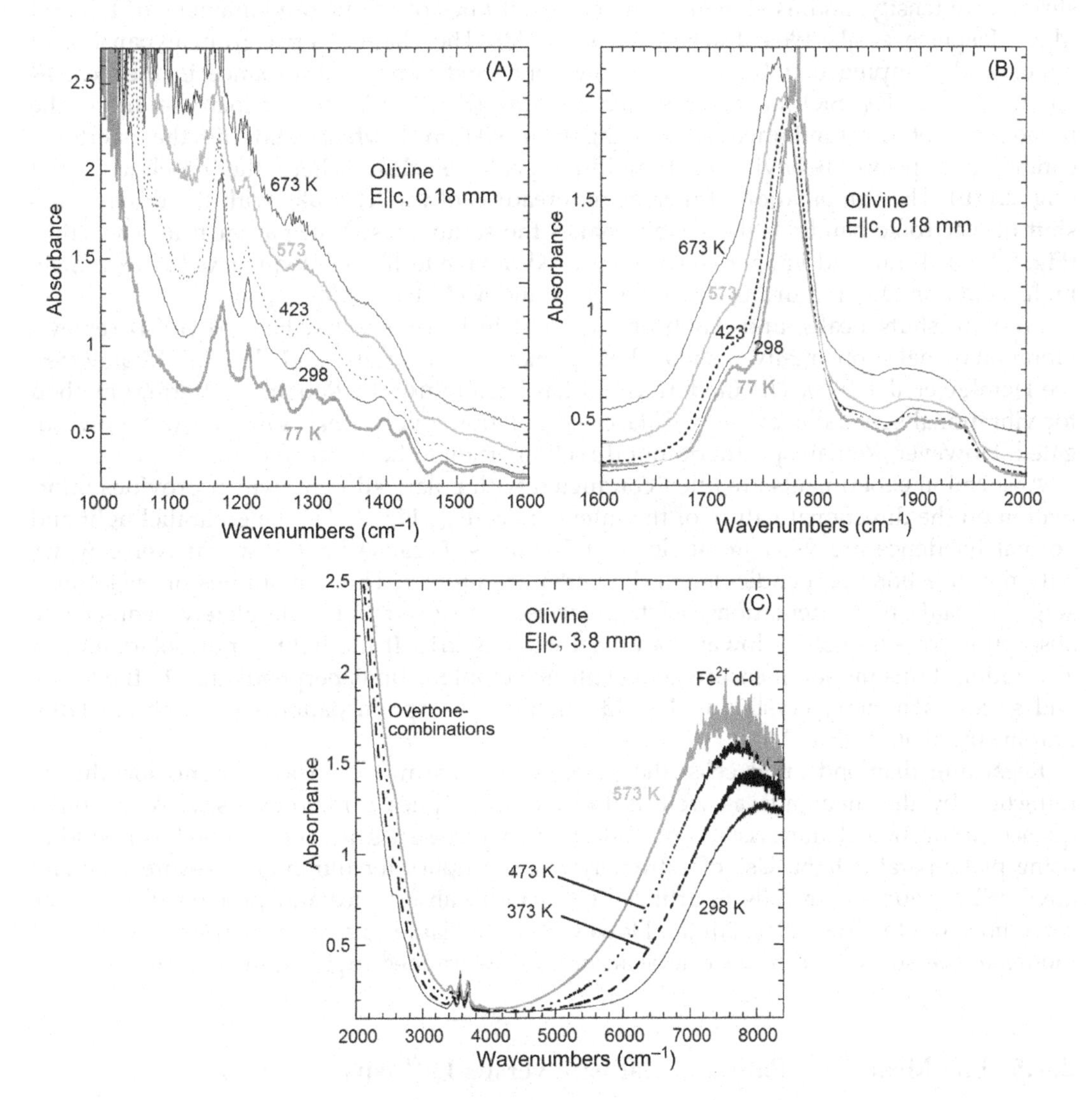

FIGURE 2.11 Near-IR absorbance spectra polarized along the *c*-axis of olivine ($Mg_{0.9}Fe_{0.1}SiO_4$), a common mineral in the lithosphere and upper mantle as a function of temperature. Panels (A) and (B) show data from one run on a thin sample (0.18 mm). Panel (C) shows another run on a thick sample (3.8 mm). These are not baseline corrected. The underlying increase in absorbance is due to the extremely strong bands in the mid-IR, because reflectivity is low and constant in the near-IR. *Y*-axis scales differ slightly, to best show the changes in the peaks.

vacuum-UV due to metal–oxygen charge transfer and so reflectivity measurements are made (see Section 2.5 for figures). Optical interactions become again small as frequency increases in the UV, until another process is activated, namely X-ray transitions. The latter are at such high energies that these transitions do not pertain to heat transfer covered in this book.

Elevating temperature broadens peaks, which shift to lower frequency, and somewhat shrink in intensity, and as shown in the classic studies of the IR fundamentals of LiF and MgO (Jasperse et al., 1964; Kachare et al., 1972). The changes arise from expansion of bonds and damping of vibrations via interconnected bonds. Absorbance in the near-IR spectra of $Mg_{0.9}Fe_{0.1}SiO_4$ at moderate temperature (Fig. 2.11A) overall increase due to the broadening of the fundamentals existing near 900 cm^{-1} which underlie the overtone-combination peaks. Behavior of individual peaks is clear where the baseline is flat (Fig. 2.11B). The combination of increasing breadths of the IR peaks with the downwards shift of electronic bands in the visible makes the sample less transparent near 4000 cm^{-1} (Fig. 2.11C). Taran and Langer (2001) measured olivine to higher frequency. For extremely high T data on O-H impurities, see Grzechnik and McMillan (1998).

Pressure shifts peaks, and has been explored in many studies. Ferraro (1984) reviews vibrational data on many solids. For planetary materials and higher frequencies, see Hemley et al. (1998). The modern trend has been to use the Raman (scattering) method for vibrational spectra, because with laser stimulation a very small sample can be investigated. However, Raman spectra do not directly pertain to heat transfer.

Absorption experiments are often conducted with polarized light, which provides information on the directional nature of the interactions (e.g., Fig. 2.11). Plane-parallel light and normal incidence are used in single-crystal studies. Because light is a transverse wave, only the directions perpendicular to the beam are probed. Thus, in studies of anisotropic single crystals two orientations, each polarized, are needed to completely characterize absorption for symmetries lower than axially symmetric. If the light is not polarized, the absorption/transmission measurements still reflect interactions perpendicular to the beam, and so low symmetry crystals will yield different absorption patterns for each crystallographic orientation (Fig. 2.6).

Regarding diamond anvil cells, the polarizations are mixed in any orientation due to refraction by the diamonds as well as the focusing optics commonly used. As a consequence, absorption characteristics and reflection characteristics are not "paired" as is studies using plane-parallel light. Use of extremely focused optics for ultrahigh pressure diamond anvil cell experiments results in artificially enhanced absorbance and problematic baseline corrections (Hofmeister, 2010, 2014a). However, for the larger diameter samples examined at moderate pressures, absorbance can be accurately determined (e.g., Taran et al., 2011).

2.4.5 The Mean Free Path and Ballistic versus Diffusive Behavior

When reflections or scattering exists, the mfp is approximated by:

$$\frac{1}{\mathrm{mfp}(\lambda)} = \frac{1}{d} + A(\lambda) \tag{2.13}$$

where d is the average grain size or average distance between reflections.

Because absorptions in the near-IR to visible are generally weak (Fig. 2.11), conditions in these regions are generally optically thin. For this reason and because our eyes sense visible light, radiative transfer is generally described in terms of interactions of a medium with light in the visible, and the qualifier of ballistic radiative transfer is omitted (Fig. 1.6). This omission has led to many misunderstandings.

Because mid-IR absorptions are generally quite strong, the light is converted to heat in the solid, which later is lost to the surroundings during re-equilibration. This process is diffusion. Hence, optically thick conditions are connected with diffusion (Chapter 3). That high energy light can transfer heat diffusively under optically thick conditions was recognized long ago (Kellett, 1952; Lubimova, 1958). Diffusion of low frequency IR radiation inside a crystal as an important means of transporting heat was only recently considered (Hofmeister, 1999; Hofmeister et al., 2014) and is explored further in this book.

2.5 REFLECTION SPECTRA

Measuring reflectivity $R = I_{\text{refl}}/I_0$ is highly advantageous because this can be quantitatively analyzed to obtain the various related parameters listed in Section 2.2 (n, k, A or α, and ɘ). Many reflectivity measurements have been made of metals, due to their high reflectivity and technological importance. Reflectivity of electrical insulators and semiconductors are also explored, because in the mid-IR, where peaks are strong, quantitative absorption measurements are difficult. The same holds for the vacuum-UV, but this region is less explored (Section 2.3). Data on optical constants of various materials, have been compiled in the book by Palik (1998), which has been updated many times.

2.5.1 Experimental Methods and Data Analysis

Data are collected in configurations such as Fig. 2.6F, where a front-surfaced mirror is used as the reference. Generally, gold is used at low frequencies and silver at high, due to their reflection characteristics. Absolute values for silver are known up to ~200 nm, and calibrated mirrors are available.

At normal incidence for unpolarized measurements:

$$R = \frac{(n - n_{\text{med}})^2 + k^2}{(n + n_{\text{med}})^2 + k^2} \tag{2.14}$$

Optical constants are extracted using Kramers-Kronig analysis, which involves complex variables (e.g., Andermann et al., 1965; Wooten, 1972). The phase angle:

$$\Theta(\nu) = \frac{\nu}{\pi}\int_0^{\infty} \frac{\ln R(w) - \ln R(\nu)}{w^2 - \nu^2} dw \tag{2.15}$$

where w is a dummy variable, is obtained by integrating overall frequencies. To avoid confusion, we note that some publications use the magnitude ($\equiv R^{1/2}$) instead of R. For non-normal incidence (Fig. 2.6F), which is needed to make polarized measurements, the equations need modification to account for the angle (Fahrenfort, 1961; Roessler, 1965).

Due to the approximations made in Kramers-Kronig analysis, optical functions and peak parameters are considered as approximate and final results are usually obtained from classical dispersion analysis (Spitzer et al., 1962). This model describes many phenomena (Wooten, 1972) and is commonly applied to vibrations. The physical picture is damped harmonic oscillators, which are described in classical mechanics. This method depicts vibrations as Lorentzian modes. This shape describes the peaks in the imaginary part of the dielectric function:

$$\varepsilon_2(\nu) = 2nk = \sum \frac{O_i \nu_i^2 \mathrm{fwhm}_i \nu}{(\nu_i^2 - \nu^2) + \mathrm{fwhm}_i{}^2 \nu^2} \tag{2.16}$$

where the sum is over the i peaks at their TO positions of ν_i, O_i is oscillator strength of each, fwhm_i is in frequency units. The real part is:

$$\varepsilon_1(\nu) = n^2 - k^2 = \varepsilon_\infty + \sum_i O_i \nu^2 \frac{(\nu_i^2 - \nu^2)}{(\nu_i^2 - \nu^2) + \mathrm{fwhm}_i{}^2 \nu^2} \tag{2.17}$$

where the dielectric constant at high frequency (ε_∞) can be approximated from measurements of n in the visible if IR vibrations are modeled. The equations can also be cast in terms of the low frequency dielectric function (ε_∞) which can be directly measured. For non-normal incidence, reflectivity calculations need to account for the angle (Jackson, 1975; Hofmeister, 1993). Classical dispersion analysis can be modified for peak shapes other than Lorentzian, such as Gaussian or Voigt shapes, which are both mixtures. However, only the Lorentzian shape is truly consistent with Kramers-Kronig analysis (Ruzi et al., 2017). Use of Voigt shapes for glasses (de Sousa Meneses et al., 2005) can be justified because reflectivity of many glasses cannot otherwise be analyzed and because each vibrational band represents a distribution of local environments as well as a statistical distribution of energies. Highly polymerized glasses do not have this problem of very low R and are fit well by Eqs. (2.16) and (2.17) (e.g., $NaAlSi_2O_8$: Merzbacher and White, 1988).

Obtaining perfect fits with three-parameters can be difficult. This situation has motivated some researchers to use four-parameter fits. Specifically, the longitudinal (LO) modes are assumed to be independent of TO modes. However, because LO and TO modes are obviously connected, the four-parameter fits do not correctly represent the physics. Imperfect fits are not caused by decoupled LO and TO modes, but rather by the presence of poorly resolved fundamentals and overtones of the acoustic modes, which create resonances (e.g., MgO). The latter is important to heat transfer (Chapter 11).

Some difficulties exist regarding measurements and processing. First, absolute values are needed, which requires that the polished surfaces of the sample and reference mirror are equally smooth and that the spectrometer and reflection accessories are well-aligned. Second, reflectance must be known over a wide range frequencies to accurately apply Kramers-Kronig analysis (Eq. (2.15)). Wooten (1972) describes extrapolations used to approximate R below and above the regions of measurement. Fortunately, the high frequency approximation mostly affects the high frequency regions. As such, measuring reflectance spectra only into the near-IR allows us to quantitatively probe the mid-IR

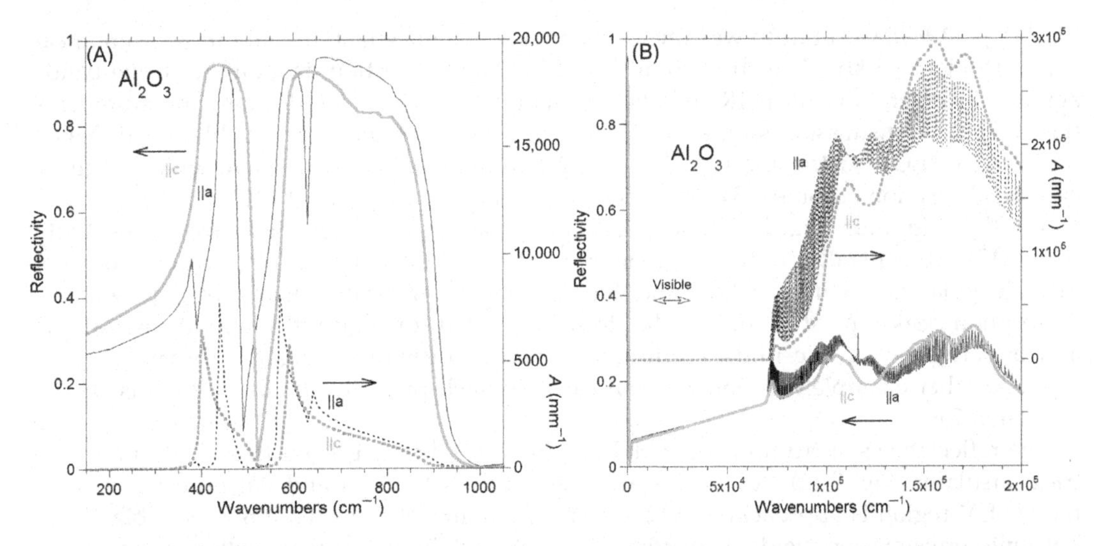

FIGURE 2.12 Reflectance spectra (upper) and absorptivity (lower) calculated from the complex index of refraction for corundum (Palik, 1998). Polarizations as labeled. (A) The IR. (B) The full range. Absorption spectra were shifted upwards for clarity. Fringing results from wide spacing of the data points.

region where the bands are strong. Similarly, the extrapolations below the data only affect the lowest frequency peak. The downwards extrapolation is generally constrained by the reflectivity (R) trend at low frequency (Fig. 2.12), in part because the extrapolation only involves a small interval in frequency. The assumed slope can be varied to explore the associated uncertainty. Third and less recognized, secondary reflections exist for thin samples in the transparent regions (Fig. 2.4A). For insulators, this is a problem for two reasons: Just above a strong IR peak, reflectivity drops sharply to near-zero (Fig. 2.12, for the peaks in near 500 and 1000 cm^{-1}). Below a strong IR peak, the reflectance gradually increases (as seen in the far-IR), but this increase is actually due to the sample absorbing whereas reflection dominates when the frequencies are between those of the TO and LO modes. Back reflection in these regions surrounding the peak perturb its shape and both Kramers-Konig and classical dispersion analyses (c.f., Hofmeister et al., 2003; Sun et al., 2008). Corundum (Fig. 2.12) does not have this problem because its peaks are close together, whereas MgO explored in the above studies has only one IR peak (with resonances). Thus, reflectivity needs to be measured under optically thick conditions (large samples and/or non-normal incidence). However, thin black samples can be studied (e.g., Sun et al. 2008), because electronic screening prevents back reflections.

2.5.2 Examples of Reflection Spectra

Corundum has been studied in great detail. Optical functions compiled in Palik (1998) were used to calculate R and A for the two polarizations (Fig. 2.12). The IR reflections of Al_2O_3 are relatively simple due to the high symmetry of the crystallographic structure

(Barker Jr., 1963). Corundum was one of the first solids for which the vibrations were measured (Coblentz, 1908). The absorption (Fig. 2.10A) and reflection data on the strong bands agree. However, the near-IR bands are not revealed by reflection measurements (Fig. 2.12A). Transmission studies are required, even throughout the visible. Weak bands intrinsic to Al_2O_3 exist in the UV (40,000–52,000 cm^{-1}, not shown) above which frequencies the absorption becomes extreme, requiring reflection data (Fig. 2.12B).

The thin film data (Fig. 2.10) and extractions of absorption coefficients from reflectivity (Fig. 2.12) are generally in good agreement, noting that absorption of powders does not provide polarization information and highly converging beams also mix polarizations. Absorption peak tops are rounded due to unaccounted for high reflectivity of the sample and imperfections in the films: corundum is hard, so uniform films are difficult to produce. Another example (the soft mineral calcite, which provides uniform films) is shown in Section 2.6.

The reflectance spectrum of the metal Fe (Fig. 2.13) differs greatly from that of an electrical insulator (Fig. 2.12). All metals are characterized by high and flat reflectivity, until the vis-UV region is approached, where a small amount of transmission is possible. Small but finite transmission yields distinctive colors for certain metals (e.g., yellow of gold) and pertains to radiative diffusion at high temperatures (Chapter 11).

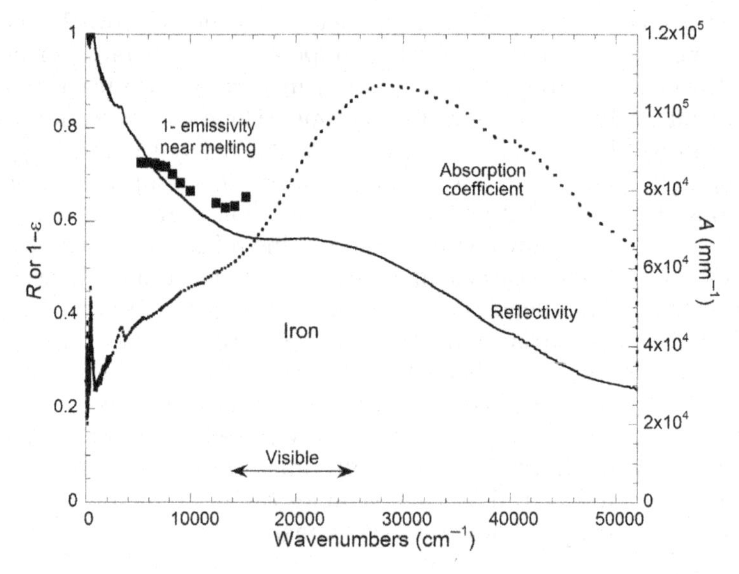

FIGURE 2.13 Reflectance spectra of Fe from the far-IR to the UV. The sample is pure electrolytic iron. The absorption coefficient was calculated from Kramers-Kronig analysis. Our results gave similar optical constants to those of Moravec et al. (1976) and Ordal et al. (1983), and to the compilation of Palik (1998) except at very low frequency, where our data are noisy. The reflectivity values agree, although previous studies did not constrain the band at 3456 cm^{-1}. Also shown are emissivities in the red spectral region and below obtained for iron near its melting point by Watanabe et al. (2003). Their emission spectra changed with time in these difficult experiments, which was attributed to oxygen and impurities.

2.6 ESCAPE OF INTERIOR PRODUCED EMISSIONS

Contradictions have accompanied emission spectroscopy from its inception, as indicated by comparing Stewart's and Kirchhoff's disparate conclusions (Chapter 1, Section 1.3.2). Coblentz (1908) was the first to measure emission spectra of minerals; he probed large samples in the transparent near-IR region at ~900K. In modern planetary science, emission spectra of large grains has been misinterpreted as being identical to emissivity due to several interwoven problems accompanying these difficult experiments, see Hofmeister (2014a). Details that will prove useful for radiative transfer models are given below. In summary:

1. Emission is measured, not emissivity ($=I_{emit}/I_0$). This is analogous to absorption spectroscopy where absorptivity cannot be directly measured due to reflections existing at interfaces. As in transmission measurements, the outcome of emission measurements depends on how high the reflectivity is, and whether conditions are optically thin, thick, or distorted.
2. Optical thickness determines the depth from which light escapes in an emission experiment.
3. As the interior emissions cross the surface, light is reflected once (Fig. 2.4). Back reflections reduce the emissions by a factor of $1-R$ (Fig. 2.4). This behavior is seen in emission spectra of partially transparent minerals (e.g. the database of Christensen et al., 2000), but this has been misinterpreted as emissivity because $\varepsilon = 1-R$ for metals and semiconductors. Dielectrics have spectral characteristics unlike metals (cf. Figs. 2.12 and 2.13), requiring a different analysis (Bates, 1978).
4. Isothermal conditions needed for Kirchhoff's law are not met in electrical insulators (Brewster, 1992). Silicate and many other minerals are dielectrics with low lattice thermal conductivity and low absorption coefficients which impede radiative transfer. The combination of these physical properties in ceramics and glasses (which have similar spectral properties to oxide minerals) causes the length scale of significant temperature change to be similar to the length scale of the interaction with light (i.e., the mean free photon path), and thus $\varepsilon \neq 1-R$.
5. Emission experiments center on transfer of radiant heat from the sample to the detector, and so the geometry of this radiative transfer and whether emissions interchange with the surroundings have an effect. Thermodynamic laws must be considered (McMahon, 1950; Gardon, 1956).

2.6.1 Theory of Emissions From Nonopaque Media

Kirchhoff's law for a body in thermodynamic equilibrium ($a = \varepsilon$) is only *strictly* true when there is no net heat transfer to or from the surface, or the surface is nonexistent, as in gas. Kirchoff's law, Eq. (2.11), being applicable to certain situations, provides the starting point for discussion.

The role of grain size when no reflections exist, is obtained by combining Eqs. (2.11) and (2.12):

$$\varepsilon(\nu, d) = 1 - \exp[-dA(\nu)] \tag{2.18}$$

The total *emissions* (E_{emit}), if no reflections occur are described by:

$$E_{emit} = \varepsilon I_{BB} \tag{2.19}$$

From Eqs. (2.19) and (2.10), emissions and emissivity will be high where transitions occur, which is consistent with physical principles. Absorption occurs when an energy absorbing transition is possible. Thus, if a vibrational mode is excited, it will emit energy near the transition frequency in order to return to the ground state from the excited state. However, measurements of emission spectra are complicated by the presence of back reflections (Fig. 2.4). Hence, in all cases where the sample butts against air or vacuum, emitted light goes as

$$E_{emit} = \varepsilon(1 - R)I_{BB} \quad \text{for a surface against ''space''} \tag{2.20}$$

Grain-size is crucial because this affects the amount of light reflected and that emitted, analogous to transmission measurements (Section 2.4). Thus, the emissions will depend on optical thickness. In the planetary science literature, large grains are often measured, which should be optically thick, so Christensen et al. (2000) and Salisbury et al. (1991), for example, assumed transmittivity is zero (in Eq. (2.7)) to arrive at $\varepsilon + R \cong 1$. However, this approximation is only valid for materials that very strongly and uniformly absorb, such as metals and semiconductors. For dielectrics (e.g., partially transparent minerals), thermodynamic considerations (McMahon, 1950; Gardon, 1956) extend Kirchhoff's law:

$$\varepsilon + R + \Upsilon \cong 1 \tag{2.21}$$

Bates (1978) derived an exact result by considering forward and backwards scattering at an interface:

$$\varepsilon + R + \Upsilon = 1 + \varepsilon R\Upsilon + R\Upsilon \tag{2.22}$$

Although the cross-products are small, they are nonzero. To understand what emission spectra are actually measuring, we consider optically thin and thick conditions. First, we rearrange the terms in Eq. (2.22):

$$\varepsilon = 1 - R - \Upsilon + r\Upsilon + \varepsilon R\Upsilon \tag{2.23}$$

Although transmittivity Υ is small for optically thick conditions, it is non-negligible, making ε larger than $1 - r$ for this case (usually large grains). Further manipulation provides the exact formula, which is relevant to optically thick or optically distorted conditions:

$$\varepsilon = \frac{1 - R - \Upsilon + R\Upsilon}{1 + R\Upsilon} \tag{2.24}$$

(Hofmeister, 2014a). For optically thin conditions, Eq. (2.11) ($\varepsilon \cong a < 1$) is a reasonable approximation.

2.6.2 Emission Results Under Optically Thin Conditions at Various Temperatures

At room temperature, emissivity similar to absorptivity was recorded in experiments by Low and Coleman (1966) for grain-sizes and mineral compositions similar to those of Christensen et al. (2000). In detail, the emissivity peaks of Low and Coleman (1966) have high values where the absorptivity is also high, consistent with Kirchhoff's result. However, in the spectra of Christensen et al. (2000), their emission values are lowest where that of the absorption is highest. In a nutshell, this planetary data base provides spectra with peaks pointing in the "wrong" direction (see Figure 8 in Hofmeister, 2014a).

The difference in results is tied to the experimental conditions. Low and Coleman (1966) placed a cold dry ice block nearby, but separate from the sample which is held at slightly lower temperature (20°C) than the detector at 24°C (Fig. 2.14). The sample radiates its heat to the dry ice block, creating a thermal gradient at the surface, which make the conditions effectively optically thin. In the planetary database, the sample radiates to the detector only and the light arising in the interior is back reflected at the surface. Conditions are optically thick. Internal surface reflections affect the planetary database, causing a minimum in emissivity where the absorption is highest (discussed in detail below).

At very high temperatures, the ice block is not needed, because the flux from the sample is large and a thermal gradient exists at the surface. Consequently, thin films of molten nitrates (Bates and Boyd, 1973) have peaks that resemble absorption spectra ($a \cong \varepsilon$) because optically thin conditions are met. As film thickness increases, the height of the most intense peak decreases relative to the weaker peaks (e.g., Bates and Boyd, 1973) due to R being largest for the intense peak. This behavior parallels the dependence of absorptivity on thickness in transmission experiments (Fig. 2.8). It is far more difficult to obtain optically thin conditions for strong peaks than for weak.

Clouds of mineral dust in space are subject to very low temperatures. Thermal equilibrium exists, due to the dust being a suspension in vacuum: no conductive mechanism exists and temperatures are governed solely by radiative transfer from the central star. Starlight warms the grains mostly in the visible to UV range, which then re-radiate the heat at low frequencies, due to low temperatures of space. Far-IR spectra from dust grains in the proto-planetary nebula NGC-6302 have a spectrum that closely resembles those of

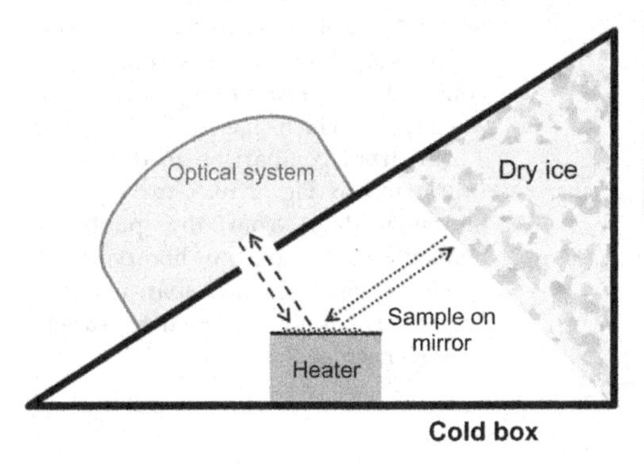

FIGURE 2.14 Schematic of ambient temperature experimental setup of Low and Coleman (1966, modified after their Fig. 1). The presence of cold material in the box minimizes the effect of laboratory background radiation. The spectra thus represent the net differences in emissions between the sample and detector.

oxide minerals (e.g., Kemper et al., 2002). Forsterite is robustly identified. Dust grains in circumstellar environments are considered to be small, ~μm sizes. Thus, optically thin conditions exist, particularly in the far-IR in which peaks are weak, and Kirchhoff's law is upheld.

2.6.3 Emission Spectra Under Optically Thick Conditions

Metals are essentially opaque. Their emissivity equals 1-R theoretically and experimentally, as shown in Fig. 2.13.

Electrically insulating minerals are partially transparent. Sub-mm grains that meet optically thick conditions and will have $\alpha = \varepsilon = 1$ in thermal equilibrium. Inserting $\varepsilon = 1$ in Eq. (2.10) shows that emissions from large grains will go as $I_{BB}(1 - R)$, due to the reflection at the surface. Thus, for the $d \sim 0.8$ mm grains studied by Christensen et al. (2000) less light is received at frequencies where transitions occur, that is, peaks point down as seen in their spectra (Fig. 2.15). For calcite, the correlation of the emissions with reflectivity, but not with the absorbance spectrum, is clear in the details (e.g., at 700 and 900 cm^{-1}). Emission spectra for large grains predominantly record the effect of back reflections (Section 2.6.1), and are unrelated to emissivity.

To understand peak shapes in emission spectra under optically thick conditions, one must recognize that reflection spectra are determined using plane polarized light, which distinguishes TO from LO modes (Fig. 2.6, see discussions in Wooten, 1972; Burns, 1990). At frequencies between the TO and LO modes, R determined in reflectivity measurements is large and more or less constant for strong peaks but for weak peaks, R is low and the shape is pointy (Fig. 2.15). Different shapes occur for strong and weak peaks in reflection measurements because the separation of the TO and LO mode for any given peak depends on its oscillator strength (Wooten, 1972). The separation also makes reflectivity high and broad for the strong modes. Strong peaks are pointy in absorption-transmission

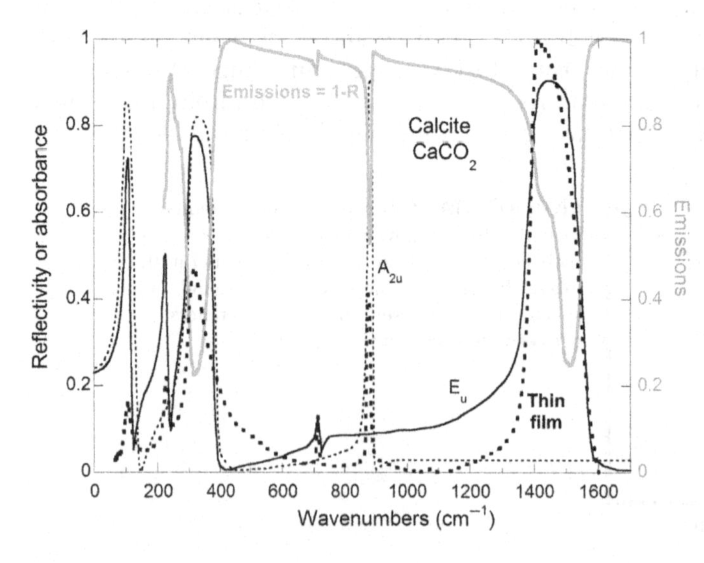

FIGURE 2.15 Calcite emissions versus reflections and absorptions. Polarized reflectivity from Hellwege et al. (1970): Solid line = E_u symmetry where light is polarized along the a-axis and fine dotted line = A_{2u} symmetry polarized along the c-axis). Thick dotted curve = absorption data collected from a thin film in a diamond anvil cell, which has a thickness of 1.2 μm from comparison to data from thin slabs in Fig. 2.10. Gray curve = emission data from the publically available ASU spectral library, #503, single crystal of Iceland spar, which is the clear, doubly refracting variety (Christensen et al., 2000).

measurements (Fig. 2.7, bottom of Fig. 2.15) because only the TO components are sampled. In contrast, blackbody radiation is not plane polarized, so TO and LO modes are not fully distinguished and polarizations are mixed. Thus, peaks in emission spectra of large grains are sharper than in reflection data, particularly for the most intense peaks (Fig. 2.15). Quartz data behave similarly (Hofmeister et al., 2000). The correlation of these emission spectra with 1-*R* is most apparent for the cubic minerals and the larger grain sizes.

That $\varepsilon = 1$ for large grains has not been recognized in the remote sensing literature (e.g. Salisbury et al., 1991; Christensen et al., 2000); instead, the factor of (1-*R*) has wrongly been considered as equal to the emissivity. Their reasoning was that $\tau = 0$ for an opaque or optically thick material, which ignores previous results and theory (McMahon, 1950; Gardon, 1956; Bates, 1978).

The planetary database (Christensen et al., 2000) also presents data on smaller mineral grains heated to 80°C, which shed light to the detector. Their spectrum #611 of calcite with grain size of 63–90 μm has a similar (reflection) peak at 1500 cm^{-1}, which is the strongest band. The weaker band at 880 cm^{-1} is seen as a reflection feature for grain sizes >200 μm. Thermal equilibrium does not exist. It is also relevant that emission experiments are not taken from mirror polish surfaces, but typically from grains, and thus scattering alters *R* through its dependence on angle. The main effect is reduction of the difference in reflection at peaks versus valleys. Consequently, a continuum of behaviors is exhibited. Emissions, but not emissivity, are provided in this database.

References

Andermann, G., Caron, A., Dows, D.A., 1965. Kramers-Kronig dispersion analysis of infrared reflectance bands. J. Opt. Soc. Am. 55, 1210–1216.

Arp, H.C., 2003. Catalogue of Discordant Redshift Associations. Apeiron, Montreal, Canada.

Barker Jr., A.S., 1963. Infrared lattice vibrations and dielectric dispersion in corundum. Phys. Rev. 132, 1474–1481.

Barr, E.S., 1963. Men and milestones in optics. II. Thomas Young. Appl. Optics 2, 639–647.

Bates, J.B., 1978. Infrared emission spectroscopy. Fourier Transform IR Spect. 1, 99–142.

Bates, J.B., Boyd, G.E., 1973. Infrared emission spectra of molten salts. Appl. Spect. 27, 204–208.

Bell, E., 1967. Light and matter. In: Flugge, S., Genzel, L. (Eds.), Encyclopedia of Physics. Springer-Verlag, Berlin, Germany, pp. 1–58.

Benedict, W.S., Plyler, E.K., 1951. Absorption spectra of water vapor and carbon dioxide in the region of 2.7 microns. J. Res. Natl. Bur. Stand. 46, 246–265. Available from: https://archive.org/details/jresv46n3p246.

Bernath, P.F., 2002. The spectroscopy of water vapour: experiment, theory and applications. Phys. Chem. Chem. Phys. 4, 1501–1509.

Bernath, P.F., 2016. Spectra of Atoms and Molecules, third ed. Oxford University Press, Oxford, UK.

Berreman, D., 1963. Infrared absorption at longitudinal optic frequency in cubic crystal films. Phys. Rev. 130, 2193–2198.

Bertie, J.E., Lan, Z., 1996. Infrared intensities of liquids XX: the intensity of the OH stretching band of liquid water revisited, and the best current values of the optical constants of $H_2O(l)$ at 25°C between 15,000 and 1 cm^{-1}. Appl. Spectrosc. 50, 1047–1057.

Brewster, M.Q., 1992. Thermal Radiative Transfer and Properties. John Wiley & Sons, New York, NY.

Burns, G., 1990. Solid State Physics. Academic Press, San Diego, CA.

Chalmers, J.M., Griffiths, P.R., 2006. Handbook of Vibrational Spectroscopy, vol. 1. John Wiley and Sons, New York, NY. Available from: https://doi.org/10.1002/0470027320.

Christensen, P.R., Bandfield, J.L., Hamilton, V.E., Howard, D.A., Lane, M.D., Piatek, J.L., et al., 2000. A thermal emission spectral library of rock-forming minerals. J. Geophys. Res. 105, 9735–9739.

Coblentz, W.W., 1908. Supplementary investigations of infrared spectra. The Carnegie Inst. Publ. No. 97, Washington, DC.

Cotton, F.A., 1971. Chemical Applications of Group Theory, second ed. Wiley-Interscience, New York, NY.

Dardono, S., Jaworowski, M., 2010. In situ spectroscopic ellipsometry studies of trivalent chromium coating on aluminum. Appl. Phys. Lett. 97. Available from: https://doi.org/10.1063/1.3511472. no. 181908.

Debye, P., 1914. Vortrage über die kinetische Theorie der Materie und der Electrizität. B.G. Teuber, Berlin, Germany.

de Sousa Meneses, D., Gruener, G., Malki, M., Echegut, P., 2005. Causal Voigt profile for modeling reflectivity spectra of glasses. J. Non-Crystal. Solids 351, 124–129.

Fahrenfort, J., 1961. Attenuated total reflection: a new principal for the production of useful infra-red reflection spectra of organic compounds. Spectrochim. Acta. 17, 698–709.

Fateley, W.G., Dollish, F.R., McDevitt, N.T., Bentley, F.F., 1972. Infrared and Raman Selection Rules for Molecular and Lattice Vibrations: The Correlation Method. Wiley Interscience, New York, NY.

Ferraro, J.R., 1984. Vibrational Spectroscopy at High External Pressures: The Diamond Anvil Cell. Academic Press, Orlando, FL.

French, R.H., Winey, K.I., Yang, M.K., Qui, W., 2007. Optical properties and van der Waals–London dispersion interactions of polystyrene determined by vacuum ultraviolet spectroscopy and spectroscopic ellipsometry. Aust. J. Chem. 60, 251–263.

Gardon, R., 1956. The emissivity of transparent materials. J. Am. Ceram. Soc. 39, 278–287.

Grechko, M., Boyarkin, O.V., Rizzo, T.R., Maksyutenko, P., Zobov, N.F., Shirin, S.V., et al., 2009. State-selective spectroscopy of water up to its first dissociation limit. J. Chem. Phys. 131, no. 221105.

Griffiths, P., de Haseth, J., 1986. Fourier Transform Infrared Spectrometry. John Wiley & Sons, New York, NY.

Grzechnik, A., McMillan, P.F., 1998. Temperature dependence of the OH^- absorption in the SiO_2 glass and melt to 1975 K. Am. Mineral. 83, 331–338.

Halliday, D., Resnick, R., 1966. Physics. John Wiley & Sons Inc, New York, NY.

Hellwege, K.H., Lesch, W., Plihal, M., Schaack, G., 1970. Zwei-Phononen-Absorptionsspektren und dispersion der schwingungszweige in kristallen der kalkspatstruktur. Zeit. Physik 282, 61–86.

Hemley, R.J., Mao, H.K., Cohen, R.E., 1998. High-pressure electronic and magnetic properties. Rev. Mineral. 37, 591–638.

Hirschmugl, C., 2004. An introduction to infrared spectroscopy for geochemistry and remote sensing. In: King, P. L., Ramsey, M.S., Swayze, G. (Eds.). Molecules to Planets: Infrared Spectroscopy in Geochemistry, Exploration Geochemistry and Remote Sensing, Short Course. In: Raeside, R. (Ed.) . Mineral. Assoc. Canada, Ontario, Canada, vol. 33, pp. 1–16.

Hofmeister, A.M., 1993. IR reflectance spectra of natural ilmenite: comparison with isostructural compounds and calculation of thermodynamic properties. Eur. J. Miner. 5, 281–295.

Hofmeister, A.M., 1999. Mantle values of thermal conductivity and a geotherm from phonon lifetimes. Science 283, 1699–1706.

Hofmeister, A.M., 2010. Scale aspects of heat transport in the diamond anvil cell, in spectroscopic modeling, and in Earth's mantle: implications for secular cooling. Phys. Earth Planet. Inter. 180, 138–147.

Hofmeister, A.M., 2014a. Carryover of sampling errors and other problems in far-infrared to far-ultraviolet spectra to associated applications. Rev. Mineral. Geochem. 78, 481–508.

Hofmeister, A.M., Bowey, J.E., 2006. Quantitative IR spectra of hydrosilicates and related minerals. Mon. Not. R. Astron. Soc. 367, 577–591.

Hofmeister, A.M., Keppel, E., Bowey, J.E., Speck, A.K., 2000. Causes of artifacts in the infrared spectra of powders. In: Salama, A., Kessler, M.F., Leech, K., Schulz, B. (Eds.), ISO Beyond the Peaks: The 2nd ISO Workshop on Analytical Spectroscopy. European Space Agency, Noordwijk, The Netherlands, pp. 343–346.

Hofmeister, A.M., Keppel, E., Speck, A.K., 2003. Absorption and reflection spectra of MgO and other diatomic compounds. Mon. Not. R. Astron. Soc. 345, 16–38.

Hofmeister, A.M., Pitman, K.M., Goncharov, A.F., Speck, A.K., 2009. Optical constants of silicon carbide for astrophysical applications. II. Extending optical functions from IR to UV using single-crystal absorption spectra. Astrophys. J. 696, 1502–1516.

Hofmeister, A.M., Dong, J.J., Branlund, J.M., 2014. Thermal diffusivity of electrical insulators at high temperatures: evidence for diffusion of phonon-polaritons at infrared frequencies augmenting phonon heat conduction. J. Appl. Phys. 115. Available from: https://doi.org/10.1063/1.4873295. no. 163517.

Hollas, M.J., 1996. Modern Spectroscopy, third ed. Wiley, New York, NY, ISBN 0471965227.

Hunter, W.R., 1982. Measurement of optical properties of materials in the vacuum ultraviolet spectral region. Appl. Opt. 21, 2103–2114.

Jackson, J.D., 1975. Classical Electrodynamics. John Wiley and Sons, New York, NY.

Jasperse, J.R., Kahan, A., Plendl, J.N., Mitra, S.S., 1964. Temperature dependence of infrared dispersion in ionic crystals LiF and MgO. Phys. Rev. 146, 526–542.

Kachare, A., Andermann, G., Brantley, L., 1972. Reliability of classical dispersion analysis of LiF and MgO reflectance data. J. Phys. Chem. Solids. 33, 467–475.

Kayser, H., 1900. Handbuch der Spectroscopie, vol. 1. S. Hirzel, Leipzig, Germany.

Kayser, H., 1908. Handbuch der Spectroscopie, vol. 3. S. Hirzel, Leipzig, Germany.

Kayser, H., Koner, H., 1934. Handbuch der Spectroscopie, vol. 7. S. Hirzel, Leipzig, Germany.

Keefe, C.D., Wilcox, T., Campbell, E., 2012. Measurement and applications of absolute infrared intensities. J. Mol. Struct. 1009, 111–122.

Kellett, B.S., 1952. Transmission of radiation through glass in tank furnaces. J. Soc. Glass Tech. 36, 115–123.

Kemper, F., Jaeger, C., Waters, L.B.F.M., Henning, T., Molster, F.J., Barlow, M.J., et al., 2002. Detection of carbonates in dust shells around evolved stars. Nature 415, 295–297.

Libowitzky, E., Beran, A., 2006. The structure of hydrous species in nominally anhydrous minerals: information from polarized IR spectroscopy. Rev. Mineral. Geochem. 62, 29–52.

Lindon, J., Tranter, G.E., Koppenaal, D., 2016. Encyclopedia of Spectroscopy and Spectrometry, third ed. Academic Press/Elsevier, Amsterdam, The Netherlands.

Lipson, H.G., 1960. Infra-red transmission of alpha silicon carbide. In: O'Connor, J.R., Smiltens, J. (Eds.), Silicon Carbide. Pergamon Press Inc., New York, NY, pp. 371–375.

Low, M.J.D., Coleman, I., 1966. Measurement of the spectral emission of infrared radiation of minerals and rocks using multiple-scan interferometry. Appl. Opt. 5, 1453–1455.

Lubimova, H., 1958. Thermal history of the earth with consideration of the variable thermal conductivity of the mantle. Geophys. J. Royal Ast. Soc. 1, 115–134.

Malus, E.L., 1809. Sur une Propriété de la Lumière Refléchié. Memoires de Physique et de Chimie de la Société d'Arcueil 2, 143–158.

Malus, E.L., 1811. Théorie de la double réfraction de la lumière dans les substances cristallines. Mémoires présentés à l'Institut des sciences par divers savants 2, 303–508.

Matsushima, F., Odashima, H., Iwasaki, T., Tsunekawa, S., Takagi, K., 1995. Frequency measurement of pure rotational transitions of H_2O from 0.5 to 5 THz. J. Mol. Struct. 252/253, 371–378.

Maxwell, J.C., 1865. A dynamical theory of the electromagnetic field. Proc. R. Soc. Lond. 13, 531–636.

McGucken, W., 1969. Nineteenth-Century Spectroscopy. The Johns Hopkins Press, Baltimore, MD and London, UK.

McMahon, H.O., 1950. Thermal radiation from partially transparent reflecting bodies. J. Opt. Soc. Am. 40, 376–380.

Merzbacher, C.I., White, W.B., 1988. Structure of Na in aluminosilicate glasses: a far-infrared reflectance spectroscopic study. Am. Mineral. 73, 1089–1094.

Moravec, T.J., Rife, J.C., Dexter, R.N., 1976. Optical constants of nickel, iron, and nickel-iron alloys in the vacuum ultraviolet. Phys. Rev. B13, 3297–3306.

NASA Astrobiology Institute, 2017. Molecular Database. <http://depts.washington.edu/naivpl/content/molecular-database> (accessed 29.03.17).

Nakamoto, K., 1978. Infrared and Raman Spectra of Inorganic and Coordination Compounds, third ed. John Wiley & Sons, New York, NY.

Ordal, M.A., Bell, R.J., Alexander Jr., R.W., Long, L.L., Querry, M.R., 1983. Optical properties of fourteen metals in the infrared and far infrared; Al, Co, Cu, Au, Fe, Pb, Mo, Ni, Pd, Pt, Ag, Ti, V, and W. Appl. Opt. 22, 1099–1119.

Palik, E.D., 1998. Handbook of Optical Constants of Solids. Academic Press, San Diego, CA.

Planck, M., 1901. Über das Gesetz der Energievertilung im Normalspektrum. Ann. Physik (Ser. 4) 4, 553–563.

Robitaille, P.M., 2009. Kirchhoff's law of thermal emission: 150 years. Prog. Phys 4, 3–13.

Roessler, D., 1965. Kramers-Kronig analysis of non-normal incidence reflection. Brit. J. Appl. Phys. 16, 1359–1366.

Rossman, G.R., 2014. Optical spectroscopy. Rev. Mineral. Geochem. 78, 371–398.

Ruzi, M., Ennis, C., Robertson, E.G., 2017. Spectral curve fitting of dielectric constants. AIP Advances 7, no. 015042. <https://doi.org/10.1063/1.4975398> (accessed 30.05.17).

Salisbury, J., Walter, L., Vergo, N., D'Aria, D., 1991. Infrared (2.1–2.5 um) Spectra of Minerals. Johns Hopkins University Press, Baltimore, MD.

Sawyer, R.A., 1963. Experimental Spectroscopy, third ed Dover Publications Inc, New York, NY (comparison with the 1944 edition reveals the rapid development of IR instrumentation after World War II).

Schermaul, R., Learner, R.C.M., Newnham, D.A., Williams, R.G., Ballard, J., Zobov, N.F., et al., 2001. The water vapor spectrum in the region 8600–15000 cm^{-1}: experimental and theoretical studies for a new spectral line database I. Laboratory Measurements. J. Molec. Spectr. 208, 32–42.

Shankland, T.J., Nitsan, U., Duba, A.G., 1979. Optical absorption and radiative heat transport in olivine at high temperature. J. Geo. Res. 84, 1603–1610.

Siegel, R., Howell, J.R., 1972. Thermal Radiation Heat Transfer. McGraw-Hill, New York, NY.

Sloan, G.C., Price, S.D., 1995. Silicate emission at 10 microns in variables on the asymptotic giant branch. Astrophys. J. 451, 758–767.

Sloan, G.C., Kraemer, K.E., Goebel, J.H., Price, S.D., 2003. Guilt by association: the 13 micron dust emission feature and its correlation to other gas and dust features. Astrophys. J. 594, 483–495.

Sommer, L., 1989. Analytical Absorption Spectrophotometry in the Visible and Ultraviolet: The Principles. Elsevier, New York, NY.

Spitzer, W.G., Miller, R.C., Kleinman, D.A., Howarth, L.E., 1962. Far-infrared dielectric dispersion in $BaTiO_3$, $SrTiO_3$, and TiO_2. Phys. Rev. 126, 1710–1721.

Subramanian, M.A., Shannon, R.D., Chai, B.H.T., Abraham, M.M., Wintersgill, M.C., 1989. Dielectric constants of BeO, MgO and CaO using the two-terminal method. Phys. Chem. Min 16, 741–746.

Sun, T., Allen, P., Stahnke, D., Jacobsen, S., Homes, C., 2008. Infrared properties of ferropericlase $Mg_{1-x}Fe_xO$: experiment and theory. Phys. Rev. no. 134303, 77.

Tan, H.P., Maestre, B., Lallemand, M., 1991. Transient and steady-state combined heat transfer in semitransparent materials subjected to a pulse or step irradiation. J. Heat. Transfer 113, 166–173.

Taran, M.N., Langer, K., 2001. Electronic absorption spectra of Fe^{2+} ions in oxygen-based rock-forming minerals at temperatures between 297 and 600 K. Phys. Chem. Mineral 28, 199–210.

Taran, M.N., Ohashi, H., Langer, K., Vishnevskyy, A.A., 2011. High-pressure electronic absorption spectroscopy of natural and synthetic Cr^{3+}-bearing clinopyroxenes. Phys. Chem. Mineral 38, 345–356.

Tennyson, J., Bernath, P.F., Brown, L.R., Campargue, A., Császár, A.G., Daumont, L., et al., 2014. A database of water transitions from experiment and theory (IUPAC Technical Report). Pure Appl. Chem. 86, 71–83.

Townes, C.H., Schawlow, A.L., 2012. Microwave Spectroscopy. Dover Publications Inc, Mineola, NY.

von Bahr, E., 1913. Ultra-rot Absorption von Gasen. Ber. Duet. Physikal. Gas 15, 710–730.

Warren, S.G., Brandt, R.E., 2008. Optical constants of ice from the ultraviolet to the microwave:a revised compilation. J. Geophys. Res. 113. Available from: https://doi.org/10.1029/2007JD009744. no. D14220.

Watanabe, H., Susa, M., Fukuyama, H., Nagata, K., 2003. Phase (liquid/solid) dependence of the normal spectral emissivity for iron, cobalt, and nickel at melting points. Inter. J. Thermophys. 24, 473–488.

Weiss, R.J., 1996. Brief History Of Light And Those That Lit The Way. World Scientific Publishing, Singapore.

Wilson, E.M., Wenger, J.M., Venables, D.S., 2016. Upper limits for absorption by water vapor in the near-UV. J. Quant. Spectr. Rad. Trans. 170, 194–199.

Wooten, F., 1972. Optical Properties of Solids. Academic Press, Inc, San Diego, CA.

Wozniak, B., Dera, J., 2007. Light absorption by water molecules and inorganic substances dissolved in sea water. In: Mysak, L.A., et al., (Eds.), Atmospheric and Oceanographic Sciences Library. Springer, New York, NY, pp. 11–81.

Young, T., 1804. Bakerian lecture: experiments and calculations relative to physical optics. Phil. Trans. Roy. Soc. 94, 1–16.

Websites

Websites listing many spectra websites:
http://depts.washington.edu/naivpl/content/spectroscopy-links (focuses on molecules)
http://www.astro.uni-jena.de/Laboratory/Database/jpdoc/f-dbase.html (optical constants)
A brief explanation of molecular spectroscopy:

http://www.pci.tu-bs.de/aggericke/PC4e/Kap_III/Molekuelspektren.htm
Microwave spectra of rotations (diatomic, triatomic and hydrocarbon molecules):
https://physics.nist.gov/PhysRefData/MolSpec/freqsearch.html https://www.nist.gov/pml/molecular-microwave-spectral-databases
Molecular optical spectra over many frequency ranges:
http://depts.washington.edu/naivpl/content/molecular-database (provides line lists and absorption cross sections). https://spec.jpl.nasa.gov/ (technical repository in progress)
UV-visible spectra of gaseous molecules:
http://satellite.mpic.de/spectral_atlas
Spectral databases of organic compounds:
http://sdbs.db.aist.go.jp/sdbs/cgi-bin/cre_index.cgi (infrared, Raman, 1H and 13C nuclear magnetic resonance, electron spin resonance, and electron impact mass spectra).
http://webbook.nist.gov/chemistry/(infrared, UV-vis and other spectra).
Optical constants over a long wavelength range for diverse materials of interest to astronomy and planetary science:
http://www.astro.uni-jena.de/Laboratory/Database/jpdoc/f-dbase.html
Absorption spectra (mostly IR and visible; some polarized) of minerals:
http://minerals.gps.caltech.edu/
Database of Raman spectra, IR total reflectance of powder, X-ray diffraction, and chemistry data for minerals:
http://rruff.info/
Thermal infrared emission spectra (typically 2000–220 cm^{-1}) of solids, mostly minerals:
http://tes.asu.edu/spectral/library/ (Emissivity is reported but the data actually represent 1-reflectivity)
Blackbody calculator:
http://www.spectralcalc.com/blackbody_calculator/blackbody.php

CHAPTER

3

The Macroscopic Picture of Diffusive Heat Flow at Low Energy

OUTLINE

"Heat acts in the same manner in a vacuum, in elastic fluids, and in liquid or solid masses, it is propagated only by way of radiation, but its sensible effects differ according to the nature of bodies." ***Joseph Fourier (1822), p. 39 of 1955 reprinted translation.***

The flow of heat is difficult to comprehend because temperature, not heat, is monitored, and because multiple processes usually exist on the microscopic scale, yet in the

Measurements, Mechanisms, and Models of Heat Transport
DOI: https://doi.org/10.1016/B978-0-12-809981-0.00003-6

macroscopic picture, no process is specified. Nonetheless, the macroscopic viewpoint is both simple and illuminating, and thus is the best place to start.

Interestingly, studies of macroscopic heat flow preceded the development of classical thermodynamics. Newton's law of cooling describes how a body achieves radiative equilibrium with the surroundings. Newton's model does not address the flow of heat through a body. The model for heat transport through a material was developed almost entirely through the theoretical and experimental efforts of Jean Baptiste Joseph Fourier (1768–1830). This pursuit, along with Egyptology, were his hobbies during ~1804–1824 while he was employed as a public servant. This inception of heat flow rests on the caloric viewpoint, supported with steady-state experiments. For detailed discussions of the historical development, see Grattan-Guinness and Ravetz (1972), Herivel (1975), and Narasimhan (1999).

Fourier's discovery that heat flow is a time-dependent phenomenon played a rather minor role in the development of thermodynamics, although it should have been given more consideration (Chapter 1, Section 1.2). Neither have implications of his findings been evaluated subsequent to the development of classical thermodynamics. Such dynamic behavior is much more difficult to understand than steady-state. Also, because time-dependent experimental data were not available until well after his development of a heat flow model, misconceptions have arisen. For example, modern sources state that three mechanisms for heat transfer exist: conduction, radiation, and convection. This misconstrued triad originated with Maxwell (1871), due to incomplete knowledge during his lifetime. Yet, as Nusselt (1915) discussed, conduction and radiation are the only modes of heat transfer, whereas convection is a not separate mechanism because it is always involves conduction. The dimensionless ratio of convective over conductive heat flux being named for Nusselt makes one wonder why his finding has been ignored, but this can be explained by Maxwell's stature. Convection is a system-wide response and is labeled as a mode by Incropera and Dewitt (2005).

Because some of the terms used are vague, two definitions are in order:

- Conduction involves diffusion of energy across some material, which involves particle interactions.
- Ballistic transfer involves energy crossing a material (e.g., space) with negligible interactions.

Neither the type of energy nor the carrier is specified above because it actually is unnecessary with these definitions. These different processes are illustrated schematically in Fig. 1.6 for separated, distinct molecules (a gas), and in Fig. 3.1 for a continuum (solids and highly viscous liquids). Even with the above definitions, some discussion of light as a carrier of energy is needed. This situation arises because conduction via diffusion is associated with low energies (which can include infrared (IR) light; Chapter 2, Fig. 2.2 and Chapter 11; Hofmeister et al., 2014) whereas ballistic transport typically involves electromagnetic waves in the visible region, which is usually denoted as light. Electromagnetic radiation can be addressed macroscopically to a good extent, but not entirely: even in Newton's time, the microscopic nature of light was debated. Although we must discuss the existence of photons here, but do not need to address their properties or quantum interactions.

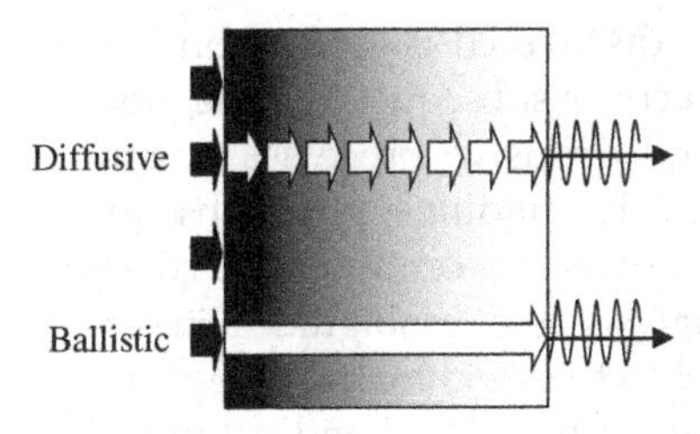

FIGURE 3.1 Schematics illustrating heat being transported by diffusive and ballistic end-members in a continuous medium with a horizontal temperature gradient. Black arrows: heat supplied to the left side. White arrows: various heat exchanges. Squiggle arrows: loss of radiant heat from the right side to the surroundings. Top, diffusion consists of a series of nearly identical exchanges across the body. Bottom, ballistic transport consists of a single exchange from the hot to cold surfaces, without participation of the body.

The term "radiative" transfer usually means ballistic transfer in the visible region because this situation is familiar (e.g., Earth is warmed by sunlight): this is the process that Nusselt had in mind, as well as Fourier, whose studies were motivated by the unexplained variations in terrestrial heat with latitude (Herivel, 1975). Regarding conduction, the view at Nusselt's time was that this process involved motion of heat, which energy source originated in the fluctuating motions of atoms. This atomistic description of heat flow is too restrictive, given that diffusion of high-frequency light also exists (e.g., Kellett, 1952), and moreover, that a continuum of behaviors from diffusive to ballistic is possible regarding interaction of electromagnetic radiation with a material (Chapter 1, Section 1.3). The term *heat conduction* exists throughout the literature: the process being described is diffusion of energy at fairly low temperatures and energies, with the implicit assumption that ballistic exchange is not important.

Understanding any phenomena requires laboratory measurements. Regarding heat flow, time and length scales in experiments are relatively short and temperatures are fairly low. Mostly, lengths of a few mm and temperatures below 1300K have been explored. Most experiments on heat flow concern steady-state conditions, which depict the system after it has undergone re-equilibration, and thus do not reveal the dynamics involved in this phenomenon. In addition, physical interfaces are known to impede heat transfer. Furthermore, materials that are electrical insulators are commonly transparent in the near-IR and visible (Figs. 2.10 and 2.12), so ballistic transfer is generally present, but can go unrecognized, particularly at low temperature (e.g., perovskites: Hofmeister, 2010a). Materials that electrically conduct (metals) do not transmit visible light, but instead have free electrons that can participate in heat flow: their relative contribution to heat conduction in alloys has been long debated (e.g., Klemens and Williams, 1986). Opportunity for misunderstanding abounds.

To delay discussing microscopic mechanisms, the details of heat conduction in metals are covered later in the book. Because ballistic transfer in electrical insulators is insidious, efforts need to be taken to ensure that we compare Fourier's model to experimental data that indeed record only diffusion of heat. As shown later in this chapter, a time-dependent (transient) method is essential, which furthermore avoids ballistic radiative transfer gains and physical contact losses. The only technique which meets these three criteria is laser-flash analysis (LFA), the gold-standard of industry (e.g., Vozár and Hohenauer, 2003/2004, 2005). Chapter 4 covers LFA and other methods, focusing on common approaches, recent developments, and the experimental limitations and uncertainties. Some data are provided and compared in Chapter 4.

The present chapter revisits Fourier's work. What Fourier discovered is correct, but some unrecognized restrictions apply to certain situations. We also cover subsequent efforts, not all of which are correct (e.g., Meissner, 1915), with ballistic and diffusive processes in mind. Section 3.1 discusses work leading to Fourier's discovery and his findings, plus subsequent adaptations. The far-reaching consequences of his discoveries over two centuries are covered by Narasimhan (1999). Section 3.2 covers dimensional analysis and ambiguities in time-independent measurements. Section 3.3 links Fourier's model to thermodynamics. Section 3.4 discusses the overextended connection with flow of electricity. Section 3.5 provides sum rules from an adiabatic assessment of heat flow which conserve energy.

3.1 FOURIER'S MACROSCOPIC MODEL OF HEAT TRANSFER

Two competing theories regarding the nature of heat existed in Fourier's time: one was the dynamical theory of heat (after Leibnitz), which involved atomic motions. But because the concept of the atom was unclear, and Lavoisier's work supported heat as a fluid (the caloric of Morveau), the mechanistic hypothesis (after Newton) was favored in the early 1800s, despite Rumford having already established through canon boring that heat is not a substance. The caloric picture was not entirely discarded, but morphed (see, e.g., the discussion of Laplace's role, below). Fourier considered both types of heat transfer to be radiative, based on the caloric model and his observations. Because Fourier's experiments were steady-state, for which heat in = heat out, these experiments *can* be accurately described by independent conservation of this imponderable fluid. Because spectroscopy was a later development, Fourier was not aware that ballistic and diffusive modes of heat transport can coexist inside a body. However, he recognized that receipt of heat by a body (from another body or across space) is distinct from how this heat crosses the body, and treated these processes in different ways. This division stems from Fourier accepting that molecules were somehow involved in heat transfer, and from his observations and key, early efforts.

3.1.1 Key Findings of Fourier's Predecessors

Fourier's model rests on four developments in physical science (Narasimhan, 1999).

- Invention of the closed tube mercury thermometer by Fahrenheit in 1714 lead to his temperature scale (Fegley, 2015). This discovery is important because measuring temperature is essential to deciphering heat and its motion.
- Through observations of ice melting, Black deduced circa 1760 that heat differs from temperature. This precipitated development of a calorimeter by Lavosier and Laplace in 1783, which allowed them to measure latent- and specific heat, with reference to behavior of water. Their work provides a foundation for thermodynamics and popularized the "caloric" model of heat as an imponderable fluid.
- Some measurements of conduction were made by Lambert in 1779, and by Dalton in 1799 of liquids. Experiments on a thin metal bar by Biot in 1804 led him to recognize that accumulation of heat at some place along the bar led to an increase in its temperature. Because Biot was of the Leibnitz school, heat transfer was considered to

involve fluctuating motions of many atoms or molecules. The atomistic view, as voiced by Laplace, was that heat conduction involved particles interacting through attractive and repulsive forces (Purrington, 1997). This many-body problem is not tractable. Additionally, Biot's idealization did not involve distance, and so he did not recognize the importance of a temperature gradient (Grattan-Guinness and Ravetz, 1972).

- Mathematical advances in differential equations, particularly by Daniel Bernoulli, provided the necessary tools. Many notables were involved, such as Laplace, Green, d'Alembert, Euler, and Lagrange, see for example, Narasimhan (1999).

3.1.2 How Fourier Visualized Heat Transfer

Although Fourier's publications are in the realm of mathematical physics, his goal was to explain observations. Fourier conducted experiments in heat flow for 2 years prior to preparing his first manuscript; he repeated almost all previous measurements and added some of his own design.

Fourier recognized that the process by which heat received by a body is incorporated in its interior differs from the process of heat energy being transmitted through the body. His attempt to quantify both processes was presented to the French Academy in 1807, but was rejected primarily based on the responses of Laplace and Lagrange (Narasimhan, 1999). Upon winning a prize regarding terrestrial heat flow in 1811, he appealed the decision on the 1807 manuscript. Because Lagrange would not change his stance, Fourier expanded his ~1807 research into his 1822 book. Fourier's final paper on the subject in 1824 mostly summarizes his previous work, but it also contained his new idea that the temperature of space is due to starlight (Herivel, 1975). We will return to Fourier's discussion of how a body receives and sheds its heat later in the book. This chapter concerns flow of heat through a body, for which independent conservation of energy is valid.

In formulating his model of heat transfer, Fourier initially pursued the action-at-a-distance approach of Biot, but soon realized that the heat diffusion problem involves a continuum. By taking an empirical, observational approach that idealized how matter behaves macroscopically, Fourier avoided discussing the nature of heat and microscopic interactions.

Fourier defined flux ($\Im$ = energy per area per time), a difficult concept, and furthermore defined the important physical property of thermal conductivity (K), considering K to be a constant specific to each material (Herivel, 1975). Fourier visualized the heat conduction/diffusion problem in terms of three components (Fig. 3.2): heat transport through the space occupied by the body of interest; heat storage within a small element of the body (i.e., its heat capacity), and boundary conditions (assumed to be known from independent means). Regarding the physics of transfer, Fourier assumed that the amount of heat transferred is *proportional* to the temperature difference between the elements, which he referred to as molecules (p. 456 in Fourier, 1955 reprint). He viewed the process as the hot molecule giving a small amount of its heat to the cold. Basically, his assumptions are equivalent to the less controversial version of the second law: Fourier devised

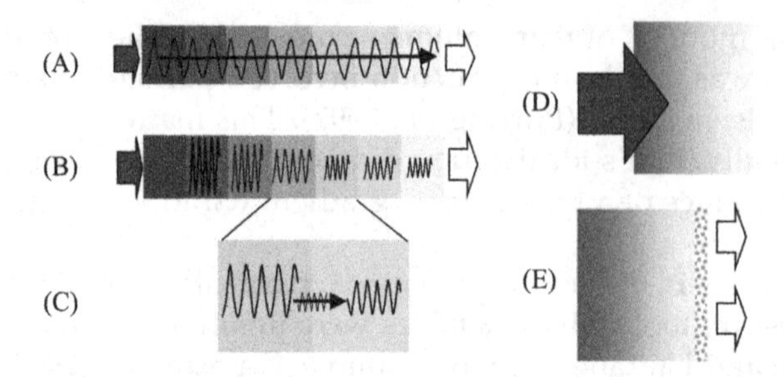

FIGURE 3.2 Fourier's visualization of heat flow across a bar. (A) Flow of heat (squiggle with arrow) across a bar with an end that is kept hot by an input (black arrow) while the other end sheds its heat radiatively to the surroundings (gray arrow). (B) Inside the bar, envisioned to consists of elements, each with a temperature (gray color) and a heat content (squiggle arrows of different size). (C) Expanded view of two elements, showing that the hotter element gives a small amount of its heat to the colder element. Fourier referred to this as conductability, but sometimes also as permeability. (D) Entry of heat into the bar. Some amount of the heat (tip of black arrow) enters into the bar, which Fourier called penetrability. (E) Exit of heat from the bar. Fourier considered the surface (speckles) to have its own conductivity, after which the heat was radiated to the surroundings, in accord with Newton's law of cooling.

the mathematical constraints. The proportionality relationship follows Newton's analysis of radiative cooling.

3.1.3 Fourier's Mathematical Model of Heat Diffusion

Using the above depiction, Fourier formulated the differential equations pertaining to flow across the interior. The methodology involved linearization of the transfer. Fourier began with Cartesian geometry, but considered cylindrical and spherical–polar systems as well. Modern expressions of his equations are:

$$\Im = -K\nabla T \tag{3.1}$$

and

$$\rho c_P \frac{\partial T}{\partial t} = \nabla \cdot K\nabla T \tag{3.2}$$

where ρ is density, c_P is specific heat at constant pressure (heat per unit mass per degree). The latter equation differs from Fourier's by allowing that K may depend on temperature. Presumably, heat is not consumed in deforming the solid, and so the heat capacity and thermal conductivity are derived under constant strain; but, more practically, constant pressure describes the laboratory conditions (see Carslaw and Jaeger, 1959). Eq. (3.2) is obtained from Eq. (3.1) by conserving energy. This requirement makes heat capacity important to heat transfer, as was recognized by Fourier (also see Section 3.5).

Eq. (3.2) is commonly presented in a simpler form, referred to as the heat equation:

$$\frac{\partial T}{\partial t} = D\nabla^2 T \tag{3.3}$$

where thermal diffusivity is related to K via:

$$K = \rho c_P D \tag{3.4}$$

Eq. (3.4) is a definition. Regarding Eq. (3.3), we have assumed that either K and D are independent of T (as Fourier did) or that the temperature changes are sufficiently small that their variation with T is unimportant (as in laser-flash measurements). The heat equation, too, should be attributed to Fourier, as it is entirely consistent with his derivation.

Parabolic Eq. (3.2) expresses the conservation of heat-energy per unit volume over an infinitesimally small volume lying in the interior of the body. In solving Eq. (3.2), boundary conditions and initial conditions must be imposed. Fourier expounded upon these stipulations at length, discussing several different problems in different geometries. For steady-state ($\partial T/\partial t = 0$) and constant K, Eq. (3.2) reduces to Laplace's equation, meaning that only Eq. (3.1) needs to be solved. The focus on steady-state has led to misunderstandings in both heat flow and thermodynamics as discussed in several chapters.

Fourier's invention of linearizing his equations allowed him to utilize the principle of superposition, that is, that particular solutions can be summed to form general solutions (Grattan-Guinness and Ravetz, 1972). Yet, utilizing trigonometric functions as solutions greatly contributed to denial of publication of his 1807 paper by the French Academy.

Fourier also stated "Heat is the origin of all elasticity: it is the repulsive force which preserves the form of solid masses, and the volume of liquids." However, Fourier's model depicts heat flowing in rays, that is, as flow of the imponderable fluid (Müller and Müller, 2009). He did not pursue this interesting idea, which has some merits in view of the existence of radiative pressure and that light and heat are the same phenomenon.

3.2 DIMENSIONAL ANALYSIS, LUMPED FACTORS, AND THE MEANING OF FOURIER'S HEAT EQUATIONS

Dimensional analysis is utilized to address complex problems which may not be tractable, to help formulate a problem mathematically, and to understand the implications of a mathematical formulation in a simple way (e.g., to estimate results). Various approaches exist, depending on which of the above goals is being pursued. Dimensional analysis permeates the field of fluid dynamics which describes the combined flow of heat and mass.

In the conceptual approach of Buckingham (1914), the number of pertinent dimensionless groups (termed "π's") is ascertained from the difference between the numbers of relevant variables and distinct unit-dimensions. How the variables are determined is not specified, but this leads to the four π's in fluid dynamics, which then combine to produce the four well-known dimensionless numbers of Grashof (Gr), Nusselt (Nu), Prandtl (Pr), and Rayleigh (Ra), for example, Hofmeister and Criss (2018). This method is useful for deriving equations.

A more common approach, that is more relevant to understanding and using existing equations, is to simplify the governing equations using the unit dimensions of the parameters (e.g., Bridgman, 1927). Tritton (1977) uses this approach to obtain Gr, Nu, Pr, and Ra and other dimensionless parameters in fluid dynamics.

3.2.1 Dimensional Analysis of the Heat Equation

Because heat flows down a temperature gradient, the one-dimensional form of Eq. (3.3) describes the essentials of heat diffusion. For Cartesian coordinates, prevalent in laboratory experiments, the relevant equation is:

$$\frac{\partial T}{\partial t} = D\frac{\partial^2 T}{\partial z^2} \tag{3.5}$$

where we have assumed D is constant over the temperature range of interest. All solutions to Eq. (3.5) have the form:

$$D \sim \frac{L^2}{\zeta} \tag{3.6}$$

where ζ is a thermalization time (e.g., Carslaw and Jaeger, 1959). Because diffusion consists of repeated events, one can view the process as motion of a heating front (Fig. 3.3A).

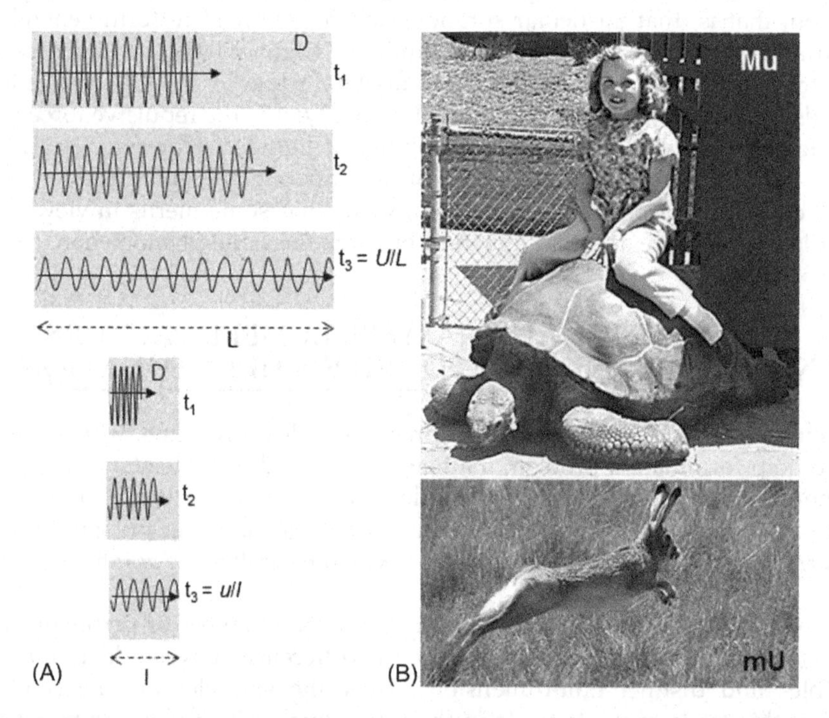

FIGURE 3.3 Schematics illustrating dimensional analysis of Fourier's equations. (A) The heat equation describes motion of a heating front. Different length blocks of the same material have the same D, but much greater time elapses before heat crosses the long bar (top) than the short bar (bottom). Thus, the characteristic speeds differ for the bars with different lengths. However, the speed and length are interdependent, through their relationship to the diffusivity, either in the heat (Eq. (3.3)), or its linearization (Eqs. (3.6)–(3.8)). (B) Momentum ($=Mu = mU$) analogy to the flux equation, which describes how heat is conducted via the parameter $K = Dc = dC$ (where use of lower and upper case suggests magnitude, not different variables). Heat is represented by mass in this analogy, illustrating that a value of K does not tell us if much heat moves slowly (tortoise, top) or if a little bit of heat moves fast (hare, bottom). The girl in the picture is the author in 1961, taken at the Toledo Zoo in Ohio.

Dimensional analysis furthermore tells us that the heating front moves at a linearized speed of

$$u \sim \frac{L}{\zeta} \tag{3.7}$$

(Hofmeister, 2010b). Combining the above linear approximations gives:

$$u \sim \frac{D}{L} \tag{3.8}$$

In summary, because cooling depends on time, each process is associated with a length scale and also with a characteristic speed pertaining to that scale.

If it were possible to conduct the experiments over lengths L comparable to the mean free path, which describes distances between exchange of heat, then u would be the average carrier velocity. Thus, high D correlates with high carrier speed. For an experiment at a given L, mechanisms with substantially different D have cooling fronts that progress at substantially different rates (speeds) and can be distinguished, and in theory quantified, in transient experiments (Mehling et al., 1998; Criss and Hofmeister, 2017; Chapter 4). For these reasons, time-dependent experiments are mandated to reveal the rate and mechanism by which energy diffuses through a material, independent of the amount of heat that is being moved, which is related to the heat capacity.

3.2.2 The Flux Equation and Ambiguities in Thermal Conductivity

When temperature is *not* a function of time, Eq. (3.2) (or Eq. (3.3)) reduces to Fourier's flux (Eq. (3.1)). The flux equation describing steady-state is simple, but this simplicity comes at the price of lost information.

Thermal conductivity is a product (Eq. (3.4)). Regarding the temperature response, density has a very small effect, even over temperature differences of $\sim$1000K, because thermal expansivity is small, $\sim 10^{-5}$ K^{-1} for the linear response of many oxides (e.g., Fei, 1995). The two other properties (D and C) respond to temperature in the opposite directions. Importantly, heat capacity is a complicated function and commonly depends much more strongly on temperature than does D. A case in point is disordered substances, for example, glasses made from natural lavas (Hofmeister et al., 2016) or feldspars (Branlund and Hofmeister, 2012), for which K increases with temperature above ambient conditions, in contrast to behavior of most crystalline solids, where K decreases above room temperature. The concept of a glass-like thermal conductivity (Cahill et al., 1992) is due to heat capacity overtaking the quite weak response of thermal diffusivity as temperature increases.

Thermal diffusivity governs how fast thermal fields change, whereas thermal conductivity describes the amount of heat-energy which is moving down a thermal gradient. This product is inherently ambiguous because large K could represent much heat moving slowly, or a small amount of heat moving very fast, or some heat moving at a moderate pace (Fig. 3.3B provides an analogy). Thus, considering steady-state behavior, which involves K alone (Eq. (3.1)), ambiguously describes the physical process, that is, the mechanism.

To drive this important point home, we plot each of D and K against mean atomic weight for the elements and some simple electrical insulators (Fig. 3.4). Thermal diffusivity varies by less than three orders-of-magnitude, whereas thermal conductivity varies by over five. At a single temperature, density is an important factor, making K high for metals but low for the noble gases (Fig. 3.4). The importance of density is also evident in the trends shown.

As recognized by Transtrum et al. (2015), ambiguous behavior is a hallmark of models which contain lumped (multiplied) factors, whereby data can be fit if even if the physics is incorrect. Transtrum et al. (2015) focused on computer models with many variables, and many different trade-offs, referring to these as sloppy models, although in a positive way, since reproducing and extrapolating data is useful in many applications. The same sloppiness can exist with 2 parameters (Figs. 3.3B and 3.4; Criss and Hofmeister, 2017) as with 10, and thus Transtrum et al.'s caution pertains to diverse efforts in the physical sciences. A good fit to the data is insufficient to verify the correctness of a mathematical model, if unconstrained variables are involved, or if trade-offs exist among constrained variables, even if the models are very useful and appear to be predictive.

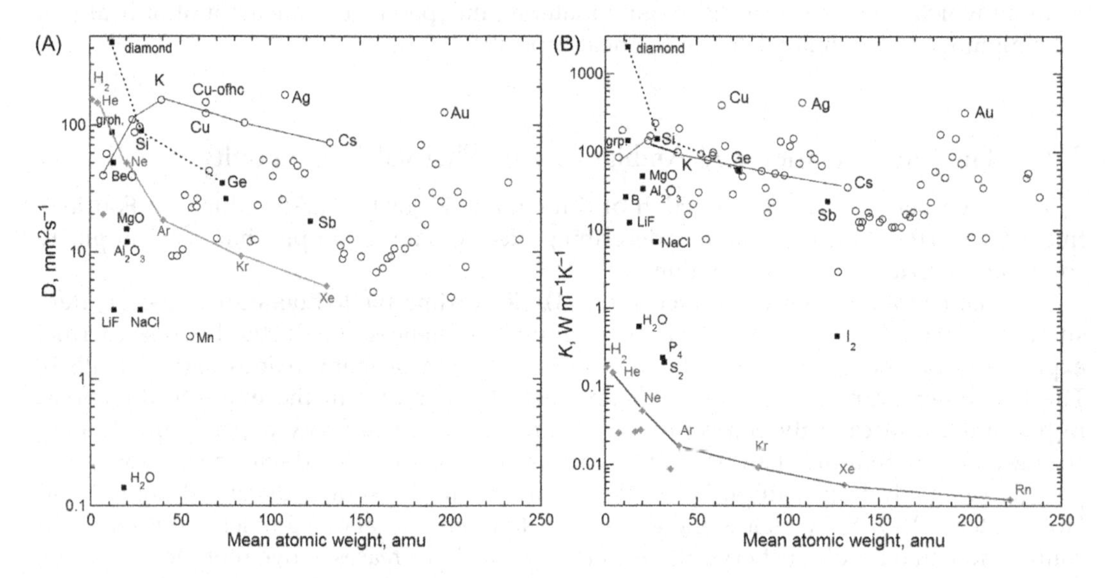

FIGURE 3.4 Thermal diffusivity (A) and thermal conductivity (B) of elements and simple compounds at ambient conditions plotted against mean atomic weight. Circles: metals; squares: nonmetals; gray diamonds: gases. Solid, dotted, and gray lines link elements in columns on the periodic table. Data from compilations of Touloukian et al., 1970; Touloukian et al., 1973; Hofmeister et al., 2014 and various websites. Some points from the compilation of gas data by Kestin et al., 1984; others are new measurements of Criss and Hofmeister, 2017. Many of the measurements are quite old; others have been repeated many times. Trends are shown for noble gases in gray; alkali metals in black; and nonmetallic simple elements in the dotted pattern. Select substances are labeled.

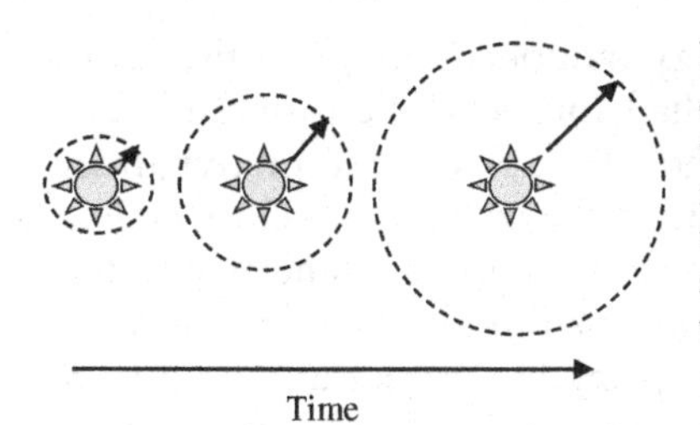

FIGURE 3.5 Schematic of thermal diffusivity involving pure energy. Sun: energy source (essentially a point). Arrow indicates one of the many rays. Dotted circle: a front of photons during a radially symmetric flare from a star. The flare is described the growth of the surface with time. The progression of the front is the key feature of heat diffusion, whereas thermal conductivity measures the product of the heat with its progression outwards.

3.2.3 The Meanings of Thermal Diffusivity and Thermal Conductivity

The physics of thermal diffusivity can be explained from several perspectives: Inverting Eq. (3.4) provides information on macroscopic behavior:

$$D = \frac{K}{\rho c_P};\ \text{diffusivity} = \frac{\text{conductivity}}{\text{storativity}} \tag{3.9}$$

Thus, D measures the ability of a material to conduct thermal energy relative to its ability to store thermal energy (Incropera and Dewitt, 2005). When D is large, the material responds rapidly to thermal inputs, whereas when D is small, the response is sluggish and achieving equilibrium takes a long time. Regarding heat flow, storativity is identical to the volumetrically based heat capacity: the single word is used to distinguish the product from C_V. The RHS of Eq. (3.9) is more general than heat flow. For example, the hydrologic diffusivity is used to explain the rate of change of hydrologic head (Davisson and Criss, 2017).

Diffusivity can be understood from a thought experiment. In Fig. 3.5, we consider the pure energy of light, and a star as a point source. At some initial time, the star emits a burst of energy, as a spherically symmetric flare (dotted circle). At later times, the flare reaches greater distances. The diffusion of this energy is described as the area of the flare and how this increases with time. For this reason, the units of D are area over time. Whether the photons in the flare are very high energy ultraviolet light or low energy IR light is a distinct issue from the time-evolution of the front.

Lastly, the microscopic meaning of D is revealed by dimensional analysis (Section 3.2.1). When heat (or matter: Section 3.3.4) diffuse, the efficiency is determined by the distance between exchange events and the speeds at which heat moves. Thus, thermal diffusivity isolates the time dependence of heat transport, whereas thermal conductivity combines the amount of heat moving with the evolutionary process. Both aspects are important, but more information exists when diffusivity and storability are separately determined. The lumped parameter is K (Eq. (3.4)) because the units, physics, and behavior of D are simpler (Eq. (3.6); data are discussed in Chapter 4).

3.3 HEAT FLOW AND THE THERMODYNAMIC FRAMEWORK

Heat flow is a confusing phenomenon because we record its motion by measuring temperature, which is not the same entity. For maximal clarity, we start with a simple scenario and work-up in complexity. Fourier's laws conserve energy independently, and thus concern only heat, not work, which involves externally applied forces. Three different basic situations are possible, which are related to the various idealizations in thermodynamics described in Chapter 1, Sections 1.1 and 1.2; while situations are considered as reversible vs irreversible (Fig. 1.1), and changes (regarding heat) are classified as isothermal, adiabatic, or isentropic (Fig. 1.3). In a nutshell, heat transfer models are interwoven with conceptualizations of classical thermodynamics, which are incomplete. Importantly, and for example, to understand heat and its flow, we consider the constant volume approximation (cf. Chapter 1, Section 1.2.3), even though a material will expand as heated. For incremental changes in temperature, constant volume suffices, although in practice, constant pressure describes most laboratory situations (but not free expansion: Chapter 1, Section 1.2.6).

3.3.1 Thermal Equilibrium

Thermal equilibrium seems rather boring insofar as temperature does not change and is equal between bodies. However, is this state actually static? All objects at finite temperature radiate heat (Chapter 1), which is a dynamic process (Fig. 3.6A). Emissions necessarily cool the body, as was known by Newton, and cause the heat death required by the second law, unless either a compensating internal energy source exists or heat is externally applied to the body. For the situation of Fig. 3.6A to be truly static, the bodies and the surroundings must be at the same temperature, so all fluxes are in balance. No changes occur

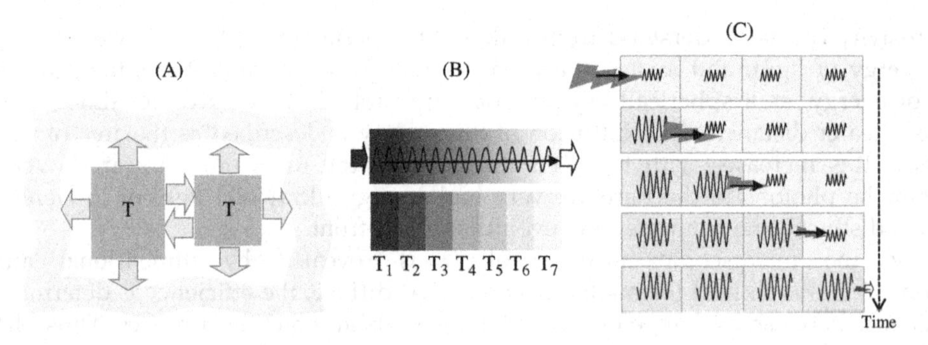

FIGURE 3.6 Schematics illustrating types of heat flow. (A) Thermal equilibrium between two bodies at identical, isothermal temperatures (medium gray). Surface conditions can be often neglected, but emissions of heat to the surroundings (light gray wavy box) always exist. If the surroundings possess the same temperature, conditions are static, even though heat is continually being exchanged. (B) Steady-state flow of heat across a bar. Top shows the heat in motion across a bar which is held at high temperature on its left end, and cools to the surroundings on the right end. Bottom shows the bar as a series of thin slices, each with constant temperature. Heat is transferred at each intersection between slices, but time is unimportant, so each slice is in quasiequilibrium. With thinner and thinner slices, continuity and smoothness are achieved, while local equilibrium exists in each slice. (C) Transient flow of heat across a bar. Top: initial condition where a finite energy pulse (lightning bolt) is applied to the left end of an isothermal bar. As time progresses, the pulse loses some energy to each of the segments that it passes through, so the segment becomes hotter (higher temperature = darker gray). Bottom: time when the pulse reaches the far (right) end of the bar. The pulse has very little energy left, but some of its energy must leave the bar because the bar is now at higher temperature and must emit more heat to the surroundings.

so conditions are isothermal and reversible. More realistically, heat flows to and from the surroundings (e.g., Earth's 'ambient' temperature results from sunlight warming it).

The existence of thermal emissions is well-known, but the connection of the radiative losses with diffusive transfer of heat across the body has not been fully elucidated. The situation exists because (1) radiative losses from the sample and diffusive progression inside the sample are assumed to be two distinct processes and (2) spectroscopic and heat transfer studies are not fully unified.

Distinct processes were indeed envisioned by Fourier (see Fig. 3.2), where he specified that input, output, and heat transfer across a body were all some type of radiative transfer. Fourier discusses receipt of radiant heat in terms of permeability, which role was elucidated by McMahon and Gardon a century later (Chapter 2, Section 2.6). Furthermore, Fourier dealt with radiative emissions by considering object surfaces to be special part of the body with their own conductivity.

In understanding material properties from a macroscopic perspective, we can largely ignore the role of the surface and therefore, of the emissions. But in understanding radiative transfer or microscopic behavior, we cannot (Chapters 5 and subsequently).

3.3.2 Steady-State

Heat flow under steady-state conditions is correctly modeled by the caloric model and Fourier's Eqs. (3.1) and (3.2), because no work is preformed, and so heat-energy is conserved separately (Fig. 3.6B). Under steady-state, heat may enter or leave the system, but a balance exists, resulting in no net change of temperature with time. Steady-state under Cartesian symmetry can be viewed as a series of slices in *local* equilibrium, altogether constituting a temperature gradient. The steady-state condition exists piecewise in local equilibrium, and can be described as quasi-equilibrium. The term *local* connotes limited space (Fig. 3.3) which is tied to short time intervals.

Thermodynamics can explain a steady-state situation, which is time-independent, with only the addition of Fourier's concepts of heat flux and the property of thermal conductivity (Eq. (3.1)) to the laws. Conditions are not reversible, but if the system is not changing, it can be adiabatic (net flow is zero) and isentropic (the state does not change; see Chapter 1, Section 1.2). The system is closed, but not isolated because heat can flow in and flow out, unaccompanied by the flow of mass. Ignoring small changes in volume is consistent with local equilibrium existing. In actuality, the assumption is constant strain (Carslaw and Jaeger, 1959). If deformation occurs in an experiment, not only is energy consumed in this process, but the consumption is time-dependent. Meeting this stipulation is difficult in high-pressure studies, and the repercussions of its violations have been overlooked (Chapter 4).

Because heat is being lost to the surroundings, the second law holds, in the original form of Clausius. This is important, because temperatures are constrained to only flow down the gradient. For Cartesian systems, this stipulation reduces heat flow in steady-state to an one-dimensional problem (the matrix can be diagonalized). In real bodies, edge effects occur, which are complicated, involving interfaces and radiation. But, the one-dimensional nature of heat flow remains, as exemplified by, laser-flash experiments and other transient techniques (Chapter 4).

Limited information can be obtained from steady-state experiments (due to trade-offs between D and C composing K, which is analogous to mass and velocity composing momentum: Fig. 3.3). Furthermore, because establishing a temperature gradient across a material depends on both entry and exit of heat (Fig. 3.6), surface behavior is important, and cannot be overlooked. From this problem arises the concept of interface resistance, which traces to Fourier's notions (Fig. 3.2). Lastly, a dynamic situation precedes data collection in steady-state conditions, so whether quasiequilibrium has been reached is always a concern.

3.3.3 Transient Conditions

When imbalances in temperature or heat content exist, the flow of heat is decidedly time-dependent. Fig. 3.6C shows a simplified version of the dynamic behavior during transient measurements in a bar. The dynamic process of thermal re-equilibration can be described as a series of heating steps in an adiabatic process. The actual behavior can be quite complicated. First, different mechanisms may operate at short times than at long times because each mechanism has an associated thermalization time and speed (Section 3.2). For electrical insulators, heat can initially be transported by ultrafast photons (i.e., internal ballistic radiation at high energy), and then slower, lower energy vibrations could diffuse. For metals, heat can initially be diffused by fast electrons, and then slow vibrations could complete the process (Criss and Hofmeister, 2017). If a change in mechanism occurs during a heating or cooling event, then the transport of heat cannot be represented by a single value of K. Fourier's equations hold, but care must be taken in applying them. In a nutshell, much more information exists in transient measurements, requiring a considerably more complicated analysis (see Chapter 4).

The measurements made by Fourier and the many data collections subsequently performed mostly concerned steady-state conditions, which cannot describe the progressive process of equilibration, and cannot differentiate mechanisms. Additionally, time-dependent methods mostly explore conditions nearing steady-state, while overlooking initial, highly transient behavior. Focus on behavior near equilibrium has created misunderstandings of heat transfer some 200 years after Fourier's important discoveries. But, we now have the means to probe all aspects of heat transfer, as covered in detail below and in the following chapters.

3.3.4 Comparison With Fick's Equations Reveals the Importance of Space

The macroscopic model for diffusion of mass (Fick, 1855) postdated studies of heat diffusion and was based on Fourier's model. However, an important difference can be gleaned by examining Fick's equations governing diffusion of mass concentrations:

$$\Im_{\mathrm{m}} = -D_{\mathrm{m}}\frac{\partial \xi}{\partial x} \quad \text{and} \quad \frac{\partial \xi}{\partial t} = D_{\mathrm{m}}\frac{\partial^2 \xi}{\partial x^2} \tag{3.10}$$

where $\Im_{\mathrm{m}}$ is mass flux (kg m^{-2} s^{-1}), ξ is concentration (kg m^{-3}), and D_{m} is the mass diffusion constant (m^2 s^{-1}). Eq. (3.10) presumes that temperature is held constant. Mass concentrations are dissipated, as matter flows outward to occupy a greater volume. Mass flow, analogous to heat flow, proceeds down the gradient in concentration. Note that Fourier's and Fick's equations are cast in terms of intensive variables, expect for the spatial derivatives, which are extensive and thereby make the fluxes extensive properties. However, the behaviors of heat and mass during their flow are not precisely analogous:

Only two physical properties (ξ and D_m) plus flux govern diffusion of mass in Eq. (3.10), whereas three parameters (T, $C = \rho c$, and D) plus flux govern diffusion of heat as in Fourier's Eqs. (3.1) and (3.3). Time-dependent flow of matter involves mass and space (volume or distance) only, because concentration being the density of a particular kind of mass links mass to space occupied and because these equations presume temperature is not changing. In contrast, time-dependent flow of heat involves energy, space, and temperature, because the amount of heat depends on the two key, independent state variables of volume (or mass, both are extensive and describe occupation of space) and temperature. For this reason, one more parameter (essentially, heat capacity) appears in the equations governing flow of energy than mass.

Fourier recognized the extreme importance of heat capacity to the flow of heat. This key aspect has been insufficiently accounted for in subsequent analyses. In part, problems have arisen because our understanding of heat stems from classical thermodynamics, which rests on modeling heat at constant volume (e.g., Chapter 1). Although this restriction permitted exploring the nature of heat, it cannot possibly be met during time-dependent heat flow. Hence, in understanding time-dependent heat flow, thermal diffusivity (or thermal conductivity) and heat capacity must be considered together (Section 3.5). In certain situations, a model can be constructed based on thermal diffusivity alone, such as in LFA (Parker et al., 1961) or Angstrom's method (Touloukian et al., 1973).

3.4 ERRORS STEMMING FROM EQUATING ELECTRICAL FLOW WITH HEAT FLOW

Flow of electrical charge and heat are both described by continuum theories. Ohm stressed the mathematical nature of his proof of his law for electrical currents and emphasized his reliance on Fourier's macroscopic theory of heat conduction (Purrington, 1997). Ohm's approach is reasonable because Fourier's model assumes no mechanical work is performed and that no internal heat is generated. The forward extrapolation of Ohm is valid. But because situations can exist where heat-energy is not conserved independent of work, the reverse extrapolation of the equations for electrical flow, which conserve charge, to heat flow is problematic. As shown in Section 3.4, mass and heat flow cannot be equated because the flow of heat is related to both volume and temperature, while the flow of mass depends only on volume. The same caveat holds for electrical current, because the defining equations governing the motion of charge are more similar to those which govern mass flow than heat flow.

Interestingly, Maxwell (1888, pp. 52–53) cautioned against carrying the analog of heat and current too far, listing many reasons. Maxwell noted that steady flow of heat requires both a continuous supply and a continuous loss (Fig. 3.2A), whereas an electrified body placed in a perfectly insulating medium could in theory remain electrified forever without any supply from external sources. He noted that there is nothing in the electrostatic system that can be described as flow. In contrast, Fig. 3.3A shows that heat flows even during thermal equilibrium. Lastly, certain bodies may be strongly electrified without undergoing any physical change, but in great contrast, the temperature of a body cannot be altered without affecting the physical state of the body, such as density or phase.

Despite these historical cautions, modern models of substances with multiple mechanisms assume that a simple sum pertains:

$$K_{meas} = K_1 + K_2 \tag{3.11}$$

(Uher, 2005). Meissner (1915) proposed Eq. (3.11) based on his analysis of the bulk thermal conductivity measurements of metals, for which heat can be potentially be carried by electrons in addition to "ordinary" conduction by vibrations and/or light. Meissner's equation is based on fitting a few experimental data points for a few samples near ambient temperature to $K = a + bT^{-1}$, where a and b are constants, presumed to represent different mechanisms (see e.g., Gruneisen and Goens, 1927). A derivation was neither provided in these early works nor subsequently, to the best of our knowledge.

Eq. (3.11) was probably based on analogies to electrical circuits and steady-state conditions, and likely was accepted on this basis. Many engineering problems in heat transfer are solved by adding thermal resistances to describe flow elements in parallel and by taking the inverse to describe series. However, as regards multiple mechanisms in a single sample, which are described by parallel heat flow, the validity of Eq. (3.11) is limited due to the geometrical constraints needed to "blend" mechanisms (Fig. 3.7). Limited validity can be demonstrated by summing heat currents (Kirchhoff's law) while considering the behavior prior to reaching steady-state:

$$J = -K\frac{\partial T}{\partial x} = -\left(K_1\frac{\partial T_1}{\partial x} + K_2\frac{\partial T_2}{\partial x}\right) = J_1 + J_2 \tag{3.12}$$

For Eq. (3.11) to hold at all time, all temperature gradients in Eq. (3.12) need to be identical. This is possible if one mechanism operates (i.e., either $K_1 = K_2$ or $K_2 = 0$), or if the mechanisms interact very strongly Eq. (3.13).

$$\frac{J_1}{K_1} = \frac{J_2}{K_2} \tag{3.13}$$

Instead, if the mechanisms are independent, each will set up its own temperature gradient, see Criss and Hofmeister (2017) for details.

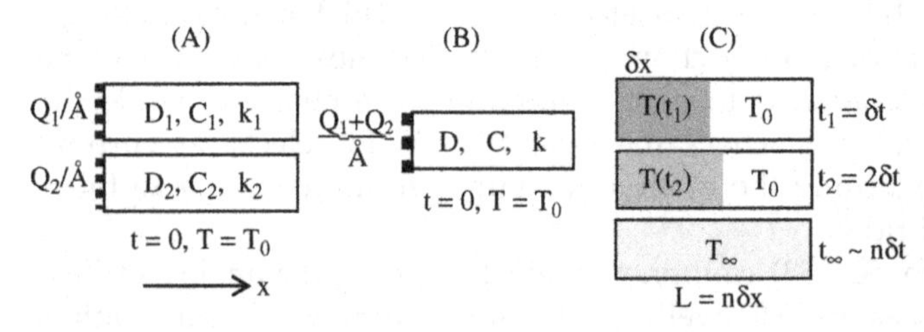

FIGURE 3.7 Schematic of heat flow in parallel. (A) Initial conditions in two parallel bars. Heavy dots represent application of the instantaneous pulse. (B) Initial conditions for a blended bar. (C) Time-evolution during an adiabatic process. *From Criss, E.M., Hofmeister, A. M., 2017. Isolating lattice from electronic contributions in thermal transport measurements of metals and alloys and a new model. Int. J. Mod. Phys. B 31, No. 175020. doi:10.1142/S0217979217502058.*

Experimental confirmation of unequal gradients is provided by femptosecond spectroscopic experiments on metals which create a high electron flux by stimulating electrons occupying a broad range of energy levels to higher energies using high-intensity, monochromatic light (usually visible), and then monitor the short-time response. Initially, the electron temperature is ~500K above ambient (Bauer et al., 2015), which is far from equilibrium.

For Eq. (3.11) to hold for two mechanisms while meeting Eq. (3.13) means that these processes cannot independently move heat. From the previous sections, including Fourier's recognition that heat capacity is important to energy conservation, Eq. (3.13) implicitly links heat capacities to thermal conductivities in a manner that is not dictated by physical properties, for example, the mechanism with higher K has higher C at all temperatures. This is not necessarily true. As indicated by Fig. 3.3B, high K means the product CD is high, but various combinations are permissible (very high C with low D or low C with very high D), which is the case for electrons (e.g., Kittel, 1971). To correctly compute the sum, energy must be conserved, which requires considering heat capacity. The following section provides details.

3.5 SUM RULES FOR HEAT FLOW THAT CONSERVE ENERGY

For heat flow, as in current flow, both parallel and series configurations exist. The series configuration applies to heat moving across layers, which occurs in many situations. The parallel configuration is especially important, because this describes multiple mechanisms in a single material (Fig. 3.7).

This section follows the derivation of Criss and Hofmeister (2017). The basis is that a rise in temperature for a material is related to the net heat received (Q) by:

$$\Delta T \cong \frac{Q}{\text{Å}LC} \tag{3.14}$$

where C = heat capacity on a per volume basis at constant pressure = ρc, Å is area, and L is length, providing volume (V = ÅL). Eq. (3.14) presumes that a single temperature field applies, and was presented by Parker et al. (1961) in order to relate the change in temperature during pulsed heating of laser-flash experiments to the heat capacity. For simplicity, the remainder of Section 3.5 utilizes C, without a subscript. Note also, that it is impossible to exactly characterize transient behavior with a single conductivity. In actuality, we are considering an equivalent conductivity ($K_{equivalent}$) which is an approximate value that conserves energy while representing the change in temperature as a simple, continuous rise as the heat traverses the sample (Fig. 3.6). The time-dependent response for multiple mechanisms with very different D, connected with very different u and C, is described by the Mehling et al. (1998) model for LFA.

3.5.1 Transport of Heat in Parallel

We consider one-dimensional flow under a nonnegligible T-gradient, which describes experiments on solids. The adiabatic approximation is used. A general rule is derived for independent mechanisms, by considering four limiting cases. Storativity (C) pertains, rather than specific heat (c), since the derivation utilizes Eq. (3.14) whereby heat is conserved.

To model transient behavior, we consider nearly instantaneous application of a prescribed amount of heat (Q) to a bar of material with initial temperature (T_0). Parallel heat flow involving simultaneous, but independent mechanisms, is described by multiple bars of equal density wherein only one mechanism operates in each bar (Fig. 3.7). This approach is essential to understand condensed mater. For metals, competing electronic and lattice mechanisms exist. For partially transparent electrical insulators (many minerals), competing radiative (photonic) and lattice (phononic) mechanisms exist.

We emphasize that distinct temperature fields are a consequence of noninteracting mechanisms at disequilibrium, which immediately follows a thermal disturbance. Such behavior is exemplified by ultrafast spectroscopic studies of electrons in metals, where applying a laser pulse immediately heats the conduction electrons to about 500K above ambient (Bauer et al., 2015). Conserving energy (e.g., Eq. (3.14)) requires that the apparent temperature field is a weighted sum of temperature fields associated with each mechanism in the differing bars:

$$CT = C_1T_1 + C_2T_2 = \sum C_iT_i \tag{3.15}$$

Assuming that the independent mechanisms obey Fourier's law leads to

$$K\frac{\partial T}{\partial x} = K_1\frac{\partial T_1}{\partial x_1} + K_2\frac{\partial T_2}{\partial x_2} + K_n\frac{\partial T_n}{\partial x_n} = \sum K_i\frac{\partial T_i}{\partial x_i} \tag{3.16}$$

where the subscripts indicate which bar the mechanisms operate in. Combining Eq. (3.15) and Eq. (3.16):

$$K\sum C_i\frac{\partial T_i}{\partial x_i} = C\sum K_i\frac{\partial T_i}{\partial x_i} \tag{3.17}$$

Solving Eq. (3.17) requires making additional assumptions, which is possible through considering a few limiting cases (Criss and Hofmeister, 2017).

3.5.1.1 *Case 1*

First, we assume that the thermal diffusivity ($D = K/C$) is similar in both bars, resulting in a single thermal rise, and an equivalent conductivity due to governance by Eq. (3.3). Hence, their temperature evolves similarly with distance and time, providing thermal equilibrium ($T \approx T_1 \approx T_2$) as re-equilibration progresses. Blending the two bars provides a single medium (Fig. 3.7B). Considering incremental changes at quasiequilibrium (Fig. 3.7C), and conserving energy yields:

$$T_1 = T_0 + \frac{Q_1}{C_1\mathring{A}\delta x_1}; \quad T_2 = T_0 + \frac{Q_2}{C_2\mathring{A}\delta x_2}; \quad T = T_0 + \frac{Q}{C\mathring{A}\delta x} \tag{3.18}$$

If $C_1 \approx C_2$ then $K_1 \approx K_2$ and $Q_1 \approx Q_2 \approx Q/2$. The last stipulation conserves energy, while also insuring that the final temperatures associated with the two mechanisms are similar. For two mechanisms, $C = C_1 + C_2$. Combining these stipulations with Eq. (3.18) gives:

$$\frac{C}{2}\delta x \approx C_1 \delta x_1 \approx C_2 \delta x_2 \tag{3.19}$$

From Fig. 3.7A and B, heat flux ($\Im = Q/(\text{Å} t_{\text{pulse}})$ where t_{pulse} is the duration of the applied heat pulse), as applied to the blended bar, equals the sum $\Im_1 + \Im_2$. Letting $\partial T/\partial x = \delta T/\delta x$ and $\partial T = \delta T \approx \delta T_i$ relates the slopes of the temperature fields such that:

$$n\frac{C_i}{C}\frac{\partial T}{\partial x} \approx \frac{\partial T_i}{\partial x_i} \tag{3.20}$$

Applying Eq. (3.20) to the summation of heat currents (Eq. (3.12)) results in:

$$K\frac{\partial T}{\partial x} = K_1\frac{\partial T_1}{\partial x_1} + K_2\frac{\partial T_2}{\partial x_2} = K_1\frac{2C_1}{C}\frac{\partial T}{\partial x} + K_2\frac{2C_2}{C}\frac{\partial T}{\partial x} = \sum K_i \frac{nC_i}{C}\frac{\partial T}{\partial x} \tag{3.21}$$

For n independent mechanisms, $C = \Sigma C_i$, so:

$$K = n\frac{\sum C_i K_i}{\sum C_i} \quad \text{for } C_i \approx C_j \text{ and } D_i \approx D_j \tag{3.22}$$

where the subscript j refers to one of the i mechanisms. For the more restrictive condition of $C_i = C_j$, then Eq. (3.22) reduces to $K = \Sigma K_i$ as in Eq. (3.11) but only under the restriction of $D_i \approx D_j$.

In Case 1 above, we have described a single rise for a T-t curve that embodies both mechanisms. When the C_i's and D_i's are each similar, the curve represents the average behavior of the various carriers. For this averaging to occur requires either accidental similarity of the physical properties or that the processes exchange energy (i.e., are *not* independent), thereby creating similar properties.

3.5.1.2 Case 2

In this variant, we assume similar D's, but do not require similar C_i values. Moreover, we allow the temperature fields to vary relative to one another. Applying Eq. (3.20) to each side of Eq. (3.17) results in:

$$K\sum C_i \frac{nC_i}{C}\frac{\partial T}{\partial x} \cong C\sum K_i \frac{nC_i}{C}\frac{\partial T}{\partial x} \tag{3.23}$$

Canceling like terms, using $C = \Sigma C_i$, and rearranging Eq. (3.23) gives:

$$K \cong \sum K_i C_i \frac{\sum C_i}{\sum C_i^2} \tag{3.24}$$

If $C_i = C_j$, then Eq. (3.24) reduces to $K = \Sigma K_i$ as did Eq. (3.22). Eq. (3.24) relates K to C by assuming a relationship between the spatial derivative of independent temperature fields and heat capacity.

Eqs (3.22) and (3.24) for noninteracting mechanisms, but with similar diffusivities, diverge from Meissner's sum, Eq. (3.11). The weighting factors revealed by our derivations account for energy conservation and effects of transient phenomena in heat flow. Eq. (3.11) incorrectly describes heat transfer, except for special cases involving equivalent diffusivities.

3.5.1.3 *Case 3*

This case is more realistic than the two previous. For simplicity, only two mechanisms are considered. We assume that the specific heats of each mechanism differ significantly, such that $C_1 << C_2 \approx C$ resulting in $D_1 >> D_2$, for $K_1 >$ or $\sim K_2$. This important case represents behavior of metals at ordinary temperatures (i.e., T above some tens of Kelvins), with subscript 1 representing the electrons, and subscript 2 representing the lattice. Such disparate thermal diffusivity values will cause the temperature–distance evolution associated with each process to differ greatly (as was observed for metals: Criss and Hofmeister, 2017). Over a relatively short time, fast process 1 will equilibrate to its final T_∞ (Fig. 3.7C). Because bar 1 representing process 1 is isothermal, $\partial T_1/\partial x_1 = 0$. However, over this same time interval, T_2 in bar 2 still changes with distance, providing a heat current of $K_2\partial T_2/\partial x_2$. Summing currents gives $K\partial T/\partial x = K_2\partial T_2/\partial x_2 + 0$ and thus, $K = K_2$.

In more detail, because energy is conserved, T_∞ for process 1 equals that of process 2 and of the blend, which is computed from Eq. (3.12) by setting $\delta x = \delta x_1 = \delta x_2 = L$. Consequently, $Q_1/(C_1V) = Q_2/(C_2V) = Q/(CV)$ from Eq. (3.18). (Note that although fast process 1 in the blended bar might initially absorb more heat than expected from thermal equilibrium, i.e., $Q^* > Q_1 = QC_1/C_2$, carriers of type 1 must subsequently transfer this excess (Q^*-Q_1) to carriers of type 2, in order to attain isothermal conditions at very long times.) Because $Q_1 = QC_1/C_2$, Q_1 is negligible compared to Q_2, then $K_1\partial T_1/\partial x_1$ is negligible compared to $K_2\partial T_2/\partial x_2$. Summing heat currents gives

$$K = K_2, \quad \text{for} \quad C_1 < C_2 \tag{3.25}$$

Importantly, the weighted sum of Eq. (3.24) reduces to Eq. (3.25) if $C_1 << C_2$, showing that Eq. (3.24) is more general than the restricting conditions used in the derivation would suggest. In contrast, Eq. (3.22) forces the temperature fields to be the same, and thus only occurs under special circumstances for independent mechanisms, but it is generally applicable for a single mechanism with multiple carriers (Section 3.5.2).

For Case 3, wherein $D_1 >> D_2$, mechanism 1 proceeds over a much faster rate, based on Eqs. (3.3) and (3.8) because L is the same for both processes (Fig. 3.7). Hence, $u_1 >> u_2$. But, mechanism 1 carries little heat ($C_1 << C_2$). Hence, during re-equilibration, as shown in *T-t* curves, the fast mechanism is expressed as a small rise at short times, and thus, the fast mechanism 1, which carries little heat, has little or nothing to do with long-term behavior.

3.5.1.4 *Case 4*

For completeness (i.e., to consider all possible values of K), we now assume that $K_1 < K_2$ while $C_1 < C_2$. In this variant, both the amount of heat and how fast it is transported by mechanism 1 are negligibly small, so that this mechanism has a negligible effect on heat transfer. The material property is again represented by Eq. (3.25).

3.5.2 Importance of Heat Capacity

Storativity (or specific heat, if density differences need to be considered) is an equally important parameter as thermal conductivity in describing heat transfer in a material with multiple mechanisms or in parallel transport. This finding is consistent with the form of Fourier's original Eq. (3.2). As a consequence, the slower process dominates should it have a significantly higher specific heat. Importantly, because our analysis covers the initial conditions up to thermal equilibration, it describes the approach to steady-state, and thus describes diverse experimental measurements of thermal conductivity. Our model is not limited to transient phenomena, but describes heat conduction when energy is conserved. However, our analysis does not cover the very short period of time after a pulse is applied when the very fast carriers equilibrate: to understand this requires detailed modeling of Eqs. (3.2) or (3.3), discussed later in the book.

3.5.3 Transport of Heat in Series

The time-evolution of heat flow of Eq. (3.10) for two and three layers has been modeled (Lee et al., 1978) and is utilized in laser-flash measurements (Chapter 4, Section 4.2). Here, we are interested in ($K_{equivalent}$) for a multilayer sample (Fig. 3.8) under steady-state conditions. For simplicity, we assume the properties K_i and C_i are independent of temperature over any given layer i. For a series configuration with temperature dependent properties, the layer width can be made small to account for this behavior.

The applied flux $\Im$ pertains to each interface. For each layer, conserving energy gives:

$$\Delta T_i \cong \frac{Q}{\mathring{A} L_i C_i} \tag{3.26}$$

Fourier's law holds across all layers and across each layer individually, yielding:

$$-\Im = K\frac{dT}{dx} = K\frac{\Delta T}{L} = K_i\frac{\Delta T_i}{L_i} \tag{3.27}$$

Combining the above gives many similar equations for different layers denoted i and j:

$$\frac{K_j}{L_j^2 C_j} = \frac{K_i}{L_i^2 C_i} = \frac{K}{L}\sum\frac{1}{L_i C_i} \tag{3.28}$$

Inverting the summation, and then using the equivalence of the products gives:

$$\frac{L}{K} = \frac{L_j^2 C_j}{K_j}\sum\frac{1}{L_i C_i} = \sum\frac{L_i}{K_i} \tag{3.29}$$

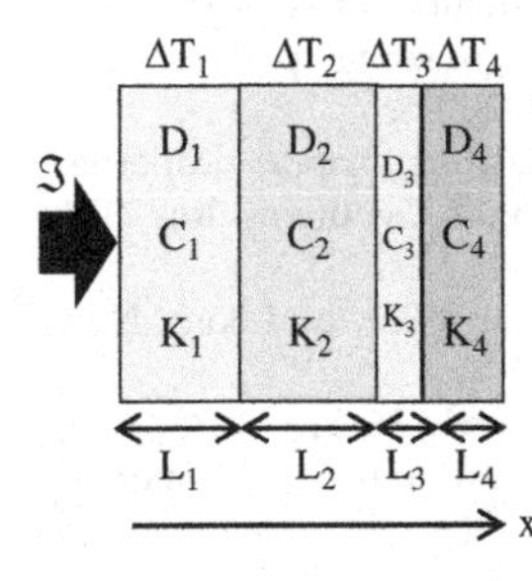

FIGURE 3.8 Schematic of heat flow in series. Heavy dots represent application heat (constant flux). Each layer has different physical properties and width, L. Interfaces can be represented as thin layers.

Our result agrees with previous analyses made by analogy to electrical resistors in series. Eq. (3.29) shows that K is a type of harmonic mean, which is always less than the simple mean. In this geometry, the temperature field is constrained and independent of time for each block. The consequence is that low conductivity layers dominate, even if thin. Again, storativity pertains.

References

Bauer, M., Marienfeld, A., Aeschlimann, M., 2015. Hot electron lifetimes in metals probed by time-resolved two-photon photoemission. Prog. Surface Sci. 90, 319–376.

Branlund, J.M., Hofmeister, A.M., 2012. Heat transfer in plagioclase feldspars. Amer. Mineral. 97, 1145–1154.

Bridgman, P., 1927. Dimensional Analysis. Yale University Press, New Haven, CT.

Buckingham, E., 1914. On physically similar systems; Illustrations of the use of dimensional equations. Phys. Rev. 4, 345–376.

Cahill, D.G., Watson, S.K., Pohl, R.O., 1992. Lower limit of thermal conductivity of disordered solids. Phys. Rev. B 46, 6131–6140.

Carslaw, H.S., Jaeger, J.C., 1959. Conduction of Heat in Solids, second ed. Oxford University Press, New York, NY.

Criss, E.M., Hofmeister, A.M., 2017. Isolating lattice from electronic contributions in thermal transport measurements of metals and alloys and a new model. Int. J. Mod. Phys. B 31, No. 175020. Available from: https://doi.org/10.1142/S0217979217502058.

Davisson, M. L., Criss, R.E., 2017. Observational Hydrology. http://www.observationalhydrology.com/ (accessed 16.10.17).

Fegley Jr., B., 2015. Practical Chemical Thermodynamics for Geoscientists. Academic Press\Elsevier, Waltham, MA.

Fei, Y., 1995. Thermal expansion. In: Ahrens, T.J. (Ed.), Mineral Physics and Crystallography: A Handbook of Physical Constants. American Geophysical Union, Washington, DC, pp. 29–44.

Fick, A., 1855. Über Diffusion. Ann. der Physik 94, 59–86.

Fourier, J.B.J., 1822. Théorie Analytique de la Chaleur. Chez Firmin Didot, Paris. (Translated in 1955 as The Analytic Theory of Heat (A. Freeman, Trans.). Dover Publications Inc., New York, NY.

Grattan-Guinness, I., Ravetz, J.R., 1972. Joseph Fourier, 1768–1830. MIT Press, Cambridge, MA, p. 516.

Gruneisen, E., Goens, E., 1927. Investigation of metal crystals. V. Electrical and thermal conductivity of single and polycrystalline metals of regular systems. Zeit. Phys. 44, 615–642.

Herivel, J., 1975. Joseph Fourier, The Man and the Physicist. Clarendon, Oxford, England.

Hofmeister, A.M., 2010a. Thermal diffusivity of perovskite-type compounds at elevated temperature. J. Appl. Phys. 107, No. 103532.

Hofmeister, A.M., 2010b. Scale aspects of heat transport in the diamond anvil cell, in spectroscopic modeling, and in Earth's mantle: implications for secular cooling. Phys. Earth Planet. Inter. 180, 138–147.

Hofmeister, A.M., Criss, E.M., 2018. How properties that distinguish solids from fluids and constraints of spherical geometry suppress lower mantle convection. J. Earth Sci. 29, 1–20. Available from: https://doi.org/10.1007/s12583-017-0819-4.

Hofmeister, A.M., Dong, J.J., Branlund, J.M., 2014. Thermal diffusivity of electrical insulators at high temperatures: evidence for diffusion of phonon-polaritons at infrared frequencies augmenting phonon heat conductionJ. Appl. Phys. 115, No. 163517. Available from: https://doi.org/10.1063/1.4873295.

Hofmeister, A.M., Sehlke, A., Avard, G., Bollasina, A.J., Robert, G., Whittington, A.G., 2016. Transport properties of glassy and molten lavas as a function of temperature and composition. J. Volcanol. Geotherm. Res. 327, 380–388.

Incopera, F.P., DeWitt, D.P., 2005. Fundamentals of Heat and Mass Transfer, fifth ed. John Wiley and Sons, New York, NY.

Kellett, B.S., 1952. Transmission of radiation through glass in tank furnaces. J. Soc. Glass Tech. 36, 115–123.

Kestin, J., Knierrim, K., Mason, E.A., Najafi, B., Ro, S.T., Waldman, M., 1984. Equilibrium and transport properties of the noble gases and their mixtures at low density. J. Phys. Chem. Ref. Data 13, 229–303.

Kittel, C., 1971. Introduction to Solid State Physics, Fourth ed. John Wiley and Sons, New York, NY, pp. 224–226, 239–265.

Klemens, P.G., Williams, R.K., 1986. Thermal conductivity of metals and alloys. Int. Metals Rev. 31, 197–215.

Lee, T.Y.R., Donaldson, A.B., Taylor, R.E., 1978. Thermal diffusivity of layered composites. Thermal Conduct. 15, 135–148.

Maxwell, J.C., 1871. The Theory of Heat. Longmans, Green, and Co, London, UK (Posthumus editions, with corrections and additions by Lord Rayleigh were published in 1891 and 1908).

Maxwell, J.C., 1888. An Elementary Treatise on Electricity. Henry Frowde, London, UK.

Mehling, H., Hautzinger, G., Nilsson, O., Fricke, J., Hofmann, R., Hahn, O., 1998. Thermal diffusivity of semitransparent materials determined by the laser-flash method applying a new mathematical model. Int. J. Thermophys. 19, 941.

Meissner, W., 1915. Thermische und elektrische Leitfähigkeit einiger Metalle zwischen 20° und 373° absolut. Ann. der Physik 47, 1001–1058.

Müller, I., Müller, W.H., 2009. Fundamentals of Thermodynamics and Applications. Springer, Verlag, Berlin, Germany.

Narasimhan, T.N., 1999. Fourier's heat conduction equation: history, influence, and connections. Rev. Geophys. 37, 151–172.

Nusselt, W., 1915. Das grundgesetz des wärrmeüberganges. Susndh. Ing. Bd. 38, 477–482, 490–496.

Parker, W.J., Jenkins, R.J., Butler, C.P., Abbot, G.L., 1961. Flash method of determining thermal diffusivity, heat capacity, and thermal conductivity. J. Appl. Phys. 32, 1679.

Purrington, R.D., 1997. Physics in the Nineteenth Century. Rutgers University Press, New Brunswick, NJ.

Touloukian, Y.S., Powell, R.W., Ho, C.Y., Klemens, P.G., 1970. Thermal Conductivity: Non-Metallic Solids. IFI/Plenum, New York, NY.

Touloukian, Y.S., Powell, R.W., Ho, C.Y., Nicolaou, M.S., 1973. Thermal Diffusivity. IFI/Plenum, New York, NY.

Transtrum, M.K., Machta, B.B., Brown, K.S., Daniels, B.C., Myers, C.R., Sethna, J.P., 2015. Perspective: sloppiness and emergent theories in physics, biology, and beyond. J. Chem. Phys. 143. Available from: https://doi.org/10.1063/1.4923066.

Tritton, D.J., 1977. Physical Fluid Dynamics. Van Nostrand Reinhold, New York, NY.

Uher, C., 2005. Thermal conductivity in metals. In: Tritt, T.M. (Ed.), Thermal Conductivity. Kluwer Academic/Plenum Publishers, New York, NY, pp. 21–91.

Vozár, L., Hohenauer, W., 2003/2004. Flash method of measuring the thermal diffusivity: a review. High Temp. High Press. 35/36 (3), 253–264.

Vozár, L., Hohenauer, W., 2005. Uncertainty of thermal diffusivity measurements using the laser flash method. Int. J. Thermophys. 26, 1899–1915.

CHAPTER

4

Methods Used to Determine Heat Transport and Related Properties, With Comparisons

OUTLINE

Measurements, Mechanisms, and Models of Heat Transport
DOI: https://doi.org/10.1016/B978-0-12-809981-0.00004-8

When one observes a spread in reported data for any material, it is readily apparent... that many workers have not examined adequately the errors involved in their technique. *D. McElroy and J. P. Moore (1969).*

4.1 OVERVIEW

Heat transport properties cannot be directly measured, but rather are inferred from changes in temperature via a plethora of methods. The dynamic nature of the phenomena makes these measurements fairly uncertain: even with a modern apparatus, accuracies better than 5% are difficult to achieve (see e.g., the review by Zhao et al., 2016). Many approaches are unique to individual laboratories (e.g., Laubitz, 1969; Berman, 1976). Yet, many of the physical principles upon which ~1800s experiments are based underlie modern studies (Reif-Acherman, 2014). The following commonalities persist: it is advantageous to keep the temperature difference small, which simplifies solving and applying Fourier's equations. Linear responses are assumed, after Newton's and Fourier's theoretical approaches. Unidirectional heat flow in Cartesian geometry is commonly explored, in which case results for anisotropic solids are given with respect to the crystallographic axes (early work was the domain of mineralogists: Narasimhan, 2010). Some methods are radial, in which case the material is analyzed as if its properties are uniaxial or isotropic. Fortunately, the fact that heat flows down the thermal gradient makes one-dimensional experiments sufficient to constrain heat transport properties.

Crucial differences among the experiments arise in how the heat that is applied depends on time, because this naturally determines the outcome. Hence, all experiments are classified in one of three categories: *steady-state, periodic temperature,* or *transitory (transient) temperature* methods (Danielson and Sidles, 1969). Although the latter two categories have been grouped together as nonsteady-state, these differ in regards to not only what is being measured but moreover how it is being measured (Fig. 4.1). To guide the reader in evaluating the copious literature, this chapter focuses on principles of operation associated with each category, and on the methods considered to be most accurate.

Traditional books begin by discussing the common and straightforward approach of measuring heat flow under steady-state conditions. However, steady-state conditions only describe the behavior of a system under quasiequilibrium (Chapter 3) and do not provide information on how heat interacts with a material. Although periodic measurements involve time, these do not ascertain thermal evolution, but instead compare behavior during a temperature excursion to a mean temperature (Berman, 1976). Hence, mechanisms of heat transfer cannot be ascertained from experiments under these two categories. Yet, from a technological viewpoint, modern periodic methods are of great importance, since these are used to evaluate heat transport across interfaces and in thin films and coatings (Zhao et al., 2016). The originator of periodic methods was Fourier (Narasimhan, 2010).

All bodies emit radiation to the surroundings. Isothermal conditions describe the delicate balance where the incoming heat offsets the outgoing thermal emissions, and thus

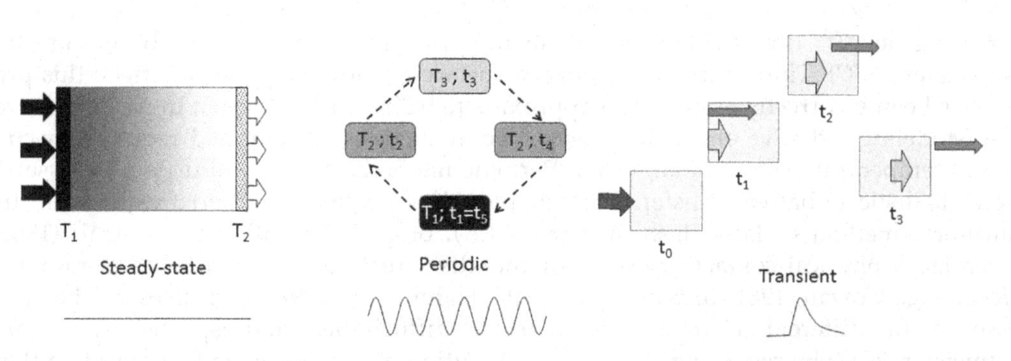

FIGURE 4.1 Categories of methods, which are distinguished based on the dependence of applied heat on time in any given experiment. Principles of operation are emphasized. Arrows indicate heat applied or emitted. Gray colors suggest temperatures of the medium. LHS: steady-state methods, showing time-independent application of heat and the resulting unwavering thermal gradient. Middle: periodic methods, showing cyclical application of heat and a cyclical response. The thermal response is viewed in terms of averages temperature differences in the cycle, not as temperature changing with time. RHS: transitory methods, showing heat applied as a pulse at some starting time (t_0). Measuring how the temperature on the side opposite to the perturbation evolves with time reveals the progression of heat across the sample. An example of two independent mechanisms is shown.

constitute the end-member case of steady-state conditions. This fact underscores the incomplete nature of steady-state measurements. Additionally, many techniques provide thermal conductivity (Eq. (3.4)):

$$K = \rho c_P D = C_P D \tag{4.1}$$

Because thermal conductivity is the product of heat capacity on a per volume basis (storativity) with thermal diffusivity, K blends the amount of heat being carried with how efficiently this heat is being transported (Fig. 3.3B; Section 3.4.2 in Chapter 3). Thus, these two components of heat transport cannot be distinguished in many experiments.

Note that comparing thermal conductivity to thermal diffusivity presents little, if any problems, mainly because ρ (or its temperature response, thermal expansivity, α_P), and the various representations of heat capacity (e.g., specific heat, c_P) are static properties. Maglić et al. (1984, 1992) and Czichos et al. (2006) review methods for measuring ρ, α_P, and c_P. Measurements for heat capacity for solids, including minerals, are plentiful (e.g., Berman and Brown, 1985; Fei, 1995): accuracies are below 3% and often below 1%. Density is more accurately measured near ambient temperature. Fortunately, ρ does not vary much with T for hard materials, so using constant $\alpha_P \sim 10^{-5}\,K^{-1}$ for the linear effect, is reasonable under most circumstances. For glasses and liquids, ρ and c_P are generally available or can be estimated with reasonable accuracy (e.g., Richet, 1987; Lange, 1997). Uncertainties lie in the heat transport measurements, as follows:

First, utilization of thermal contacts, which produce thermal losses at interfaces, artificially reduce thermal conductivity (Fried, 1969): the amount is significant in contact measurements of hard, oxide minerals (e.g., Hofmeister, 2007a). Second, if equilibrium is not truly achieved, and/or if the sample is partially transparent, ballistic radiative transfer across the sample exists to varying degrees, producing a spurious augmentation of thermal conductivity that strongly increases with temperature due to blackbody intensity increasing as T^3 (for a comparison using crystalline quartz, see Branlund and Hofmeister, 2007). Third, radiation from the surface is known to have an effect: this process has been incorrectly assumed to approximate ballistic effects, even in modern literature. Modeling radiative effects in steady-state methods is difficult and requires spectral data at temperature (Kunc et al., 1984). Periodic methods reduce, but do not necessarily avoid, ballistic radiative transfer. This chapter thus begins with, and focusses on, the transitory, method of laser-flash analysis (LFA), originating with Parker et al. (1961), which lacks physical contacts, permits quantitative treatment of both types of radiative effects (e.g., Cowan, 1961; Blumm et al., 1997; Hahn et al., 1997), and distinguishes processes with different characteristic thermal diffusivities and speeds (Criss and Hofmeister, 2017) by recording the thermal evolution of the sample in response to a thermal perturbation.

Because one goal of the companion book is an improved understanding of the thermal state and thermal evolution of the Earth and other immense bodies, this chapter mostly concerns measurements of solids (Sections 4.2–4.4) and focuses on methods applied to bulk material (Table 4.1). For details on diverse approaches, see the volume edited by Tye (1969). This work and the extensive compilation of Touloukian et al. (1970, 1973)

TABLE 4.1 Summary of Important Modern Methods for Measurement of Heat Transport Properties

Name	Type[a]	Property	Phase[b]	Notes[c]	Max. T	First Use[d]	Accuracy[e]
Laser-flash analysis	T	D	s,l	©,®,$	~3000K	1961	2%–3%
Ångström's	Pw	D	s	P	~1500K	1861	~5%
3ω	Pw	K	s,l,g,tf	P,$,©	~750K[f]	1908	<5%
Thermoreflectance	Pp	$\text{¢} = CK$	s,l,tf	P,$,©	~600K[f]	1986	5%–10%
DC hot wire	Sa	K	s,l,g		~1800K	1888	1%–15%
AC hot wire	Pw	D	s,l,g		~1800K	1931	1%–15%
Stip or planar wave	Pw	D,K	s,l	P	<1200K	1991	Varies
Electrical heating	Sa	K	metals		~3000K	1898	2%–10%
Guarded hot plate	Sa	K (low)	s		<1200K	1912	2%–5%
Comparative	Sc	K/K_{std}	s,l,g		~1200K	1789	10%–20%

[a]*T: transient; P: periodic (w: waves; p: pulses); S: steady-state (a: absolute; c: comparative).*
[b]*s: solid; l: liquid; g: gas; tf: used for thin films and also to determine interface resistance.*
[c]*P: pressure studies undertaken; ©: contact free, but a coating is required; ®: ballistic internal radiation (important above 300K) can be removed; $: surface radiation is low or presents no problem. Systematic uncertainties associated with electrical insulators have not been investigated for all methods.*
[d]*Mostly from Reif-Ackerman (2010), who provides a history with detailed references and pictures of apparatii. Also see Narasimhan (2010) and Zhao et al. (2016) who focusses on measurement principles, provide schematics, and references to standard procedures published by American Society for Testing Materials.*
[e]*Nominal for the technique from the above references. This figure addresses random errors, but not necessarily systematic errors, because benchmarking is frequently against metals.*
[f]*Most studies are near room temperature. Thermoreflectance imaging, which is a different type of measurement, is performed at higher temperatures.*
Notes: This summary was distilled from the reviews and volumes cited in Section 4.1, which cover many additional methods.

remain relevant because much data, particularly those on elements, are of this vintage and because many historic methods are still used. Regarding more recent review volumes, Maglić et al. (1992) provide a guide to performing experiments: whereas Tritt (2004) and Ventura and Perfetti (2014) give succinct summaries. However, none of these works sufficiently cover ballistic radiative transfer. The underlying reason is that the focus, historically and currently, is on thermal conductivity of metals or metalloids (reviewed by Reif-Acherman, 2014), not on thermal diffusivity of electrical insulators. Plus, in materials science, the focus is moderate to cold temperatures and steady-state, where radiation is not so important.

Concerns about the spread in values between laboratories persist (e.g., Laubitz, 1969; Touloukian et al., 1970, 1973). The variations in insulators are most simply explained by differing proportions of systematic errors with opposing signs (Section 4.5). As discussed by Fried (1969), interfaces offer thermal resistance, which can be described as surface roughness, whereby pockets of air separate the materials. These "gaps" effectively reduce the contact area, and thus impede heat flow. However, quantifying contact losses at interfaces theoretically is very difficult, and so experimental approaches are used in Section 4.5, after Fried (1969). Radiative effects are discussed in detail in Sections 4.2 and 4.5.

In solids, heat and mass move independently. Oddly, all microscopic models are based on behavior of gas, in which heat and mass are transported inseparably. Because gases are prone to convective instabilities even in low temperature gradients, the relevant measurements are steady-state. Furthermore, the fundamental microscopic model of transport, that is, the kinetic theory of gas, ties heat conduction to each of mass diffusion and viscosity, which comprise the remaining transport properties. To evaluate this model in Chapter 5 requires considering the data. Therefore, methods used to determine heat conduction, mass diffusion, and viscous drag of gases are summarized in Section 4.6. Liquids that flow under any stress are also covered. Chapter 10 covers methods for silicate liquids, which require stress to flow.

4.1.1 Synopsis of Key Methods for Measuring Heat Transport

Section 4.2 focusses on the transient method of LFA, which is demonstrably accurate, reliable, and versatile. Recent reviews are Maglić and Taylor (1992), Vozár and Hohenauer (2003/4), and Corbin and Turriff (2012). The LFA method, developed by Parker et al. (1961) and Rudkin et al. (1962), is robust, yet simple. The large number of scientific papers dealing with applications and continued developments of the laser-flash method (e.g., Chen et al., 2010) testify to the quality of the underlying principles. Other transient methods exist (see e.g., Danielson and Sidles, 1969) but do not have all the advantages of LFA, and so are not commonly utilized.

Section 4.3 focusses on periodic methods, which have been misleadingly designated as transient due to time being a component of the measurement. However, the raw data do not describe thermal evolution, but instead examine excursions from a mean temperature (Berman, 1976). The method originating with Ångström (1861) is commonly used in Earth science (e.g., Xu et al., 2004), despite the presence of contact losses and radiative transfer gains. The 3ω technique is popular in materials science below ~300K, as it is viewed as being free of radiative effects (Ventura and Perfetti, 2014). Hence, comparison of results from these methods with LFA helps to quantify the two contrasting systematic errors. Although thermoreflectance (Paddock and Easley, 1986) is mostly applied to thin films, this relatively new method is discussed here mostly to elucidate its limitations with regard to current use in geophysically relevant pressure studies. The rapid, oscillating changes for the above nonsteady-state methods being unlike those in slowly evolving geologic terranes or planetary bodies is a concern. Do the results represent mechanisms present near equilibrium, or is a mixture of mechanisms present?

Section 4.4 summarizes steady-state methods focusing on metals and alloys (e.g., Flynn, 1969) because conventional techniques provide accurate data for these materials, making these important in benchmarking. Accuracy is possible because (1) metals adhere well to metal thermocouples and power sources, thus minimizing thermal losses at interfaces, (2) metals are opaque, except as thin films, so spurious radiative transfer is not a concern, and (3) the metal sample itself can be heated electrically. Therefore, some high-pressure methods for metals are covered here. Note that values reported for K of metals are often calculated from measurements of electrical conductivity (Ventura and Perfetti, 2014). This is a potential problem because recent LFA measurements (Criss and Hofmeister, 2017) show that electrons transport heat in metals only over short-time intervals (transiently), whereas the long-term, quasiequilibrium mechanism involves vibrations.

4.1.2 A Brief Discussion of Topics not Covered

Measurements of hard electrical insulators (mantle phases) under high pressure are inherently suspect because avoiding contact losses is crucial for accuracy in any study (Fried, 1969), but is impossible for pressure studies. A pressure transmitting medium is always present which not only conveys heat from the sample to the apparatus (e.g., to the diamond anvils) but also its geometry varies as compression increases in the experiments. Pressure studies of insulators via traditional methods applied to bulk materials are not discussed in detail here, not only due to these unavoidable problems but also because the strong dependence of thermal diffusivity on thickness (Chapter 7) precludes obtaining data relevant to planetary interiors from tiny samples that can be studied at very high pressure. Hofmeister (2009, 2010a), Hofmeister et al. (2007), and Hofmeister and Branlund (2015) discuss several problems and limitations in studies of electrical insulators at pressure. Ross et al. (1984) review diverse materials: thermal transport in metals at low pressure are reasonably constrained by the hot wire/hot strip techniques (e.g. Andersson and Bäckström, 1986) because metal–metal contact losses and ballistic radiative transfer gains are unimportant.

Extensions of the LFA method in materials science are summarized by Vozár and Hohenauer (2003/4). Adaptations in the geosciences have some problems. For example, the experiments of Schilling (1999), Ray et al. (2006), and a few others utilize thermocouples, which are known to induce errors (Taylor and Maglić, 1984; Lee and Hasselman, 1985). More importantly, the pulse used in these studies is very broad, which is not taken into account. Radiative transfer is incorrectly addressed in these studies, as discussed by Hofmeister et al. (2007). Another case is the "flash" method of Beck et al. (2007) and Goncharov et al. (2009) which measures the temperature from a metal film only while it is remotely heated, to purportedly constrain diffusion through the surrounding medium. These experiments involve ballistic transfer of visible light over the scale of the apparatus, rather than diffusion of low energy heat over a specified distance, as discussed by Hofmeister (2009, 2010a).

Methods unique to cryogenic studies are not covered because cryogenic data relevant to planetary studies (on icy material composing cold objects in the outer Solar System) are not yet available, and because the cryogenic data that we do discuss are obtained using common methods. For comprehensive reviews, see White (1969) and Ventura and Perfetti (2014). In brief, ballistic effects are present (Hofmeister, 2010b; Hofmeister and Whittington, 2012) but have been overlooked because it is assumed that this effect is mitigated by the low intensity of blackbody radiation at low temperature. However, low emission intensity is compensated for by high sample transparency at low temperatures, which stems from narrow peaks widths, depopulated states, and transitions and absorptions in solids mainly occurring in the mid-infrared (IR) which is higher than the blackbody peaks at cryogenic conditions (Chapter 2). Laser-flash experiments (which remediate ballistic transfer) are conducted routinely with commercial apparatus down to 150K (e.g., Aggarwal et al., 2005) and occasionally to near liquid nitrogen temperatures (e.g., Kogure et al., 1986; Hemberger et al., 2010) and in principle can be extended to ~9K temperatures with liquid He-cooled detectors and cryostats. Below 1K, the recently developed method of optical heating and magnetization thermometry method (Hao et al., 2004) is promising. Their data on silica glass are in agreement with steady-state measurements of Lasjaunias et al. (1975) but not with higher T cryogenic data obtained using periodic methods, suggesting that radiative transfer exists in the later experiments (see Whittington and Hofmeister, 2012, and Section 4.5).

For additional optical techniques with a focus on thin films and nanosamples, see Borca-Tasciuc and Chen (2004). Although behavior over small scales and short times is of great technological importance, large bodies involve slow cooling. However, small distances do pertain to polycrystalline substances (rocks). Grain-boundary effects, which are a type of interface resistance, are discussed later, in Chapter 7.

Photoacoustic and photothermal methods are not covered. These are not commonly used.

4.2 THE TRANSITORY METHOD OF LASER-FLASH ANALYSIS

4.2.1 Principles and Essentials

In LFA experiments, the front side of a typically disk-shaped sample is heated by a short high-power light pulse. The temperature increase generated by the light pulse propagates through the thickness of the disk and leads to a small temperature change of the entire sample. The temperature increase on the rear surface is measured as a function of time as the small addition of heat crosses a well-defined length. Thermal diffusivity is extracted from these raw data based on solutions to the heat equation. Because the geometry is conducive for one-dimensional flow and the change in temperature during the experiment is purposefully kept small, Eq. (3.3) reduces to

$$\frac{\partial T}{\partial t} = D(T)\frac{\partial^2 T}{\partial z^2} \tag{4.2}$$

A heat pulse is applied remotely to a thin layer on the front (bottom) surface of the slab with length L, parallel faces, and area Å using a laser (or sometimes an ultraviolet (UV) lamp). The subsequent time-evolution of the emissions from the rear (top) surface are recorded remotely, using an IR detector (Fig. 4.2). Section 4.2.2 provides further experimental details. Because small amounts of heat are applied and temperature changes are small, emissions are proportional to temperature (Vozár and Hohenauer, 2003/2004). Thus, the raw data are referred to as a temperature–time (T–t) curve. Its shape determines the thermal diffusivity in accord with Eq. (4.2). For a diffusional process, an asymmetric "S" shape is observed (Fig. 4.3) which results from the rear (top) surface first being heated as the laser pulse crosses, and subsequently shedding heat to the surroundings as the sample re-equilibrates with the surroundings. Improvements in the LFA technique over the 55 years since its development have led to a highly accurate (±2%) and robust, yet simple, method for evaluating transient heat conduction (Maglić and Taylor, 1992; Vozár and Hohenauer, 2003/4; Corbin and Turriff, 2012). For a brief summary, see ASTM (2013). The major sources of errors early on (finite pulse time, heat losses to the surroundings, nonuniform heating during the pulse, penetration of the pulse, radiative transfer internally) have been addressed through advances in laboratory techniques, instrumentation, and theoretical models. For the early history, see Taylor and Maglić (1984).

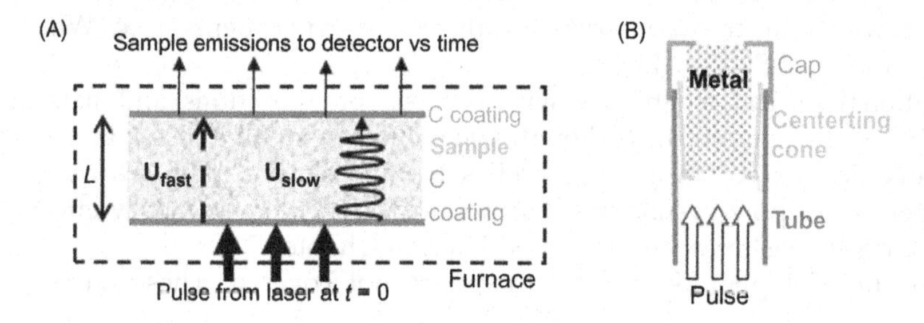

FIGURE 4.2 Schematics of LFA experiments. (A) Operation essentials. Dashed frame indicates furnace surrounding the sample. Speckled rectangle with thickness L shows the sample edge-on, which has (often) graphite-coated faces. Arrows show receipt of energy from the laser and travel of emissions from sample to detector. Dashed and squiggle arrows inside the sample indicate heat transfer processes traveling at relatively fast and slow speeds. (B) Detailed cross-section of the cylindrical sample and holder. Solids: three graphite parts, distinguished by shade. Stipple: metal sample. Arrows: path of the laser, which is blocked from reaching the detector, which therefore only collects emissions from the faces of the sample and cap. The base of the centering cone may be heated by the laser pulse, but its increase in temperature must be transferred radially through the tube and cap, and then vertically to be sensed. Thermal contacts between the graphite parts and metal are poor so any such conductive transfer is greatly delayed, and cannot possibly produce the rapid rise observed for certain metals. Source: *From Criss, E.M., Hofmeister, A.M., 2017. Isolating lattice from electronic contributions in thermal transport measurements of metals and alloys and a new model. Int. J. Mod. Phys. B 31, No. 175020. doi:10.1142/S0217979217502058.*

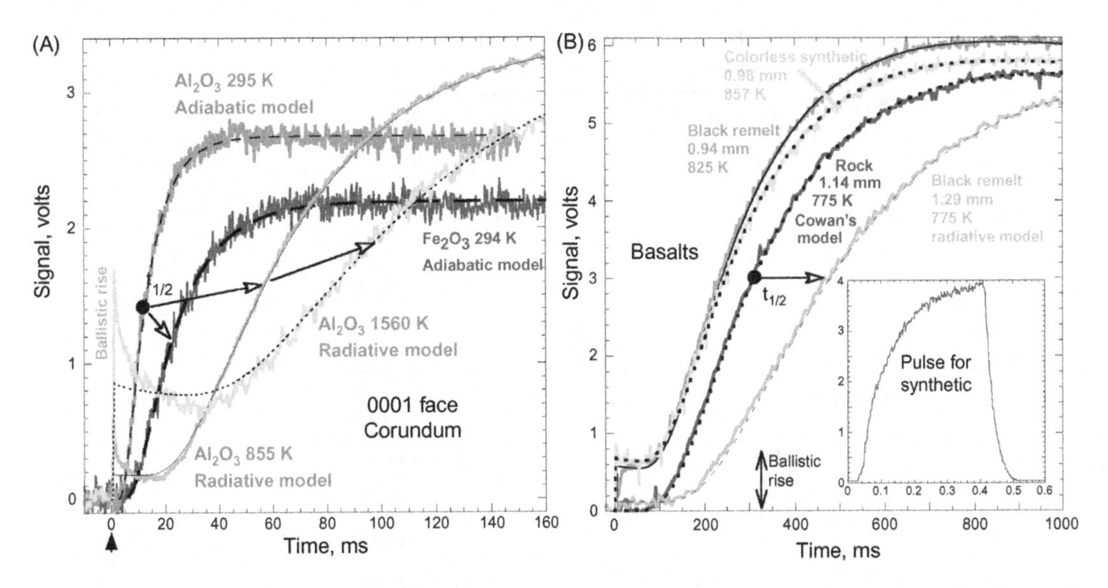

FIGURE 4.3 Temperature–time curves of partially transparent electrical insulators in crystalline (A) and glassy (B) states. The laser pulse is shown in the inset. Signal voltage is arbitrary and was scaled to best depict the raw data but with similar voltages at the maxima. Labels give sample name, thickness, temperature, and model used to fit the raw data, which are shown as the thick, gray curves of different shades with varying amounts of noise. Black lines with various patterns: model fits. Arrow below the *x*-axis indicates firing of the laser pulse at $t = 0$ ms. Black dot: position of $t_{1/2}$ on various curves. Open arrows connect $t_{1/2}$ positions for different samples at similar conditions (e.g., between corundum and its structural analog hematite at room temperature, which has slower D due to larger lattice spacings and slower speeds), or between different conditions for the same sample (e.g., corundum as T increases). In part A, curves were measured of the *c*-face of single crystals, where Al_2O_3 is 1.00 mm thick and Fe_2O_3 is 0.60 mm thick. In part B, colorless synthetic basaltic glass is paired with a similar composition (excepting Fe cations) of black, remelted Hawaiian basalt, showing similar thermal evolution, including amounts of radiative transfer, despite the extreme color difference. The latter occurs due to the near-IR being the relevant spectral region for ballistic transport, rather than the visible. Part B also compares the almost black source rock and its remelted glass, showing dissimilar thermal evolution and radiative transfer, due to the small crystals scattering light of various frequencies while more efficiently diffusing heat internally. For descriptions of samples and results for thermal diffusivity, see Hofmeister (2014) and Hofmeister et al. (2016).

4.2.1.1 *The Basic, Adiabatic Model*

The physics is embodied in a simple relationship based on adiabatic conditions, after Parker et al. (1961):

$$D = 0.138785\frac{L^2}{t_{1/2}} = b_j\frac{L^2}{t_j} \tag{4.3}$$

where L is sample length, and $t_{1/2}$ is the time taken for the rear surface to reach half of the maximum temperature (Figs. 4.2A and 4.3). Note that $t_{1/2}$ is a measurement time that grows with length, which is related to much shorter lifetimes through dimensional analysis as discussed in Chapter 3. The half-time is convenient to measure: other fractions (j) with other numerical factors (b_j, see Table 4.2) are equally valid.

TABLE 4.2 Values[a] of Geometrical Factor b_j in Eq. (4.3)

j	b_j
0.1	0.066108
0.2	0.084252
¼	0.092725
0.3	0.101213
⅓	0.106976
0.4	0.118960
½	0.138785
0.6	0.162236
⅔	0.181067
0.7	0.191874
¾	0.210493
0.8	0.233200
0.9	0.303530

[a] *After Maglić and Taylor (1992), in their Table 1. Reprinted with permissions from Springer-Nature.*

Eq. (4.3) comes from an one-dimensional heating model that uses a Fourier series (Carslaw and Jaeger, 1959). Assuming $T=0$ for the front surface, the temperature at rear surface at L is described by:

$$T(L,t)=\frac{Q}{C\rho L\mathring{A}}\left[1+2\sum_{n=1}^{\infty}(-1)^n\exp\left(\frac{-n^2\pi^2Dt}{L^2}\right)\right] \tag{4.4}$$

where $Q=$ pulse energy, $C=$ heat capacity on a per volume basis $=\rho c=$ storativity, and $\mathring{A}$ is area. The numerical factor b_j in Eq. (4.3) results from the summations when evaluating the maximum temperature reached by the back surface (see 298K curves in Fig. 4.3A). Because time and distance can be measured accurately, D can be measured accurately.

The rise in temperature (for a single carrier and adiabatic conditions) is described by:

$$\Delta T\cong\frac{Q}{\mathring{A}LC} \tag{4.5}$$

(Parker et al., 1961). Constant pressure is assumed, although constant strain is actually required. Pulse energy is kept sufficiently low so that $\Delta T<4°C$. Consequently, conditions are not far from equilibrium. The temperature dependence of D is then determined by simply varying furnace temperature. Heat capacity can be determined from ΔT using additional information on Q. But because the efficiency of energy absorption affects Q, this determination assumes *only* one carrier exists. Two carriers (e.g., phonons and electrons in a metal) will produce different rise heights depending on uptake proportions, which related to their respective storativities, and the speed with which each carrier diffuses heat across the sample (Criss and Hofmeister, 2017).

The adiabatic model (Eq. (4.3)) reasonably describes data when the flux is relatively low and the diffusivity is fairly high. At high temperature, the adiabatic model adequately fits $T-t$ curves for many metals (see e.g., Ni and Al curves in Criss and Hofmeister, 2017). Emissions are low for metals from the far-IR to almost the visible, due to their high reflectivity and low emissivity (ε, see Chapter 2), and thus the radiative cooling is reduced. Additionally, these have fairly high thermal diffusivity and high heat capacity, so that the warming of the back surface is fast compared to its cooling and the sample soaks up heat strongly for its size, compared to electrically insulating solids, which meet these conditions infrequently. Specifically, radiation heat losses depend on emissivity and roughly go as $\varepsilon L/(DC_P)$, per Cowan (1963) and Cape and Lehman (1963). Details follow.

4.2.1.2 *Advantages of LFA*

The primary theoretical advantage of LFA is that the thermal evolution of the sample is monitored, and so Eq. (4.2) directly applies, plus the dimensional analysis of Chapter 3 is relevant. Thus, the behavior of the sample during transport of heat is evident in the raw data (see below).

Key experimental advantages are that neither heat input nor absolute temperature need be quantified, and physical contacts, which produce heat losses, are avoided. Furthermore, samples can be fairly small ($\sim$6–20 mm diameter and $\sim$1 mm thickness) while baseline corrections remove effects of stray light and background radiation from hot components supporting the sample (Fig. 4.2). Requirements are straightforward: it is important to have parallel faces, to avoid strong temperature gradients across the sample (i.e., minimal heating by the pulse), and for the pulse to be absorbed by a thin coating. Meeting these conditions permits treating D as a constant over the temperature change during the experiment (Eq. (4.1)).

4.2.1.3 *Departures From Adiabatic Conditions and Instantaneous Pulses*

Commonly, the detector signal decreases with time after reaching a maximum (Fig. 4.4) due to the sample cooling by radiation from its surface of the "extra" thermal energy of the pulse to the slightly cooler surroundings. This behavior and effects arising from the finite lateral extent of the disk require using more complicated models than Eq. (4.3). Details on commonly used models are given in Section 4.2.4. To summarize, the simple model of Cowan (1963) for heat losses (e.g., by radiative cooling) converges rapidly and is popular. Clark and Taylor (1975) modeled the heating part of the curve to addressed nonuniform pulse using averaging schemes of fractional times (in Table 4.2). These schemes can be blended. Improvements in the essentially two-dimensional model of Cape and Lehman (1963) by Josell et al. (1995) and Blumm and Opfermann (2002) have made this highly accurate. This development and computational advances have largely replaced the averaging schemes, although Cowan's model remains popular due to rapid convergence combined with reasonable accuracy.

Cowan's model was used to fit most of our data on metals, polycrystals, and grainy rocks (Chapter 7), which do not have ballistic radiative transfer across the sample.

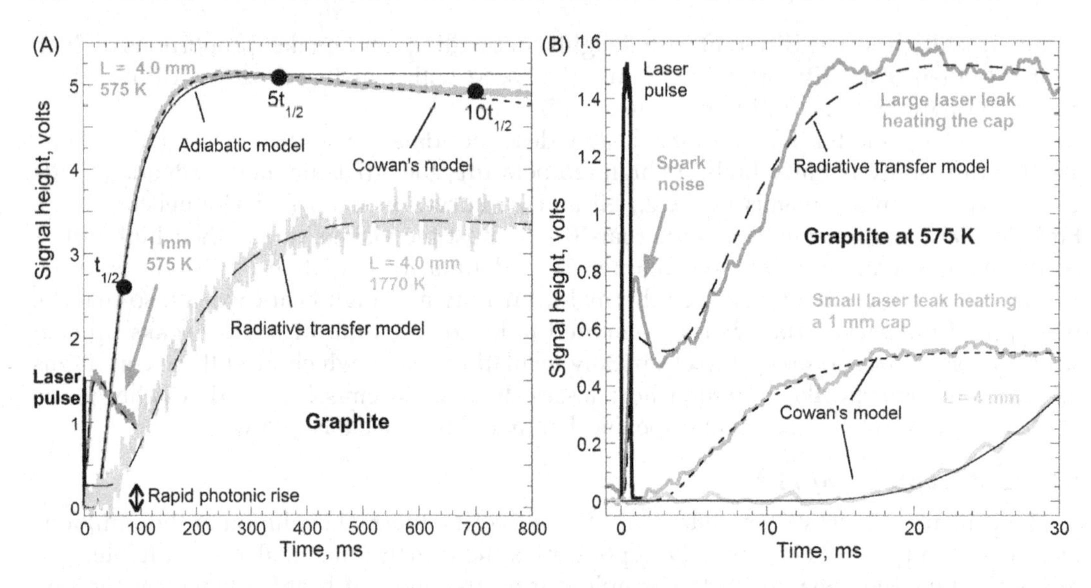

FIGURE 4.4 Temperature–time curves for graphite. Symbols as described in Fig. 4.2. (A) Comparison of data from samples with 1 and 4 mm thicknesses at 575K and of low and high temperatures for the 4 mm sample. The latter shows that radiative transfer exists at high T even for a dark and grainy sample. Fits of adiabatic and Cowan's models to one dataset are shown. These are marked with the half-time used in Parker et al.'s adiabatic model and with the cooling times used in Cowan's model. Double arrow near the axis indicates the initial rise due to photons. (B) Expanded view of the same T–t curves at 575K. The thin samples are actually the cap (Fig. 4.1), which was heated by the laser due to a small metal sample not completely covering the aperture of the sample cone. This light leakage is known as "spark noise" in older literature (Maglić and Taylor, 1992), where a tiny amount of laser light was leaked to set "$t = 0$." For the small leak, the cap intercepted all the laser light, and the data were fit by Cowan's model, like the thick sample. For the large leak, some of the laser light is transmitted ballistically to the detector. Due to the pulse having a finite width as shown, the signal rises with time as the laser fires. Here, the radiative model is needed. For results on D of the 4 mm thick graphite, see Hofmeister et al. (2014). For further discussion of the experiments on light leakage, see Criss and Hofmeister (2017).

Cowan's model also describes grainy or very dark colored electrical insulators, which respectively scatter or strongly absorb in the near-IR, if sufficiently thick (Fig. 4.3). The low temperature T–t curves for partially transparent electrical insulators, including glasses, are also well-described (Fig. 4.2; curves are also shown in Pertermann et al., 2008; Hofmeister, 2010b; Yu and Hofmeister, 2011). In actuality, very small amounts of ballistic transfer exist at 298K, but are at the noise level in the detector response at low T in our apparatus, which uses an InSb detector and is optimal in the near-IR.

Eq. (4.3) and other models for data analysis (see Section 4.2.4 for details) require that the pulse width be significantly shorter than the time heat takes to cross the sample. Errors associated with finite duration of spikey laser pulses are small and corrected through models, beginning with Cape and Lehman (1963). Blumm and Lemarchand (2002) further account for greater than ideal heating by high-energy pulses, and correct for effects of a strong temperature dependence of thermal diffusivity of the sample. Ohta et al. (2002) also address the latter problem. Small corrections for beam size are provided by Thermitus (2010).

Models commonly used (Section 4.2.4) assume that the samples are coated and that the pulse is absorbed by the coating. For uncoated samples, where the laser pulse penetrates, amended models were provided by Tischler et al. (1988) and McMasters et al. (1999).

4.2.2 Effects of Radiative and Electronic Transport Process on $T-t$ Curves

The models discussed above describe radiative losses from the sample as it cools down to temperature of the surroundings. Radiative processes also occur internally in partially transparent samples. Such ballistic transport added a spurious augmentation to the thermal diffusivity of partially transparent samples in many LFA experiments prior to about 1994. This problem no longer exists. First, the effect is reduced by applying a thin metal coating by sputtering and then overcoating with graphite (Degiovanni et al., 1994). Second, remnants of ballistic transfer across the sample (which is distinct from radiation from the surfaces to the surroundings discussed above) are removed by modeling. Assumptions underlying the model are described by Blumm et al. (1997). Hahn et al. (1997) fully developed this model, which was tested by Hofmann et al. (1997) and Mehling et al. (1998).

Distinguishing ballistic radiative contributions which elevate the temperature on the back side (Fig. 4.2) from diffusion of heat is possible because these mechanisms are largely independent in materials denoted as "diathermic." The underlying cause is characteristic speeds differing for the two processes. Ballistic radiative transfer across the sample occurs at c/n where n is the index of refraction. Diffusion of the lattice energy is much slower, and to first-order has a characteristic speed resembling that of acoustic modes, as long known and covered in Chapter 11. Ballistic and lattice heat transfer are visually discernable in the temperature–time curves (Figs. 4.3 and 4.4).

The formulation detailed by Hahn et al. (1997) and summarized by Mehling et al. (1998) is based on optically thin conditions in the near-IR and accounts for absorbance being frequency-dependent, although values of optical properties are not needed. Blumm et al. (1997) and Hofmann et al. (1997) established model accuracy (Section 4.2.4.4).

The characteristics of the photonic signal in diathermics are unique. One identifying factor is that its maximum height increases greatly with T for such partially transparent insulators (Fig. 4.3, also see $T-t$ curves in Hofmeister, 2006; 2010a,b: Pertermann and Hofmeister, 2006; Branlund and Hofmeister, 2007; Hofmeister and Pertermann, 2008; Pertermann et al., 2008; Yu and Hofmeister, 2011). Another clue is that the photonic shape is not the characteristic, diffusional "S" following the pulse. Instead, the signal height increases strongly *only* while the laser pulse is applied: after the pulse, decay is quite rapid as the sample re-equilibrates with the surroundings via radiation from the surface (Fig. 4.3A, light gray curve). The photonic increase is not instantaneous because the pulse is gradually applied (inset in Figs. 4.3B and 4.4B). But because photons are ultrafast, the time dependence of photonic rise is unaffected by either changing sample length (e.g., from ~0.1 to 10 mm) or by temperature: the photonic rise always overlaps with the pulse.

Semiconductors have similarly shaped photonic signals, but a different temperature response exists. For uncoated Si, the intensity was very high at ambient T, and decreased from 800 to 1000K (see $T-t$ curves in Milošević, 2010). The behavior is connected with the optical spectra of Si differing by broadly absorbing and reflecting across the IR region, whereas the insulators absorb in the IR, but transmit in the near-IR (Chapter 2). A model with coupling of high and low energy processes, such as that of Lazard et al. (2004) may be needed to fully understand semiconductor behavior.

Sequential rises with time are also observed for metals (Fig. 4.5; many other examples are given by Criss and Hofmeister, 2017). The signal differs from that arising from radiative transfer, which was avoided in the experiments. Both phonons and electrons diffuse heat in a metal, as first proposed by Koenigsberger circa 1900. From dimensional analysis of Eq. (3.3) (or Eq. (4.2)), different heat transport mechanisms, if these have sufficiently different thermal diffusivities, will be manifest in the temperature–time history as sequentially occurring rises for any given length. This deduction is evident in Eq. (4.3), and is documented for photons, as discussed above. Specifically, for an experiment at a given L, mechanisms with substantially different D will have different $t_{1/2}$, and should be distinct in the T–t curves, so long as interactions between carriers are weak, and the apparatus has sufficient resolution to capture the more rapid mechanism. Carrier speeds for electron–electron and phonon–phonon mechanisms differ by $\times 1000$. So, roughly, D_{ele} should be about $1000 \times D_{lat}$, and so the two independent mechanisms are distinguishable in T–t curves (e.g.,

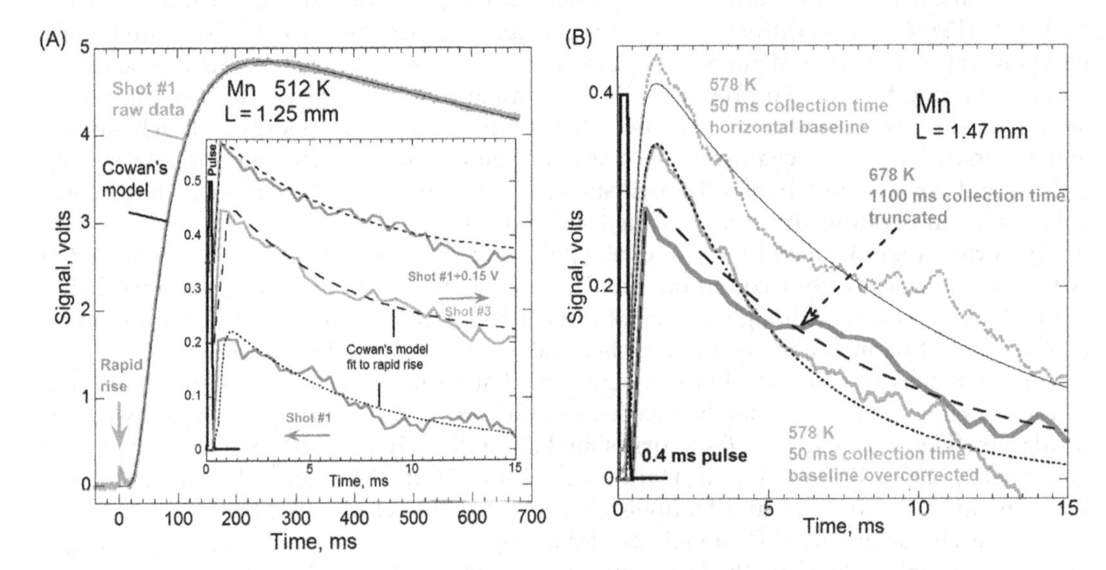

FIGURE 4.5 Temperature–time curves for thin samples of the element Mn, showing slow and rapid rises connected with diffusion of slow lattice vibrations and fast electrons. (A) Mn at 512K with $L = 1.25$ mm. The fits (thin black) to the raw data (thick gray) are over the entire curve (large panel), or only over the rapid rises (inset). Thick black line: laser pulse. Because the slow rise begins at 20 ms whereas 50 ms is the shortest collection duration possible with our software, we could not record the rapid rise at better resolution. Nonetheless, the rapid rise differs insignificantly between the three shots acquired under the same conditions, and has characteristics unlike ballistic radiative transfer. (B) Slightly thicker sample of Mn, comparing data from short and long collection times, using a narrow pulse. Linear baseline corrections have a small effect on most shots. Short shots used a horizontal baseline correction. Source: *From Criss, E.M., Hofmeister, A.M., 2017. Isolating lattice from electronic contributions in thermal transport measurements of metals and alloys and a new model. Int. J. Mod. Phys. B 31, No. 175020. doi:10.1142/S0217979217502058.*

Fig. 4.5). Electrons are fast, but carry negligible heat, so only a small portion of the applied heat is carried to during this transient behavior.

Resolving the rapid signal generally requires long lengths. Previous LFA measurements of metals did not resolve the signals for several reasons. First, spark noise (light leakage: Fig. 4.4) exists in early studies (Maglić and Taylor, 1992) and would mask the rapid rise. Second, typical thickness were 1–4 mm (Parker et al., 1961; Monaghan and Quested, 2001) which provide D from ~2 to ~170 mm^2 s^{-1} that is consistent with steady-state measurements (Touloukian et al., 1973; Henderson et al., 1998a,b; Hust and Lankford, 1984; Gorbatov et al., 2012), and so a rapid signal was not sought. The transient signal is observed in Mn with a typical ~mm length because this metal is disordered, which is associated with low thermal diffusivity, irregardless of whether electrons or phonons are involved. Criss and Hofmeister (2017) in exploring longer lengths and short collection times while taking extra steps to avoid spark noise, found that rapid and slow rises were present in metals and alloys if their electronic heat capacity was large relative to that of an average metal. Penetration of the blackbody radiation (sample reflectivity) is also relevant. Independent responses of electrons and phonon to thermal events were demonstrated earlier in fs-spectroscopy, where the electron population is ~500K hotter than the lattice temperature (see review by Bauer et al., 2015). The consistent behavior observed and implications for mechanisms are discussed later in the book.

4.2.3 Details on LFA Experiments

For a summary of experimental conditions that produce highly accurate results and models that address departures from the optimal ranges of parameters, see Corbin and Turriff (2012). For a summary of experimental uncertainties (with reference to a steel standard), see Vozár and Hohenauer (2005). For measurement techniques, see Maglić and Taylor (1992).

The main components of a LFA are a controlled atmosphere furnace, a high-energy pulsed laser (or sometimes a UV lamp), and an IR detector. The apparatus used in our laboratory to 2200K is described by Bräuer et al. (1992). The instrument described by Taylor (1980) reaches 3000K. Baba and Ono (2001) describe an instrument with a radiometer that reaches 1500K. Additional commercial instruments exist (e.g., Min et al., 2007), along with many instruments unique to laboratories (see the list in Vozár and Hohenauer, 2005).

A sample in the form of a small, thin slab with parallel faces (~0.3–~15 mm thick by 6–15 mm diameter in our laboratory) is held by its edges at some temperature provided by the surrounding the furnace (Fig. 4.1). Neither thermocouples nor heater contact the sample. Top and bottom surfaces of the sample are graphite coated (thickness ~1 μm) to absorb laser light, thereby shielding the detector, and to enhance intensity of the emissions. Graphite coatings also buffer oxygen fugacity to C–CO, preventing oxidation of Fe^{2+}.

At moderate temperature, graphite coatings alone reduce the ballistic radiative transfer to an extent that accurate results are obtained from the model of Hahn et al. (1997), see Mehling et al. (1998) Essentially, the bottom coat converts the narrow laser pulse to a broad band spectrum of lower intensity, for which the energy is distributed across absorbing and transmitting spectral regions. As temperature increases, the peak of the blackbody moves into the more transparent near-IR region. The position, plus the higher intensity, results in more ballistic transfer. Thicker graphite coatings are needed, but at some point the thickness of the film has an effect, if the sample is fairly thin (e.g., Lim et al., 2009) and must be accounted for in analyzing the $T-t$ curves. Sputtering on a thin metal coating (as is common in scanning electron microscopy studies) prior to spraying graphite on the sample reduces the radiative transfer due to the generally high reflectivity of metals. However, problems accompany use of thick coatings when their thermal diffusivity is much different than D of the sample (series configuration with reciprocals adding: Chapter 3). Gold and silver have high thermal diffusivity (127 and 170 $mm^2 s^{-1}$, respectively, Touloukian et al., 1973) and so for glasses which have very low D, a very thin coating of Au must be used. For glasses, thickness is far less critical if coatings are made using metals with lower D, such as nickel, palladium, or platinum (all are below $\sim 25\ mm^2 s^{-1}$: Touloukian et al., 1973). The above-mentioned metals have low work functions and are easily sputtered using relatively inexpensive equipment (see e.g., VEM, 2017). Refractory sprays and molybdenum have been used for samples that react with graphite (e.g., Shibata et al., 2000; Criss and Hofmeister, 2017): the important issue are integrity of the coating, keeping this thin, and nonreactivity with the sample or atmosphere. Notably, oxygen impurities in the purge gas or water impurities in the sample itself can remove graphite coatings at high temperature.

For a robust fit to the $T-t$ curves, the experimental duration should be ~ 10 half-times. However, very long durations are often accompanied by fluctuations or instabilities, attributed to the electronics of the detection system or fluctuations in the furnace stability (baseline drift). Therefore, thin samples are used to provide short collection times when ascertaining low D, as in glasses or disordered silicates. For example, D being $\sim 2\ mm^2 s^{-1}$ requires L below ~ 1 mm, and a thickness to diameter ratio of 1:8 (e.g., Pertermann et al., 2008). Regarding the contrasting situation of high D, the shortest duration available in any given apparatus requires some minimum L for any given D. Also, rise times should exceed the $\sim$ms duration of the purse (shown in Fig. 4.2B, inset). Thus, to quantify large D, large L is needed (Eq. (4.2)). This requirement can be problematic at very high D, such as diamond, due to either sample availability or cost, or to sample chambers limiting the lateral dimension of the sample.

Sources of experimental uncertainty in laser-flash measurements are implicit in simple models for temperature–time curves arising from phonon transport alone (Eq. (4.2) for opaque samples, or very low temperatures). The strong dependence of the half-time on thickness in Eq. (4.2) remains the main source of uncertainty, because measurement time is accurately determined. The reliability of each data point is insured by requiring that calculated and measured temperature–time curves match. Accuracy for the technique is $\sim 2\%$, determined through benchmarking against metals. These opaque and soft materials lack radiative transfer and have good thermal contact, allowing calibration against results from conventional methods for standard reference materials (e.g., Henderson et al., 1998a,b). Graphite was once used, but this material is not sufficiently uniform in porosity and/or impurity distribution for accuracy to be guaranteed.

A small source of uncertainty is the actual sample temperature determination. Calibrations can be made against the Curie transition in metals and melting points.

Another source of error is detector noise, typically at low temperature where detectors are less sensitive. This can be reduced through a larger number of data collections at temperature setpoints. Typically, 3 data collections at higher temperature suffice, and 6–10 at room temperature.

4.2.4 Further Information on Commonly Used Thermal Models

The adiabatic model was covered in Section 4.2.1. Many complicated models are reviewed by Vozár and Hohenauer (2003/2004), who cover data reduction in detail. In summary, the experimental data (the emission signal vs time) are fit numerically to a model using well-known, least-squares minimization methods. Which model is best can be ascertained statistically, based on a correlation coefficient: this approach accounts for random errors. Differences among models are quite small when the experiment is reasonably close to the ideal conditions described above and internal radiative transfer is not present, see for example, the comparison of models using experiments on steel by Massard et al. (2004) who provide a succinct summary of the models. This present section is also a summary, with a focus on the basics and some updates.

4.2.4.1 *Pulse Corrections*

Corrections for effect of finite pulse width began with square wave idealizations of pulse shape (Cape and Lehman, 1963). Azumi and Takahashi's (1981) discussion of the fairly realistic triangular case underlies Blumm and Lemarchand's (2002) approach to the actual pulse shape, as recorded by a sensor. We begin with the effect of a triangular pulse on adiabatic heating case to illustrate the principles:

If the pulse has finite width (t_{pulse}), arrival of the heat is spread out over an interval, so the rise time is delayed: this can be understood by considering the pulse to consist of a series of spikes, where "$t = 0$" is shifted with each successive spike. In addition, the adiabatic case (Eq. (4.3)) will involve proportionality constants different than those provided in Table 4.2, because the terms in the summation of Eq. (4.4) are changed due to each spike providing its own increase in temperature. Addressing both effects for the triangular pulse gives:

$$D = b_j \frac{L^2}{\left[c_j t_j - t_{pulse}\right]} \tag{4.6}$$

where the correction factors b_j and c_j depend on the particular rise height (j) used in the analysis and on the pulse characteristics. Table 4.3 gives one example. Other shapes will change these correction factors.

The above analysis was extended by Watt (1966) to address behaviors other than adiabatic and pulses with any shape. If the shape is defined by $\varphi(t)$, and the solution for a delta function pulse is T_δ, then the correct solution is:

$$T(t) = \frac{\int_0^t \varphi(u) T_\delta(t-u) du}{\int_0^t \varphi(u) T_\delta du} \tag{4.7}$$

where u is a dummy variable.

Blumm and Lemarchand (2002), Blumm and Opfermann (2002), and Ohta et al. (2002) address the effect of pulse energy on various models of the T–t curves when thermal diffusivity depends strongly on T. It is best to keep the heating by the laser pulse small.

TABLE 4.3 Values[a] of Geometrical Factors b_j and c_j in Eq. (4.6) for $j = ½$ and a Triangular Pulse Occurring Over $t = 0$ to t_{pulse} With a Maximum at $t = f\, t_{pulse}$

f	b_j	c_j
0.15	0.34844	2.5106
0.28	0.31550	2.2730
0.29	0.31110	2.2454
0.3	0.30648	2.2375
0.5	0.27057	1.9496

[a]*After Table 2 in Maglić and Taylor (1992).*

4.2.4.2 Cowan's Model for Heat Losses From the Surfaces

Cowan (1963) focused on cooling of the sample by radiation to the surroundings, subsequent to pulsed heating. This model is based on his earlier and more extensive mathematical analysis of the pulse method (Cowan, 1961). A more restricted cooling model was provided by Parker and Jenkins (1962).

The main effects of cooling are to shift the maximum temperature to shorter times, to reduce the maximum temperature attained, and to return the sample to the surrounding temperature rapidly (Fig. 4.4A). If radiative losses are linearized, after Newton's law of cooling, which is reasonable because changes in T are small, the decrease in temperature is exponential with time. The rear surface temperature is derived from this assumption using the adiabatic solution for warming. Using Eq. (4.5) for $T_{max}-T_0$ and setting the initial temperature to 0 (which is equivalent to the baseline correction of the emissions in experiments) yields:

$$\frac{T(L,t)}{T_{max}} = \exp\left(\frac{-bDt}{L^2}\right) + 2\sum_{n=1}^{\infty}(-1)^n \exp\left(\frac{-n^2\pi^2 Dt}{L^2}\right) \tag{4.8}$$

where the cooling parameter b varies from 0 (no radiative loss) to order of 1. Cowan's theoretical $T-t$ curves used the approximation:

$$b \approx 2.3\times10^5 L\left[1+\frac{c_{x=L}}{c_{x=0}}\right]\left(\frac{T}{1000}\right)^3\frac{\varepsilon}{K} \tag{4.9}$$

where the units of $K = \rho cD$ are ergs cm^{-1}K^{-1} and emissivity is non-dimensional. The ratio of the specific heats is very close to unity, because the temperature changes are kept low. The ratio of unity is particularly true at high T, where radiative cooling is strongest, because heat capacity of solids approaches the Dulong-Petit limit. The curves calculated by Cowan (1963) showed that this factor has a weak affect. For this reason, his model addresses cooling from any cause (e.g., contact of the edges with a holder). Cowan's results using $c_{x=L}/c_{x=0} = 1$ are shown in Fig. 4.6A. In our presentation, T_{max} for all curves was set to unity, in accord with detector signal in experiments being arbitrary.

Temperature–time curves are modeled by fitting the signal to Eq. (4.8), with b as a fitting parameter. A fairly accurate value can be obtain using the procedure of Cowan (1963), where the temperatures (signal height) in the cooling part of the curves are compared to the temperature (signal height) at the half-time and various multiples (e.g., $5t_{1/2}$ or $10t_{1/2}$, as shown in Fig. 4.4A). Because the T–t curves are referenced to $T=0$ at $t=0$, the relevant parameters are the ratios:

$$\frac{T(5t_{1/2})}{T(t_{1/2})} \text{ and/or } \frac{T(10t_{1/2})}{T(t_{1/2})} \tag{4.10}$$

The theoretical curves of Cowan (1963), shown in Fig. 4.5B and using $c_{x=L}/c_{x=0}=1$, provide $Dt_{1/2}/L^2$ graphically from these ratios with a high degree of accuracy. This accuracy stems from small temperature changes resulting from the pulse.

4.2.4.3 *Cape and Lehman's Two-Dimensional Model*

Cape and Lehman (1963) devised and analytically solved a two-dimensional model for LFA experiments that includes the effects of face and edge heat losses and a nonideal pulse. They assumed a square wave shape and that the pulse energy was uniformly distributed over the front surface and was fully absorbed. Improvements have been made by Josell et al. (1995) and by Blumm and Opfermann (2002). The modern versions of Cape and Lehman's model are considered to be highly accurate, if internal radiative transport is not present. The fairly complicated mathematics are not presented here, because we have found that if the temperature–time curves are well-described by Cape and Lehman's

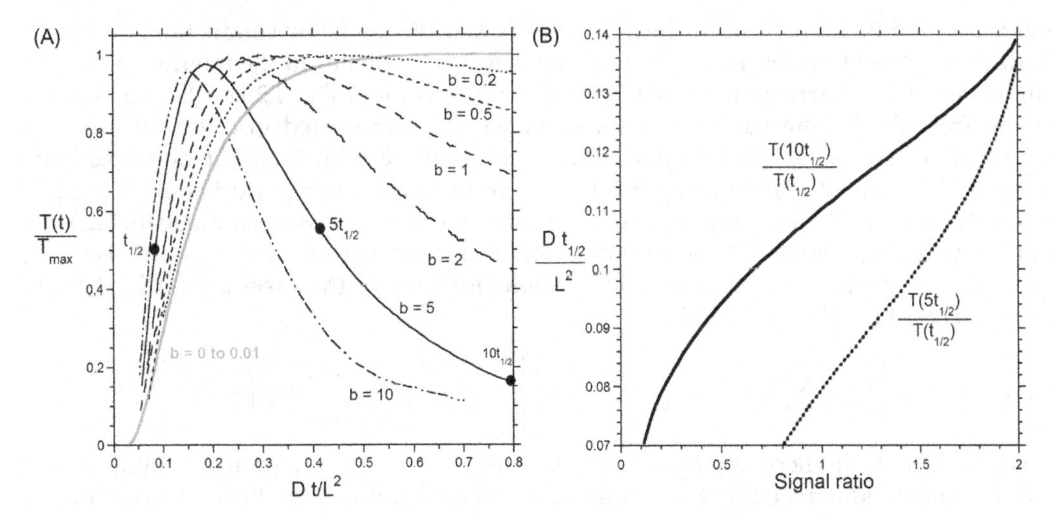

FIGURE 4.6 Cowan's model, assuming using $c_{x=L}/c_{x=0}=1$, appropriate for small temperature changes used in experiments. (A) Theoretical cooling curves. For all curves the maximum temperature is set to unity, because the signal voltage is recorded, which is proportional to temperature. Cooling parameter b is assumed to be constant over data collection. For low values of b, the cooling is not observable over the time scale considered (gray curve). Dots: the half-time and its multiples. (B) Theoretical graph for ascertaining thermal diffusivity, from length and half-time multiples. The ratios describe the cooling portion of the curve, as shown by the dots in Fig. 4.3 and part A.

model, then they are also well-described by Cowan's model. However, our natural samples are not pristine or pure and so yield a small range of D-values reflecting specific imperfections and impurities in the sample of interest.

4.2.4.4 Models Addressing Internal Radiative Transfer

Two types of models exist, which address either the end-member case of ballistic transport where the medium participates negligibly (Blumm et al., 1997; Hahn et al., 1997), or coupled radiation and conduction (e.g., Srinivasan et al., 1994; Andre and Degiovanni, 1995; Liu and Tan, 2004; Lazard et al., 2004). Models of coupled phenomena commonly provide extensive numerical simulations, but rarely analyze T–t data. Lazard et al. (2004) tested their model against silica glass, which is highly transparent, in which case the simpler ballistic model suffices. For the above reasons, and because many Earth materials are diathermic (e.g., partially transparent oxides) where the very fast speed of radiative transfer at high frequencies decouples this process from lattice conduction, this subsection only discusses ballistic models, whose purpose is to extract D from LFA experiments (e.g., Hofmann et al., 1997; Mehling et al., 1998).

Blumm et al. (1997) describe a simpler, forerunner of the ballistic model. With decoupled mechanisms, fluxes add, so Fourier's equation under small temperature changes (Eqs. (3.1) and (3.2) combined) becomes:

$$\rho c_P \frac{\partial T}{\partial t} = -\left(\frac{\partial \Im_{dif}}{\partial x} + \frac{\partial \Im_{bal}}{\partial x}\right) \tag{4.11}$$

where $\Im_{dif}$ and $\Im_{bal}$ represent the heat fluxes due to diffusion (slow conduction) and ballistic (fast radiation) transfer, respectively. Note that $\Im_{dif}$ originates in the interior so this flux can be modeled as arriving from just below the top surface of Fig. 4.2 ($x = L$), whereas $\Im_{bal}$ arrives from the bottom surface ($x = 0$), never having participated in diffusion. Also, the top graphite coat is assumed to not appreciably interfere with $\Im_{bal}$. For decoupled processes with a small radiative component, $\Im_{bal}$ can be treated as a perturbation. The signal is then the sum of the temperature rises from the top surface (assumed adiabatic) and the bottom, which is assumed to contribute a small fraction (χ) of the long-term rise T_{max}. Blumm et al. (1997) used the adiabatic solutions for each of the cases $x = L$ (Eq. 4.4) and $x = 0$ (Parker et al., 1961) to provide:

$$T(L,t) = T_{max}\left[1 + 2\sum_{n=1}^{\infty}(-1)^n \exp\left(\frac{-n^2\pi^2 Dt}{L^2}\right)\right] + \chi T_{max}\left[1 + 2\sum_{n=1}^{\infty}\exp\left(\frac{-n^2\pi^2 Dt}{L^2}\right)\right] \tag{4.12}$$

Eq. (4.12) is fit numerically to the signal, using χ as a fitting parameter. Blumm et al. (1997) found good agreement when comparing thermal diffusivity obtained from conventional models for soda-lime and pure silica glasses with a gold undercoat and to D obtained from Eq. (4.12) for glasses with only a graphite top coat. Importantly, optical properties are not needed. The fraction represents several factors, such as absorption and re-emission of $\Im_{bal}$ by the coating at $x = L$ and/or partially losses of the flux from $x = 0$ inside the sample. When $\Im_{bal}$ is a perturbation and the temperature changes (heat from the pulse) are small, such nuances are unimportant.

Hahn et al. (1997) improved the accuracy of the ballistic model by including the T^3 dependence of the radiative heat losses. Effects of both graphite layers, modeled as surfaces with constant emissivity, were included. Green's functions were used to solve the heat equation. Their analytical solution was cross-checked against the solution of a related, but simpler, problem by Carslaw and Jaeger (1959). Hahn et al. (1997) suggest that a flat signal following the radiative increase is associated with complete absorption of the pulse by the bottom coating, whereas a peak immediately following the pulse is associated with incomplete absorption by the bottom coating, so that some of the laser signal heats the top coat. Incomplete absorption exists for damaged coatings or very thin coatings. Generally, the incomplete absorption is neglected in analyzing data, as can be inferred from Figs. 4.3 and 4.4.

Hofmann et al. (1997) summarize Hahn et al.'s (1997) model and tested it against float glass (a complex silicate glass) and crystalline CaF_2, which, respectively, are transparent above $\sim 2000\ cm^{-1}$ and above $\sim 1000\ cm^{-1}$. In addition, CaF_2, absorbs weakly near $800\ cm^{-1}$ (the IR region), which makes fluorite difficult to measure with periodic or steady-state methods. By using numerical simulations that included spectral properties and by conducting experiments with various coatings, Hofmann et al. (1997) showed that applying the model to coated samples provided accuracy within 1% at high temperature, and that the model is valid for materials that are approximately diathermic. Mehling et al. (1998) discuss the model further, conducted experiments on coated and uncoated silicate glass of a different chemical composition plus fused silica. They show that a graphite coating is sufficient: a metal undercoating is not needed.

As discussed by Hahn et al. (1997), the ballistic model is valid when the sample is either fully transparent, or strongly scattering, or strongly absorbing. Most geologic samples fall in one of these three categories in the frequency range of interest, that is, that of the blackbody at the temperature of the furnace. Of the hundreds of samples studied in our laboratory, only a few departed from the model. These departures are minor (Fig. 4.7A). Smooth variations with temperature show that the differences are not caused by light leakage or some other artifact. The effect on $D(T)$ is also minor, as exemplified by comparing $D(T)$ of rare-earth garnets ($X_3Ga_5O_{12}$, where $X = Nd$, Sm, or Eu to the trends for transparent $Gd_3Ga_5O_{12}$) (Hofmeister, 2006). Existence of many narrow bands between 2000 and $>21{,}000\ cm^{-1}$ at the 1–2 mm thicknesses studied differ from Hahn et al.'s categories. The other departure was for black spinels (Hofmeister, 2007b). For one of these hercynites, $Mg_{0.77}Fe^{2+}{}_{0.22}Zn_{0.01}(Al_{1.84}Fe^{3+}{}_{0.15}Ti_{0.01})O_4$, the modeled differed from the theoretical fit below 700K, but was a good match above 900K. Both samples have low frequency IR fundamentals and a transparent region at about $1500\ cm^{-1}$ (Fig. 4.7B). The peak of the blackbody crosses the transparent region during the temperatures examined. Silicates with high frequency Si–O absorptions, broad absorptions of Fe^{2+} over the near-IR, and/or narrow Fe^{3+} or O–H absorptions in the visible and IR, respectively (all of which are common in geologic samples), present T–t curves similar to the ballistic model. To understand the behavior requires delving into diffusive radiative transfer (Chapters 8 and 11).

4.2.4.5 *Models of Layered Substances*

Heat transfer across layers or interfaces is of considerable technological interest. Examples are composite materials, and thermal barrier coatings. Vozár and Hohenauer (2003/4) review the early work in LFA on layered substances. The basics stem from two different PhD theses,

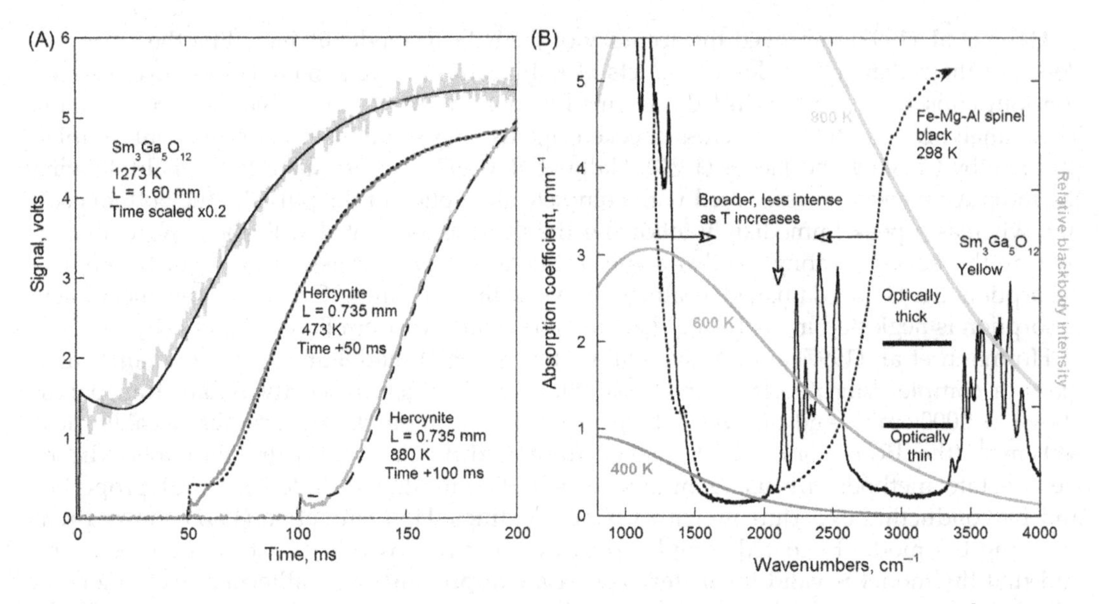

FIGURE 4.7 LFA and spectral data for yellow rare-earth garnet and black hercynite spinel: $Mg_{0.77}Fe^{2+}{}_{0.22}Zn_{0.01})(Al_{1.84}Fe^{3+}{}_{0.15}Ti_{0.01})O_4$. (A) Expanded view of the radiative effects on the thermal evolution following the laser pulse. The variance of curve for hercynite at low temperature from Hahn et al. (1997) model decreased smoothly as temperature increased. For $Sm_3Ga_5O_{12}$, the variance began at moderate temperature. The shape for both samples is similar, and was observed for other rare-earth garnets and black spinels at different temperatures. The resulting D values are perturbed at levels near the uncertainty of ~2% (Hofmeister, 2006, 2007b). (B) Comparison of near-IR spectra at 298K with Planck curves. Bands become broader and less intense as T increases, whereas the Planck curves become more intense and shift towards the visible. Heavy black lines show the limits of optically thick or thin conditions, which are approximate. For black hercynite, the gap in the transparent region is reduced gradually as T increases while the Planck curve intensity increases where the sample is opaque. When optical thickness becomes large everywhere (apparently above ~900K), the $T-t$ curves match the ballistic model. For yellow garnet at low temperatures, the Planck curve mostly overlaps with the opaque IR region. As temperature increases, the Planck curve increasing "samples" the spikey near-IR features which alternate with spectral gaps. High T will smear out the narrow bands into broad envelopes with absorption coefficients between optically thick and thin conditions, which does not meet model assumptions.

which yielded the seminal papers of Lee and Taylor (1976) and Lee et al. (1978). Chen et al. (2010) provide an updated summary and devise a model for a coating on a substrate of known thermal diffusivity, which constrains both the interface resistance and the coating value. Campbell (2010) addresses thin films, arguing that the three-layer model is needed due to interface resistance. Layer models are discussed further in Section 4.6.1.

4.3 PERIODIC METHODS

Imposed, periodic temperature variations have a long history. Pulsed power, AC hot wire, two strop, and transient plane source techniques are commonly used to probe heat transfer properties of bulk samples, whereas 3ω and thermoreflectance are currently applied to thin films (Reif-Acherman, 2014; Zhao et al., 2016). These methods do not require attaining steady-state conditions. But, neither do they resolve behavior as a function of time: that is, the thermal evolution of the sample is not monitored as in LFA (Berman, 1976). Generally, K or D is assumed to be constant over a temperature cycle (Fig. 4.1).

One concern is behavior arising when multiple mechanisms exist, since independent mechanisms are associated with different values of D, different storativities, different carrier speeds and operate over different time-scales and lengths. For metals, fast electrons and slow vibrations both transport heat, whereas for electrical insulators, ultrafast photons and slow vibrations are present. Thus, proportions of the different mechanisms will depend on how large the period of the temperature variation is with respect to the time-scale of the process and the size of the sample ($D \sim L^2/t$). Almost all studies assume that quasiequilibrium exists, and that some effective thermal conductivity describes the situation.

The volumes of Tye (1969) and Maglić et al. (1984, 1992) cover many different periodic methods. This section focuses on those commonly used in modern studies or relevant to the planetary sciences.

4.3.1 Angstrom's Method and Its Modifications

Ångström's (1861) method is used in geoscience in its modified form (e.g., Xu et al., 2004). To understand the principles, we first describe the original experiment:

A sinusoidal source of heat at frequency ω is applied to one end of a rod-shaped sample, and measurement of the phase and/or attenuation of the sinusoidal temperature wave is recorded with distance (x) along the rod (Fig. 4.8A). In principle, thermal diffusivity can be determined from measurements either of the attenuation or the wave or of its phase lag, based on an one-dimensional solution to Fourier's heat equation (Berman, 1976). Specifically, the difference ΔT at the end of the rod ($x = 0$) between the actual temperature and mean temperature is described by:

$$\Delta T(0, t) = T^* \sin(\omega t) \tag{4.13}$$

where T^* is the maximum excursion (the amplitude). If no heat is lost from the sides of the rod, then the temperature difference along the rod varies in time in accord with:

$$\Delta T(x, t) = T^* \exp\left[-x\sqrt{\frac{\omega}{2D}}\right] \sin\left[\omega t - x\sqrt{\frac{\omega}{2D}}\right] \tag{4.14}$$

Importantly, it is assumed that one value of D represents all mechanisms present, or that only one mechanism is present. Ideally, the attenuation (Λ) and phase lag (Φ) are:

$$\Lambda(x) = \exp\left[x\sqrt{\frac{\omega}{2D}}\right]; \quad \Phi(x) = x\sqrt{\frac{\omega}{2D}} \tag{4.15}$$

Diffusivity is then obtained from any of:

$$D = \frac{\omega x^2}{2[\ln \Lambda(x)]^2}; \quad \text{or} = \frac{\omega x^2}{2\Phi(x)^2}; \quad \text{or} = \frac{\omega x^2}{2\Phi(x)\ln \Lambda(x)} \tag{4.16}$$

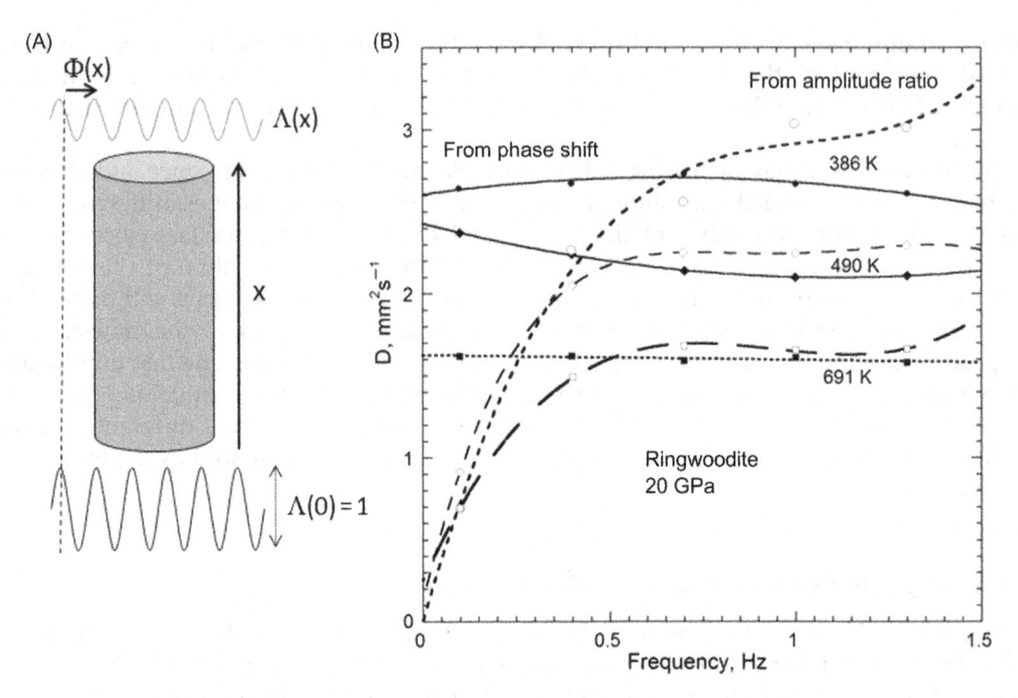

FIGURE 4.8 Ångström's and closely related methods. (A) Principles of operation. Samples are long cylinders if thermally conductive, or short if insulating. An AC voltage is applied to metal at the base ($x = 0$). At some point (x) along the length or at the end (L), the decayed amplitude of the signal and its phase shift are recorded. (B) Thermal diffusivity inferred using different frequencies by Xu et al. (2004) for $(Mg_{0.9}Fe_{0.1})_2SiO_4$ in the spinel structure. Different shaped symbols are for different temperatures. Filled symbols use the phase shift whereas open symbols use the amplitude decay. Dotted line: linear fit. Solid lines: second-order polynomial fits. Dashed lines: third-order polynomial fits.

The LHS is considered to be the most accurate and to be valid when heat is lost from the sides (Berman, 1976). Other methods of analysis are covered by Danielson and Sidles (1969).

Correction terms address boundary conditions such as surface losses from the sides via radiation to the surroundings. Experimental difficulties center on establishing a purely sinusoidal temperature variation and in dealing with the gradual climb in the base temperature with time (Berman, 1976). Contact losses exist because thermocouples are used, as well as ballistic radiative transfer across the sample. This classical method has been applied mainly to metals, which fortunately lack these problems.

The "modified" Ångström's method refers to measurements of slabs, rather than long rods (e.g., Bellings and Unsworth, 1987). Eq. (4.16, LHS) is used. A flat geometry is advantageous for low thermal diffusivity samples. Zhao et al. (2016) list recent studies, involving efforts to simultaneously measure heat capacity, and note that the surface loss corrections do not much affect diffusivity, but are important for heat capacity. Statements that radiative transfer effects are reduced by comparing amplitudes (e.g., Manthilake et al., 2011) refer to surface radiation, and not to internal, ballistic processes. This deduction is evident in Eq. (4.11). In analyzing data from slabs, different frequencies are explored to improve accuracy (e.g., Vandersande and Pohl, 1980). Raw data in Fig. 4.7B show that thermal diffusivity depends on frequency, and suggest that the RHS side of Eq. (4.11) reduce uncertainties through averaging. Different behaviors at different temperature could be ballistic effects for the partially transparent samples studied.

The pulsed power method of Maldonado for room temperature and below is closely related to Ångström's. Summaries are given by Tritt and Weston (2004) and Zhao et al. (2016).

4.3.2 AC Hot Wire, Hot Strip, and Needle Point Methods

Some techniques provide both k and D; the AC hot wire and hot strip methods have uncertainties of $\pm 6\%$ and 30%, respectively (Hammerschmidt and Sabuga, 2000). The needle point method is similar to the hot wire, but is not accurate when applied to dissolved minerals, see Hofmeister et al. (2007). In all three methods, the metal serves as both a heater and a thermometer. Several chapters in the volumes of Tye (1969) and Maglić et al. (1984, 1992) discuss the details of these and related techniques. A brief summary is provided here.

Early studies used a hot wire with direct current to measure radial heat transport in metals in steady state. Data analysis is similar to the "absolute" methods of Section 4.4. Currently, alternating currents are used, and so all methods are periodic, where data analysis is similar to Ångström's. Xu et al. (2004) actually used the hot wire method. The difference is geometrical: the hot wire is inside the sample, whereas in the methods detailed in Section 4.3.1, the electrical leads are at the ends of the sample in a Cartesian geometry.

The hot strip method differs by having a planar source, and is denoted as a transient plane source method, even though this is actually a periodic method. Equations vary from Ångström's method due to geometrical considerations. Variations include a two strip method (Andersson and Bäckström, 1986) which is more accurate and has been used at pressure.

4.3.3 The 3ω Method

The 3ω method for thermal conductivity is based on the electrical resistivity of a metal being temperature-dependent. In contrast to the methods of Sections 4.3.1 and 4.3.2, where the temperature is measured after heat passes through a sample over a well-defined distance, the 3ω method is based solely on the response of the heating wire to periodic temperature changes of the surroundings. Contact losses are minimized by use of sputtering, but the results depend on a thermal model of the wire and substrate.

The method traces to ~1910 when O.M. Corbino detected the presence of a third harmonic component in the voltage when he applied an alternating current through a heater. At this time, H. Eberling used this approach to investigate filaments in light bulbs. The 3ω technique became popular after Cahill and Pohl (1987) provided an approximate analytical solution for the self-heating of a narrow metal line deposited on a substrate. Today, considerably efforts are directed towards technologically important thin films and nanomaterials, see Borca-Tasciuc and Chen (2004), Reif-Acherman (2014), and Ventura and Perfetti (2014) for summaries and updates.

Essentials of the 3ω method are as follows: applying an AC current $J(t) = J_0\cos(\omega t)$ to a metal line produces temperature oscillations at an angular frequency of 2ω. The metal serves as both heater and temperature sensor (Fig. 4.8A and B). Changing temperature affects the resistance of the metal line of length L. A thermal model is essential to relate the periodically changing temperature to the electrical characteristics of the wire and the experimental parameters. Assuming some average temperature pertains to the applied voltage (v_0) for the resistive metal line gives the solution to the one-dimensional heat equation as:

$$\Delta T = \frac{dQ/dt}{\pi L K_{sub}} \int_0^\infty \frac{\sin(ux)}{(ux)^2\sqrt{x^2+q^2}} du \tag{4.17}$$

where Q is the heat applied, L is the length of the metal line and x is its half-width, K_{sub} is the thermal conductivity of the substrate, u is a dummy variable, and $q = (2i\omega/b)^{1/2}$ contains the frequency dependence of the oscillation at the surface, and b is a constant. Eq. (4.17) was only recently solved analytically (Duquesne et al., 2010). Under the assumptions of isotropic thermal conductivity, effectively infinite length of the metal line, and the substrate being effectively semi-infinite, thermal conductivity of the substrate is reasonably constrained from:

$$K_{sub} = \frac{v_0^2 \left(\frac{\partial R}{\partial T}\right) \ln\left(\frac{\omega_2}{\omega_1}\right)}{4\pi R_0 L \; (v_{3\omega 1} - v_{3\omega 2})} \tag{4.18}$$

where R_0 is resistivity of the metal at the base temperature, with a known temperature derivative $\partial R/\partial T$ considered to be constant over the range of temperatures explored, and two different applied frequencies are used (Cahill and Pohl, 1987). For each frequency, the corresponding harmonic voltage (v) is determined from the experiment (Fig. 4.9). Because the temperature changes are small, the linear derivative of the resistivity is sufficient, but contributes substantially to the uncertainty in this technique. Note that the dimensions of the wire are important.

Recent improvements in the model allow for two-dimensional conduction of finite thickness heaters and sample anisotropy. For a summary, see Zhao et al. (2016).

Advantages are the ability to probe small samples, insensitivity to blackbody radiation due to the small surface area of the metal heater, and fairly rapid measurement times of minutes. Importantly, the radiative calculation depicts steady-state surface radiation from the wire, not transient ballistic transfer of energy within the sample. At the temperatures typically accessed (mostly <400K; e.g., Langenberg et al., 2016), ballistic internal transport and surface radiation are both low. Because the wire is deposited on the sample, contact resistance is minimized. Disadvantages are microfabrication being required for the wire and that the method provides K, and thus does not separate the efficiency of heat transport from the amount carried.

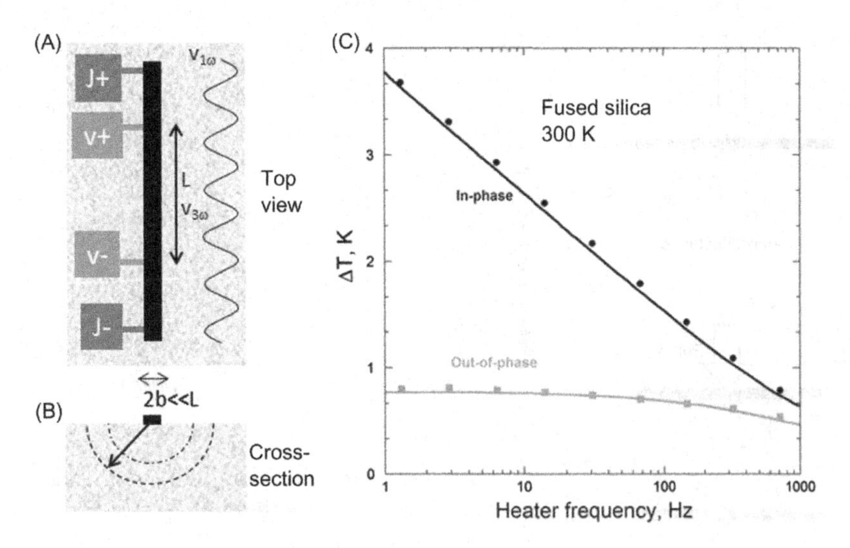

FIGURE 4.9 The 3ω method, as applied to a substrate. (A) Conditions of operation in from a top view of the sample (gray speckle). Black: a thin wire with width of $2b$. An AC current is applied to the outer leads, whereas the voltage is measured over a shorter length $L>>2b$. The signal for the 3ω affect is very weak. (B) Principles of operation, suggest by a cross-section. Heat from the wire moves radially (arrow). It is assumed that the length L is effectively infinite to approximate heat transfer from the wire as one-dimensional (i.e., radial) into the substrate. This transfer depends then on $2b$ and the power applied divided by the length of the wire. (C) Periodic temperature differences in the 3ω method. Dots: measurements of vitreosil which probably has ~150 ppm OH impurities. Curves: the model of Cahill (1990), which differs negligibly from Eq. (4.17), obtained using $D = 0.83\ mm^2\ s^{-1}$, which is appropriate for fused silica with low water contents, but not dry (Hofmeister and Whittington, 2012). Out-of-phase temperatures are derived from the imaginary part of the solution (note q is an imaginary number) and in-phase temperatures are derived from the real part. Source: *Part (C) is reproduced from Figure 6 in Cahill, D.G., 1990. Thermal conductivity measurement from 30 to 750K: the 3ω method. Rev. Sci. Inst. 61, 802, with the permission of AIP Publishing.*

4.3.4 Thermoreflectance

The thermoreflectance technique, developed by Paddock and Easley (1986) and Young et al. (1986), is widely used in materials science near ambient temperature to study thermal transport across interfaces. This technique is applied to thin films, epitaxial layers, and nanomaterials with a recent focus on small-scale differences in advanced materials; see reviews by Borca-Tasciuc and Chen, 2004; Reif-Acherman, 2014; Zhao et al., 2016.

In contrast to the periodic methods of Sections 4.3.1 to 4.3.3, thermoreflectance measurements and analyses are not simple and are not related to a measureable length scale. The experiments always involve layers (Fig. 4.10A). Several properties of the layers are relevant, depending on the circumstances: thermal conductivity K, storativity C_V, thermal effusivity ϵ:

$$\epsilon^2 = \rho c K = \rho^2 c^2 D \tag{4.19}$$

as well as thermal resistance (or its inverse of conductance, which is energy per area per degree) of the interfaces. The challenge of the technique is modeling the behavior in order to extract several properties of interest (Zhao et al., 2016).

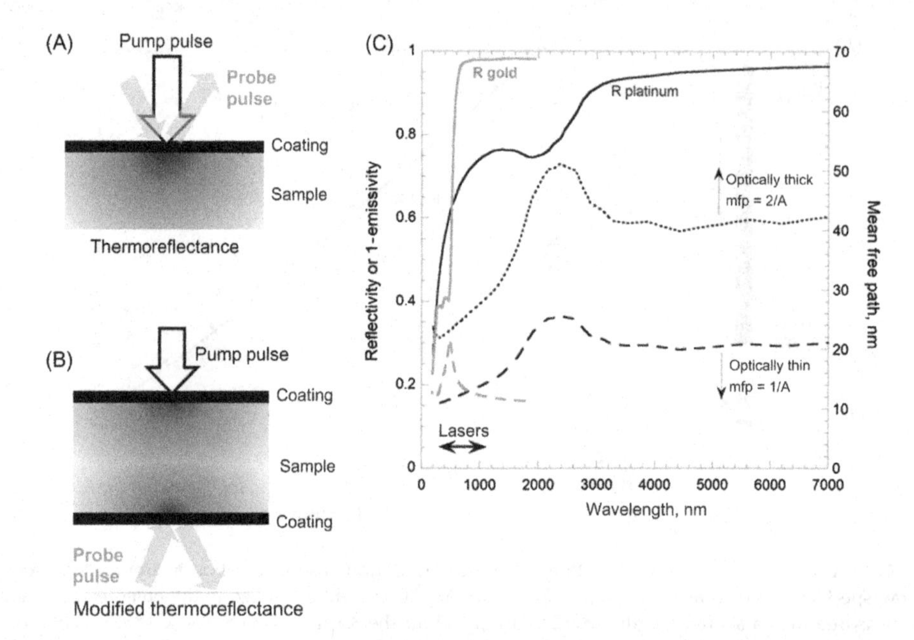

FIGURE 4.10 Thermoreflectance principles and relevant spectral data. (A) Configuration (cross-section) used in materials science, where the same laser is used to heat the metal coating and to measure the change in reflectivity (beams are split with a delay). The angle of the probe beam is exaggerated. Heat from the hot spot diffuses through the metal coating into the sample, as indicated by the shading: this is three-dimensional. (B) Geometry used by Yagi et al. (2011) and others in diamond anvil cells. Heating and sampling are on opposite sides. Their thermal model is one-dimensional and of three layers, and differs by considering a length over which heat moves. (C) Reflectivity and mean free path calculated from the tabulated optical constants of gold and platinum of Palik (1998).

Although this method probes short times using fast detection systems, experimental conditions are quasisteady-state, resulting from sequential application of multiple pulses. Hence, thermoreflectance is not a transient technique, although it involves short measurement times. For this reason, mechanism proportions from thermoreflectance data are deduced from a model (Schmidt et al., 2008), as is the case when analyzing K (e.g., Klemens and Williams, 1986). Exploring individual pulses in a modified approach is discussed below.

The essence of the method is sketched in Fig. 4.10A: Heating is applied remotely by a fast, pulsed laser near visible wavelengths (e.g., 800 nm, with pulse occurring at a frequency of 80 MHz or 12.5 ns between pulses) to a thin metal film (e.g., ~80 to 100 nm of Al, Pt) that is deposited on a sample. Spot sizes are roughly 10 microns, and each pump pulse heats the surface by several Kelvins (e.g., Lyeo and Cahill, 2006). Reflections from this surface are monitored remotely as a function of time using probe pulses also from the laser, but with a smaller spot size and after a short delay. Reflectivity of the metal surface is assumed to depend linearly on the temperature. Strain has an effect, and can be used to ascertain thickness.

Measurements are made and analyzed in either a time-domain or a frequency-domain. Zhao et al. (2016) outline both approaches and provide some of the equations need for this complex analysis. Averages are used, which reduces noise. Equations in the frequency-domain are simpler, such that the response of the surface to the pulses is given by:

$$Z(\omega_0, \tau_d) = v_{inplane} + iv_{outplane} = \beta \sum_{j=-\infty}^{\infty} H(\omega_0 + j\omega_s)\exp(ij\omega_S\tau_d) \quad (4.20)$$

where v is the response in and out of the plane, which is a complex function, H is the response of the surface to heating by a continuous plane wave modulated at a frequency of $\omega_0 + j\omega_S$, and τ_d is the delay time between the laser stimulation (the pump pulse) and sampling (the probe pulse). The response H at any given frequency is related to the K_{in}, K_{out}, and C_V of each layer and the thermal conductance between each pair of layers. Typically, v_{in}/v_{out} is modeled as a function of frequency or the time delay. Thermal conductivity of some materials depends on frequency (i.e., on delay time), which is of interest, being unexpected from the standard model (Koh and Cahill, 2007).

Thermoreflectance has been benchmarked against bulk materials, mostly metals. For large substrates, thermal conductivity can be determined by using a low frequency and small spot size for deeper heating (Fig. 4.10A) whereas heat capacity is determined using a high frequency and large spot size for shallow heating, and assuming isotropic properties. Assumptions exist regarding the depth and spread into the substrate that sustains a temperature gradient. Note that the laser beam is attenuated in the metal films: As shown in Fig. 4.10C, the mean free path is smaller than the film thickness. This feature and use of spots during laser stimulation and sampling makes the diffusion of heat into the sample two-dimensional in thermoreflectance.

The interface is important. Measured conductance of the interface between sputtered Al on Al_2O_3 and other metal–insulator pairs is large ~150 MW m^{-2}K^{-1} (e.g., Lyeo and Cahill, 2006). Since K = 25 W m^{-1}K^{-1} for ceramic corundum (200 W m^{-1}K^{-1} for Al), the sputtered interface is equivalent to an additional layer of 100 nm of corundum (or of 1000 nm of Al). This thickness is significant because sputter coats are in this range of thickness.

Yagi et al. (2011) modified the thermoreflectance technique for use in diamond anvil cells. In these experiments, the metal coating on one side is heated with a pulsed near-IR laser, while another coating on the opposite side is sampled by reflecting a continuous wave laser near 800 nm (Fig. 4.10B). Recent studies use thinner gold coatings and probe lasers emitting at 532 nm (e.g., Okuda et al., 2017). The method of analysis differs by being based on an one-dimensional model of three layers, requiring inputs of the diffusivity of the metal, and also using a fitting parameter to describe the effusivity of NaCl which surrounds the sample. This approach has not been benchmarked at low pressure.

On the one hand, the well-defined length over which heat is transferred ties this experiment (Yagi et al., 2011; Okuda et al., 2017) to long-standing measurement principles and Fourier's equations. On the other hand, heat from the lased spot flows in three dimensions, not one-dimension as modeled. The signal from the pump pulse is unlikely to cross the micron-sized samples, given the low emissivity and high reflectivity of relevant metals (Fig. 4.10C). Moreover, the small amount of energy in the pulses (0.5 μJ over a 50 μm spot), which would only raise the metal temperature within a 5 nm depth by ~0.1K. For deeper penetration, the increase in T is proportionately reduced. Thus, the changes recorded by the probe laser (Fig. 4.10B) record heat transfer associated with the probe pulse, not due to the pump pulse. Lastly, interface resistance needs to be addressed in layered models (Campbell, 2010; Zhao et al., 2016; Section 4.2.4). However, we have some concerns about the materials science approach, given the high reflectivity of metals, application of small amounts of heat as visible radiation (increases are considered to be only 1–2K), and the fs-spectroscopic studies of electron–lattice disequilibrium following stimulation of metals and metal films by visible light (e.g., Bauer et al., 2015). Evaluation requires discussing radiative transfer (Chapter 8) and D depending on length (Chapter 7), and is postponed.

4.4 STEADY-STATE METHODS

Steady-state methods require information on both the flux and the temperature gradient and yield K. In one dimension, Eq. (3.1) becomes:

$$\Im = \frac{1}{\mathring{A}}\frac{\partial Q}{\partial t} = -K(T)\frac{\partial T}{\partial x} \tag{4.21}$$

where $-\partial Q/\partial t = \text{power} = \Pi$. Boundary conditions and sample geometry influence the results. These experiments are represented by a mathematical expression for the power:

$$\Pi = -\frac{\partial Q}{\partial t} = -\frac{\int_{T_0}^{T_1} K(T)dT}{\int \frac{1}{\mathring{A}}dx} \cong \frac{\mathring{A}}{L}\int_{T_0}^{T_1} K(T)dT \tag{4.22}$$

The RHS assumes that the geometry ($\mathring{A}/L$) does not change in the experiment, which is a limitation in high-pressure studies. Generally, the temperature is maintained by an electrical heat source and the gradient is not determined; rather, the difference (ΔT) across some length is determined from thermocouples. The length does not need to involve the entire samples and radial geometries are used.

Steady-state methods were importantly historically (Reif-Acherman, 2014) and circa 1950–1970, when many measurements were made. For details, see various chapters in Tye (1969) or in Maglić et al. (1984, 1992); for a summary, see Zhao et al. (2016). For a synopsis of low temperature techniques, see Ventura and Perfetti (2014).

The methods are characterized as (1) absolute, where temperatures across a set length are measured, not that the results are without uncertainty, and (2) comparative, where a standard and sample are measured simultaneously in either parallel or series configurations. For both categories, physical contacts, interface resistance, conduction of heat through the thermocouples, and radiative or convective losses from the surface are involved. Common features are being limited to ~1200K (Table 4.1). Experimental challenges involve minimizing and estimating these losses while accurately measuring ΔT, over the long durations needed to reach steady-state for the large sample size required for such methods. Special efforts are needed for small samples (Zawilski et al., 2001). Measurement uncertainties do not fully address effects of these systematic errors, as suggested by the comparisons in Section 4.5.

Because K is measured, not D, and times are long, steady-state methods reflect mechanisms operating under near-equilibrium conditions and do not differentiate rate from the amount of heat carried.

4.4.1 Absolute Methods

In two-probe potentiometric methods, L is the sample length, whereas in four-probe methods, interior points are used. This method is best suited for wires or rods (i.e., metals) and is done in a vacuum to minimize side losses to radiation. If losses are negligible, $K(T)$ is calculated from Eqs. (4.21) and (4.22). Thus, either small temperature differences are used, or a function must be deduced for K vs T.

Berman (1976) emphasizes the problems with interface resistance in longitudinal measurements. For the description of combined conductive and radiative heat transfer, models have been developed over the past two decades which are applicable to stationary measurement systems, such as the guarded hot plate (Kunc et al., 1984).

To reduce surface radiative losses from the sample to the surroundings, radial geometries and internal heating are used (e.g., Flynn, 1969). However, this approach cannot avoid ballistic transport within the sample. Infinite length is assumed, requiring large samples.

4.4.2 Comparative Methods

This approach eliminates measuring the power transmittal. The basic equation is:

$$\frac{K}{K_{\text{stnd}}} = \frac{\mathring{A}}{\mathring{A}_{\text{stnd}}} \frac{L}{L_{\text{stnd}}} \frac{\Delta T}{\Delta T_{\text{stnd}}} \tag{4.23}$$

Comparative methods, where a standard is an intrinsic part of the measurement, are viewed as involving lower uncertainties (Powell, 1969). Laubitz (1969) discusses errors associated with sample conductivity mismatches. A valid comparison further requires similar spectroscopic characteristics and similar interface resistance, so that the two common systematic errors play a minor role in the comparison. Due to these requirements infrequently being met, we share Laubitz's concerns.

4.5 COMPARISON OF HEAT TRANSFER DATA FOR ELECTRICALLY INSULATING SOLIDS FROM DIFFERENT METHODS

Random and systematic errors (plus limitations of some kind) are inherent to experiments. The comparisons made here involve high purity samples, because impurities greatly reduce K and D from end-member values, as is long known. It would be advantageous to compare samples near the middle of binary solid solutions (also known as mixed crystals) because changes in chemical composition have a weak affect in this situation. However, for any given solid solution, usually only one method has been applied. In addition, single-crystal data are considered here because grain boundaries can provide thermal resistance.

It is long known that ballistic radiative transfer at high temperature is a problem in many experiments. The comparisons provided below are near ambient temperature because this temperature is accessed in diverse methods (Table 4.1) and because radiative transfer is relatively low.

The focus is on bulk materials and conventional, contact methods commonly used to study planetary materials. Although rocks have grains and grain boundaries, scales are relatively large, and so bulk materials are of the greatest interest to planetary science. We do not make a detailed comparison with results from thermoreflectance because this technique involves very small distances and very short times (Section 4.3.4).

The recognized sources of systematic errors in heat transport measurements are contact resistance and radiative effects (internal ballistic gains and surface cooling). Because metals adhere well to each other, contact resistance arising from use of thermocouples or heaters does not affect these measurements. Metals are also opaque, with mean free paths ~20 nm in the visible (Fig. 4.1C), so ballistic transport of high frequency light will not affect measurements of bulk samples. Thus, conventional, contact measurements and LFA of the main, slow near-equilibrium mechanism in metals are in good agreement, as shown by Criss and Hofmeister (2017), see Chapter 9.

4.5.1 Consequences of Time and Length Scale Limitations

A filter exists in all measurements, namely the length- and time-scales of the experiment. Fig. 4.11 summarizes the conditions of the methods listed in Table 4.1 and Sections 4.2–4.4. Also shown are lines of constant $D = 0.3L^2/t$, from Parker's adiabatic model at $t = t_{0.9}$ (Table 4.2) to approximate measurement times which exceed the characteristic lifetimes of diffusion (see Chapter 3). The approximation for thermal diffusivity of ~1 $mm^2\,s^{-1}$, which represents complex electrical insulators (minerals), crosses most experiments. Faster insulators (diamond is highest of all solids at 453 $mm^2\,s^{-1}$) and metals lie between the lines approximating $D = 1$ and 1000 $mm^2\,s^{-1}$. Obviously, most techniques are geared towards measuring typical values of thermal diffusivity.

Importantly, Fig. 4.11 implicitly assumes that the measured diffusivity pertains to the mechanism operating near the equilibrium or quasiequilibrium conditions that are explored in most of these measurements. Because LFA records temperature against time, unlike the other measurement categories (Fig. 4.1), this method that can discern mechanisms operating following the initial application of heat, during far from equilibrium conditions, as well as D-values for mechanisms operating near equilibrium, simply by varying length and data acquisition intervals of the measurements (Criss and Hofmeister, 2017). Other transitory methods exist (Danielson and Sidles, 1969), but have longer measurement times than LFA, and thus cannot explore fast, transient phenomena (Fig. 4.11).

4.5.2 Competing Effects of Contact Losses and Ballistic Gains

Quantifying contact losses at interfaces theoretically is very difficult, and so experimental approaches are used (e.g., Fried, 1969; Campbell, 2010; Zhao et al., 2016). However, experimental comparisons have pitfalls, due to ballistic radiation offsetting contact losses.

Fig. 4.12 compares materials that are known their high purity, several of which are relevant to planetary science. Because natural garnets tend not to be pure, yttrium aluminum garnet is shown instead. Several factors exist that make it difficult to decipher effects of the competing systematic errors. Also, the two contact methods include both steady-state

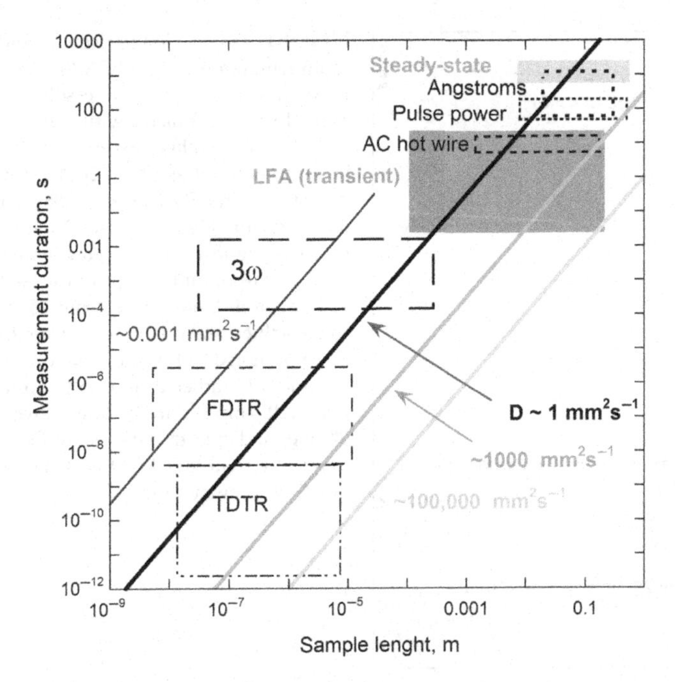

FIGURE 4.11 Dependence of the time required to make various measurements of heat transport properties to the length over which heat moves during the experiment. Values are mostly from Zhao et al. (2016), who provide a similar comparison with additional techniques, but do not differentiate between method categories nor make a comparison with theory. Light gray: steady-state methods. Dark gray: the transient method of LFA. White boxes with patterned outlines: periodic methods, where TDTR is time-domain thermoreflectance and FDTR is frequency-domain thermoreflectance. Gray lines: approximate values of thermal diffusivity from an adiabatic model, considering an appropriately long lifetime (see text). The heavy gray lines are compatible with low *D* of slow lattice vibrations, high *D* for fast electrons, and extremely high *D* of ultrafast photons, as suggested by the dimensional analysis of Chapter 3. The light gray line approximates very porous insulting materials.

and periodic approaches, and so have variable amounts of ballistic radiative transfer. However, soft solids with low frequency modes have fairly flat lying trends. Hard solids ($H > 7$), have D that decreases as the number of physical contacts increase. Hence, contact resistance is the important factor for silicate minerals near 298K. The behavior of MgO is a puzzle, since this is moderately hard but D increases with the number of contacts. This trend can be explained by large amounts of radiative transfer, particularly in the cryogenic experiments of Slack (1962). But then, why do not the other transparent solids have similar trends? Additional comparisons are needed.

Fig. 4.13 depicts the effect of hardness on D obtained at 298K by various methods for materials with the B1 structure. The silver halides are both opaque and soft, in which case contact methods yield the same values as LFA. Possible, the silver halides have a small amount of each type of systematic error, close to the level of uncertainties. Softer materials which are partially transparent yield higher D from steady-state contact methods, showing that ballistic transfer has a slight effect at 298K. Data on very soft alkali halides also exist from periodic methods, but pressed powders were used (see summary by Yu and

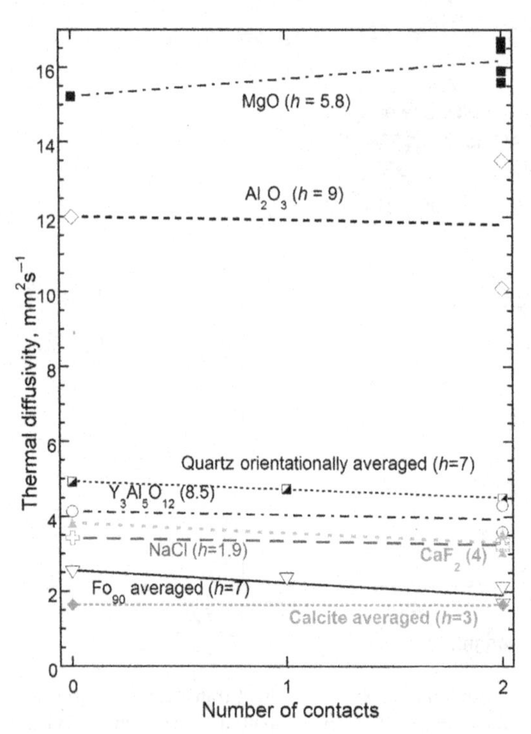

FIGURE 4.12 Comparison of thermal diffusivity at room temperature from LFA (0 contacts) to conventional measurements of crystalline solids, which mostly have two physical contacts, with thermocouples or heaters. All are end-members except for olivine. Data compiled by Hofmeister et al. (2007, 2014). MgO data compiled by Andersson and Bäckström (1986). High D from contact methods for MgO and Al_2O_3 are from Slack (1962), whereas the low value is an average of various sources from the web, and is referred to as the accepted value. The range is large for K (from 25 to 50 W $m^{-1}K^{-1}$) but some values are miscopied or for higher T than 298K, but not noted. LFA data on calcite are unpublished. Hardness is indicted and soft substances are shown in gray. All samples are colorless except for olivine. YAG, olivine, and quartz have many IR modes at fairly high frequency, whereas MgO and the soft substances have low frequency modes.

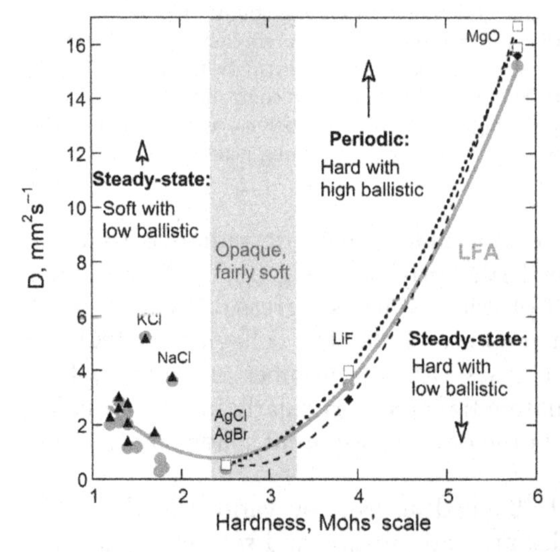

FIGURE 4.13 Thermal diffusivity of single-crystals with the rocksalt structure at 298K as a function of hardness and of the type of method. Gray: LFA data (Yu and Hofmeister, 2011; who compiled results from conventional sources). Periodic studies used Angstrom's method, except for MgO data were from the strip method of Andersson and Bäckström (1986), who compiled earlier data. Most steady-state measurements are "absolute." Triangles: soft alkali halides that were not included in the second-order polynomial fits shown. For MgO, the thickness used for LFA was 1 mm, whereas the steady-state and periodic methods used 3–4 and 7 mm samples, respectively.

Hofmeister, 2011). Harder, LiF has lower D in steady-state, indicating that contact resistance is important. However, this datum for LiF was corrected for high radiative transfer by ~30% by Berman and Brock (1965). The data for the other transparent alkali halides may have been corrected, but the amounts were not reported. Regarding MgO, the steady-state measurements are from Slack (1962), who also reported high values for sapphire (Fig. 4.12). Values on the web include 13.6 $mm^2 s^{-1}$, but this seems to be for ceramic. Thus, for transparent crystalline LiF and MgO, radiative transfer is very important and overshadows contact resistance. Length may be an issue because this affects transmission and thus ballistic transport (Chapter 7).

4.5.3 Effects of Spurious Radiative Transfer Alone

Coatings are deposited on samples by sputtering or spraying for the 3ω and LFA methods, so interface resistance should be unimportant in both techniques. Data on substrates, not thin films, are compared in Fig. 4.14, to further minimize effects of interfaces and sample size. Results from the 3ω method are 10% higher than from LFA, on average. Because density and heat capacity are well established for these materials, this increase is unconnected with converting from K to D. The samples are colorless and all have ballistic transfer at high T. Although the amounts of ballistic transfer at room temperature are low, and difficult to discern in the LFA curves (see Figure 2 in Hofmeister, 2010b), the amounts are not zero. Ballistic transfer exists at all finite temperatures, but whether it can be resolved in the $T-t$ curves depends on experimental parameters, such as the signal-to-noise ratio.

Materials which have large amounts of ballistic transfer in contact measurements (Figs. 4.12 and 4.13) should also have large amounts in 3ω measurements, that is, more than the ~10% average. Materials with low D and complex chemistry and structure (e.g., silicates and several oxide perovskites) are less impacted (Fig. 4.14). Smaller thicknesses used in Angstrom's method have less ballistic transport and should give lower, not higher, values

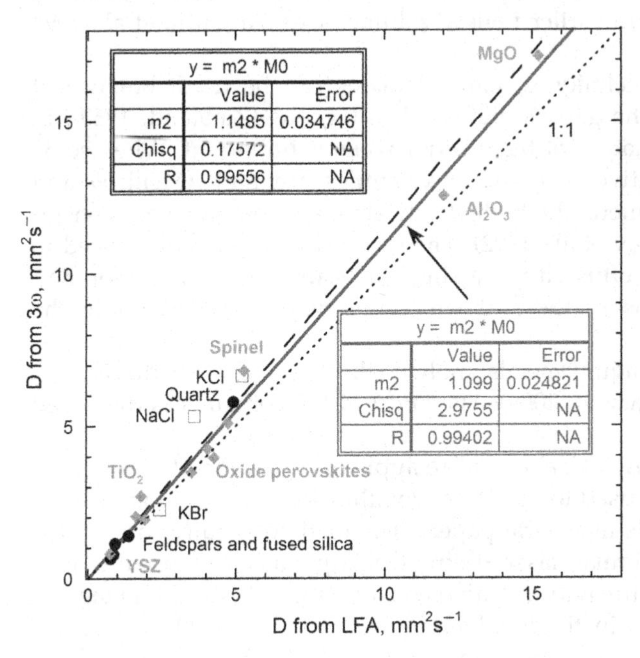

FIGURE 4.14 Comparison of results from the 3ω and laser-flash analysis methods at ambient temperature. Feldspars, fused silica, YSZ, alkali halides (listed by major component) from Cahill et al. (1992), where the latter were extrapolated from measurements at lower temperature and were not included in the fit to this dataset (dashed line). Gray triangles from Langenberg et al. (2016) for 0.5 mm thick samples. LFA from ~1 mm thick crystals are directional averages of compiled data (Hofmeister et al., 2014). TiO_2 is strongly anisotropic (1.40 and 2.56 $mm^2 s^{-1}$, perpendicular and parallel to the c-axis, respectively) as is quartz, whereas all other samples are cubic (MgO, spinel, glass, alkali halides, $SrTiO_3$, $KTaO_3$, Y-stabilized cubic zirconia, and LSAT perovskite) or weakly anisotropic. Synthetic spinels are disordered to varying degrees, so this comparison has some uncertainties.

than that for the ~1 mm samples used in LFA (Chapter 7). Spurious radiative effects may be larger than out ~10% estimate.

Ballistic radiative transfer exists in all periodic methods of partially transparent materials because the temperature is constantly changing (Fig. 4.1). Most of the energy is transferred via diffusion, but a small component is not. Ballistic transfer exists in steady-state measurements probably because equilibrium was not reached.

We did not find a sufficient amount of data to make either a length or a high-temperature comparison. Fig. 4.14 suggests that higher temperatures in 3ω measurements of partially transparent solids will overestimate *K* (and *D*) substantially, as observed for Ångström's and AC hot wire techniques (see e.g., figures in Hofmeister et al., 2007).

4.6 METHODS FOR MEASURING TRANSPORT PROPERTIES OF FLUIDS

Measurements of transport properties of fluids, which like those of solids, involve variations on a theme. The term "liquid" in this section denotes dense matter that flows under any stress. Silicate glasses are discussed in Chapter 10: for these, methods for *D* and *K* of solids apply. The term gas is used here to connote rarified matter, for example, fluids above the critical point.

4.6.1 Thermal Conductivity and Thermal Diffusivity

Many of the methods for solids, listed in Table 4.1, are also applied to liquids. For a recent summary of methods applied to liquids, see Nunes et al. (2010). Details of many experimental approaches are covered in earlier review volumes (see Vargaftik et al., 1994 and Section 4.1).

Among the conventional contact techniques, parallel plates can be used, but use of coaxial cylinders, with the liquid in the gap, is highly advantageous (Ziebland, 1969). In this radial geometry, only surface radiation of the exterior walls is important. When coaxial cylinders are used in a comparative method, uncertainties are high (McElroy and Moore, 1969). Hot wire and hot strip methods (both steady-state and periodic) are considered to be the most accurate (Phylippov et al., 1992). The two strip variant can be used to obtain *K* and *C*, and therefore, *D* and effusivity in a single measurement (Andersson and Bäckström, 1986). Many modifications of these above techniques can be found in the literature.

Transient LFA has been applied to liquid metals, molten silicates, and other fluids (e.g., Büttner et al., 1998; Stankus and Savchenko, 2009). The effect of the container is discussed below.

Periodic methods in addition to those of Table 4.1 are applied. Optical methods include forced Rayleigh scattering, which was used to ascertain *D* within 4%–11% of molten alkali halides to ~1200K and of other liquids in several papers. The method requires dye and is detailed by Nagasaka et al. (1992). A similar, laser grating technique has been used to measure thermal diffusivity and mass diffusivity simultaneously, which takes advantage of the fact that motions of the molecules in fluids advect much of the heat. However, this method is comparative, which reduces accuracy (Nunes et al., 2010).

Overall, methods providing K and D for fluids have similar limitations and systematic errors as those applied to solids. An advantage is that contact resistance is reduced because "gaps" are not present. However, the presence of the container affects many studies, requiring that conduction across the container be modeled. Ballistic radiation is more common than in solids, due to lack of grain boundaries. Due to this problem, some researchers process their data using coupled radiative–conductive models. As shown in Section 4.5, ballistic transfer exists in both periodic and steady-state methods. Random experimental errors are associated with nonisothermal fluid or probes; end effects, imperfect oscillations and imperfect geometry of the probe (Phylippov et al., 1992).

For laser-flash studies of liquids in containers, three-layer models are used, after Lee et al. (1978). This model does not include ballistic internal radiation. Liquid metals are opaque substances. Transparent Al_2O_3 containers are used. Thus, if the $D(T)$ data for sapphire are of the actual container, which many include some radiative transfer, then diffusion across the liquid metal can be accurately modeled. For other liquids, metal containers are used for which $D(T)$ is well established. Graphite is also possible, if values for the specific material are used, to eliminate variability known to exist in this material. Ballistic transport will exist across many liquids. To the best of our knowledge, a model to extract D that addresses both layering and ballistic transfer has not yet been developed. Heat transport properties of partially transparent silicate melts at high T may require experimental innovations, so that Hahn et al.'s (1997) model can be used.

Currently our laboratory measures D from plates of silicate glass as the material passes through the glass transition, up until the temperature where the material either flows or crystallizes (e.g., Pertermann et al., 2008). A wide temperature range is possible for high-viscosity silica-rich melts, but not for low viscosity basalts, which flow, or for samples with large amounts of water, which outgasses (e.g., Hofmeister et al., 2016).

For gases, fewer methods are available than for liquids or solids (Table 4.1). As in measurements of condensed matter, many variations exist on common themes (Phylippov et al., 1992). Gases are optically thin, so ballistic transport is possible. But because the experiments are conducted under low temperature gradients and mostly steady-state conditions to avoid convective instability, internal radiation is greatly reduced. Surface radiation still exists, but this is ammenable to modeling. Meeting stability requirements means that most studies of gas provide thermal conductivity.

4.6.2 Mass Diffusivity

Fick's Eqs. (3.10) parallel those of Fourier, but are simpler because only two physical properties pertain, concentration and mass diffusivity (D_m). With mass flux being directly and simply related to D_m, fewer types of measurements are needed to constrain mass transport.

Suárez-Iglesias et al. (2015) review data and methods for self-diffusion for 360 fluids, citing 575 papers, and tabulated numerical results from this immense body of work. Comparisons of datasets are included, along with discussion of uncertainties. Many materials such as noble gases and water have been studied multiple times, which leads to accurate values. Literature citations are provided of the diverse techniques, which are performed as a function of T and/or P. As in measurements of heat transport properties, the methods for mass diffusion are variations on a few themes. Two basic approaches exist:

Tracer methods were used exclusively before 1950, and are still used for rarified gas. These techniques introduce a small amount of an isotopic variant of the molecules of the fluid of interest. Various techniques are used, such as chromatography or interferometry. Most procedures are absolute (not requiring a calibration). In using Fick's Eqs. (3.10), an isotopic correction is made to obtain the mass self-diffusity:

$$D_{m,11} = D_{m,12}\sqrt{\frac{2M_2}{M_1 + M_2}} \tag{4.24}$$

where subscript 1 refers to the main species and 2 refers to the trace and M is the molecular mass. Eq. (4.24) pertains to low density, but is generally valid. Data on vapors and gases are in generally good agreement of 5%–10% (Suárez-Iglesias et al., 2015).

Nuclear magnetic resonance provides more accurate values of D_m, and does not require the above mass correction. However, reasonably large viscous resistance, and thus high density, is required, so that the gases are studied must be either under compression or held at cryogenic temperatures. A great number of liquids have been studied using virtually the same spin-echo technique as the original studies of Hahn, Carr, and Purcell (Suárez-Iglesias et al., 2015). The NMR method requires use of a standard. Water is most frequently used, due to the availability of very precise and reliable data from tracer studies at moderate temperatures and atmospheric pressure (under saturated conditions). Additional data on water that are less accurate but cover a wide T and P range, or data on benzene, which is the second most frequently studied liquid, are used to extend the conditions or to provide cross-checks.

Neutron scattering is also used. Suárez-Iglesias et al. (2015) point out that only the data on hydrogen are a valid measurement. These are within 10% of tracer methods. Otherwise, this technique is comparative with considerable modeling.

4.6.3 Viscoscity

Because viscosity of liquids, dilute and dense gas, and mixtures are important to engineering applications, considerable data exist. This property is governed by entirely different equations. For Newtonian fluids, which are the simplest flow behavior and appropriate to many materials, shear stresses (σ_{xy}) arise during viscous drag:

$$\sigma_{xy} \equiv \upsilon\rho\frac{\partial u_x}{\partial y} = \frac{F_{\text{drag}}}{\text{Å}} \tag{4.25}$$

where υ is kinematic viscosity, Å is area, u is speed, and y is the direction perpendicular to the flow that is along x. Thus, if stress (or force per area) is applied in a controlled manner, measuring velocity for a parcel fluid of known density yields its kinematic viscosity. Instruments are known as viscometers. Many reviews exist in earth science for methods applied to dense matter: these deferred to Chapter 10, where the data are discussed.

As with mass diffusivity, viscosity becomes increasingly difficult to measure as density decreases. Calibrations are based on condensed matter: unsurprisingly, the common standard is water, which is 100 times more viscous than gas at ambient conditions. For gases, measurement techniques are based on pressure-driven or shear-driven gas flow. Capillary tubes are often used and laminar flow is generally assumed. Minimizing or avoiding friction is important for accuracy. The coefficient of viscosity is derived either from the mass flux, or the time for objective to fall a specific distance, or the damping of some oscillating resonator. Even for commonly studied gases, such as CO_2, pressures in the experiments are above ambient and temperatures are below 700K. To explore rarified conditions, recent developments utilize laser spectroscopic methods (Gao et al., 2016) or frictionless diamagnetic levitation of plates which are spun inside the medium (Shimokawa et al., 2016).

References

Aggarwal, R.L., Ripin, D.J., Ochoa, J.R., Fan, T.Y., 2005. Measurement of thermo-optic properties of $Y_3Al_5O_{12}$, $Lu_3Al_5O_{12}$, $YAlO_3$, $LiYF_4$, $LiLuF_4$, BaY_2F_8, $KGd(WO_4)_2$, and $KY(WO_4)_2$ laser crystals in the 80–300 K temperature range. J. Appl. Phys. 98, 103514.

Andersson, S., Bäckström, G., 1986. Techniques for determining thermal conductivity and heat capacity under hydrostatic pressure. Rev. Sci. Instrum. 57, 1633–1639.

Andre, S., Degiovanni, A.A., 1995. Theoretical study of the transient couple conduction and radiation heat transfer in glass: phononic diffusivity measurements by flash technique. Int. J. Heat. Mass. Trans. 38, 3401–3414.

Ångström, A.J., 1861. Neue methode, das wärmeleitungsvermögen der körper zu bestimmen. Ann. Phys. 114 (4), 513–530 (Berlin, Ger.) (Translation: New method of determining the thermal conductibility of bodies. Philos. Mag. 25 (166) (1863) 130–142).

ASTM, 2013. ASTM E1461-13, Standard Test Method for Thermal Diffusivity by the Flash Method. ASTM International, West Conshohocken, PA. Available from: www.astm.org.

Azumi, T., Takahashi, Y., 1981. Novel finite pulse-width correction in flash thermal diffusivity measurement. Rev. Sci. Instru. 52, 1411–1413.

Baba, T., Ono, A., 2001. Improvement of the laser flash method to reduce uncertainty in thermal diffusivity measurements. Meas. Sci. Technol. 12, 2046–2057.

Bauer, M., Marienfeld, A., Aeschlimann, M., 2015. Hot electron lifetimes in metals probed by time-resolved two-photon photoemission. Prog. Surface Sci. 90, 319–376.

Beck, P., Goncharov, A.F., Struzhkin, V.V., Militzer, B., Mao, H.K., Hemley, R.J., 2007. Measurement of thermal diffusivity at high pressure using a transient heating technique. Appl. Phys. Lett. 91, 181914.

Belling, J.M., Unsworth, J., 1987. Modified Angström's method for measurement of thermal diffusivity of materials with low conductivity. Rev. Sci. Inst 58, 997–1102. Available from: https://doi.org/10.1063/1.1139589.

Berman, R., 1976. Thermal Conduction in Solids. Clarendon Press, Oxford, UK.

Berman, R., Brock, J.C., 1965. The effect of isotopes on lattice heat conduction. I. Lithium fluoride. Proc. R. Soc. London A 289, 46–65.

Berman, R.G., Brown, T.H., 1985. Heat capacity of minerals in the system Na_2O-K_2O-CaO-MgO-FeO-Fe_2O_3-Al_2O_3-SiO_2-TiO_2-H_2O-CO_2: representation, estimation, and high temperature extrapolation. Contrib. Mineral. Petrol. 89, 168–180.

Blumm, J., Lemarchand, S., 2002. Influence of test conditions on the accuracy of laser flash measurements. High Temp. High Pres. 34, 523–528.

Blumm, J., Opfermann, J., 2002. Improvement of the mathematical modeling of flash measurements. High Temp. High Pres. 34, 515–521.

Blumm, J., Henderson, J.B., Nilson, O., Fricke, J., 1997. Laser flash measurement of the phononic thermal diffusivity of glasses in the presence of ballistic radiative transfer. High Temp. High Pres. 34, 555–560.

Borca-Tasciuc, T., Chen, G., 2004. Experimental techniques for thin-film thermal conductivity characterization. In: Tritt, T.M. (Ed.), Thermal Conductivity. Kluwer Academic/Plenum Publishers, New York, NY, pp. 205–237.

Branlund, J.M., Hofmeister, A.M., 2007. Thermal diffusivity of quartz to 1000 degrees C: Effects of impurities and the α-β phase transition. Phys Chem. Minerals. 34, 581–595.

Bräuer, H., Dusza, L., Schulz, B., 1992. New laser flash equipment LFA 427. Interceram 41, 489–492.

Büttner, R., Zimanowski, B., Blumm, J., Hagemann, L., 1998. Thermal conductivity of a volcanic rock material (olivine-melilitite) in the temperature range between 288 and 1470 K. J. Volcan. Geotherm. Res. 80, 293–302.

Cahill, D.G., 1990. Thermal conductivity measurement from 30 to 750 K: the 3ω method. Rev. Sci. Inst 61, 802–808.

Cahill, D.G., Pohl, R.O., 1987. Thermal conductivity of amorphous solids above the plateau. Phys. Rev. B 35, 4067–4073.

Cahill, D.G., Watson, S.K., Pohl, R.O., 1992. Lower limit of thermal conductivity of disordered solids. Phys. Rev. B. 46, 6131–6140.

Campbell, R.C., 2010. Approximations in the use of two and three layer analysis models in flash diffusivity measurements. Thermal Conduct. 30, 328–338.

Cape, J.A., Lehman, G.W., 1963. Temperature and finite-pulse effects in the flash method for measuring thermal diffusivity. J. Appl. Phys. 34, 1909–1913.

Carslaw, H.S., Jaeger, J.C., 1959. Conduction of Heat in Solids, second ed. Oxford University Press, New York, NY.

Clark III, L.M., Taylor, R.E., 1975. Radiation loss in the flash method for thermal diffusivity. J. Appl. Phys. 46, 714–719.

Chen, L., Limarga, A.M., Clarke, D.R., 2010. A new data reduction method for pulse diffusivity measurements on coated samples. Comput. Mater. Sci. 50, 77–82.

Corbin, S.F., Turriff, D.M., 2012. Thermal diffusivity by the laser flash method. In: Kaufmann, E.N. (Ed.), Characterization of Materials. John Wiley and Sons, Inc, New York, NY, pp. 1–10. Available from: https://doi.org/10.1002/0471266965.

Cowan, D.R., 1961. Proposed method of measuring thermal diffusivity at high temperatures. J. Appl. Phys. 32, 1363–1370.

Cowan, D.R., 1963. Pulse method of measuring thermal diffusivity at high temperatures. J. Appl. Phys. 34, 926–927.

Criss, E.M., Hofmeister, A.M., 2017. Isolating lattice from electronic contributions in thermal transport measurements of metals and alloys and a new model. Int. J. Mod. Phys. B 31. Available from: https://doi.org/10.1142/S0217979217502058. No. 175020.

Czichos, H., Saito, T., Smith, L., 2006. Springer Handbook of Materials Measurement Methods. Springer, Würzburg, Germany.

Danielson, G.C., Sidles, P.H., 1969. Thermal diffusivity and other non-steady-state methods. In: Tye, R.P. (Ed.), Thermal Conductivity, vol. 2. Academic Press, London, UK, pp. 149–201.

Degiovanni, A., Andre, S., Maillet, D., 1994. Phonic conductivity measurement of a semi-transparent material. In: Tong, T.W. (Ed.), Thermal Conductivity, vol. 22. Technomic, Lancaster, PN, pp. 623–633.

Duquesne, J.Y., Fournier, D., Frétigny, C., 2010. Analytical solutions of the heat diffusion equation for 3ω method geometry. J. Appl. Phys. 108, 086104.

Fei, Y., 1995. Thermal expansion. In: Ahrens, T.J. (Ed.), Mineral Physics and Crystallography. A Handbook of Physical Constants. American Geophysical Union, Washington, DC, pp. 29–44.

Flynn, D.R., 1969. Measurement of thermal conductivity by steady-state methods in which the sample is heated directly by passage of an electric current. In: Tye, R.P. (Ed.), Thermal Conductivity, vol. 1. Academic Press, London, UK, pp. 241–300.

Fried, E., 1969. Thermal conduction contribution to heat transfer at contacts. In: Tye, R.P. (Ed.), Thermal Conductivity, vol. 2. Academic Press, London, UK, pp. 253–275.

Gao, R.-K., Sheehe, S.L., Kurtz, J., O'Byrne, S., 2016. Measurement of gas viscosity using photonic crystal fiber. 30th International Symposium on Rarefied Gas Dynamics. AIP Conf. Proc. No. 1786. https://doi.org/10.1063/1.4967601

Goncharov, A.F., Beck, P., Struzhkin, V.V., Haugen, B.D., Jacobsen, S.D., 2009. Thermal conductivity of lower-mantle minerals. Phys. Earth Planet. Inter. 174, 24–32.

Gorbatov, V.I., Polev, V.F., Korshunov, I.G., Taluts, S.G., 2012. Thermal diffusivity of iron at high temperatures. High Temp. 50, 292–294.

Hahn, O., Hofmann, R., Raether, F., Mehling, H., Fricke, J., 1997. Transient heat transfer in coated diathermic media: a theoretical study. High Temp. High Press. 29, 693–701.

Hammerschmidt, U., Sabuga, W., 2000. Transient hot wire (THW) method: uncertainty assessment. Int. J. Thermophys. 21, 1255–1278.

Hao, Y.-H., Neuwmann, M., Enss, C., Fleischmann, A., 2004. Contactless technique for thermal conductivity measurement at very low temperature. Rev. Sci. Instr. 75, 2718–2725.

Hemberger, F., Göbel, A., Ebert, H.P., 2010. Determination of the thermal diffusivity of electrically nonconductive solids in the temperature range from 80K to 300K by laser-flash measurement. Int. J. Thermophys. 31, 2187–2200.

Henderson, J.B., Giblin, F., Blumm, J., Hagemann, L., 1998a. SRM1460 series as a thermal diffusivity standard for laser flash instruments. Int. J. Thermophys. 19, 1647–1656.

Henderson, J.B., Hagemann, L., Blumm, J., 1998b. Development of SRM8420 series electrolytic iron as a thermal diffusivity standard. Netzsch Applications Laboratory Thermophysical Properties Section Report No. I-9E.

Hofmeister, A.M., 2006. Thermal diffusivity of garnets at high temperature. Phys. Chem. Minerals 33, 45–62.

Hofmeister, A.M., 2007a. Pressure dependence of thermal transport properties. Proc. Natl Acad. Sci. USA 104, 9192–9197.

Hofmeister, A.M., 2007b. Thermal diffusivity of aluminous spinels and magnetite at elevated temperature with implications for heat transport in Earth's transition zone. Am. Mineral. 92, 1899–1911.

Hofmeister, A.M., 2009. Comment on "measurement of thermal diffusivity at high pressure using a transient heating technique" in [Appl. Phys. Lett. 91, 181914 (2007)]. Appl. Phys. Lett. 95, 096101.

Hofmeister, A.M., 2010a. Scale aspects of heat transport in the diamond anvil cell, in spectroscopic modeling, and in Earth's mantle. Phys. Earth Planet. Inter. 180, 138–147. Available from: https://doi.org/10.1016/j.pepi.2009.12.006.

Hofmeister, A.M., 2010b. Thermal diffusivity of perovskite-type compounds at elevated temperature. J. Appl. Phys. 107, 103532. Available from: https://doi.org/10.1063/1.3371815.

Hofmeister, A.M., 2014. Thermal diffusivity and thermal conductivity of single-crystal MgO and Al_2O_3 as a function of temperature. Phys. Chem. Miner. 41, 361–371.

Hofmeister, A.M., Branlund, J.M., 2015. Thermal conductivity of the Earth. In: Schubert, G., Price, G.D. (Eds.), Treatise in Geophysics, vol. 2. Mineral Physics, second ed. Elsevier, Amsterdam, The Netherlands, pp. 584–608.

Hofmeister, A.M., Pertermann, M., 2008. Thermal diffusivity of clinopyroxenes at elevated temperature. Eur. J. Mineral. 20, 537–549.

Hofmeister, A.M., Pertermann, M., Branlund, J.M., 2007. Thermal conductivity of the Earth. In: Schubert, G. (Ed.), Treatise in Geophysics, vol 2. Mineral Physics (edited by G.D. Price). Elsevier, The Netherlands, pp. 543–578.

Hofmeister, A.M., Whittington, A.G., 2012. Effects of hydration, annealing, and melting on heat transport properties of fused quartz and fused silica from laser-flash analysis. J. Non-Cryst. Solids 358, 1072–1082.

Hofmeister, A.M., Dong, J.J., Branlund, J.M., 2014. Thermal diffusivity of electrical insulators at high temperatures: evidence for diffusion of phonon-polaritons at infrared frequencies augmenting phonon heat conduction. J. Appl. Phys. 115. Available from: https://doi.org/10.1063/1.4873295. No. 163517.

Hofmeister, A.M., Sehlke, A., Avard, G., Bollasina, A.J., Robert, G., Whittington, A.G., 2016. Transport properties of glassy and molten lavas as a function of temperature and composition. J. Volcanol. Geotherm. Res. 327, 380–388.

Hofmann, R., Hahn, O., Raether, F., Mehling, H., Fricke, J., 1997. Determination of thermal diffusivity in diathermic materials by laser-flash technique. High Temp.- High Press. 29, 703–710.

Hust, J.G., Lankford, A.B., 1984. Update of thermal conductivity and electrical resistivity of electrolytic iron, tungsten, and stainless steel. National Bureau of Standards Spec. Pub. 260–290, 1–71.

Josell, D., Warren, K., Czairliyan, A., 1995. Correcting an error in Cape and Lehman's analysis for determining thermal diffusivity from thermal pulse experiments. J. Appl. Phys. 78, 6867–6869.

Klemens, P.G., Williams, R.K., 1986. Thermal conductivity of metals and alloys. Int. Metals Rev. 31, 197–215.

Kogure, Y., Mugishima, T., Hiki, Y., 1986. Low-temperature thermal diffusivity measurement by laser-flash method. J. Phys. Soc. Jpn. 55, 3469–3478.

Koh, Y.K., Cahill, D.G., 2007. Frequency dependence of the thermal conductivity of semiconductor alloys. Phys. Rev. B 76, 075207.

Kunc, T., Lallemand, M., Saulnier, J.B., 1984. Some new developments on coupled radiative-conductive heat transfer in glasses: experiments and modeling. Int. J. Heat Mass Trans. 27, 2307–2319.

Lange, R.A., 1997. A revised model for the density and thermal expansivity of K_2O-Na_2O-CaO-MgO-Al_2O_3-SiO_2 liquids from 700 to 1900 K, extension to crystal magmatic temperatures. Contrib. Mineral. Petrol. 130, 1–11.

Langenberg, E., Ferreiro-Vila, E., Leborán, V., Fumega, A.O., Pardo, V., Rivadulla, F., 2016. Analysis of the temperature dependence of the thermal conductivity of insulating single crystal oxides. APL Mater. 4, No. 104815.

Lasjaunias, J.C., Ravex, A., Vandorpe, M., 1975. The density of low energy states in vitreous silica: specific heat and thermal conductivity down to 25 mK. Solid State Commun. 17, 1045–1049.

Laubitz, M.J., 1969. Measurement of the thermal conductivity of solids at high temperature by using steady-state linear and quasi-linear heat flow. In: Tye, R.P. (Ed.), Thermal Conductivity, vol.1. Academic Press, London, UK, pp. 111–183.

Lazard, M., André, S., Maillet, D., 2004. Diffusivity measurement of semi-transparent media: model of the coupled transient heat transfer and experiments on glass, silica glass and zinc selenide. Int. J. Heat Mass Trans. 47, 477–487.

Lee, H.L., Hasselman, D.P.H., 1985. Comparison of data for thermal diffusivity obtained by laser-flash method using thermocouple and photodector. J. Am. Ceram. Soc. 68, C12–13.

Lee, H.J., Taylor, R.E., 1976. Thermal diffusivity of dispersed composites. J. Appl. Phys. 47, 148–151.

Lee, T.Y.R., Donaldson, A.B., Taylor, R.E., 1978. Thermal diffusivity of layered composites. Thermal Conduct. 15, 135–148.

Lim, K., Kim, S., Chung, M., 2009. Improvement of the thermal diffusivity measurement of thin samples by the flash method. Thermochim. Acta 494, 71–79.

Liu, L.H., Tan, H.P., 2004. Transient temperature response in semitransparent variable refractive index medium subjected to a pulse irradiation. J. Quant. Spectrosc. Radiat. Transfer 83, 333–344.

Lyeo, H.-K., Cahill, D., 2006. Thermal conductance of interfaces between highly dissimilar materials. Phys. Rev. B 73, 144301.

Maglić, K.D., Taylor, R.E., 1992. The apparatus for thermal diffusivity by the laser-flash method. In: Maglić, K.D., Cezairliyan, A., Peletsky, V.E. (Eds.), Compendium of Thermophysical Property Measurement Methods, vol. 2, Recommended Measurement Techniques and Practices. Plenum Press, New York, NY, pp. 281–314.

Maglić, K.D., Cezairliyan, A., Peletsky, V.E. (Eds.), 1984. Compendium of thermophysical property measurement methods. vol. 2, Recommended Measurement Techniques and Practices. Plenum Press, New York, NY.

Manthilake, G.M., deKoker, N., Frost, D.J., McCammon, C.A., 2011. Lattice thermal conductivity of lower mantle minerals and heat flux from Earth's core. Proc. Natl Acad. Sci. 108, 17901–17904.

Massard, H., Pinto, C.S.C., Couto, P., Orlande, H.R.B., Cotta, R.M., 2004. Analysis of flash method physical models for the measurement of thermal diffusivity of solid materials. Proceedings of the 10th Brazilian Congress of Thermal Sciences and Engineering, No. CIT04–0537.

McElroy, D.L., Moore, J.P., 1969. Radial heat flow methods for the measurement of the thermal conductivity of solids. In: Tye, R.P. (Ed.), Thermal Conductivity, vol.1. Academic Press, London, UK, pp. 185–239.

McMasters, R.L., Beck, J.V., Dinwiddie, R.B., Wang, H., 1999. Accounting for penetration of laser heating in flash thermal diffusivity experiments. J. Heat Trans. 121, 15–21.

Mehling, H., Hautzinger, G., Nilsson, O., Fricke, J., Hofmann, R., Hahn, O., 1998. Thermal diffusivity of semitransparent materials determined by the laser-flash method applying a new mathematical model. Inter. J. Thermophys. 19, 941–949.

Milošević, N., 2010. Application of the laser pulse method of measuring thermal diffusivity to thin alumina and silicon samples in a wide temperature range. J. Therm. Sci. 14, 417–423.

Min, S., Blumm, J., Lindemann, A., 2007. A new laser flash system for measurement of the thermophysical properties. Thermochimica Acta 455, 46–49.

Monaghan, B.J., Quested, P.N., 2001. Thermal diffusivity of iron at high temperature in both the liquid and solid states. ISIJ Int. 41, 1524–1528.

Nagasaka, Y., Nakazawa, N., Nagashima, A., 1992. Experimental determination of the thermal diffusivity of molten alkali halides by the forced Rayleigh scattering method. I. Molten LiCl, NaCl, KCl, RbCl, and CsCl. Int. J. Thermophys. 13, 555–574.

Narasimhan, T.N., 2010. Thermal conductivity through the 19th century. Phys. Today 63, 36–41. Available from: https://doi.org/10.1063/1.3480074.

Nunes, V.M.B., Lourenco, M.J.V., Santos, F.J.V., Lopes, M.L.S.M., Nieto de Castro, C.A., 2010. Accurate measurement of physicochemical properties on ionic liquids and molten salts. In: Gaune-Escare, M., Seddon, K.R. (Eds.), Molten Salts and Ionic Liquids. John Wiley and Sons, New York, NY, pp. 229–263.

Ohta, H., Shibata, B.H., Waseda, Y., 2002. Evaluation of the effective sample temperature in thermal diffusivity measurements using the laser flash method. Int. J. Thermophys. 23, 1659–1668.

Okuda, Y., Ohta, K., Yagi, T., Sinmyo, R., Wakamatsu, T., Hirose, K., et al., 2017. The effect of iron and aluminum incorporation on lattice thermal conductivity of bridgmanite at the Earth's lower mantle. Earth Planet. Sci. Lett 474, 25–31.

Paddock, C.A., Easley, G.L., 1986. Transient thermoreflectance from thin metal films. J. Appl. Phys. 60, 285–290.

Palik, E.D., 1998. Handbook of Optical Constants of Solids. Academic Press, San Diego, CA.

Parker, W.J., Jenkins, R.J., 1962. Thermal conductivity measurements on bismuth telluride in the presence of a 2 MeV electron beam. Adv. Energy Convers. 2, 87–103.

Parker, W.J., Jenkins, R.J., Butler, C.P., Abbot, G.L., 1961. Flash method of determining thermal diffusivity, heat capacity, and thermal conductivity. J. Appl. Phys. 32, 1679.

Pertermann, M., Hofmeister, A.M., 2006. Thermal diffusivity of olivine-group minerals. Am. Mineral. 91, 1747–1760.

Pertermann, M., Whittington, A.G., Hofmeister, A.M., Spera, F.J., Zayak, J., 2008. Thermal diffusivity of low-sanidine single-crystals, glasses and melts at high temperatures. Contrib. Mineral. Petrol. 155, 689–702. Available from: https://doi.org/10.1007/s00410-007-0265-x.

Phylippov, L.P., Kravchun, S.N., Tleubaev, A.S., 1992. Apparatus for measuring thermophysical properties of liquids by AC hot-wire techniques. In: Maglić, K.D., Cezairliyan, A., Peletsky, V.E. (Eds.), Compendium of Thermophysical Property Measurement Methods, Vol. 2, Recommended Measurement Techniques and Practices. Plenum Press, New York, NY, pp. 375–405.

Powell, R.W., 1969. Thermal conductivity determinations by thermal comparator methods. In: Tye, R.P. (Ed.), Thermal Conductivity, vol. 2. Academic Press, London, UK, pp. 275–338.

Ray, L., Förster, H.J., Schilling, F.R., Förster, A., 2006. Thermal diffusivity of felsic to mafic granulites at elevated temperatures. Earth. Planet. Sci. Lett. 251, 241–253.

Reif-Acherman, S., 2014. Early and current experimental methods for determining thermal conductivities of metals. Int. J. Heat Mass Trans. 77, 542–563.

Richet, P., 1987. Heat capacity of silicate glasses. Chem. Geol. 62, 111–124.

Ross, R.G., Andersson, P., Sundqvist, B., Bäckström, G., 1984. Thermal conductivity of solids and liquids under pressure. Rep. Prog. Phys. 47, 1347–1402.

Rudkin, R.L., Jenkins, R.J., Parker, W.J., 1962. Thermal diffusivity measurements on metals at high temperatures. Rev. Sci. Instrum. 33, 21–24.

Schilling, F.R., 1999. A transient technique to measure thermal diffusivity at elevated temperature. Eur. J. Mineral. 11, 1115–1124.

Schmidt, A.J., Chen, X., Chen, G., 2008. Pulse accumulation, radial heat conduction, and anisotropic thermal conductivity in pump-probe transient thermoreflectance. Rev. Sci. Instrum. 79, 114902.

Shibata, H., Ohta, H., Suzuki, A., Waseda, Y., 2000. Applicability of platinum and molybdenum coatings for measuring thermal diffusivity of transparent glass specimens by the laser flash method at high temperatures. Mater. Trans., JIM 41, 1616–1620.

Shimokawa, Y., Matsuura, Y., Hirano, T., Sakai, K., 2016. Gas viscosity measurement with diamagnetic-levitation viscometer based on electromagnetically spinning system. Rev. Sci. Instrum. 87, No. 125105.

Slack, G., 1962. Thermal conductivity of MgO, Al_2O_3, $MgAl_2O_4$, and Fe_3O_4 crystals from 3° to 300°K. Phys. Rev. 126, 427–441.

Srinivasan, N.S., Xiao, X.G., Seetharaman, S., 1994. Radiation effects in high-temperature thermal diffusion measurements using the laser-flash method. J. Appl. Phys. 75, 2325–2331.

Stankus, S.V., Savchenko, I.V., 2009. Laser flash method for measurement of liquid metals heat transfer coefficients. Thermophys. Aeromech. 16, 589–592.

Suárez-Iglesias, O., Medina, I., Sanz, M., Pizarro, C., Bueno, J.L., 2015. Self-diffusion in molecular fluids and noble gases: available data. J. Chem. Eng. Data 60, 2757–2817. Available from: https://doi.org/10.1021/acs.jced.5b00323.

Taylor, R., 1980. Construction of apparatus for heat pulse thermal diffusivity measurements from 300–3000K. J. Phys. E: Sci. Instrum. 13, 1193–1199.

Taylor, R.E., Maglić, K.D., 1984. Pulse method for thermal diffusivity measurement. In: Maglić, K.D., Cezairliyan, A., Peletsky, V.E. (Eds.), Compendium of Thermophysical Property Measurement Methods. Vol. 1, Survey of Measurement Techniques. Plenum Press, New York, NY, pp. 305–336.

Thermitus, M.A., 2010. New beam size correction for thermal diffusivity measurement with the flash method. In: Gaal, D.S., Gaal, P.S. (Eds.), Proceeding of the 30th International Thermal Conductivity Conference and the 18th International Thermal Expansion Symposium. DEStech Publications, Lancaster, PA, pp. 217–225.

Tischler, M., Kohanoff, J.J., Ranguni, G.A., Ondracek, G., 1988. Pulse method of measuring thermal diffusivity and optical absorption depth for partially transparent materials. J. Appl. Phys. 63, 1259–1264.

Touloukian, Y.S., Powell, R.W., Ho, C.Y., Klemens, P.G., 1970. Thermal Conductivity: Non-Metallic Solids. IFI/Plenum, New York, NY.

Touloukian, Y.S., Powell, R.W., Ho, C.Y., Nicolaou, M.S., 1973. Thermal Diffusivity. IFI/Plenum, New York, NY.

Tritt, T.M. (Ed.), 2004. Thermal Conductivity: Theory, Properties, and Applications. Kluwer Academic/Plenum Publishers, New York, NY.

Tritt, T.M., Weston, D., 2004. Measurement techniques and considerations for determining thermal conductivity of bulk materials. In: Tritt, T.M. (Ed.), Thermal Conductivity. Kluwer Academic/Plenum Publishers, New York, NY, pp. 187–203.

Tye, R.P. (Ed.), 1969. Thermal Conductivity, vols. 1–2. Academic Press, London, UK.

Vandersande, J.W., Pohl, R.O., 1980. Simple apparatus for the measurement of thermal diffusivity between 80–500 K using the modified Angstrom method. Rev. Sci. Instrum. 51, 1694–1699.

Vargaftik, N.B., Phylippov, L.P., Tarzimanov, A.A., Totskii, E., 1994. Handbook of Thermal Conductivity of Liquids and Gas. CRC Press, Boca Raton, FL (translator: Yu.A. Gorshkov).

VEM, 2017. Thin Film Evaporation Guide. https://www.vem-co.com/guide (accessed 01.11.17).

Ventura, G., Perfetti, M., 2014. Thermal Properties of Solids at Room and Cryogenic Temperatures. Springer, Dordrecht, The Netherlands.

Vozár, L., Hohenauer, W., 2003/2004. Flash method of measuring the thermal diffusivity: a review. High Temp. High Press. 35/36 (3), 253–264.

Vozár, L., Hohenauer, W., 2005. Uncertainty of thermal diffusivity measurements using the laser flash method. Int. J. Thermophys. 26, 1899–1915.

Watt, D.A., 1966. Theory of thermal diffusivity by pulse technique. Br. J. Appl. Phys. 17, 231–240.

White, G.K., 1969. Measurement of solid conductors at low temperature. In: Tye, R.P. (Ed.), Thermal Conductivity, vol.1. Academic Press, London, UK, pp. 69–109.

Xu, Y., Shankland, T.J., Linhardt, S., Rubie, D.C., Langenhorst, F., Klasinski, K., 2004. Thermal diffusivity and conductivity of olivine, wadsleyite, and ringwoodite to 20 GPa and 1373 K. Phys. Earth Planet. Inter. 143–144, 321–326.

Yagi, T., Ohta, K., Kobayashi, K., Taketoshi, N., Hirose, K., Baba, T., 2011. Thermal diffusivity measurement in a diamond anvil cell using a light pulse thermoreflectance technique. Meas. Sci. Technol. 22, 024011.

Young, D.A., Thomsen, C., Grahn, H.T., Maris, H.J., Tauc, J., 1986. Heat flow in glasses on a picosecond timescale. In: Anderson, A.C., Wolfe, J.P. (Eds.), Phonon Scattering in Condensed Matter V, 1986. Springer-Verlag, Berlin, Germany, pp. 49–51.

Yu, X., Hofmeister, A.M., 2011. Thermal diffusivity of alkali and silver halides. J. Appl. Phys. 109, 033516. Available from: https://doi.org/10.1063/1.3544444.

Zawilski, B.M., Iv, R.T.L., Tritt, T.M., 2001. Description of the parallel thermal conductance technique for the measurement of the thermal conductivity of small diameter samples. Rev. Sci. Instrum. 72, 1770–1774.

Zhao, D., Qian, X., Gu, X., Jajja, S.A., Yang, R., 2016. Measurement techniques for thermal conductivity and interfacial thermal conductance of bulk and thin film materials. J. Electron. Packag. 138, 040802.

Ziebland, H., 1969. Experimental determination of the thermal conductivity of fluids. In: Tye, R.P. (Ed.), Thermal Conductivity, vol. 2. Academic Press, London, UK, pp. 185–239.

CHAPTER

5

Reconciling the Kinetic Theory of Gas With Gas Transport Data

OUTLINE

Measurements, Mechanisms, and Models of Heat Transport
DOI: https://doi.org/10.1016/B978-0-12-809981-0.00005-X

Use of the computer substitutes specific knowledge for understanding. You understand a subject when you have grasped its structure, not when you are merely informed of specific numerical results. **H.R.** ***Post, Inaugural lecture, Chelsea College, London 1974 as quoted by C. Truesdell (1984).***

On a microscopic scale, a system or a material consists of an enormous number of atoms or molecules, each of which can exist in ground and/or excited states with different energies. The micro- and macroscopic approaches differ greatly (Zemansky and Dittman, 1981) because the former requires many assumptions about the existence and nature of matter, and utilizes many quantities which cannot be directly measured. The constraints available from macroscopic measurements represent spatial or temporal averages of these quantities which involve a large number of particles. Through this connection, depiction of the locally occurring processes can be verified (or refuted) if the ambiguities inherent to complex models are taken into account.

For the microscopic picture, the relevant branch of modern science is statistical mechanics. Regarding modern heat transfer, the central concepts stem from classical efforts to describe the behavior of gases and interaction of this type of matter with heat. These ideas were developed before the structure of the atom was understood and before light and heat were recognized to be the same phenomenon. Given this history, imperfections in the old ideas are expected. The present chapter demonstrates that the kinetic theory of gas, which underlies all microscopic models of heat flow, has several shortcomings.

Modern modeling has not uncovered these historical errors, despite recent computational advancements that have promoted the popularity of the microscopic approach. A significant problem is that when parameters are multiplied, calculations can appear to agree with data even if the physics is incorrect (Transtrum et al., 2015; Fig. 3.5). The numerous, unmeasurable quantities incorporated in microscopic models provide considerable opportunity to "match" data using incorrect equations, all while greatly increasing the complexity of the overarching models.

Historic and modern microscopic models are based on gas behavior which is less complex than behavior of condensed matter. Matter and heat-energy move largely together in gases, but not in solids. The linkage in gases between mass flow and energy transfer permitted development of thermodynamic principles based on extremely limited information. Furthermore, thermodynamics has been centered on the ideal gas, which consists entirely of hypothetical point masses, because this assumption reduces mathematical complexity. The importance of the ideal gas to thermal physics is exemplified by its temperature being numerically equivalent to Kelvin's absolute temperature scale. Yet, the ideal gas is unlike any real material, and so this simplification comes at the cost of lost information and possible errors.

Importantly, the kinetic theory of gas, in assuming perfectly elastic collisions, cannot describe changes in temperature with time (Criss and Hofmeister, 2017). This peculiarity stems from historical developments: this kinetic model portrays conditions in gas experiments of the 1800s, which were steady-state with very low temperature gradients. These conditions were selected to avoid convective motions (e.g., Purrington, 1997). Importantly, steady-state conditions cannot reveal how systems progress towards equilibrium, which is the essence of heat transfer. In addition, point masses cannot collide. The latter inconsistency was circumvented by Maxwell and Boltzmann through considering atoms as small hard spheres. However, an additional process is associated with spheres of finite size: namely, frictional dissipation (viscous drag) during near misses. This type of nonelastic behavior was not included in the classical models.

Besides incorrectly portraying collisions as elastic, the kinetic theory of gas does not account for the one-dimensional nature of heat flow (Chapter 3) nor does it allow for photon participation in heat transfer. Yet, blackbody radiation ubiquitously emanating from all bodies (Chapter 1, Section 5.1.4) means that photons continuously leave any surface. The kinetic theory of gas (Fig. 5.1A and B) considers heat as being carried solely by translations of atoms (or molecules). These shortcomings require revision of the historical approach.

Brush (1976) provides details on the development of the kinetic theory of gas. Purrington (1997) provides a concise summary. This chapter focuses on the formulations of Clausius and Maxwell, which underlie modern models, and expands upon the findings of Criss and Hofmeister (2017) that were introduced in Chapter 3. Section 5.1 describes the

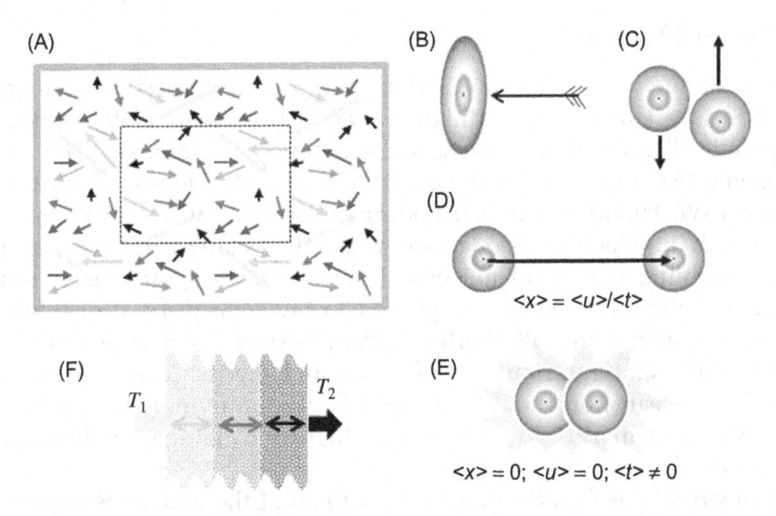

FIGURE 5.1 Schematics illustrating the kinetic theory of gas. (A) Body of gas with one kind of atom, but with a distribution of translational velocities. The average velocity is related to the temperature. The dotted perimeter box indicates a representative subvolume of the material. Collisions with the walls of the container (gray box) are viewed to constrain gas temperature, whereas ubiquitous radiation from the walls is not considered in the kinetic theory of gas. (B) Analogy of collisions with archery. A very tiny bullseye is difficult to hit. A point, which actually has no area, cannot be struck by an arrow if this also has no area (i.e., is a line segment). (C) Near misses of gas atoms with finite size. (D) Depiction of collisions with no interactions: for this case, distance, speed and time are simply related. (E) Depiction of a collision involving interaction of the particles: for this case, the duration of the interaction is important. (F) Gas subject to a low temperature gradient (to avoid convection) in steady-state. Temperatures at the ends are maintained by the walls. The gas can be considered as a series of narrow strips. As the width of each strip becomes infinitesimal, each strip can be considered as being in local thermal equilibrium with a characteristic velocity and temperature.

underpinnings of the kinetic theory of gas. Section 5.2 provides the mathematics for the historical model, which is based on monatomic gas and uses average properties. Effects of inelastic collisions, viscous drag, and energy reservoirs in addition to translations are addressed. Section 5.3 qualitatively accounts for the effects of rotations and vibrations of molecules in a gas. Section 5.4 compares the model to data on monatomic gases and gases with increasingly complex molecules, in order to evaluate its accuracy and the need for amendments.

5.1 COLLISIONS IN MONATOMIC GAS AND IMPLICATIONS FOR HEAT TRANSFER

Visualizing and quantifying the behavior of atoms or molecules in gas during collisions depends on a description of the atomic (or molecular) structure. The kinetic theory of gas assumes elastic collisions of point masses but incorporates cross-sections for head-on collisions of small hard spheres. Problems associated with this inconsistent depiction and general implications stemming from the inelastic nature of collisions are covered below.

5.1.1 Timeline of Discovery

The structure of the atom was unveiled in part through spectroscopic studies from ~1860 to ~1913 (see Chapter 2, Partington, 1960; McGucken, 1969; Purrington, 1997). During this period all aspects were debated, sometimes vociferously. The turning point was J.J. Thomson's 1897 discovery of the electron in the cathode ray apparatus developed by Plücker circa 1859. Thomson's crucial finding obliterated the vortex theory of the atom championed by Kelvin, which envisioned atoms as interlocking rings composing a continuous, nonrigid medium, but it did not end debates regarding atoms and structure. For example, Mach and Ostwald remained opposed to the concept of atoms, which contributed to Boltzmann's suicide in 1906 (Porter, 1998). Discovery of radioactivity by Becquerel in 1896 and the important contributions of Marie and Pierre Curie lead to Rutherford's suggestion of the nuclear atom in 1911. Proposals of atomic models shortly followed in 1913 by Niels Bohr and in 1916 by G.N. Lewis. In 1919, I. Langmuir distinguished ionic and covalent bonds.

The kinetic theory of gas was developed prior to all of the above discoveries, including discovery of the electron. Model development began in ~1738 with Daniel Bernoulli's explanation of Boyle's law (Purrington, 1997). At this time, Leonhard Euler, who was as much a physicist as a mathematician, derived an equation of state (EOS) for gas from kinetic theory (Truesdell, 1984). Contributions by Herapath and Waterston were overshadowed by the key papers of Clausius (1857) and Maxwell (1867), which lead to Boltzmann's tome (translated by Sharp and Matschinsky, 2015). Clausius supported the concept of molecules, whereas Maxwell waffled on atomic theory, describing the constituents of gases as particles. These particles were initially considered point masses.

Eventually, Boltzmann proposed that molecules were rigid spheres which collided elastically. However, Maxwell was greatly concerned about the hypothesis of elastic collisions up to his death in 1879 (Niven, 1890). This concern was justified, as discussed below.

The main goal early-on was to address the data on specific heats of gases, compiled in 1828 by Dulong and Petit. As the work progressed, the focus turned to heat transfer. This shift occurred through the interactive debates of Clausius and Maxwell, and also through their responses to critics. However, when Boltzmann developed his work, Clausius was in ill health and refused to communicate with Maxwell, who was focused on electromagnetism anyways. Moreover, Maxwell was active for only about a year after publication of Boltzmann's work. As stated by Gibbs, Clausius' approach was mechanical whereas Maxwell and Boltzmann emphasized statistics. Yet, as stated by Maxwell, neither he nor Boltzmann understood each other's way of thinking (Purrington, 1997).

The present chapter is semi-classical, and considers continuum behavior, although discretization is introduced at the end of this chapter to understand conditions under which this behavior becomes important. A continuum approach is justified because (1) the translations of atoms in monoatomic gas are not discretized and (2) a blackbody curve (Fig. 2.3) is described mathematically by a continuous function. Moreover, the classical, continuum formulation of Wien (Eq. (2.4)) is a good approximation overall: fits at various temperatures and in frequency and wavelength domains are shown in Halliday and Resnick (1966) and in Chapters 2 and 8.

5.1.2 Pseudopoint Masses and Elastic Collisions

The kinetic theory of gas involves a statistical analysis which assumes:

1. Some average speed (u) describes motions of the gas molecules, and their temperature.
2. The molecules are very small compared to the volume occupied by the gas.
3. Some average distance (the mean free path, Λ) describes travel of molecules between collisions.
4. Time *during* collisions is negligible compared to the lifetime (τ) *between* collisions.
5. Collisions are perfectly elastic (hard sphere molecules).
6. Thermal conductivities sum (underlies computing an average).

In summary, assumptions #1 and #3 are valid, but the formula pertains to spheres and limits energy to translations. Problems with assumption #6 were addressed in Chapter 3 and are discussed further below. Assumption #5 is invalid for the reasons described below, which makes assumption #4 problematic. Assumption #4 also implicitly assumes that forces act only during collisions, which requires that assumption #2 be valid, and that forces are short-range, which describes neither Coulombic nor gravitational interactions. Assumption #2 seems problematic from our addition to the third law in Chapter 1 which states that empty space cannot exist.

Point masses have no area and so cannot collide (Fig. 5.1B). This fault is known and has thus far been addressed by considering atoms to be hard, small spheres, which requires introducing an additional parameter (a cross-section) into the kinetic theory of gas, and also assumes that the atoms are small compared to the volume per atom in space (Halliday and Resnick, 1966). Reif (1965) discusses a simple model for collisions of hard spheres. This fix is insufficient. Although not previously recognized, spheres having finite size will also interact if they touch while passing (drag during near misses: Fig. 5.1C) and thus viscosity involves more than the simple transfer of momentum of point masses envisioned by Maxwell (e.g., Reif, 1965). This modification (Sections 5.2.3 and 5.2.5) allows for inelastic near misses affecting viscosity, but does not remedy the crucial problem of inelastic collisions being required for temperature to evolve with time.

Although elastic collisions can move a parcel of heat from one end of the gas container to the other, they cannot alter the average molecular velocity, which is connected with the temperature of the gas (Clausius, 1857). For isothermal gas, the average velocity describes all properties (Fig. 5.1A). If a thermal gradient exists, which is necessarily imposed by exterior constraints, the material can be viewed as strips of matter in local equilibrium (Fig. 5.1F). Each narrow strip is described by a constant temperature. The temperature of the strips does not change and the system does not evolve if the sole mechanism for heat exchange is the elastic collisions of the atoms themselves.

Atoms are neither point masses nor hard, elastic spheres. Atoms instead have small, dense nuclei surrounded by light electrons, which can be depicted as shells of negative charge around a dense, positively charged nucleus. It follows that descriptions of atomic or molecular collisions must address the finite size and structural properties of atoms, which were debated after the lifetimes of Clausius, Maxwell, and Boltzmann. Neither light nor the relative motions of the atoms composing the molecules can be ignored regarding heat and its transport.

5.1.3 Inelastic Collisions of Deformable Atoms Provide a Photon Gas

As a faster molecule approaches a slower molecule, the strong Coulombic repulsions between their surrounding clouds of negative changes (Fig. 5.2A) increase in magnitude. The shapes of the clouds (or shells) change, that is, the atoms deform (Fig. 5.2A and B). The ease with which a molecule deforms is known as polarizability. Deformation of atoms during collisions in the gaseous state is studied both experimentally and theoretically (e.g., Brouard et al., 2016). Also, polarization of a molecule in response to visible light is essential for a vibration to be stimulated during Raman spectroscopy (e.g., Ferraro et al., 2003). Thus, inelasticity must affect gas transport properties.

Let us suppose the slower molecule is the locus of deformation: During the interaction, some kinetic energy of the faster molecule is consumed in deforming the slower molecule. In this brief interval, the faster molecule decelerates and reverses direction, while the slower molecule gains both kinetic and internal energy as it enters an excited state. Re-equilibration releases energy as heat: specifically, a low energy photon is emitted from the slower atom. A second photon is likely emitted from the faster atom, because this atom too would be deformed, consuming some kinetic energy. This semi-classical picture conserves energy and momentum. The deformation would be small and the photons would

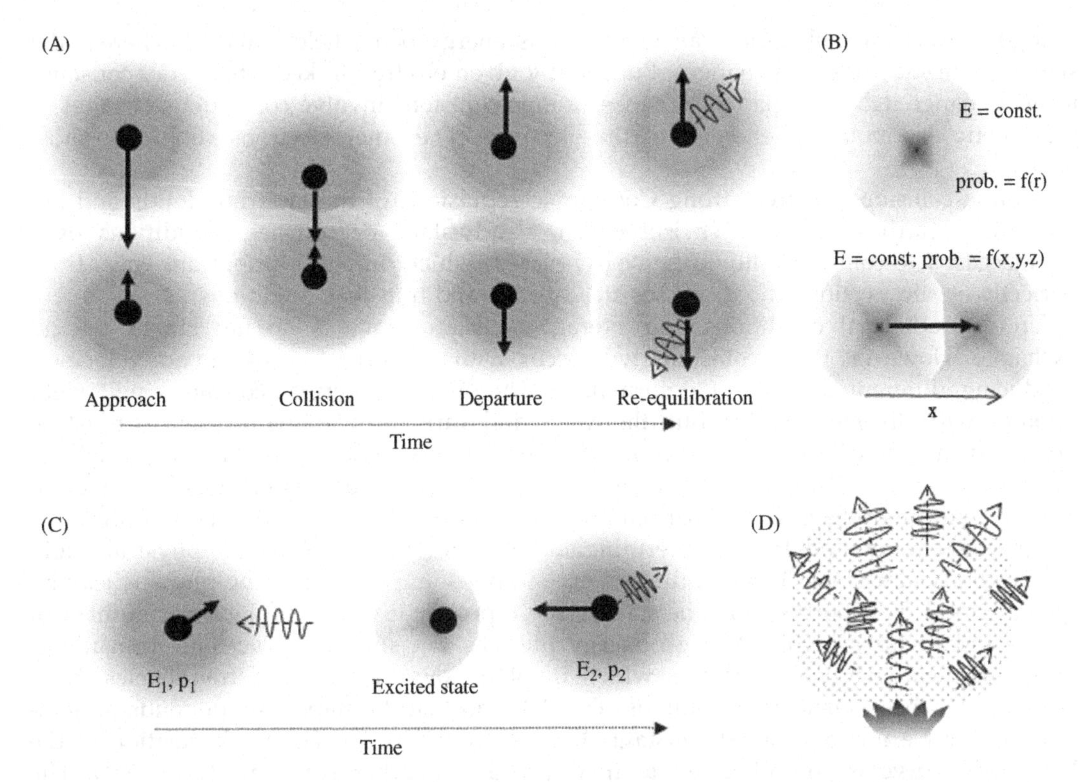

FIGURE 5.2 Schematic of dynamic interactions of finite size atoms in monatomic gas. (A) The progression of an inelastic collision with time. Black: nuclei. Gray: shells of electrons, forming a shielding cloud. Squiggle arrow: thermal photon. (B) Comparison of an isolated gas atom (top) with a colliding pair (bottom). The outer electrons of the isolated atom are in a spherically symmetric s-shell, which has a characteristic energy, but a radial probability distribution. For the colliding atoms, the reduction in distance along the x-direction increases the Coulombic repulsion and breaks the spherical symmetry. One can view this as the shells retaining a constant energy, but then the shape can no longer spherical, that is, the probability function has changed. (C) Interaction of a low energy photon with a gas atom as a function of time. The photon has momentum, and provides radiation pressure to the atom. Since pressure equals force/area, a force is applied, which can deform the electron cloud, due to low electron mass and outward location. Largely, the photon–atom collision is elastic, so the distributions of energy and momentum are little changed. (D) Steady-state conditions in a body of gas. Applied heat (the fire) establishes T and a distribution of molecular velocities. The velocity distribution produces a thermal photon gas with its distribution linked to the velocities. Thermal photons escape, providing blackbody emissions, with this same distribution. To maintain constant T of the matter gas, heat must be continually supplied.

be quite low energy. Furthermore, some of the photons released during re-equilibration would interact with other atoms, see below. This interaction is essential because otherwise a monatomic gas could not be heated by incoming light.

As indicted in Fig. 5.2B, the process of collisions does not involve a transition of an electron between levels, which typically involves energies in the visible region and discrete transitions. Importantly, the collision of a pair of atoms is a symmetry breaking operation, and so characteristics of the electronic ground state changes with time as the atoms approach each other. Because the change in time (or position) is continuous, so is the change in electron shells. One can view this as energy being held constant whereas the shape of the shell or cloud changes. But for any given electron to keep its energy constant, it must change its orbital path. Therefore, atomic collisions involve continuous changes in the kinetic and potential energies of the electrons. The outer electrons should be most affected.

Atomic collisions involve strong Coulombic repulsions of the electron clouds and are inelastic, regardless of whether isothermal, steady-state, or transient conditions exist for the gas. The collisions should be nearly elastic, which differs in important ways from perfectly elastic. Section 5.1.5 discusses steady-state and transient conditions.

Under isothermal conditions, the average collision provides no net kinetic energy exchange. However, these collisions will occur with a distribution of impact velocities, producing different degrees of deformation. Therefore, isothermal collisions will yield photons with diverse energies. But, the range and variation of photon energies must be described by a distribution that is compatible with the statistics of the kinetic energies of the gas molecules. Therefore, the average energy of the emitted photons must be linked to the temperature of the gas. In a thermal gradient, this would involve local equilibrium.

The existence of a photon gas intermingled with matter gas explains the physical origin of blackbody emissions. These emissions are simply some portion of the photons leaving the gas cloud. This portion must be statistically representative due to the sheer number of atoms and photons involved. High energy photons can stimulate electronic transitions. However, heat-energy is germane, which involves low energies and frequencies. Note that the blackbody curve is continuous (Fig. 2.3) and that its shape only depends on temperature and a few additional constants (Eqs. (2.3) and (2.4)). The peak position of the blackbody curve is controlled by temperature and another constant (Eq. (1.25)). The shapes from classical (Wien, 1893) and quantum physics (Planck, 1901) are quite similar (cf. Eqs. (2.3) and (2.4)). However, the basic shape predicted from the Rayleigh-Jeans formula is quite different, and does not correspond well with observations. The merits and limitations of the derivations of these formulae (mentioned in Chapter 1) are revisited in Chapter 8 which further addresses discrete (high energy) transitions.

Photon–photon collisions can be dismissed because photons lack mass; this dismissal is consistent with photons being points so that these collision cross-sections are negligibly small. However, the photon points can collide with the atoms of finite area (Fig. 5.2C). Photons have energy and momentum and therefore apply radiation pressure to a surface. Obviously, a photon striking an individual atom contributes a small force, which can redirect the translational motion of the impacted atom, and alter its kinetic energy. In this manner, the statistical distributions of the internal photon gas and the translations of the atoms are interdependent.

For a gaseous body such as a raincloud or nebula, some photons will escape from its surface, while other photons arrive from the surroundings. In this manner, the temperature of the gas is established (Fig. 5.2D). Under radiative equilibrium (i.e., isothermal conditions), photon influx equals photon outflux. Although atoms (or molecules) may also diffuse along the boundary, these are bound by gravity, surface tension, the relationship of the surface to phase equilibria, or by a container. Thus, matter escapes far less easily from a cloud than photons. To first order, heat flux balance involves only photons, due to their immense speed. This flux constitutes the blackbody emissions. The radiated photons have the same energy distribution as the retained photons because all photons have the same high speed (i.e., higher energy photons are not preferentially lost). The photon energy distribution (Fig. 2.3) is a direct consequence of the molecules having a certain distribution of kinetic energy and momentum. Distinct transitions at certain frequencies exist but are in addition because these are superimposed on the blackbody curve, see e.g., Chapters 2 and 8.

Thus, photons coexist with gas molecules in a nebula, cloud, or container (Fig. 5.2D). The statistical distribution of kinetic energy amongst the molecules determines the statistical distribution of the frequencies of the thermal photon assemblage: the photons represent the heat content of the gas, as reflected in the total radiance from the material (Eq. (1.24)) or the luminosity. Because the blackbody curve is described by a single temperature (Wien, 1893; Planck, 1901), the statistics of the photon gas is also represented by that same temperature. Temperature and heat are inseparable, but are not identical.

Importantly, gases do not reflect light strongly, in contrast to the surface of a condensed matter (Chapter 2). A small amount of reflection could occur if the indices of refraction of the gas and its surroundings differ, but this is mitigated by the indistinct surface of any gas. Water clouds in Earth's atmosphere do not provide a counter example, because the condensed droplets are the entity providing reflections. Internally to a gas, reflection is imperceptibly small, which makes the spectra, thermodynamics, and transport properties of a matter gas simple.

5.1.4 The Photon Gas and (Macroscopic) Thermodynamic Laws

Photons are not part of classical thermodynamics, but can explain several of the laws of thermodynamics when radiative transfer is taken into account (Text Box 5.1; Chapter 1). Importantly, inelastic collisions are required for thermal evolution. If instead collisions were perfectly elastic, the gas could maintain its temperature for an indefinite period of time, and so once a gas were heated, the heat source can be shut off and the gas would not cool. It should also be clear that Fourier's equations, which conserve heat-energy (the caloric model), are fully compatible with a moving photon gas.

Although this section considers monatomic gases, the findings must extend to other types of matter. Ubiquitous occurrence of the photon gas is compatible with Maxwell's (1871) remark: "All heat is of the same kind." A flux of photons on a surface of a material provides heat to that material. A flux of photons from the surface sheds heat to the surroundings. If the inbound flux is less than the outbound flux, the body cools.

5.1.5 The Internal Photon Gas and Microscopic Heat Conduction

As illustrated in Fig. 5.2A, a photon gas is produced by inelastic collisions of atoms in a gas. Under isothermal conditions discussed above, nothing stops the photons from leaving the matter gas, although departure may involve a series of interactions with atoms. In order to maintain isothermal conditions, heat must be added (e.g., radiation from container walls or more distant surroundings). The added photons (and the coexisting gas) must interact with the colliding molecules (Fig. 5.2C), otherwise, the average velocity of

TEXT BOX 5.1

THERMODYNAMIC LAWS RECAST IN TERMS OF MACROSCOPIC PHOTON BEHAVIOR.

First Law	Third Law
Photons are energy and can do work, or can be produced by work, and thus must be accounted for in conservation laws.	Photons are everywhere.

First + Third Law Corollary

A photon gas resides inside every assembly of matter due to atomic collisions being inelastic.

Second Law

Because the internal photon gas of a medium is not bound by gravity or other forces, it flows outward providing emissions, and as such prohibits reversible processes. The strong growth of emitted flux with temperature constrains the direction of heat flow.

Zeroth Law

Thermal equilibrium requires the same fluxes and thus the same temperature, allowing equilibrium to be communicated among multiple bodies.

the atoms would degrade progressively with each collision and the matter gas would cool regardless of heat received. Participation of the photon gas in heat transfer is essential.

Steady-state is closely related to isothermal conditions. The gradation of temperature under steady-state (Fig. 5.3A) can be viewed as a series of isothermal strips (Fig. 5.3B). However, no matter how many strips are considered, there still exists a temperature difference across each interface, and therefore each strip radiates to its cooler neighbor while receiving flux from its hotter neighbor (Fig. 5.3C). Actually, any given strip radiates in both directions, but the net flux is from hot to cold. The gas atoms are not physically contained within the strip. However, due to each strip having a representative temperature, and due to mean free paths being short ($\sim$200 pm; Section 5.2), the gas in each strip is effectively localized under local thermal equilibrium. As is long known, this thermal gradient must be very, very weak, or the gas would convect. Short times may be required to sustain any finite gradient. Under such circumstances, the ensemble of the strips (Fig. 5.3D) consists of gas atoms that fluctuate locally in space while the net flux of the photon gas is down the temperature gradient and on to the near and far surroundings. Although the gas molecules might provide a net flow which describes the gradient, this behavior requires that a return flow exists, that is, convection occurs. For this reason, gases can sustain only low temperature gradients in steady-state. Local fluctuations better describe the exchange of energy between molecules, and are required to provide the balance of energetics between

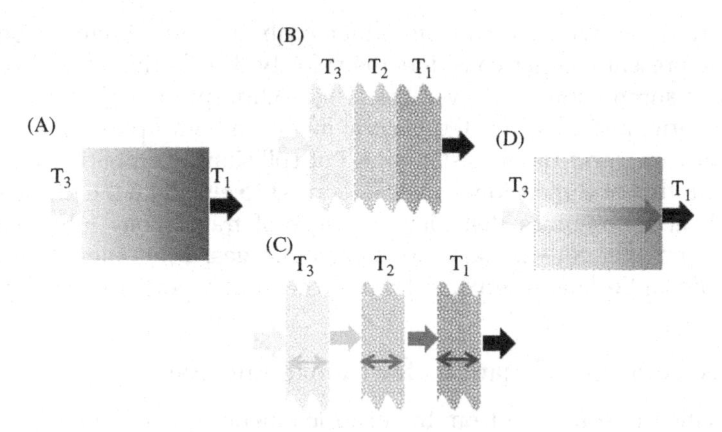

FIGURE 5.3 Steady-state conditions and behavior of the photon gas. (A) Schematic of a gas subject to a temperature gradient. The hot end is held at T_3, while the cold end is at lower temperature T_1. The temperatures cannot differ greatly or the gas will convectively circulate. (B) Division of the gas into isothermal strips. Three strips are shown, but the divisions can be many more to provide local equilibrium. (C) Pull-apart view of the interfaces. Each of the hotter strips must radiate to its colder neighbor. (D) Continuum view. With a large number of strips, the temperature changes continuously, and the photon gas flows continuously.

the matter gas and the photon gas. But, the photon gas differs substantially because it cannot be retained, and photons travel at speeds many orders of magnitude faster than molecular.

Thus, measurements of thermal conductivity in some matter gas under steady-state conditions actually monitor the continuous flow of the photon gas. Thermal conductivity of the gas (Eq. (3.1)) thus reflects its ability to convey a prescribed flux of photons.

Although the existence of a photon gas is based on behavior of colliding atoms in gas, this finding also applies to condensed matter. Otherwise, Fourier could not have devised a formula for steady-state macroscopic heat flow on the basis of the caloric model (Chapter 3). The importance of the photon gas to heat transport is underscored by laser-flash analysis experiments. In this transient method, a small increment of heat is remotely applied to one side of a sample and how sample temperature evolves is determined by sampling the blackbody emissions from the opposite side as a function of time (see Chapter 4). Therefore, the flow of photons describes all heat transfer, whether diffusive or ballistic, and regardless of whether the flow depends on time or is static, or involves high or low energy. This hypothesis will be evaluated through measurements as the book unfolds.

5.2 THE KINETIC THEORY OF MONATOMIC GAS

The existing model for heat transport, centers on translational motions of atoms in a gas. Although summations are the basis of the model, this does not require discretization, since integration is equivalent to summation with infinitely small steps. Moreover, averages are used in many cases, which pertain to either continuous or discrete distributions. For these reasons, the present chapter considers the velocity distribution to be a continuum, but for simplicity uses summations and averages. This section presents heat transfer equations in their simplest form, and modifies the historic model to both uphold the second law and to address viscous drag and the inelastic nature of collisions. Comparing the results of the kinetic theory with that of dimensional analysis is used to elucidate the physics.

The historical model assumes that kinetic energy of translations fully accounts for the thermal energy, and therefore pertains to monatomic gas. Molecules rotate and vibrate, which yield additional thermal reservoirs. Their effect on transport is covered in Section 5.3.

5.2.1 Previous Transport Equations for Monatomic Gas

The basic equation of transport from the historic kinetic theory of gas is:

$$K_{\text{gas}} = \frac{1}{3} N c_{\text{atom}} < u > \Lambda = \frac{1}{3} C < u > \Lambda_{\text{elast}} \tag{5.1}$$

where K is thermal conductivity, N is the number of atoms, c_{atom} is the heat capacity per atom (or molecule), the brackets designate an average, u is speed, and Λ is the mean free path. The subscript "elast" emphasizes that elastic collisions are presumed. The factor of $1/3$ originates in the assumption that in some volume, $1/3$ of the molecules are moving in each of the x, y, and z directions. However, this partitioning is recognized as being approximate. Reif (1965) and others provide a step-by-step derivation, which is not repeated here.

The above situation purportedly describes steady-state conditions with a low temperature gradient, which is required to prevent convection of gas. During steady-state conditions, temperature does not depend on time. As such, one might envision slight variations in T existing in all three directions. However, each representative volume (dotted box in Fig. 5.1A) will have a representative temperature. This realization has several implications:

Recasting the average of Eq. (5.1) to a summation yields:

$$K_{\text{gas}} = \frac{\sum u_i c_i \lambda_i}{3N} = \frac{\sum K_i}{3N} = \frac{\langle K_i \rangle}{3}; \text{ but } \frac{\langle K_i \rangle}{3} \equiv \frac{K}{3} \neq K \tag{5.2}$$

Thus, one supposition of the mean free gas model leads to a nonsensical result. The underlying problem is that averaging in three directions violates the second law. In a temperature gradient, heat can only move down the local gradient, which makes heat flow effectively one-dimensional: consider matrix diagonalization in Cartesian coordinates (also see Chapter 3). Therefore, translations of the molecules can only move heat in one direction, down the thermal gradient (Fig. 5.2C). Molecular collisions in the directions perpendicular to the gradient exist, but on average transmit no heat. In other words, the KE does not change and so the temperature does not change. Similarly, the net motion of photon gas is also down the thermal gradient.

The kinetic theory of gas describes the two other transport properties, mass diffusion and viscosity, which also are related to atoms moving. Derivations for viscosity and self-diffusivity provided by Reif (1965) are equivalent to Maxwell's approach. The result for the coefficient of viscosity is $\eta = 1/3 N m < u > \Lambda$, where m is atomic mass. For diffusion of mass, the formula is simpler, $D_{\text{mass}} = 1/3 u\Lambda$. From the definitions of thermal diffusivity

(Eq. (3.2); $K = C_P D$, where $C_P = \rho c_P$ is heat capacity on a per volume basis), and of kinematic viscosity ($\upsilon = \eta/\rho$):

$$D_{heat} = \upsilon = D_{mass} = \frac{1}{3} < u > \Lambda_{elast} \quad \text{(historic)} \tag{5.3}$$

The diffusivities (D_{heat}, D_{mass}) for monatomic gas are identical in the above historic formula because the atoms were presumed to carry all heat and all mass. Consequently, gases also diffuse momentum. This behavior is compatible qualitatively with experiments on gas, wherein gas viscosity *increases* with temperature (e.g., Kestin et al., 1984) because the mean free path becomes shorter (more collisions). The factor of three error exists in historic formulations for all three properties.

Importantly, diffusion of momentum does not describe all possible processes. Viscosity also represents impedance to motions, as was known to Newton, and as is evident in experiments in condensed matter, wherein viscosity *decreases* as temperature increases (see e.g., data on natural glasses and melts in Chapter 10: Dingwell, 1995). The historic kinetic theory of gas does not include deformation, which is an inelastic property, and certainly does not allow for breaking and reforming bonds, which is an essential process during physical flow of stiff liquids and weak solids (Fig. 5.4C).

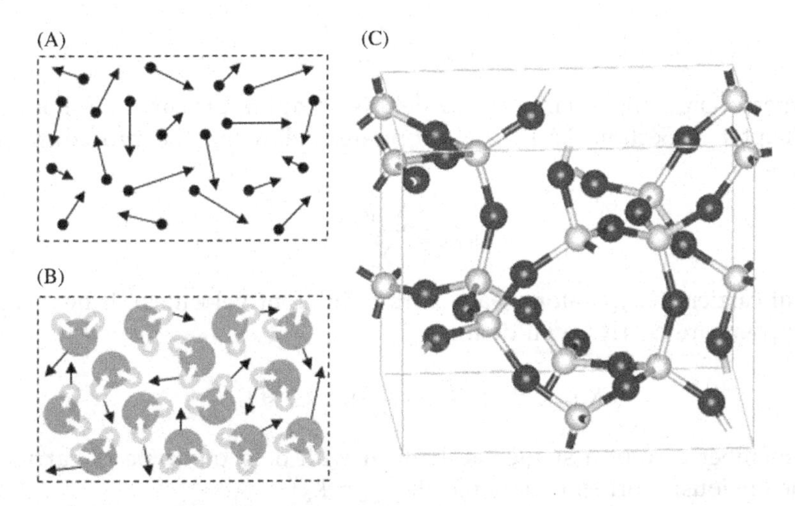

FIGURE 5.4 Schematics illustrating how transport differs among the three states of matter. Circles in various colors: atoms. Black arrows: direction of physical motion of atoms or molecules. White arrows: vibrations of atomic bonds. (A) Ideal gas, where large distances exist between atoms, so that interactions are limited to brief collisions and diffusion of mass is entirely kinetic and fully coupled to heat transfer. Motions occur even in the in the absence of organized flow. This depiction is a first approximation to real, monatomic gases. (B) Pourable liquid, where molecules are close together but motions are somewhat impeded by neighboring molecules. For simple molecules, energy exchanges are mostly kinetic, occurring via collisions, but due to proximity of molecules, energy of vibrations can be somewhat exchanged between molecules. As liquid density and the size (and complexity) of individual molecules increase, motions become more hampered and vibrations become increasingly important to the transfer of energy. (C) Ball-and-stick depiction of a crystalline solid, specifically SiO_2 with the keatite structure, taken from Creative Commons. Interatomic bonds (sticks) link constituent atoms (balls) in all directions, which severely limits motions. Vibrations are limited by symmetry. Atoms can diffuse towards defects: otherwise bonds must be broken and reformed to move atoms.

Along this vein, Eq. (5.3) was derived assuming point masses. An immediate problem, recognized historically, was that point masses cannot collide. The model morphed to consider collisions of hard spheres. Below, we show that the formulation for viscosity in the kinetic theory of gas is incomplete, because deformable spheres of finite size will not only collide, but they will also rub while passing (Fig. 5.1C).

5.2.2 Derivation of Transport Equations for Monatomic Gas by Conserving Energy

Transfer of heat down the temperature gradient in monatomic gas is analogous to transport involving parallel bars (Criss and Hofmeister, 2017) where each bar represents atoms with a particular velocity colliding with other atoms of a slightly lower velocity, and a single mechanism operates. The bars (Fig. 3.6) when "blended" provide the statistical assembly. The mechanisms are all the same, so the heat capacities are similar (although not identical). By considering the mechanisms to be independent, in which case, $C = \Sigma C_i$, Criss and Hofmeister (2017) derived Eq. (3.21) for similar heat capacities and diffusivities.

For a single mechanism, with a total thermal conductivity of K_{single}, the carriers interact. This requires a different form for the total heat capacity:

$$C_{\text{single}} = \frac{\sum C_i}{N} \tag{5.4}$$

Due to the interactions, single temperature field is an apt approximation. For this reason, Case 1 of Chapter 3, Section 3.5.1, is appropriate. Following the procedure leading to Eq. (3.21) gives:

$$K_{\text{single}} \cong \frac{\sum C_i K_i}{\sum C_i} \tag{5.5}$$

for N different carriers (i.e., N atoms each with a different velocities). If heat capacities of the various carriers are nearly equal then,

$$K_{\text{single}} = \frac{\sum K_i}{N} \equiv \langle K_i \rangle \text{ for } C_i \approx C_j \tag{5.6}$$

For the end-member case of a single mechanism with only one type of carrier, Eq. (5.6) reduces to the obviously correct equation of $K_{\text{single}} = K_i$.

Eq. (5.6) can also be derived from Eq. (3.23), which does not assume heat capacities are similar until the final step, and does not assume interactions are important or that thermal diffusivities are similar. These summations, which were derived by Criss and Hofmeister (2017) are more general than the conditions assumed in deriving them, because different approaches lead to the same result (Eq. (5.6)).

The above sum rules describe the known unidirectional nature of heat flow in solids (Fig. 3.6) during disequilibrium. The factor of ⅓ in historic models is clearly inappropriate to solids, which not only can sustain steep thermal gradients, but moreover their heat and mass transport are fully independent phenomena (Fig. 5.3) The factor of factor ⅓ should be omitted from Eqs. (4.1) to (4.3) and all other mechanistic models of microscopic processes. For gases, assuming similar heat capacities for molecules with different velocities, the correct mathematical form for the diffusivities is:

$$D_{\text{heat}} = D_{\text{mass}} = \langle u \rangle \Lambda = \frac{1}{N}\sum_{i=1}^{N} u_i \Lambda_i \tag{5.7}$$

Eq. (5.7) does not actually require monatomic gas or heat exchange by molecular collisions. Rather, the stipulation is that a single mechanism exists, which involves interacting carriers, and so the heat capacities of the individual carriers are similar (Criss and Hofmeister, 2017). For ideal monatomic gas, heat capacity is a constant (e.g., Fegley, 2015; Section 5.2.4) and therefore is the same for all atoms regardless of different velocities, so this criterion is clearly met.

5.2.3 Transport Equations for Ideal Gas From Dimensional Analysis

The averages discussed above can be derived independent of the kinetic theory by using dimensional analysis. Dimensional analysis of the one-dimensional heat equation (Eq. (3.3)) gives $D_{\text{heat}} = uL$ (Eq. (3.8)). Applying dimensional analysis to Fick's equation (Eq. (3.9), RHS), which has the same form as Fourier's equation, obviously gives $D_{\text{mass}} = uL$. These results have the same forms as taking an average of Eq. (5.7). Regarding viscosity, the historical model of gas, derived by Maxwell, considers momentum to be diffused (e.g., Reif, 1965; Cussler, 2009). Under this assumption, historic kinetic theory gives $\upsilon = D_{\text{heat}} = D_{\text{mass}}$.

Dimensional analysis of the heat equation (Eq. (3.3)) (for one-dimensional flow is consistent with the kinetic theory of gas if it is recognized that heat flow is directed down the temperature gradient. In this comparison, we do not consider the temperature dependence of the transport properties, but rather how they relate to other variables at any given temperature.

In addition, as mentioned in Section 5.1, an additional mechanism for viscosity exists when molecules have finite size. Viscosity measures resistance of a medium to flow. This concept predates the kinetic theory of gas. For Newtonian flow, which is the simplest type of flow, shear stresses (σ_{xy}) arise during viscous drag:

$$\sigma_{xy} \equiv \upsilon\rho \frac{\partial u_x}{\partial y} = \frac{F_{\text{drag}}}{\text{Å}} \tag{5.8}$$

where Å is area, u is speed, and y is the direction perpendicular to the flow (along x). Rearranging and applying dimensional analysis yields for any length-scale:

$$\upsilon = \frac{\partial y}{\partial u}\frac{m(\partial u/\partial t)}{\text{Å}(m/\text{Å}h)} = \frac{h}{u}\frac{u}{t}h = \frac{h}{t}h = u_{\text{flow}}h \tag{5.9}$$

where t is time (Hofmeister and Criss, 2018). For a microscopic scale, u = carrier speed and h = mean free path. The incorrect factor of ⅓ is not present because drag also occurs in one direction. Eq. (5.9) has the same form as Eqs. (5.7) and (3.3).

Although the velocities responsible for momentum diffusion and viscous drag must be identical, the mean free paths need not be the same because the processes differ. Rubbing of the molecules will retard diffusion of momentum. Because the average velocities are the same for both processes, mean free paths ($=u\tau$) add inversely, as occurs for lifetimes:

$$\frac{1}{\tau}=\frac{1}{\tau_{\text{mechanism-1}}}+\frac{1}{\tau_{\text{mechanism-2}}} \tag{5.10}$$

Summing reciprocals is a consequence of lifetimes being the inverse of probabilities, which sum by definition (e.g., Reif, 1965). Details are presented in Section 5.2.5, after quantifying the velocities.

5.2.4 Internal Energy and Translation Velocities of Monatomic Gas

Taken together, the experiments of Boyle, Charles, and Gay-Lussac show that the quantity PV/T is nearly constant for a fixed mass of gas (e.g., Halliday and Resnick, 1966). Their discoveries led to inferring the EOS for an ideal gas, $PV=N_M R_{gc} T$, where N_M is the number of moles. This well-known EOS is consistent with free expansion not changing the temperature of the ideal (or perfect) gas (also see Chapter 1, Section 1.2). This behavior leads to the internal energy for a collection only depending on temperature, that is, $E \propto T$ and $\partial E/\partial V|_T=0$ (Pippard, 1974). However, the proportionality constant cannot be constrained from thermodynamic relationships, although these relationships constrain many other physical properties. As Pippard (1974) stated, the thermodynamics are fully constrained by the EOS *and* a heat capacity. For similar reasons, the flow of heat also requires knowledge of the heat capacity (Criss and Hofmeister, 2017; Chapter 3, Section 3.3.4).

Formulae describing the thermodynamics of an ideal gas are very simple, for example: $\alpha_P=1/T$, $B_T=P$, and

$$C_P - C_V = N_M R_{gc} \tag{5.11}$$

Again, the thermodynamic relationships provide $E \propto R_{gc}T$, but constraining the numerical factor requires further considerations. (Below, heat capacity is provided on a per mole basis so that the equations presented are in their familiar forms.)

The historic approach to this problem of internal energy of the ideal gas underlies the kinetic theory of gas. Many textbooks (e.g., Halliday and Resnick, 1966) derive internal energy by relating pressure (=force/area) to the momentum of the gas atoms, and then taking averages. However, momentum is defined by $p=mu$, whereas energy is defined by $KE=\frac{1}{2}mu^2$. Hence, the ratio obtained depends on differences between the averages of $<u>^2$ and $<u^2>$, which obviously depend on the exact distribution of velocity over the parameter being averaged. Using $<u^2>^{1/2}$ to represent the average for momentum is inaccurate.

In the historic analyses, the average is taken over space or time, but as is implicit in Eq. (5.5) and as discussed below, the average temperature is *independent* of time. It is further assumed that the velocities have identical averages in all three directions, which is incorrect if a temperature gradient exists. The historic and textbook analyses also presume elastic collisions. Therefore, the previous derivation of internal energy for ideal gas is based on the limiting circumstances of isothermal conditions and certain velocity distributions. This unrecognized restriction clearly precludes proper formulation of time-dependent transport processes. Specifically, the historic approach requires defining a velocity distribution that adheres to $<u>^2 = <u^2>$, which then leads to an internal energy that is compatible with measured heat capacities. Obtaining these distributions was a focus of Maxwell and Boltzmann, which lead to the distribution function attached to their names. The next section presents a simple, direct approach based on classical physics, which does not require computing a statistical average.

5.2.4.1 Internal Energy of Monatomic Ideal Gas from the Virial Theorem

The Virial theorem (VT) of Clausius (1870) can be used to describe gas because restriction of matter to some volume is equivalent to forming a bound state (Hofmeister and Criss, 2016). Gas can be bound by gravity (e.g., a nebula), by a phase boundary (e.g., rainclouds), or by the walls of a container. The fluctuating motions of atoms or molecules in a gas are averaged over the space it occupies (dashed perimeter in Fig. 5.1A). Time averages are not relevant for a gas because the state variables in the EOS are macroscopic, static quantities. That is, the EOS describes a body of gas at any instant over some small space. This concept is akin to unit cells in crystallography.

An isothermal gas contained within a sphere of radius r has a surface area (Å) of $4\pi r^2$. Rotational motions of monotomic atoms pertain to interior configurations of electrons, so are not germane in our depiction of atoms as perfect spheres. Hence, the VT can be stated as $2<\mathrm{KE}> = rF$, where F is the force governing the translational motions and the energy is limited to the translations. Because pressure (P) equals F divided by area, the VT for a gas with only translational energy is:

$$2\langle \mathrm{KE}\rangle_{\mathrm{trans}} = rF = r\left(4\pi r^2 \frac{F}{\mathrm{Å}}\right) = 4\pi r^3 P = 3VP \quad (5.12)$$

(Hofmeister and Criss, 2016). Combining the EOS of ideal gas with Eq. (5.12) yields for all three translations:

$$\langle \mathrm{KE}\rangle_{\mathrm{trans}} - \frac{1}{2}m\langle u^2\rangle = \frac{3}{2}N_{\mathrm{M}}R_{\mathrm{gc}}T \quad (5.13)$$

which recapitulates the historical result for ideal, monatomic gas. From Eqs. (5.11), (4.13) and definitions, both heat capacities of an ideal gas are independent of T and P. Hence, the weighting factors of Eq. (5.5) cancel and Eq. (5.7) represents the diffusivities of a monatomic gas of hard spheres, noting that temperature independence is presumed.

Clausius (1870) did not derive Eq. (5.13) from Eq. (5.12), because he thought that the VT described the sum of the interior and exterior potentials. For a gas, these are respectively related to the interatomic forces and collisions with the walls of the container. Clausius did not realize that the VT pertains to only one (conservative) force and only one type of motion. This stipulation stems from the VT resting on the balance of linear momentum within a limited space, as demonstrated by Hofmeister and Criss (2016). Thus, the numerical factor and proportionality constant of Eq. (5.13) pertain only to translational motions of the atoms. This case describes monatomic gas, for which the internal energy equals simply the KE of translations. Polyatomic gases have vibrational and rotational motions, requiring additional Virial statements (Section 5.3).

The above derivation makes fewer restricting assumptions than the historical approach (see, e.g. Halliday and Resnick, 1966). In particular, a distribution function for the velocities is not specified. Isothermal conditions are not required, although some average temperature must describe the gas and the gas is required to be spatially contained (a bound state). Although radial symmetry was assumed, this is not essential, but rather is a convenience. The derivation does not set any limit on molecule size. It is limited to only one conservative force with one type of associated motion. Thus, inelastic production of heat is not included. For the model to reproduce measurements means that the collisions must be nearly elastic. Because only translations are considered, monatomic gas is assumed. The gas being ideal enters into the computation via the EOS, $PV = N_M R_{gc} T$, which reasonably approximates real gas, as discovered centuries ago.

5.2.4.2 Internal Energy of Monatomic van der Waals Gas From the VT

Van der Waals allowed for the effect of finite atomic volumes on the EOS. His well-known empirically deduced EOS is:

$$P = \frac{N_M R_{gc} T}{V - b} - \frac{a}{V^2} \text{ or } \left(P + \frac{a}{V^2}\right)(V - b) = N_M R_{gc} T \tag{5.14}$$

where the fitting parameter b is correction for finite atomic size and a provides a correction for intermolecular attractive forces. For one mole of atoms, b provides their volume under contact. The term a/V^2 is only important when the atoms are close together, e.g., at high pressure or low temperature, and represents an extra, internally applied pressure. For this reason, b and a are used as fitting parameters and may not be constant over a wide range of T and P conditions (see e.g., Halliday and Resnick, 1966).

Thermodynamic identities provide the difference in heat capacities (Berberan-Santos et al., 2008):

$$C_P - C_V = N_M R_{gc} \left[1 - \frac{2a}{V^3} \frac{(V-b)^3}{R_{gc} T}\right]^{-1} \tag{5.15}$$

Although the internal energy is not fully constrained, it clearly is more complicated than E of the ideal gas.

Applying the VT leads to:

$$E_{vdW} = \langle KE \rangle = \frac{1}{2} m \langle u^2 \rangle = \frac{3}{2} \frac{N R_{gc} T}{(1 - b/V)} - \frac{3}{2} \frac{a}{V} \tag{5.16a}$$

If b is small, Eq. (5.16a) can be linearized:

$$E_{vdW} = \langle KE \rangle = \frac{1}{2} m \langle u^2 \rangle \cong \frac{3}{2} N_M R_{gc} T \left(1 + \frac{b}{V}\right) - \frac{3}{2} \frac{a}{V} \tag{5.16b}$$

Essentially, at any given temperature, gas atoms that are hard spheres have higher velocities due to less volume being available, but this is offset by the change in the intermolecular forces. Eq. (5.16a) leads an exact result and also an approximation for small b:

$$C_V(T,V) = \frac{\partial E}{\partial T}\bigg|_V = \frac{3}{2}\frac{N_M R_{gc}}{(1-b/V)} \cong \frac{3}{2}N_M R_{gc}\left(1+\frac{b}{V}\right) \tag{5.17}$$

Summing Eqs. (5.17) and (5.15) gives C_P. Although C_V from Eq. (5.17) simple and similar to the ideal das, C_P depends on temperature and volume in a complex manner due to the parameter a. Crucially, the term a/V in Eq. (5.16b) is negative which leads to an impossibly negative KE as $T \to 0$K. This term is not compatible with the second and third laws.

Because collisions are the basis of kinetic theory, the effect of parameter b, which is related to the size of the atoms, should be the most important. With this simplification, the weighting factors cancel in Eq. (5.5), and $C_P - C_V = NR_{gc}$ for a monatomic gas composed of small, hard spheres. This result matches Eq. (5.11) for point masses, but reveals nothing new. Regarding the parameter a, avoiding negative KE as $T \to 0$K provides the stipulation:

$$a < \frac{N_M R_{gc} T V^2}{V-b} \lesssim N_M R_{gc} T b \tag{5.18}$$

Using this approximation removes the temperature dependence of C_P, and also the need for weighting, but makes both terms for C_P depend on volume. The limit is:

$$C_P = \frac{3}{2}\frac{N_M R_{gc}}{(1-b/V)} + \frac{N_M R_{gc}}{\left[1 - 2N_M (V-b)^2\right]} \tag{5.19}$$

5.2.5 Mean Free Paths for Collisions and Drag of Hard Spheres

5.2.5.1 Collision Probabilities

A lifetime can be estimated by considering collisions of hard spheres in "empty" space (Fig. 5.3A) by comparing the average volume available to each atom to its cross-section. The atomic cross-section is related to the physical size of the atoms. The underlying assumptions are that (1) the physical size of the atom is much smaller the separation of atoms and (2) collisions involve rigid spheres. These assumptions are the source of imperfections in the theory, as follows:

The cross-section for elastic collisions of two equal size hard spheres is $2\pi(r_{atom})^2$. As discussed by Reif (1965), the average and relative velocities enter into computing cross-sections, but these roughly cancel. Viewing the atoms as a projectile and a bullseye on a target (Fig. 5.1B), and using the EOS for 1 mole of ideal gas ($PV = N_M R_{gc} T$) yields the mean free path:

$$\Lambda_{ideal} = \frac{V^*}{2\pi r_{atom}^2} = \frac{k_B T}{2\pi r_{atom}^2 P} \tag{5.20}$$

where V^* is the volume of 1 mole of atoms. The LHS used the relationship $k_B = N_a R_{gc}$ where N_a is Avogadro's number. Under the hard sphere approximation, the radius is assumed to be constant and does not depend on V, T, or P. This simplifying assumption is suspect for two reasons. First, the location of electrons involves probabilities, so the radius is not fixed, and actually extends to infinity, with a diminishing probability. Second, from the third law, extended (Chapter 1, Section 1.2), matter fills space. Probably, the nucleus is unchanged.

However, the electron cloud expands such that the size is limited by Coulombic repulsions of other atoms in the gas, and thus the probability is null before infinity is reached.

To address soft spheres, the van der Waals EOS with $a \sim 0$ is first considered. For spherical atoms, $b = 4\pi r^3_{\text{atom}}/3$. To circumvent void space existing between spheres, tiny cubes are considered, for which $b = 8r^3_{\text{atom}}$. These two approximations to the shape provide:

$$\frac{k_B T}{2\pi r^2_{\text{atom}} P} + \frac{2}{3} r_{\text{atom}} < \Lambda_{\text{hard}} < \frac{k_B T}{2\pi r^2_{\text{atom}} P} + \frac{4}{\pi} r_{\text{atom}} \tag{5.21}$$

Because $2/3$ and $4/\pi$ bracket unity, the size to be exclused is represented simply by r_{atom}.

Regarding the effect of intermolecular forces, Eq. (5.18) indicates that another fitting parameter (ι) is needed. Hence:

$$\Lambda_{\text{vdW, heat or mass}} = \iota \frac{k_B T}{2\pi r^2_{\text{atom}} P} + r_{\text{atom}}; \quad \iota \sim < 1 \tag{5.22a}$$

The mean free path of Eqs. (5.20)–(5.22a) provides an average value, as does Eq. (5.13). The statistics are implicit in the temperature. However, is the hard sphere term actually needed? Typical values of r_{atom} are $\sim 2 \times 10^{-10}$ m (Mantina et al., 2009). The term $k_B T(2\pi r^2_{\text{atom}} P)$ is $\sim 2 \times 10^{-7}$ m at ambient conditions, and thus $\iota \sim 10^{-3}$ is required for these terms to be similar, which is unexpected. Thus, the main effect of atoms having finite size is to provide a cross-section for collisions. The hard sphere limit is negligible except at very low temperature, where the material is not a gas, but is condensed matter. The additive term r_{atom} can be dropped from Eq. (5.22a).

5.2.5.2 *Viscous Drag*

The additional process associated with near misses (Fig. 5.1C) has the same characteristic speed as diffusion (Eq. 4.8) but a different mean free path. If one visualizes two atoms passing, then the drag is described by the area of contact, πr^2_{atom}. Hence, the cross-section is not twice the area of the hard spheres as in collisions, but simply the area. The same volume pertains. Because the inverse of the mean free paths add (from Eq. (5.10)), combining the above stipulations and using results from Section 5.2.5.1 yields for both processes combined:

$$\Lambda_{\text{vdW, viscosity}} = \iota \frac{k_B T}{3\pi r^2_{\text{atom}} P}; \quad \iota \sim < 1 \tag{5.22b}$$

The correction term of r_{atom} is omitted because this is negligible at ambient conditions.

The above does not account for the expected generation of heat during drag. This omission parallels neglecting the production of heat during direct collisions.

5.2.6 Diffusivities and Kinematic Viscosity of Monatomic Gases

From dimensional analysis and the above:

$$D_{\text{heat}} = D_{\text{mass}} = \frac{3}{2} \upsilon \tag{5.23}$$

Combining Eqs. (5.7), (5.13), and (5.22a,b), and recognizing that the hard atom limit is negligibly small under most conditions, gives for small, hard monatomic spheres:

$$D_{\text{calc}} \cong \iota \frac{k_B T}{2\pi r_{\text{atom}}^2 P}\sqrt{\frac{3N_M R_{gc} T}{m}}; \quad \iota \sim 1 \tag{5.24}$$

Neither inelastic collisions, nor the existence of a photon gas were considered in deriving Eq. (5.24). The effects of atoms not being hard spheres and collisions not being elastic are next explored.

5.2.6.1 Effect of Atoms Being Fuzzy

Atomic size is the only input parameter that is uncertain in Eq. (5.24), but it is accompanied by the fitting parameter (ι) which is near unity. For the monatomic elements, atomic sizes represent atoms in very dense gas or molecular solids, whereby the atoms effectively in contact (e.g., Mantina et al., 2009), which requires either low temperatures or high compression. Regardless of the details, the radii of monatomic atoms in rarefied gas will be larger than those of compressed gas. Behavior of ions in crystals support this deduction. It is well-known that ionic radii are affected by coordination, such that their values are small for more restricted spaces and large for more open spaces in crystalline environments (e.g., Shannon and Prewitt, 1969). The importance of coordination to ionic atomic radii is analogous to the ions filling the available space. This explanation is consistent with our amendment to the third law (Chapter 1) which states that perfect vacuums do not exist. This deduction is also logical given that atoms consist of a nucleus with a cloud of moving electrons. Due to the softness of atoms, the prediction of Eq. (5.24) will exceed measured values.

Thus, r_{atom} for a neutral gas atom or molecule should depend on the space available. Because both r_{atom} and the fitting parameter ι are equivocal and combine in Eq. (5.24) as ι/r_{atom}^2, their effects cannot be distinguished in any comparison (see Fig. 3.3 or Transtrum et al., 2015). From the above discussion, and the fact that a (or ι) are only important for high density, parameter ι is set to unity in Eq. (5.24) and r_{atom} is considered to be a sloppy parameter that depends on V, and thus on T (or on T and P, if the latter pair are the independent thermodynamic variables). Dimensionally, r_{atom} is proportional to $V^{1/3}$. The same size may not pertain to the diffusivities and viscosity, since the latter involves additional, glancing collisions.

5.2.6.2 Effect of Collisions Occurring Over Non-Negligible Intervals of Time

Eqs. (5.22b) and (5.24) concern the mean free path, which is the product of the translational velocity (u) with the time *between* collisions (Fig. 5.1E). Another time interval pertains: specifically, the time *during* collisions (Fig. 5.1F). The kinetic theory of gas assumes that the latter is negligible in comparison, which is sensible for elastic collisions. However, inverse lifetimes add (Eq. (5.10)) so the "negligible" lifetime is the important lifetime even in the elastic model!

Inelastic collisions involve an excited state (Fig. 5.2) where the atoms interact over a finite duration: specifically, inelasticity is associated with a "delay" before the atoms resume their motions. Larger atoms should interact over a longer time. But the translational speed is not associated with the excited state, and so the mean free path *between* collisions is not affected by the interactions. Because the theory is cast in terms of the mean free path, an effective velocity is germane. Offhand, one would think that the effective velocity is lower than the translational velocity computed from the internal energy (Eq. (5.13)), to account for the inelastic "delay." But this deduction utilizes u from $\Lambda = u\tau$, and assumes that the lifetimes add, which is wrong: the delay has no associated velocity (Fig. 5.1).

If an effective translational velocity indeed pertains, the temperature dependence of measured transport properties will not follow the predicted power of $T^{3/2}$. Because inverse lifetimes add (Eq. (5.10)), the effective velocity is actually faster than the kinetic energy. This has two consequences. One is that velocities and thus D_{heat} are not simple sums of terms, and the other is that a stronger dependence on T should exist in the data. Furthermore, the delay should increase with atom size. Consequently, the temperature dependence departs further from the ideal gas model as atomic size increases. The mismatch arises from not including the interaction time (the inelastic delay) according to Eq. (5.10). The importance of the delay will be inferred from comparing the historical model to the data (Section 5.4).

The same "delay" may not pertain to the diffusivities and viscosity, since the latter has an addition collision mechanism (near misses). Hence, the same power laws may not hold for all three transport properties.

5.2.6.3 Summary

More complex formulae than the above are available in the literature for mass diffusivity (e.g., Cussler, 2009). However, no formula can overcome the ambiguity in the atomic radii or the problems with neglecting the small interaction time, in view of Eq. (5.10). These models focus on mass diffusion behavior of gas mixtures. This chapter instead focuses on how well the model describes all three transport properties in single-constituent gases, with the ultimate goal of understanding the limitations of extrapolating kinetic gas theory to heat transfer in condensed matter.

5.3 COMPLICATIONS WHEN GAS MOLECULES CONTAIN MULTIPLE ATOMS

The atoms or molecules constituting a gas have three translations, representing the three distinct Cartesian directions, which sum to provide one component of internal energy. Section 5.2.4 showed that the ideal gas EOS suffices to evaluate the mean velocity. For monatomic gas, all energy is translational. For gas composed of molecules, two additional energy reservoirs exist: rotations of the imperfectly spherical molecules and vibrations of the atoms defining the molecule against one another.

Symmetry (i.e., group theory) is used to ascertain the numbers and types of these interior motions (e.g., Cotton, 1971; Nakamoto, 1978). For linear molecules, including diatomics, two rotations exist that are parallel and perpendicular to the bond. For all other molecules, three rotations in perpendicular directions exist. The number of total modes of a molecule is $3j$ where j is the number of atoms in the molecule. The number of vibrations

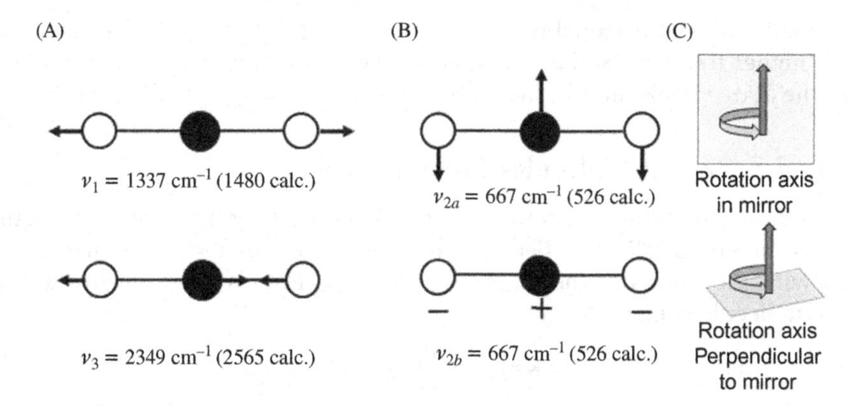

FIGURE 5.5 Normal modes of vibrations in the linear CO_2 molecule. Arrows indicate directions in the plane of the paper. + and − indicate motions out of the plane. Mode ν_1 is Raman-active only, yet has a resonance with ν_2. The others are IR-active only. See Nakamoto (1978) for a derivation, detailed discussion, and measured frequencies. Calculated frequencies from Nash (2001). (A) Stretching vibrations. (B) Bending modes, which are degenerate. These bending motions describe two distinct modes because their symmetry elements differ with respect to rotation around the short axis, as shown in (C). Only the symmetric stretch also adheres to a mirror perpendicular to the long axis, whereas both stretches are compatible with infinite symmetry of rotation about the long axis, and both bends are described by a mirror containing the long rotation axis (not shown). See Cotton (1971) for a general discussion.

is reduced from $3j$ by the number of translations (always 3) and also by the number of rotations (2, if the molecule is linear, but 3 otherwise).

The vibrational modes of any molecule are described by periodic changes in its bond lengths and bond angles, which are visualized and categorized as stretching and bending. The fundamentals are referred to as normal modes. Overtones and combinations of the fundamentals exist (examples are given in Chapter 2). Modes in which the dipole moment of the molecule changes are stimulated by absorption of light, and are denoted as infrared (IR)-active, while modes in which the polarizability changes (closely related to the shape) are simulated during inelastic scattering of high energy visible light (Raman-active). Hence, motions that are asymmetic are IR-active whereas symmetric motions are Raman-active. Notably, the perfectly symmetric stretching motion in homogeneous diatomic molecules is Raman-active only. Other vibrations can be both IR and Raman-active. Fig. 5.5 shows the four distinct internal motions of atoms in the linear CO_2 molecule and describes their connection with symmetry elements of the molecule. Modes of other molecules are shown in the above mentioned books. For actual motions, see Nash (2001). For use of group theory to determine complex motions of complex molecules, and lattice vibrations of a crystal as well, see Fateley et al. (1972).

Ascertaining frequencies requires measurements (Chapter 2) or force calculations. The basis of the latter is the simple harmonic oscillator (SHO) model. For each stretching mode (i), the force is proportional to the displacement (x) via a constant (k_i), yielding a frequency which depends on the reduced mass (m_{red}):

$$\nu_i = \frac{1}{2\pi c}\sqrt{\frac{k_i}{m_{red,i}}} \tag{5.25}$$

This simple model agrees reasonably well with spectral data (e.g., Nakamoto, 1978), but tends to give higher frequencies than the measurements. Vibrational data on molecules are compiled in the above books and in the websites listed at the end of Chapter 2.

5.3.1 Internal Energy of Molecules From the VT

Like translations, rotations are purely kinetic energy. Therefore, the VT argument of Section 5.2.4 is directly applicable. Because the van der Waals terms are negligible or can be wrapped with the radius of the atom, the ideal gas EOS is used. From Eq. (5.13), the rotations of a molecule contribute:

$$\langle \mathrm{KE} \rangle_{rot} = \frac{1}{2} l N_M R_{gc} T \tag{5.26}$$

where $l = 2$ for diatomic or linear molecules, but is 3 for all others.

Regarding vibrations, the VT provides for the SHO:

$$\mathrm{KE}_{sho} = \mathrm{PE}_{sho} \tag{5.27}$$

Because energy conservation is not invoked in deriving the VT, the total energy is then twice the kinetic energy, for details, see Hofmeister and Criss (2016). Thus, the internal energy for the sum of all vibrations is:

$$\langle \mathrm{KE} \rangle_{vib} = \frac{(3j - 3 - l)}{2} N_M R_{gc} T \tag{5.28}$$

In the above equations, the average accounts for some distribution of energies. This should be true for the rotations, due to high symmetry, but rotations are coupled with vibrations, which are discretized. Fig. 2.7 shows IR spectra of CO_2 and water vapor, where many rotational lines are superimposed on the various absorption bands. The pattern of the lines is associated with harmonic oscillations of a rigid rotator (see e.g., Wolfram Demonstrations Project, 2017). For the homogeneous diatomic molecules, rotational lines are superimposed on their Raman spectra (e.g., Compaan et al., 1994). Vibrations of homogenenous diatomics are not IR-active, and so this coupling is not observed. However, rotations of O_2 and other paramagnetic diatomics are present in microwave spectra without the vibrational modes (Brown and Carrington, 2003). Websites with spectral data on rotations and vibrations of molecules are compiled in Chapter 2.

The numbers of observed rotational lines are vast, and so the distributions are nearly continuous. Hence, the average of Eq. (5.26) is a reasonable approximation. However, not all vibrations are activated by photons and some are at too high of a frequency to be stimulated at ordinary temperature (Fig. 2.2 illustrates the connection of temperature with heat). For these reasons, the vibrational contribution to the internal energy is $\frac{1}{2} NR_{gc}T$ for each mode that is active at the temperature of interest.

The total internal energy at some temperature is thus the sum of Eqs. (5.13) and (5.26), plus a vibrational component that depends on the particular molecule and the temperature. This component is determined from the vibrational frequencies.

5.3.2 Effects of Internal Rotations and Vibrations on the Transport Properties

Translational velocities are governed by Eq. (5.13), regardless of molecular complexity. The mean free path for direct collisions (Eq. (5.22a)) pertains to spherical atoms. Linear

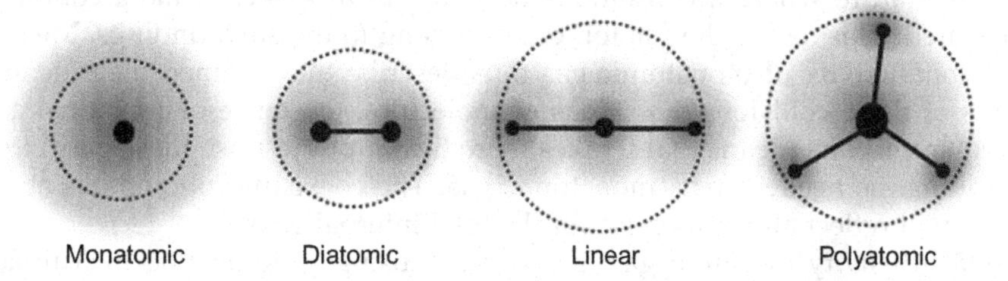

FIGURE 5.6 Schematic of size determinations for gas molecule types. Left, a monatomic molecule, where a small hard nucleus is surrounded by a soft electron cloud. The size (dotted circle) is defined by some density of the electron cloud, which decreases with distance from the nucleus. Middle left, diatomic molecules have a rigid bond which was assumed to represent the radius. Middle right, linear triatomic molecules. The sausage shape was represented by 1.5 × the average of the bond lengths. Right, the polyatomic ammonia molecule. The three H atoms are not in the same plane as the large N atom. The size for ammonia and methane molecules are reasonably approximated by a sphere. For the larger hydrocarbon molecules, the radius was computed by adding half of the C–C bond length to the C–H bond length. C_2H_2 was treated as the other hydrocarbons rather than like the linear triatomics.

molecules are certainly not spherical, but an effective radius may be represented as an average of the short and long dimensions. Fig. 5.6 explains how molecular size is estimated, to permit comparison of the kinetic theory of gas with measured transport properties.

Regarding viscosity, diffusion of momentum is impeded by drag (Fig. 5.1C). Mean free paths during glancing collisions are likely more affected by molecular shape than are direct collisions, and so the spherical approximation for υ (Eq. 5.22B) is less likely to be valid as molecule complexity and size increases, but nonetheless serves as the starting point for comparison in Section 5.4.

As discussed in Section 5.2.6.2, inelastic interactions during a collision of atoms cause a "delay." Molecules also experience deformation during collisions, which likewise contributes a "delay." Both delays affect the lifetimes inversely per Eq. (5.10). In addition, rotations and vibrations are stimulated during collisions. This interchange of translational with rotational and vibrational energy can be approximated as an elastic (conservative) process. But, this is valid only as a first-order approximation. Because these interactions occur over finite times, the presence of rotations and vibrations contribute a "delay" that augments the inelastic "delay." This energy interchange delay will increase with the number of atoms in the molecule. Thus, gases will increasingly diverge from the kinetic theory as molecular complexity increases.

5.4 EVALUATION OF THE KINETIC THEORY WITH MEASUREMENTS OF GAS

Because different methods are used to measure different properties, the identical mean free paths (or lifetimes) may not be sampled. The essentials of the experiments are as follows:

To avoid convection, heat transfer measurements of gas are made under steady-state conditions with weak temperature gradients. Thus, the measurements approximate nearly constant temperature where a continuous heat flux is supplied, while a continuous heat loss is measured. This loss is the photon gas departing to the surroundings. Measurements of thermal conductivity of monatomic gas thus describe the summed inelastic and elastic components of these collisions. For a gas of molecules, existence of rotational and vibrational energy reservoirs means that it takes more heat input to raise these gases to a certain temperature than is needed for a monatomic gas. The coexisting photon gas at any given temperature will reflect all components of the total internal energy.

Mass self-diffusivity of dilute gas is measured using isotopic tracers (Cussler, 2009). Nuclear magnetic resonance techniques are more accurate, but require substantial density, and so gases under high pressure are explored (Suárez-Iglesias et al., 2015). Thus, D_{mass} may differ from D_{heat} and υ because these measurements involve averages over the isotopes. Adding a neutron to a tiny atom should affect diffusivity in the elastic model. Yet, this effect is small, being only 4% near ambient conditions for the smallest molecules, that is, for hydrogen, deuterium, and tritium (Amdur and Beatty, 1965). Thus, isotopic differences are unimportant in our comparison of many gases.

During diffusion, each collision involves a tiny loss of velocity via conversion of the energy involved in deformation to light. If isothermal conditions are maintained during the experiment, receipt of photons from the exterior will compensate for the loss of KE during collision. If not, D_{mass} should be slightly lower than D_{heat} due to departures from perfect elastic collisions. Measurements of D_{mass} for a gas of molecules will similarly be slightly affected by inelastic losses, but should not be altered by the elastic interchanges between the three energy reservoirs during nearly equilibrium (nearly isothermal conditions), although D_{heat} may be affected.

Regarding the kinematic viscosity of gas, this quantity records both diffusion of momentum and resistance of a material to deformation (Eq. (5.8)). This property will be the most affected by inelastic collisions (Sections 5.2.3 and 5.2.5), and should be $\frac{2}{3}D_{heat}$ for monatomic gas from dimensional analysis (Eq. (5.23)), if behavior is nearly elastic. A gas of molecules could be affected more due to shape factors affecting drag.

5.4.1 Comparison of Kinetic Theory With Measurements of Heat Capacity

At 298K, the difference between C_P and C_V is very close to R_{gc} for diverse gases. Therefore, the model is compared to C_P which is directly measured. Table 5.1 shows that the internal energy equals the sum of the translational and rotational motions, with a small contribution from low frequency vibrational modes. Blackbody intensity at 298K is very low by 1500 cm^{-1} (Fig. 2.3). The energy contribution for vibrations in Table 5.1 assumes that modes below 1200 cm^{-1} participate fully in heating the gas near 298K, but that higher frequency modes are ineffective.

The above approach reasonably describes the heat capacity and thus the internal energy for all types of gases (Table 5.1). Discrepancies are associated with vibrational frequencies lying near the cut-off, which is a rough approximation. Note that diatomic gas heat capacity is affected by the vibrational mode if the frequency is sufficiently low. This is the case whether or not the vibration is active in the IR or not. The criterion is instead whether the collisional energy is sufficiently high to "match" the vibrational energy.

TABLE 5.1 Vibrational Characteristics of Gas Molecules and Comparison of Heat Capacities

Substance	Number Vibrations	Lowest Vibrations (cm^{-1})	C_P/R_{gc} @R.T.	Theory C_V/R_{gc} trans + rot + vib	Theory C_P/R_{gc} $1+C_V/R_{gc}$
He	0	n.a.	2.50	1.5 + 0 + 0	2.5
Rn	0	n.a.	2.53	1.5 + 0 + 0	2.5
H_2[a]	1[b]	4395[b]	3.47	1.5 + 1 + 0	3.5
N_2	1[b]	2360[b]	3.50	1.5 + 1 + 0	3.5
O_2	1[b]	1580[b]	3.53	1.5 + 1 + 0	3.5
F_2	1[b]	892[b]	3.73	1.5 + 1 + 0.5	4
Cl_2	1[b]	546[b]	4.08	1.5 + 1 + 0.5	4
NO[a]	1	1880	3.50	1.5 + 1 + 0	3.5
CO	1	2138	3.59	1.5 + 1 + 0	3.5
HCl	1	2991	3.41	1.5 + 1 + 0	3.5
H_2O	3	1595	4.02	1.5 + 1.5 + 0	4
H_2S	3	1183	4.16	1.5 + 1.5 + 0.5	4.5
SO_2	3	505,1114	4.93	1.5 + 1.5 + 1	5
N_2O[a]	4	590[c]	4.66	1.5 + 1 + 1	4.5
CO_2[a]	4	660[c]	4.47	1.5 + 1 + 1	4.5
NH_3	6	931	4.22	1.5 + 1.5 + 0.5	4.5
CH_4	9	1306	4.29	1.5 + 1.5 + 0	4
C_2H_2[a]	7	805[b,c],929[c,d]	5.29	1.5 + 1 + 2	5.5
C_2H_4	12	835,875,1057[d]	5.16	1.5 + 1.5 + 1.5	5.5
C_2H_6	18	275,821[c],993,1155	6.33	1.5 + 1.5 + 2.5	6.5

[a]*Linear polyatomic molecule: all diatomic molecules are linear.*
[b]*Raman active only.*
[c]*Doublet.*
[d]*Frequencies from NASA Astrobiology Institute (2017).*
Notes: Measured frequencies mostly from Nakamoto (1978) either represent the lowest mode, or list frequencies of all modes below the 1200 cm^{-1} cut-off described in the text. Data on C_P/R_{gc} are mostly from Engineering Toolbox (2017) and refer to heat capacity on a per mole basis.

5.4.2 Comparison of Kinetic Theory With Transport Measurements of Gas at Constant Conditions

All three transport properties have been determined for a modest number of chemical compounds that exist as gases near ambient conditions. Mixtures are not considered here. Table 5.2 includes all known monatomic gases, although the values for radon are approximate. Data on diatomic gases are less comprehensive, but like monatomic gases are mostly

TABLE 5.2 Thermal and Mass (Self) Diffusivities Compared to Kinematic Viscosity Data for Simple Gases at STP (0°C and Atmospheric Pressure), Unless Noted

Substance	D_{heat} mm^2 s^{-1}	D_{mass} mm^2 s^{-1}	υ mm^2 s^{-1}	r_{atom} pm	m g mol^{-1}	D_{calc} mm^2 s^{-1}	r_{calc} pm
He	158	155	105	140	4.00	394	221
Ne	49.7	46.7	33.5	154	20.18	145	263
Ar	18.1	15.5	11.8	188	39.95	69.1	367
Kr	9.43	8.22	6.32	202	83.8	41.3	423
Xe	5.40	4.79	3.60	216	131.3	28.9	499
Rn	3.8		2.1	220	222	21.4	521
H_2	135	122 ± 3	95	75	2.016	1933	284
N_2	19	17.2	13.6	110	20.18	241	392
O_2	18.7	17.4	13.6	120	32.0	189	382
F_2	16.7		12.3	140	38.0	128	387
Cl_2	5.80		3.84	200	70.9	45.8	562
CO	19.05	18.40	13.21	113	28.01	228	391
NO	17.78		13.28	115	30.01	213	398
HCl	10.06[a]	12[a]	8.17[a]	127	36.46	159	465
H_2O steam[b]	12.7[c]	17[c]	13.1[c]	96	18[b]	450[b]	569[b]
H_2S	9.96[a]		7.85[a]	133	34	467	550
SO_2	5.13[a]		3.96[a]	143	64	108	655
N_2O[d]	9.77[a]		6.87[a]	174[d]	44	50	521
CO_2[d]	8.80	9.54	6.95	173[d]	44	49	549
NH_3	14.43	15.00	11.20	109	17	359	544
CH_4	19.15	20.60	14.28	102	16	422	479
C_2H_2[d]	11.60[a]	13.3[a]	8.52[a]	166[d]	26	125	546
C_2H_4	10.63[a]	12.5[a]	7.7[a]	173	28	111	559
C_2H_6	9.63[a]	10.5[a]	8.7[a]	186	30	93	577

[a]*Value at 298K. Data on HCl are extrapolated from ~325K, and are mostly unconfirmed.*
[b]*For the polyatomics, viscosity and thermal conductivity data are at 298K from Engineer's Edge (2017) and Engineering Toolbox (2017); also see Yaws (2001). For NNO, the most recent thermal conductivity data of Richter and Sage (1959) were used.*
[c]*Data on D_{heat} and υ for water vapor (steam) was extrapolated to 298K from the trend listed in Engineer's Edge (2017) for T~100°C. For D_{mass}, the compilation of Fokin and Kalashnikov (2008) was used, which agreed with viscosity from Thermopedia (2018).*
[d]*Linear molecules.*

Notes: Gas experiments involve steady-state. Radius in pm for monatomics from van der walls radius ascertained from contacting atoms in molecular crystals (Mantina et al., 2009). For diatomics, the bond length was used (e.g., NIST, 2017), which also provided mass. Radii of polyatomics were estimated as the length of the longest bond except for linear triatomics (OCO and NNO), where 3/2 the bond lengths was used to account for their shapes being like sausages (Fig. 5.6). For HCCH, the C—C bond length was used. Transport data for noble gas are the experimental averages of Kestin et al. (1984); Suárez-Iglesias et al. (2015). Data on He from Peterson (1970). Mass diffusivity for diatomics from Winter (1951) and Hanley and Prydz (1972), and Suárez-Iglesias et al. (2015), who also cover polyatomics. For H_2, self-diffusion is the average of three isotopic tracers (Amdur and Beatty, 1965), with the range indicated. Density data from Air Liquide (2017). Viscosity data from Engineering Toolbox (2017); also see Yaws (2012). The size of the free atom was computed by comparing D_{calc} to D_{heat}.

reported at STP (0°C and 1 atm). For HCl, D_{mass} is only measured at 298K, so its other transport properties are also reported at this temperature. Most data for the simple molecules are reported at NTP (normal being 20°C) but 298K is also commonly used. Data on water vapor were extrapolated from higher temperatures. Mass diffusivity data were not found for several polyatomic gases.

Despite variable uncertainties, Table 5.2 demonstrates that thermal diffusivity and mass diffusivity are roughly equivalent for real gases, whereas kinematic viscosity is significantly lower. This comparison confirms that two mechanisms for viscosity (collisions and drag) exist.

5.4.2.1 Inelastic Effects as Revealed by the Ratios of Measured Transport Properties

By comparing the ratios of the transport properties, the problems associated with the equivocal nature of atomic radii are circumvented. Fig. 5.7A shows, within uncertainties, that υ/D_{heat} for monatomic gases under nearly isothermal conditions is nearly constant and equals ⅔, which confirms the dual nature of viscosity in gases.

The monatomics at STP are discussed first, because these are the closest to small spheres. For monatomic gas, υ/D_{heat} decreases ever so slightly with atomic radius whereas D_{mass}/D_{heat} is close to unity for He gas and decreases as the atomic radius increases. Deformation during direct collisions produces heat losses which will decrease D_{mass} and

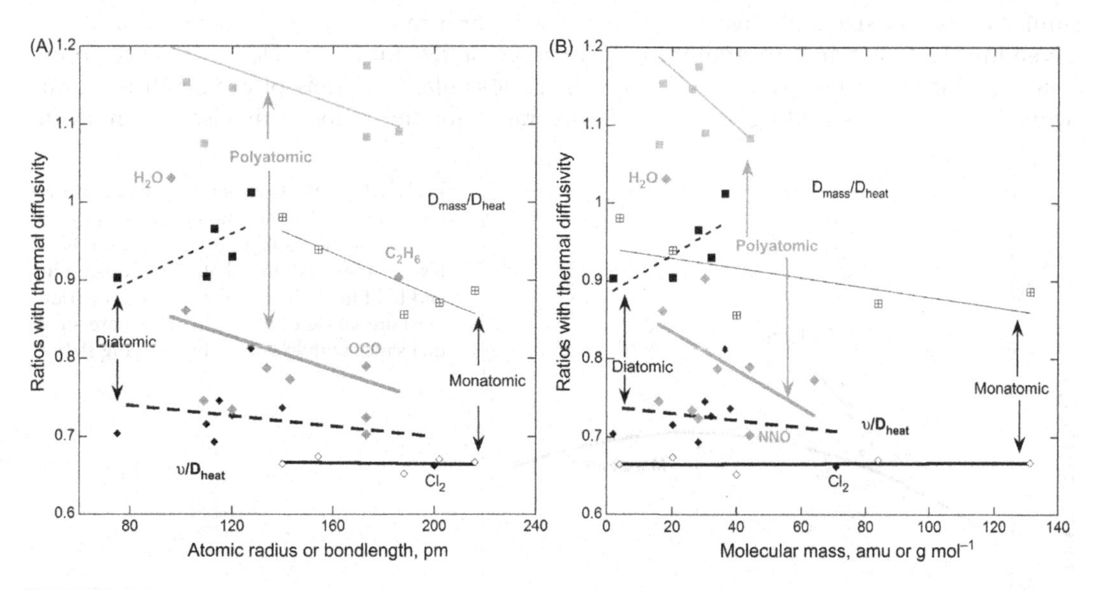

FIGURE 5.7 Dependence of ratios of transport properties on molecule size (A) or mass (B). Least squares linear fits are shown. Diamonds and heavy lines indicate υ/D_{heat}. Squares and light lines indicate D_{mass}/D_{heat}. For monatomic gas (solid black lines), atomic radii during contact are used. For diatomics (dashed lines), the total bond length is used. For other polyatomics (gray), the longest bond length is used, except for linear molecules, where 75% of the total length is used. Radon is not included because its values are approximate. Data on HCl and the polyatomics are at ambient conditions: all other data are at STP (Table 5.2). The slightly different temperature for the polyatomics and HCl (298 vs 273K) should not matter because ratios are considered (this is verified below).

υ, but does not affect D_{heat}, because this records behavior of the photon gas. Decreasing ratios of υ/D_{heat} and D_{mass}/D_{heat} are consistent with larger atoms being "softer," that is, more prone to deformation during direct collisions. However, why is the slope of υ/D_{heat} flat? Diffusion of momentum via direct collisions will decrease as inelasticity increases. However, glancing collisions during drag will increase this component of viscosity as inelasticity increases (Section 5.2.3). Different slopes for the two ratios (Fig. 5.7), and the growth of υ/D_{mass} from the ideal ratio of 0.67 for the relatively hard He gas to 0.75 for soft Xe (Fig. 5.8), are consistent with kinematic viscosity reflecting the compensating effects of deformation on direct and glancing collisions whereas D_{mass} is affected only by losses during direct collisions. The low value of υ/D_{mass} for Xe suggests that the largest atoms offer less resistance during drag, but deform greatly in direct collisions.

For the diatomic gases at STP, some scatter exists in the ratios (Figs. 5.7 and 5.8). The data in Table 5.2 were checked against several sources and should be correct. However, viscosity of chlorine gas was only determined once, ~1930, and likewise for fluorine. With this in mind, the ratios υ/D_{heat} and υ/D_{mass} are roughly constant and equal 0.72 and 0.77, respectively. In contrast, D_{mass}/D_{heat} for diatomic gases increases with size, reaching unity for large diatomic molecules, which is opposite to the trend for the monatomic gases. All ratios depend similarly on bond length and on mass, which shows that the estimates for molecule size in Fig. 5.7 are not the source of the odd increase in D_{mass}/D_{heat}. Notably, the H_2 molecule with its few electrons defines the trend, whereas the other diatomics have similar mass and size and cluster together. The highest ratio is for HCl, which is at 298K. Given the clustering and uncertainties, the average of 0.94 for D_{mass}/D_{heat} for the diatomic gases indicates that energy losses during their molecular collisions of are small but non-negligible. With this finding in mind, the averages for the ratios with viscosity indicate

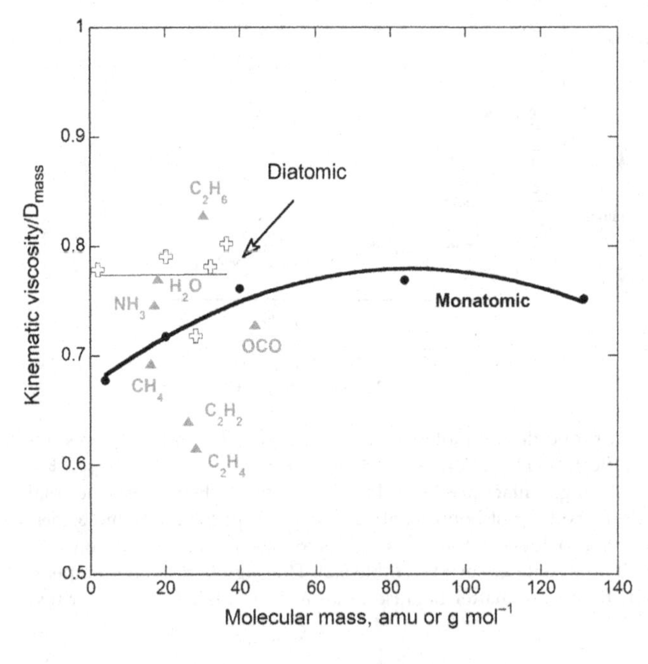

FIGURE 5.8 Comparison of measured ratios for υ/D_{mass} to atomic mass. Fits are shown for the monatomic atoms (dots hand heavy line) and the diatomics (open plus and light line). Polyatomic atoms (gray triangles) are labeled but not fit. These are small molecules with low mass but varying shape.

that an elongated shape enhances drag relative to that expected for spheres, but does not significantly impact direct collisions.

Notably, the diatomic gases have additional reservoirs for internal energy. The vibrations of the heterogeneous molecules are at too high of a frequency to absorb heat energy associated with 273 or 298K, whereas vibrations of the homogeneous diatomic molecules are IR-inactive. Thus, for the species investigated, the internal energy lies in the translations and rotations, and the photon gas represents these reservoirs. Because collisions will exchange energies between these reservoirs, equilibrium should exist. During this exchange, the colliding molecules could be delayed, which effectively reduces the efficiency of heat transport. This behavior would cause D_{mass}/D_{heat} to be larger than the ratios for monatomic gas, which is not observed. Thus, the elongated shape of diatomic molecules apparently contributes to υ/D_{heat} and of υ/D_{mass} diverging from the prediction of Eq. (5.23).

For polyatomic gases, both υ/D_{heat} and D_{mass}/D_{heat} decrease with molecule size, although this trend has considerable scatter (Fig. 5.7). These trends are consistent with deformation increasing with molecule size. Values for υ/D_{mass} are scattered about 0.73; this average is similar to the average of 0.74 for the monatomics (Fig. 5.8), consistent with shapes being both prolate and oblate. Considering the limitations in the data, drag depending on molecule shape explains variable υ/D_{mass}.

For the polyatomic gases, averages of 0.80 for υ/D_{heat} and 1.13 for D_{mass}/D_{heat} are both considerably higher than averages of the diatomic and monatomic gases (Fig. 5.7) and exceed the predictions of $^2/_3$ and 1 for elastic spheres. This behavior is consistent with vibrations of polyatomic molecules being both IR-active and at sufficiently low frequency to be simulated near 298K. Thus, D_{heat} is reduced by the delays associated with interactions of the vibrating molecules with the photon gas generated during the inelastic collisions.

5.4.2.2 Comparison of Calculated Diffusivity to Measured Transport Properties

Measured values of all three properties at STP for monatomic gas are related to the calculated values for diffusivity (Eq. (5.24)) through a power law (Fig. 5.9A). For D_{mass} and υ, the power = 4/3 within uncertainty, whereas the power for D_{heat} is only slightly smaller. For the diatomic gases at STP, the power varies from 0.82 to 0.88. For polyatomic gases at ambient conditions, the trend is flat with a power ranging from 0.26 to 0.33. Hard spheres cannot explain these trends:

Monatomic atoms are soft and expand to fill space, so r_{atom} should be roughly proportional to $V^{1/3}$ (Section 5.2). From Eq. (5.22b) and the EOS, this would weaken the temperature response for monatomic gas, not strengthen it as shown in Fig. 5.9A. The weaker dependencies on T for diatomic and monatomic gas are consistent with molecule size increasing with volume and thus with temperature, but because their diffusivities are affected by the presence of rotations and vibrations as discussed above, this finding does not overturn the conclusion based on monatomic gas behavior.

To understand the origin of the mismatch in Fig. 5.9, the radius needed for D_{calc} to match D_{heat} was computed. The computed effective radius (Table 5.2, RHS) is 2–3 times those from Mantina et al. (2009) for the monatomic atoms and likewise for estimates of molecular size. Even with this larger size, the atoms are small compared to the available

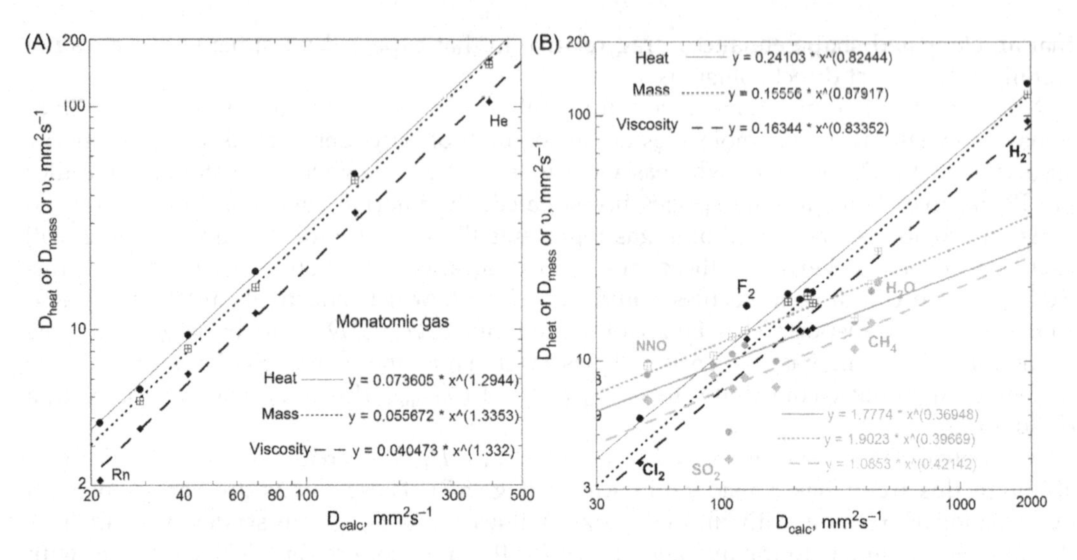

FIGURE 5.9 Comparison of physical properties at ambient condition to the kinetic theory of hard sphere atoms, Eq. (5.23) with $\iota = 1$. Circles and solid line = thermal diffusivity. Square with cross and dotted line = mass self-diffusivity. Diamond and dashed line = kinematic viscosity. Data and sources in Table 5.2, where the radius accounts for the geometry of the molecule. Least squares fits are shown. (A) Monatomic (noble) gas. For all fits, correlation coefficients are better than 0.99. (B) Diatomic gas (black symbols) and other simple gases (gray symbols). Correlations coefficients are better than 0.9 for power fits.

volume. However, if the equivocal size were the only problem, the power fits in Fig. 5.9 would be near unity, with different slopes. The model short-coming is that Eqs. (5.22a,b) may represent the mean free path, but it does not represent how the interactions during collisions perturb the lifetimes from the collisional estimate of Fig. 5.1E.

5.4.3 Implications of Temperature-Dependent Measurements on the Importance of Inelasticity to Transport

The measured transport properties on monatomic gas do not go as $T^{1.50}$, as predicted by Eq. (5.23) (Fig. 5.10). Instead, the transport properties of gas with the small atoms go as $T^{1.70}$, whereas the transport properties for intermediate size Ar goes as $T^{1.80}$, and for large Kr and Xe, the transport properties go as $T^{1.84}$. For all gases, the ratio υ/D_{heat} is virtually unaffected by temperature, and is very close to ⅔, which confirms that viscosity combines drag with diffusion of momentum. However, that the power laws differing from 1.5 shows that the mean free paths are incorrectly computed in the kinetic theory of gas.

Likewise, the measured transport properties for polyatomic gas do not agree with the kinetic theory of gas (Fig. 5.11). For the homogenous diatomic gases, the power law is $T^{1.70}$, while the ratio υ/D_{heat} is on average 0.69–0.71, but is affected by temperature. The ratio υ/D_{heat} is on average 0.73 for CO_2, 0.87 for NH_3 and 1.0 for water vapor, and depends on temperature. The power law fits are all near T^2. Unsurprisingly, as the complexity of the molecules increase, agreement with the model is degraded.

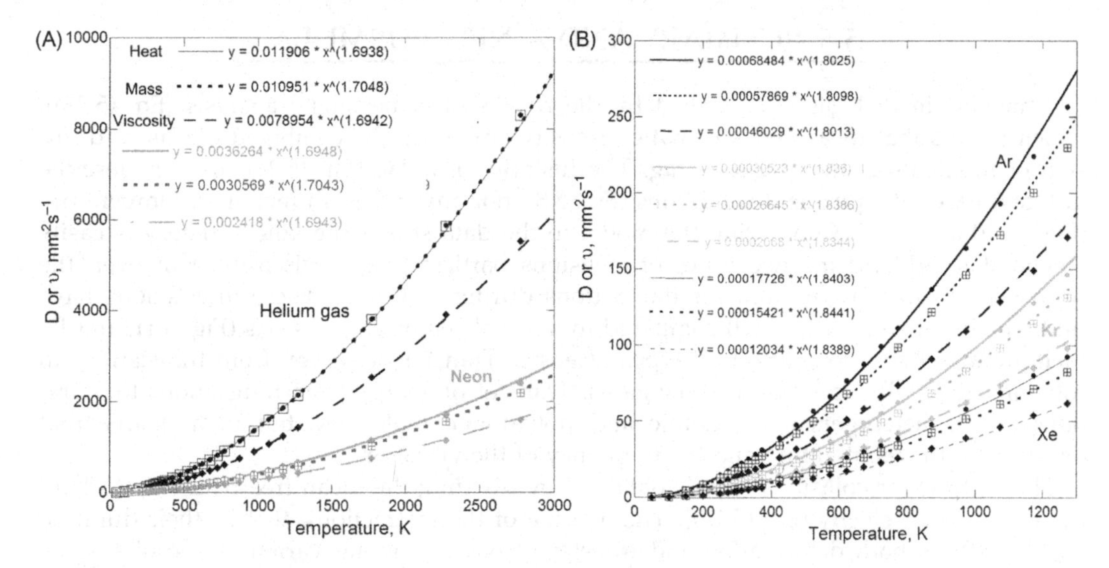

FIGURE 5.10 Dependence of measured transport properties of the noble gases on temperature. Data from the compilation of Kestin et al. (1984). For each gas, circles and solid line: thermal diffusivity. Square and dotted line: mass self-diffusivity. Diamond and dashed line: kinematic viscosity. (A) Small atoms; (B) large atoms.

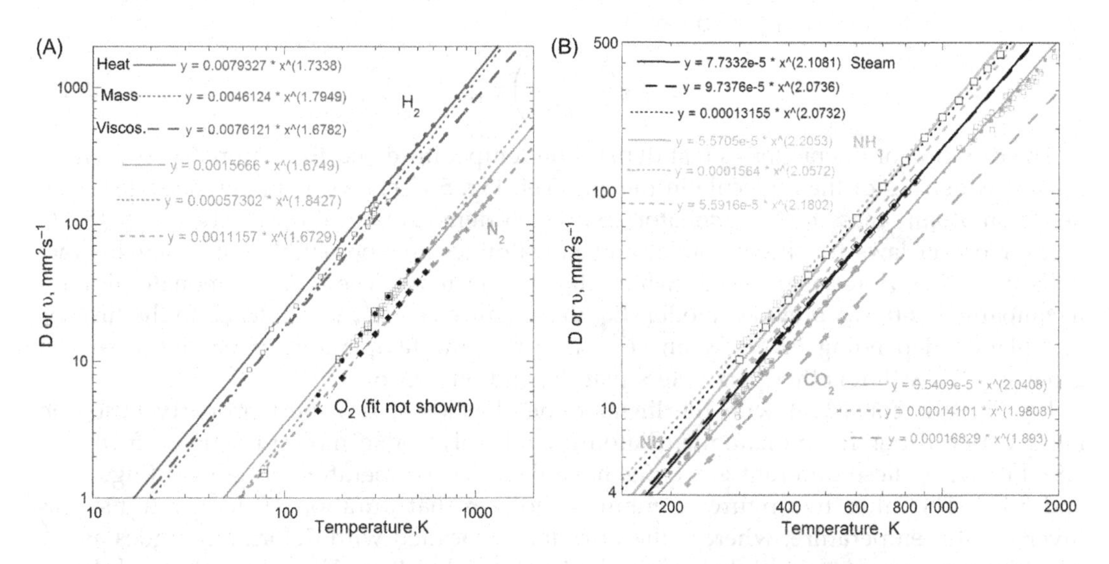

FIGURE 5.11 Dependence of measured transport properties of the polyatomic gases on temperature. Data on D_{mass} from the compilation of Suárez-Iglesias et al. (2015). Data on D_{heat} and viscosity from Engineering edge (2017); also see Thermopedia (2018). Symbols as in Fig. 5.8. (A) Diatomic gas with atoms of one kind. (B) Polyatomic gas with 2–5 small atoms per molecule.

5.5 SUMMARY AND A NEW FORMULA

Transport in real gas is described by the results of dimensional analysis (Eq. (5.23)), which proves that measured kinematic viscosity embodies the combined effects of diffusion of momentum and viscous drag. The historic formulae (Eq. (5.24)) neither describe different gases at equivalent conditions (Fig. 5.8), nor any gas as a function of temperature (Figs. 5.10 and 5.11). Comparing the model to the data shows the kinetic theory is easily amended to address the two types of collisions participating in viscosity. However, the serious problem is to account for interactions during collisions. These interactions have finite lifetimes, which are small compared to intervals between collisions (Fig. 5.1E and F). Inelastic direct collisions affect all types of gases. Transfer of energy from translations to rotations affect all except monatomic gases. Transfer of energy from translations to vibrations affects heterogeneous diatomic and polyatomic molecules, but to a degree that depends on the temperature and the frequency of the vibration.

The lifetime for collisions only is obtained by dividing the mean free path (Eq. (5.22b)) by the average velocity (Eq. (5.13)). The lifetime of the interactions, that is, their duration (τ_{delay}) reflects both deformation and transfer of energy among various reservoirs. Only one lifetime describes the delay because the various processes cannot be independent, as evidenced by rotational–vibrational coupling. Because the duration during the interactions is much shorter, and reciprocal lifetimes add (Eq. (5.10)), the shorter lifetime dominates the overall lifetime. Thus, $\tau \cong \tau_{delay}$. From Eq. (5.13), which describes the translational velocities, and the above:

$$D \cong \left(\frac{3NR_{gc}T}{m}\right)\tau_{delay} \tag{5.29}$$

The duration of interactions must depend on temperature and the symmetry and vibrational frequencies of the molecule in the gas. Yet, Fig. 5.12 shows a simple inverse dependence on atomic mass for the monatomics, which demonstrates the validity of Eq. (5.29). Weaker power laws for the diatomics and polyatomics are compatible with their interactions including vibrational effects, which depend on mass. The data on monatomics are incompatible with the previous model (Eq. (5.24)) since $m \sim r^{7.6}$, which leads to the historic formulation depending inversely on $m^{3/4}$, and that the temperature dependence should grow as $T^{3/2}$, with no other parameters that depend on T or on m.

Eq. (5.29) is consistent with the linear dependencies of transport property ratios on mass which differ for monatomic, diatomic and polyatomic molecules (Figs. 5.7B and 5.8). Likewise, these different gas types have different temperature responses (Figs. 5.10 and 5.11). The latter two figures combined indicate that duration of energy transfer is inverse with temperature, whereas the duration associated with deformation goes as $T^{1/3}$ and that these powers multiple to describe the total duration. The dependence of deformation on temperature is best explained by T being proportional to V at constant P (the ideal gas EOS). The delay depending weakly on mass is consistent with deformation of soft atoms with various shapes.

Because scattering of molecules does not describe heat transfer in gas, scattering of phonons cannot describe heat conduction in condensed matter. The phonon-scattering model

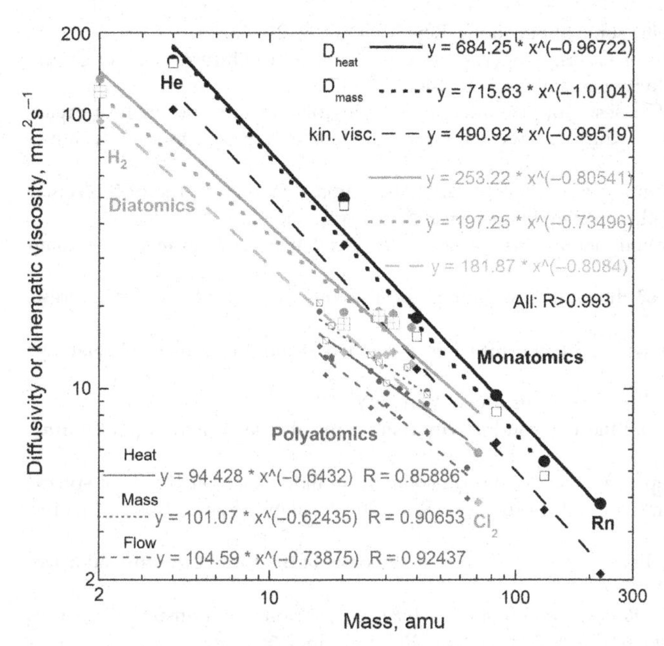

FIGURE 5.12 Dependence of measured transport properties of gases at STP on mass of their molecules. Least squares fits are shown. Dots and solid line: D_{heat}. Squares and dotted lines: D_{mass}. Diamonds and dashed lines: kinematic viscosity. Monatomics are shown in black; diatomics in gray; and polyatomics in small, dark gray symbols. Data from Table 5.2.

needs to be reformulated, and the lifetimes of interactions during collisions in gases needs further exploration. The next chapter explores liquids, because this state of matter is an intermediate situation (Fig. 5.4).

References

Air Liquide, 2017. Gas Encyclopedia. <https://encyclopedia.airliquide.com/> (accessed 25.08.17).

Amdur, I., Beatty, J.W., 1965. Diffusion coefficients of hydrogen isotopes. J. Chem. Phys. 42, 3361–3364. Available from: https://doi.org/10.1063/1.1695735.

Berberan-Santos, M.N., Bodunov, E.N., Polliani, L., 2008. The van der Waals equation: analytical and approximate solutions. J. Math. Chem. 43, 1437–1457.

Brouard, M., Chadwick, H., Gordon, S.D.S., Hornung, B., Nichols, B., Aoiz, F.J., 2016. Stereodynamics in NO(X) + Ar inelastic collisions. J. Chem. Phys. 144, 224301. Available from: https://doi.org/10.1063/1.4952649.

Brown, J.M., Carrington, A., 2003. Rotational Spectroscopy of Diatomic Molecules. Cambridge University Press, Cambridge, UK.

Brush, S.G., 1976. The kind of motion we call heat: a history of the kinetic theory of gases in the 19th century. North-Holland Pub. Co, Amsterdam, NY.

Clausius, R., 1857. On the nature of the motion we call heat. Phil. Mag. 14, Ser 4 (14), 108–127 (translated from Poggendorff's Annalen 49, p. 315, 1857).

Clausius R., 1870. On a mechanical theorem applicable to heat. Philosophical Magazine, Ser. 4. 40: 122–127 (translated from Sitz. Nidd. Ges. Bonn, p. 114 1870).

Compaan, Wagoner, A., Aydinli, A., 1994. Rotational Raman Scattering in the instructional laboratory. Am. J. Phys. 62, 639–645.

Cotton, F.A., 1971. Chemical Applications of Group Theory, 2nd ed. Wiley-Interscience, New York.

Criss, E.M., Hofmeister, A.M., 2017. Isolating lattice from electronic contributions in thermal transport measurements of metals and alloys and a new model. Int. J. Mod. Phys. B 31, 175020. Available from: https://doi.org/10.1142/S0217979217502058.

Cussler, E.L., 2009. Diffusion: Mass Transport in Fluid Systems. Cambridge University Press, Cambridge, UK.

Dingwell, D.B., 1995. Relaxation in silicate melts: some applications. Rev. Mineral. 32, 21–66.

Engineer's Edge, 2017. Thermal Conductivity of Gases; Properties of Gases at 1 atm. <http://www.engineersedge.com/heat transfer/thermal-conductivity-gases.htm> (accessed 25.08.17).

Engineering Toolbox, 2017. Heat Capacity of Gases. <http://www.engineeringtoolbox.com/specific-heat-capacity-gases-d_159.html>; Viscosity of Gases <http://www.engineeringtoolbox.com/gases-absolute-dynamic-viscosity-d_1888.html> (accessed 25.08.17).

Fateley, W.G., Dollish, F.R., McDevitt, N.T., Bentley, F.F., 1972. Infrared and Raman Selection Rules for Molecular and Lattice Vibrations: The Correlation Method. Wiley-Interscience, New York, NY.

Fegley Jr., B., 2015. Practical Chemical Thermodynamics for Geoscientists. Academic Press\Elsevier, Waltham Massachusetts.

Ferraro, J.R., Nakamoto, K., Brown, C.W., 2003. Introductory Raman Spectroscopy, second ed. Academic Press, New York, NY.

Fokin, L.R., Kalashnikov, A.N., 2008. The viscosity and self-diffusion of rarefied steam: Refinement of reference data. High Temp. 46, 614–619.

Halliday, D., Resnick, R., 1966. Physics. John Wiley & Sons Inc, New York, NY.

Hanley, H.J.M., Prydz, R., 1972. The viscosity and thermal conductivity coefficients of gaseous and liquid fluorine. J. Phys. Chem. Ref. Data 1, 1101–1113.

Hofmeister, A.M., Criss, E.M., 2018. How properties that distinguish solids from fluids and constraints of spherical geometry suppress lower mantle convection. J. Earth Sci. 29, 1–20. Available from: https://doi.org/10.1007/s12583-017-0819-4.

Hofmeister, A.M., Criss, R.E., 2016. Spatial and symmetry constraints as the basis of the Virial theorem and astrophysical implications. Can. J. Phys. 94, 380–388.

Kestin, J., Knierrim, K., Mason, E.A., Najafi, B., Ro, S.T., Waldman, M., 1984. Equilibrium and transport properties of the noble gases and their mixtures at low density. J. Phys. Chem. Ref. Data 13, 229–303.

Mantina, M., Chamberlin, A.C., Valero, R., Cramer, C.J., Truhlar, D.G., 2009. Consistent van der Waals radii for the whole main group. J. Phys. Chem. A 113, 5806–5812.

Maxwell, J.C., 1867. On the dynamical theory of gases. Phil. Trans. R. Soc. 157, 49–88.

McGucken, W., 1969. Nineteenth-Century Spectroscopy. The Johns Hopkins Press, Baltimore, MD and London, UK.

Nakamoto, K., 1978. Infrared and Raman Spectra of Inorganic and Coordination Compounds, third ed. John Wiley & Sons, New York.

NASA Astrobiology Institute, 2017. Molecular Database. <http://depts.washington.edu/naivpl/content/molecular-database> (accessed 18.09.17).

Nash, J., 2001. Vibration of Small Molecules. <http://www.chem.purdue.edu/gchelp/vibs/index.html> (accessed 19.09.17).

Niven, W.D., 1890. The Scientific Papers of James Clerk Maxwell. Cambridge University Press, Cambridge, UK.

Partington, J.R., 1960. A Short History of Chemistry, first ed. MacMillan and Co, London, UK, 1937.

Peterson, H., 1970. The properties of Helium: density, specific heats, viscosity, and thermal conductivity at pressures from 1 to 100 bar and from room temperature to about 1800 K. Risø Report No. 224 (Danish Atomic Energy Commission Establishment, Copenhagen, Denmark).

Pippard, A.B., 1974. The Elements of Classical Thermodynamics. Cambridge University Press, London.

Planck, M., 1901. Über das Gesetz der Energievertilung im Normalspektrum. Ann. Physik 4, 553–563.

Porter, N.A., 1998. Physicists in Conflict. Institute of Physics Publishing, London, UK.

Purrington, R.D., 1997. Physics in the Nineteenth Century. Rutgers University Press, New Brunswick, NJ.

Reif, F., 1965. Fundamentals of Statistical and Thermal Physics. McGraw-Hill, New York, NY.

Richter, G.N., Sage, B.H., 1959. Thermal conductivity of fluids: Nitric Oxide. J. Chem. Eng. Data 4, 36–40.

Shannon, R.D., Prewitt, C.T., 1969. Effective ionic radii in oxides and fluorides. Acta Crystallogr. Sect. B25, 925–945.

Sharp, K., Matschinsky, F., 2015. Translation of Ludwig Boltzmann's Paper "On the Relationship between the Second Fundamental Theorem of the Mechanical Theory of Heat and Probability Calculations Regarding the Conditions for Thermal Equilibrium" Sitzungberichte der Kaiserlichen Akademie der Wissenschaften. Mathematisch-Naturwissen Classe. Abt. II, LXXVI 1877, pp. 373–435 (Wien. Ber. 1877, 76:373–435). Reprinted in Wiss.Abhandlungen, Vol. II, reprint 42, p. 164–223, Barth, Leipzig, 1909. Entropy 17, 1971–2009. Available from: https://doi.org/10.3390/e17041971.

Suárez-Iglesias, O., Medina, I., Sanz, M., Pizarro, C., Bueno, J.L., 2015. Self-diffusion in molecular fluids and noble gases: available data. J. Chem. Eng. Data 60, 2757–2817. Available from: https://doi.org/10.1021/acs.jced.5b00323.

Thermopedia, 2018. Steam Tables. http://thermopedia.com/content/1150Begal (Reprinted, with permission, from NBS/NRC Steam Tables) (accessed 09.09.17; 09.01.18).

Transtrum, M.K., Machta, B.B., Brown, K.S., Daniels, B.C., Myers, C.R., Sethna, J.P., 2015. Perspective: sloppiness and emergent theories in physics, biology, and beyond. J. Chem. Phys. 143. Available from: https://doi.org/10.1063/1.4923066.

Truesdell, C., 1984. An Idiots's Fugitive Essays on Science. Springer-Verlag, New York, NY.

Wien, W., 1893. Eine neue Beziehung den Strahlen schwarze Körper zum zweiten Haupsatz den Wärmetheorie. In: Sitzungsbereichte der Koniglich Preussicchen Akademie der Wissenschaften, Berlin, p. 55–62.

Winter, E.R.S., 1951. The self-diffusion coefficients of nitrogen, oxygen, and carbon dioxide. Trans. Faraday Soc. 47, 342–348.

Wolfram Demonstrations Project, 2017. Rotational-Vibrational Spectrum of a Diatomic Molecule, Contributed by Porscha McRobie and Eitan Geva. <http://demonstrations.wolfram.com/RotationalVibrationalSpectrumOfADiatomicMolecule/> (accessed 19.09.17).

Yaws, C.L., 2001. Matheson Gas Data Book, seventh ed. McGraw-Hill, New York, NY.

Yaws, C.L., 2012. Yaws' Critical Property Data for Chemical Engineers and Chemists. Knovel, Norwich, UK and New York, NY.

Zemansky, M.W., Dittman, R.H., 1981. Heat and Thermodynamics, sixth ed. McGraw-Hill, New York, NY.

CHAPTER

6

Transport Behavior of Common, Pourable Liquids: Evidence for Mechanisms Other Than Collisions

OUTLINE

There is no water, so things are bad. If there were water, it would be better. But there is no water.

T.S. Eliot, The Waste Land (1922)

The unusual physical properties of liquid water have received considerable attention in the physical sciences (e.g., Chaplin, 2018), due to the importance of this simple fluid to life itself. From a geochemical perspective, the existence of life as we know it on Earth is predicated on its solid being less dense at the freezing point. If this behavior was reversed, as

Measurements, Mechanisms, and Models of Heat Transport

DOI: https://doi.org/10.1016/B978-0-12-809981-0.00006-1

observed for virtually all other known compounds or elements, the oceans would freeze from the bottom-up and Earth would be an ice world, not a nice world.

In great contrast, and as detailed in the present chapter, the transport properties of water are not unusual compared to diverse other substances that flow under any amount of stress (are *pourable*) near ambient conditions, except near its density maximum at 4°C in regards to pressure (Singh et al., 2017). Why? This question pertains to mechanisms, which must be understood in order to construct a viable microscopic model of heat transport in matter in states other than gaseous.

This chapter considers the state of matter commonly referred to as liquid. We probe materials that can be poured near room T and P for several reasons. First, liquids are dense, like solids, and these two phases are lumped together as "condensed matter." Yet, liquids have a critical point, such that for conditions exceeding this particular temperature and pressure, the liquid phase cannot be distinguished from a gas. Thus, the liquid state constitutes a transitional case, thereby providing a bridge between the relative simplicity of gases and the complexity of solids. Second, transport data on pourable liquids exist in abundance. There are holes and inadequacies but a wide variety of substances have been explored. Reasons include industrial needs, such as transport of foods through tubes or lubricating moving machinery. Progress of chemical reactions involving mass diffusion is another reason. Conduction of heat by fluids is of practical relevance; for example, cooling of automobile engines. The large number of measurements allow general statements to be made, while simultaneously hampering assessment of uncertainties of individual studies. Therefore, the purpose of this chapter is a general description of transport in pourable liquids as a step towards a thorough understanding the microscopic basis of heat transport in solids. Third, substances that flow under any stress are considered. Simple, Newtonian behavior dominating this category simplifies the measurements, makes available data on viscosity easy to analyze and understand, thereby permitting direct comparisons. Molecular compounds are the mainstay of our comparison, along with liquid elements, and certain oils.

This chapter does not consider data on fluids or liquids at very high temperature. Although relevant to the Earth and celestial bodies, existence of radiative transfer at high temperatures is a complicating factor. We cover silicate glasses and liquids in Chapter 10, and diffusive radiative transfer in Chapter 11. Pressure measurements of mass self-diffusivity are accurate and will be explored briefly, but data on thermal conductivity at P are uncertain by ~20% (Ross et al., 1984) or more (Hofmeister, 2010). One problem is contact losses (Chapter 4). Another is small samples (Chapter 7). Pressure determinations of viscosity are only briefly touched upon, in the context of the importance of volume to transport properties.

6.1 TRANSPORT PROPERTIES EXPECTED FOR INELASTIC COLLISIONS OF FINITE SIZE ATOMS IN LIQUIDS

Chapter 5 provides a new formula for thermal or mass diffusivity (Eq. (5.30)) that explicitly includes only one state variable, temperature. Kinematic viscosity associated with colliding spheres differs by a factor of ⅔ (Eq. (5.23)). The end result is a simple inverse dependence of the diffusivities and kinetic viscosity on atomic or molecular mass (m). For liquids, this parameterization is advantageous because the shapes of molecules vary from nearly spherical as in carbon tetrachloride to very long chains of polymers. The other key parameter tied to the substance is the time of interactions during collisions (τ_{delay}) which is certainly temperature-dependent, and likely pressure dependent, but neither was specified in the simple model presented in Chapter 5. Because the equations describing transport properties of liquids must reduce to those of the gas above the critical point, and should also be identical in the limit of point mass constituents, this section adds a few qualifiers to Eq. (5.30) based on well-known characteristics of pourable liquids, and what was learned from examining gas properties in Chapter 5. The adaptations are intended to address crucial differences existing between these states of matter.

Available data on the simplest material known, monatomic gas, showed that D_{heat} was not identical to D_{mass} due to the former being monitored via flow of the photon gas, whereas the latter involves motion of the atoms themselves, which, unlike photons, are impeded by each other. The difference between the diffusivities systematically increasing with atomic size (Figs. 5.9 and 5.10) underlies this contention. For diatomic and polyatomic gas (Fig. 5.11), the difference between D_{heat} and D_{mass} is greater. Hence, lifetimes for mass and heat transport in liquids cannot be identical. Because interactions in liquids are stronger than in gas due to proximity of the constituents, and because molecule shape and orientation during any interactions should have some effect, the lifetime associated with mass diffusion in liquids actually is a combination of several different lifetimes. Let us denote the primary collision of two atoms as being associated with the lifetime τ_{heat}. For mass transport, a secondary collision will have an effect as well. We do not need to account for the far less-frequent tertiary and higher order collisions, because these can be equivalently "lumped" with the secondary collisions. Because inverse lifetimes add, the lifetime associated with mass transport via collisions will be smaller than that with heat. For liquids, the diffusivities associated with inelastic collisions are proposed to follow:

$$D_{heat} \cong \left(\frac{3NR_{gc}T}{m}\right)\tau_{heat} = u_{heat}\tau_{heat} \tag{6.1}$$

$$D_{mass} \cong \left(\frac{3NR_{gc}T}{m}\right)\tau_{mass} = u_{mass}\tau_{mass};\ \tau_{mass} \leq \tau_{heat} \tag{6.2}$$

where N is the number of moles and R_{gc} is the gas constant. The temperature (T) or pressure dependence of the lifetimes are not specified. The RHS of these equations is general, arising from dimensional analysis, where we have indicated that each property is not only associated with a different lifetime, but also is linked to a different carrier velocity, should the mechanisms differ.

Regarding viscosity of gas, two mechanisms exist, idealized as direct and glancing collisions, which represent the forward transfer of momentum and the inhibiting effect of viscous drag. Considering the collisional cross-sections for spheres predicted that kinematic viscosity (ν) should be ⅔ of either diffusivity (Eq. (5.23)), which is observed to nearly hold. For shapes other than spheres, the cross-sections, and therefore the lifetimes of forward momentum transfer and retarding viscous drag will vary from this simple ratio. Because the data reflect both mechanisms, combined, and to preserve the point mass limit, we include the factor of ⅔ while absorbing the complexity into the lifetime:

$$\nu \cong \left(\frac{2NR_{gc}T}{m}\right)\tau_{drag} = u_{drag}\tau_{drag};\ \tau_{drag} \sim \tau_{mass} \quad (6.3)$$

Due to the competition, the size of the lifetime associated with viscous flow relative to that of the diffusivities is not obvious, but for small and round molecules the factor of ⅔ should account for the first-order differences. Use of the subscript "drag" anticipates that forward motions are greatly hindered by frictional forces. At high viscosity, diffusion of momentum may be difficult to nearly impossible.

Eqs. (6.1) to (6.3) reduce to 5.30 for equal lifetimes and velocities. This meets our stipulation that transport behavior converges above the critical point, where these phases are indistinguishable. The above formulation remains based on molecular collisions as the sole mechanism, while allowing for glancing collisions affecting kinematic viscosity. Other mechanisms are represented by the dimensional formulation, but not by the simple proportionality with temperature. Below, we test the predicted m and T dependencies against available data, recognizing that lifetimes can depend on m or T.

6.2 MEASUREMENTS OF TRANSPORT PROPERTIES FOR POURABLE LIQUIDS

Because the purpose of this chapter is to seek trends amongst the data on liquids, we rely on compilations. The transport property best constrained for liquids is D_{mass}, due to a recent, extensive, and downloadable compilation on diverse molecular compounds by Suárez-Iglesias et al. (2015). Their comprehensive review summarizes data at ambient conditions, and various temperatures and pressures, and includes molecular solids and noble gases as well. The figures in this chapter show the accurate Nuclear Magnetic Resonance (NMR) data from this review. The figures average all measurements for any given compound that were made at ambient conditions. Suárez-Iglesias et al.'s (2015) compilation for D_{mass} of diverse pure molecules in the liquid phase by can be considered as inclusive.

All transport properties of all three states of H_2O are well-studied: see the websites listed at the end of the chapter, Assael et al. (2000), Chaplin (2018), Fokin and Kalashnikov (2008), Ramires and Nieto de Castro (1995), and the compilation of Suárez-Iglesias et al. (2015). Online steam tables provide data also on water and water vapor under various conditions. For ice, data sources are James (1968), plus reviews by Weertman (1983) and Fukusako (1990).

In contrast, viscosity data shown in the figures are not from thorough and comprehensive reviews, but rather were obtained from compilations that stem from practical needs. The websites listed at the end of the chapter summarize data used by engineers. Much data exist on dynamic viscosity of diverse materials. Because density is commonly measured, kinematic viscosity is easily computed from dynamic viscosity.

For liquids, fewer data exist on heat transport. Mostly, thermal conductivity (K) is reported in the websites geared for engineering. Fortunately, heat capacity and density are commonly measured, so that most of these data on K can be converted to thermal diffusivity. However, for liquids, the transient hot-wire method is mostly used, which is considered to be accurate to 5% at ambient conditions (Watanabe, 1996; Watanabe and Kato, 2004). A small amount of radiative transfer exists at 298K, and is a contributing factor to the uncertainty. Ballistic radiation affects the temperature derivative as well. However, many liquids that pour at ambient conditions are not stable at high temperatures, and so the data are limited to low temperatures. The expectedly small amount of radiative transfer is not important to our general arguments.

Due to needs of the food industry, data on all three transport properties exist on edible oils, although not all properties have been measured for each of the various types of oil. More importantly, the chemical compositions of the oils are not precisely fixed since these are mixtures, and so the molecular mass is approximate, and the comparison is approximate. Fatty acids have well-defined compositions and tranport properties (Valeri and Meirelles, 1997), but not food oils. We include these complex compounds due to their intermediate viscosities and the number of data, which should average the uncertainties associated with compositional variations. References are: Coupland and McClements (1997), de Freitas-Cabral et al. (2011), Diamante and Lam (2014), Huang and Liu (2009), Xu et al. (2014), and Yáñez-Limón et al. (2005). Data on sugars, syrups, and other foods (Bleazard et al., 1996; Ertl and Dullian, 1973; He et al., 2006; Nguyen et al., 2012; White, 1988) were included for similar reasons.

Lubrication is essential to machinery and vehicles. Viscosity data are available for many different oils and greases. To represent this type of material, we include silicone oils, because all three transport properties are well-characterized, plus, these oils have the same building block, chained together. Many physical properties are provided by manufacturer's websites (listed at the end of the chapter). Data on low molecular mass silicone oils are summarized by Roberts et al. (2017). Thermal conductivity and heat capacity were measured by Sandberg and Sundqvist (1982). Data on large molecular mass oils are given by Koschmieder and Pallas (1974) and Thern and Lüdemann (1996), where the latter is the source of D_{mass} values.

Last but not least, we examine data on the simplest possible liquids, the elements. All three transport properties of the two metals that are liquid at or very close to ambient conditions (Ga and Hg), are well-studied (Aurnou and Olsen, 2001; Blagoveshchenskii et al., 2015; Meyer, 1961; Schriempf, 1972, 1973). The alkali metals (Cs and Rb) which melt near ambient conditions (at 29°C and 39°C, respectively) are less well-characterized. Sources are Andrade and Dobbs (1952) and the compilations of Vargaftik et al. (1994) and Iida et al. (2006). Specific heats of solid alkali metals were used which are identical at the melting point (Filby and Martin, 1966), but are affected by phase transitions (λ-shaped curves). Mercury, being further from its melting point at -38°C, should not be affected. In contrast, data on the two nonmetals, Br_2 and I_2, are rather uncertain, as discussed in the review by Touloukian and Ho (1981). Because Cl_2 is liquid at T fairly close to ambient, we include these data as well, slightly extrapolated, from the same compilation, recognizing that this is more uncertain. Mass diffusivity data have not been published, to the best of our knowledge.

A more detailed investigation than that presented here is warranted. However, the graphical analysis below and comparison with trends for gas transport properties from

Chapter 5 establish that modeling collisions during translations insufficiently describes transport properties of liquids.

6.3 TRENDS IN TRANSPORT PROPERTIES OF POURABLE LIQUIDS

6.3.1 Mass and Density Effects at Ambient Conditions

At ambient temperature and pressure, distinct trends exist for each of the three transport properties with molecular mass, although considerable scatter exists in the data (Fig. 6.1). Molecular compounds, including the complex oils and simple diatomics, define the observed trends: mass diffusivity inversely depends on molecular mass, as expected, whereas thermal diffusivity is approximately constant, and, contrastingly, kinematic viscosity is directly proportional to m.

The diatomic molecules describe a trend that is slightly steeper but lower than that of the more complex molecular compounds. The slight increase in D_{heat} with m which exists for the homogenous diatomics may be due to the large uncertainties for bromine and iodine, and with chlorine data being extrapolated, but a similar increase describes the

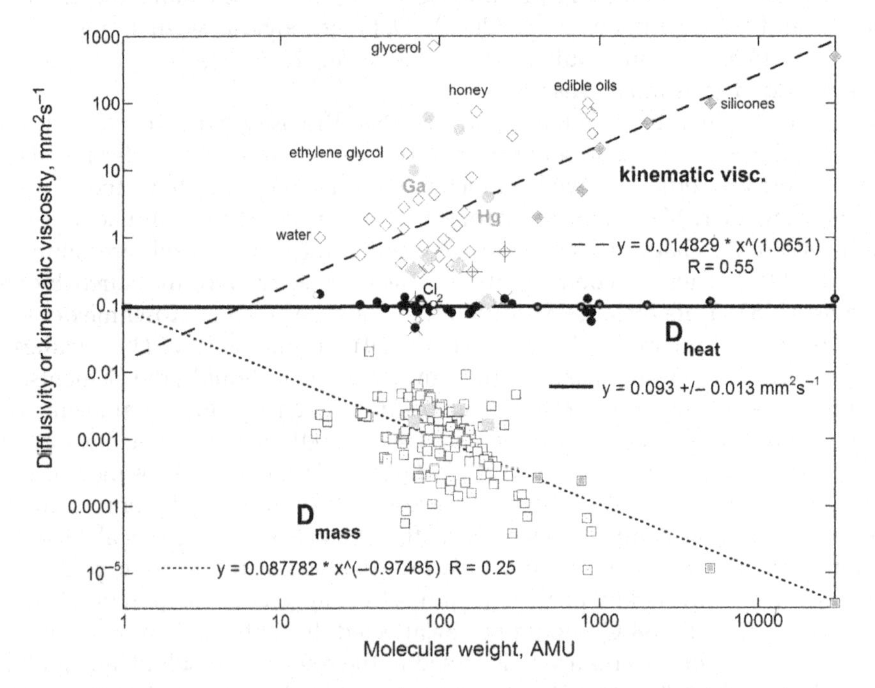

FIGURE 6.1 Dependence of measured transport properties of liquids near 298K and ambient pressure on mass of their molecules. Data on D_{mass} (squares and dotted line) mostly from the compilation of Suárez-Iglesias et al. (2015). Data on D_{heat} (dots and solid line) and on ν (diamonds and dashed line) from websites or publications as described in Section 6.2. Silicone oils (dots with gray interiors) and diatomic homogenous molecules (superimposed × or + signs) were included in the least squares fits shown. Chlorine viscosity is hidden by the D_{heat} trend. Large gray symbols: liquid metals, which were not including in the fitting. Open circles: heat transport at 300K, which was not included in the fits, but also appears to be independent of m.

accurate measurements of the silicone oils. The edible oils cluster together, which probably stems from approximated values for m. Sugary substances have larger ν than the molecular compounds, which seems connected with these compounds flowing, yet being sticky to the touch. Both food types also have higher D_{heat} but lower D_{mass} than the trends for the molecular compounds. The metallic elements have D_{mass} close to values for the molecular compounds, but possess much larger D_{heat} and lower viscosity.

Unlike the molecular compounds and diatomics, transport data for liquid metals do not depend on mass (Fig. 6.1), but instead follow an inverse power law with density (Fig. 6.2). The power-law fits are steep for thermal diffusivity, moderate for kinematic viscosity, and nearly flat for mass diffusivity. Roughly, the trend for kinematic viscosity of the metals approximates the lowest n for molecular solids while the trend for D_{mass} of the metals approximates the highest values for molecular solids. This behavior suggests an inverse correlation of D_{mass} with viscosity, which is explored further below. Fig. 6.2 further emphasizes that the thermal diffusivity of the liquid metals differs greatly from D_{heat} of the molecular compounds, and that density is not a controlling parameter for the compounds, but does have some effect.

For molecular liquids, D_{heat} is constant (Figs. 6.1 and 6.2). In round numbers, the thermal diffusivity equals 0.1 $mm^2\,s^{-1}$ within 10%, and furthermore occupies a very restricted range of 0.04–0.15 $mm^2\,s^{-1}$. Such behavior is inconsistent with molecular collisions, for which a strong mass dependence is expected. For Eq. (6.1) to be valid, requires that the lifetime be proportional to the molecular mass. This is reasonable since large molecules may become entangled with each other. Moreover, D_{heat} for very simple molecular configurations (the diatomics and the silicone oils) increases with molecular mass, suggesting that a simple dependence of lifetime on mass for all three properties does not exist. To explore this further, thermal diffusivity for restricted ranges of m was fit to power laws in density. The results of Table 6.1 show that D_{heat} increases slightly with ρ, under nearly

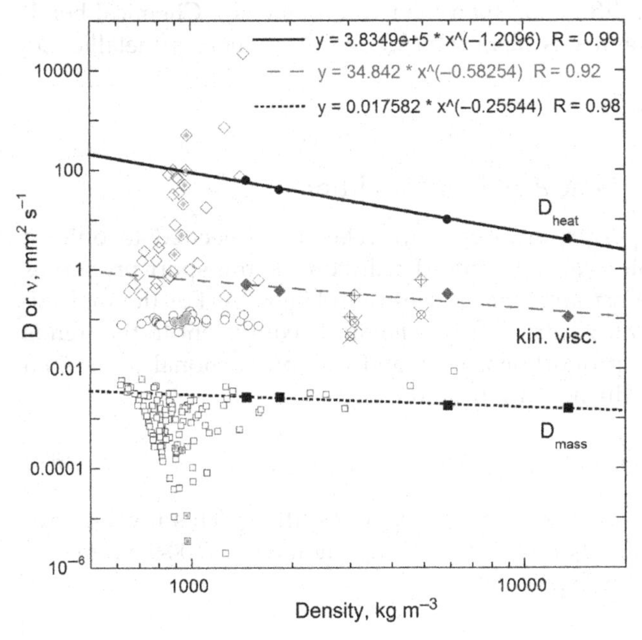

FIGURE 6.2 Dependence of measured transport properties of liquids at 298K and ambient pressure on density. Filled black symbols and lines depict liquid metals. Least squares fits are listed at the top. Other symbols as in Fig. 6.1. Data sources listed in Section 6.2.

TABLE 6.1 Fits of D_{heat} at 298–300K as a Function of Density for Molecular Liquids

Mass Range amu	No. Pts.	Density Range kg m^{-3}	Power-Law Fit	Correlation Coefficient
58–64	6	760–1160	$0.016883\ \rho^{0.251}$	0.21
70–79	9	610–1270	$0.0007126\ \rho^{0.7116}$	0.77
86–93	6	620–1280	$0.029135\ \rho^{0.1621}$	0.34
142–170	5	730–3100	$0.13734\ \rho^{-0.0673}$	0.41
253–352	3	860–4960	$0.039303\ \rho^{0.1312}$	0.92

constant molecular mass. Table 6.1 and Figs. 6.1 and 6.2 suggest that a mechanism other than inelastic collisions governs diffusion of heat in liquids composed of molecules.

The trend for D_{mass} extrapolates to the average for D_{heat} at $m = 1$ amu, which reasonably approximates a point mass atom (i.e., monatomic H). The trend for ν extrapolates to a slightly lower value at $m = 1$ amu. This behavior is consistent with our model for gas, which indicated $\nu = 2/3\ D_{heat} = 2/3\ D_{mass}$, if the cross-sections are described by inelastic collisions of ideal spherical atoms. However, the magnitudes of the material properties for liquids and gas are much different. For example, the lowest value for D_{heat} of gas is 3.8 mm^2 s^{-1} (for radon), which is $\sim 20\times$ larger than D_{heat} of liquids. The discrepancy between mass self-diffusivities is even larger. Although kinematic viscosity magnitudes are similar for liquids and gas, this is fortuitous, as follows:

The trends of Fig. 6.1 for heat transport and viscous drag are not consistent with collision based transport, although mass transport in molecular liquids appears to be associated with collisions. Liquids, like gas, have $D_{heat} > D_{mass}$, but the discrepancy in magnitude is larger and kinematic viscosity behaves much differently than the two diffusivities. Chemical bonding appears to affect drag because nonmetals have $\nu > D_{heat} > D_{mass}$ whereas metallic elements have $D_{heat} > \nu > D_{mass}$.

6.3.2 Ratios of Transport Properties at Ambient Conditions

Understanding how transport properties of liquids are related are needed to build a model. To eliminate the effects of molecular mass and density on the transport properties at STP, mass diffusivity is plotted against kinematic viscosity in Fig. 6.3A. For the molecular compounds, D_{mass} inversely correlates with ν. This behavior is equivalent to the trends in Fig. 6.1: because D_{mass} is inversely proportional to m and ν is proportional to m, then $D_{mass} = \text{constant}/\nu$. Setting the power during fitting as unity gives:

$$D_{mass} = \frac{3}{2}\frac{1}{1000\nu} \tag{6.4}$$

where the constant of 3/2 is uncertain by 5% from least squares fitting. This inverse provides a simple and testable prediction. For bromine, D_{mass} should be 0.00495 mm^2 s^{-1} at STP and for iodine, 0.00025 mm^2 s^{-1} is expected.

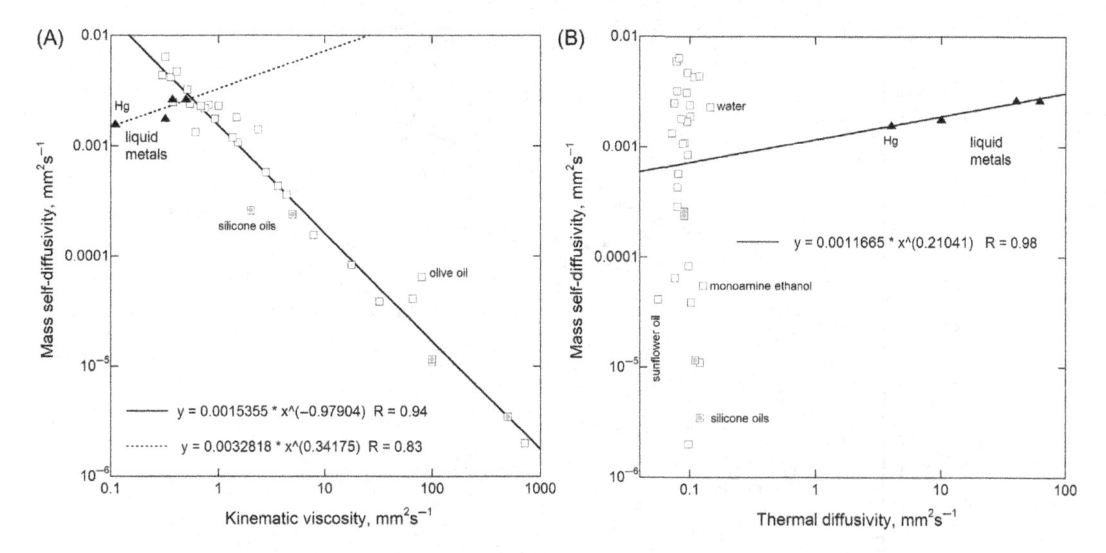

FIGURE 6.3 Cross correlations of transport properties for pourable liquids. (A) Dependence of D_{mass} on ν. (B) Dependence of D_{mass} on D_{heat}. Squares and solid line: molecular compounds. Gray interiors show the subset of silicone oils. Triangles and dotted line: liquid metals. Section 6.2 lists data sources.

For molecular compounds, the two diffusivities are uncorrelated (Fig. 6.3B). This result corroborates our inference of different mechanisms controlling their heat and mass transport.

The direct correlations of D_{mass} with ν and of D_{mass} with D_{heat} (Fig. 6.3) for the metallic elements are directly related to the trends in Fig. 6.2. Their viscosity represents a substantial component of forward diffusion of momentum, but with more inhibiting drag than occurs for the gases. Likewise, heat transfer in liquid metals involves collisions, but the mechanism differs substantially from that in gases.

To probe the lifetimes, ratios of the transport properties are useful (Fig. 6.4). With this approach, a much smaller database is explored, but one that involves well-studied substances. The power laws for the ratios are consistent with the fits in Fig. 6.1 for the molecular compounds and with the metals not depending on atomic mass. For tiny molecules in this subset of the data described in Section 6.2, the ratios converge to $1/2$. For collisions of spheres, ratios of unity and $2/3$ are expected. Agreement is reasonable, given that neither D_{heat} nor ν behave as expected for simple, collisional transport.

Ratios of liquid metal transport properties follow power laws in density (not shown) as expected from Figs. 6.2 and 6.3. These do not project to an intercept. For liquid metals, $D_{\text{mass}}/D_{\text{heat}}$ is proportional to density, where ν/D_{heat} goes as $\rho^{0.63}$.

6.3.3 The Temperature Dependence of Transport Properties

Well-studied substances are considered here: water, simple molecular solids, liquid Hg metal, and silicone oils. Because the data on the homogenous diatomics are not very accurate (Touloukian and Ho, 1981) and because these behave similar to more complex molecular compounds, we do not examine the T dependence for this category in detail. For

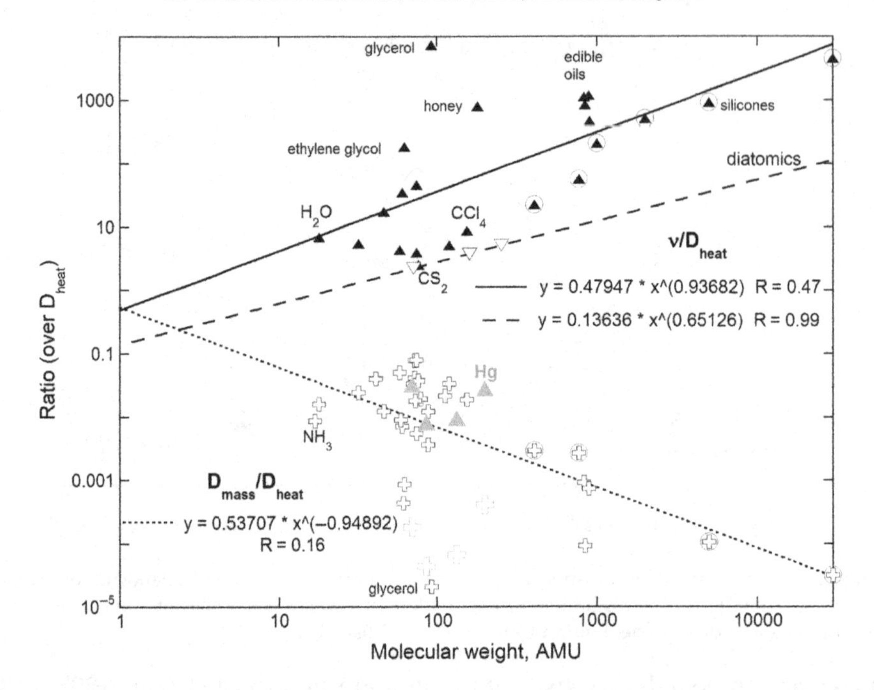

FIGURE 6.4 Ratios of the transport properties as a function of mass. Triangles = kinematic viscosity divided by thermal diffusivity: open for the diatomic nonmetals and gray for the elemental metals. Plus signs: mass diffusivity over thermal diffusivity (gray for liquid metals). Circles: silicone oils. Several compounds are labeled. Section 6.2 lists data sources.

bromine, D_{heat} is nearly constant ($=0.082\ mm^2\ s^{-1}$) from freezing to boiling whereas viscosity decreases strongly: $\nu(T) = 737200T^{-2.58}\ mm^2\ s^{-1}$.

6.3.3.1 Water, With a Comparison to Ice and Steam

For liquid water, the three transport properties depend on T in much different ways (Fig. 6.5). Thermal diffusivity weakly increases with temperature, whereas kinematic viscosity strongly decreases, while mass diffusivity strongly increases. Projected trends for D_{heat} and ν converge, but both always greatly exceed D_{mass}. For D_{mass}, essentially the same second-order polynomial in T describes both the liquid and solid phases. For each of water and ice, D_{mass} was not well-described by a power law. For each of water and ice, D_{heat} weakly depends on T and either a linear or a power-law fit suffices for the small ranges of T examined. These measurements involve non-steady-state techniques (hot-wire and Ångstrom's) and are likely influenced by ballistic transport, due to partial transparency of H_2O in the infrared (Fig. 2.7). However, near room temperature ballistic augmentation is small, so the order-or-magnitude difference in D_{heat} represents the effect of freezing on H_2O. Similarly, rheologic behaviors of water and ice differ. Water is a Newtonian fluid: nonetheless, v depends on T in a complex manner. For an accurate fit, four parameters are required for an exponential form (Fig. 6.5), while five parameters are needed for a polynomial fit (not shown). Viscosity of ice is not shown because its behavior non-Newtonian

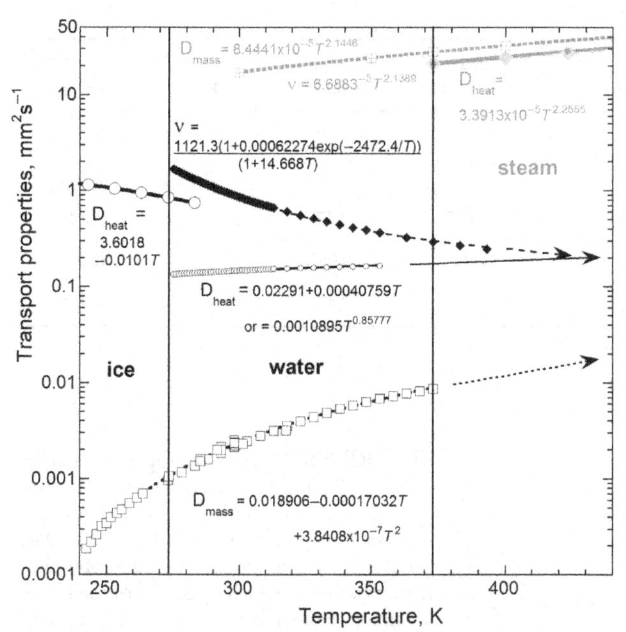

FIGURE 6.5 Transport properties of water and other phases of H_2O as a function of temperature. Least squares fits have correlation coefficients of 0.99. Arrows: extrapolations for water trends to high T. Thermal diffusivity of water (small circles) and ice (large circles) depend linearly on T, but differ in slope and are discontinuous at freezing. For D_{mass}, one trend describes both solid and liquid phases (open squares), but the gas (gray squares) differs greatly, although the fit is also a second-order polynomial. For gas, thermal diffusivity (gray dots and line, with fit listed) and kinematic viscosity (gray diamonds and dashes) are described by similar fits that parallel steam mass diffusivity. Section 6.2 lists data sources; see also Chaplin (2018).

(Weertman, 1983), and thus the creeping behavior of ice is mechanistically distinct. Although creep has been considered in Earth science to be mathematically equivalent to Newtonian viscous flow, this parallelism ignores the grain-size dependence of creep and the often used formulations in the engineering literature (Meyers and Chawla, 2009), see discussion of Hofmeister and Criss (2018).

Transport properties for steam are much higher than those of water or ice, and follow simple power laws in T. The observed behavior (Fig. 6.5, gray symbols) with T follows that of the small molecules CO_2 and NH_3 (Fig. 5.11), whereby D_{mass} is larger at high T and the powers in T are similar. The behavior and mechanisms of transport in steam are much simpler than in the two condensed phases of H_2O.

6.3.3.2 Small Molecule Liquids

In comparing the transport properties of molecular compounds, we opted for convenience: this section considers all small molecules on the viscopedia website for which kinematic viscosity as a function of temperature was listed. Mass diffusivity for all of these (Table 6.2) were compiled by Suárez-Iglesias et al. (2015). Thermal diffusivity was obtained from various publications and websites described in Section 6.2.

Fig. 6.6 shows that small molecules have similar values exist for each of the three transport properties. Moreover, the overall behavior is like that of water and bromine, where ν decreases with temperature, while D_{heat} is nearly constant, but D_{mass} increases as T increases. As observed for H_2O, D_{mass} is continuous across the freezing point and is described by a polynomial when the data cover more than a narrow range of 100K. However, the strong curvature seems to be associated with the freezing point. Unlike

TABLE 6.2 Properties of Small Molecule Liquids

Name	Formula	m amu	ρ^a kg m^{-3}	D_{heat} mm^2 s^{-1}	T_{freeze} K	T_{boil} K	Volume m^3 mol^{-1}
Benzene	C_6H_6	78	871	0.094[a]	279	353	0.0896
Toluene	C_7H_8	92	862	0.102[a]	178	384	0.106
Methanol	CH_4O	32	787	0.105[a]	175	338	0.0407
Hexane	C_6H_{14}	86	655	0.089[a]	178	342	0.131
Pentane	C_5H_{12}	72	621	0.078[b]	143	309	0.116

[a]*At 298K.*
[b]*At 300K.*

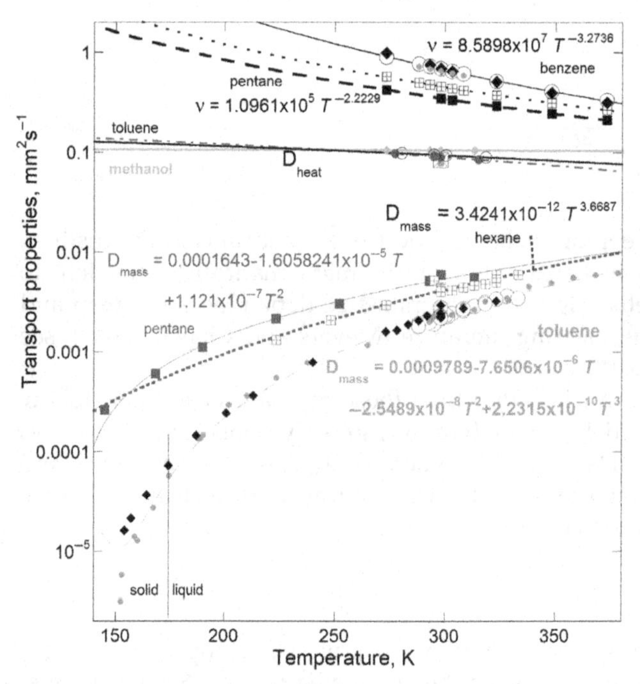

FIGURE 6.6 Transport properties of small molecule liquids as a function of temperature. Circles: benzene; dots: toluene; diamonds: methanol; open squares: hexane. Filled squares = pentane. Power laws describe kinematic viscosity, which was determined over a small temperature range. For hexane, $\nu = 463{,}300T^{-2.4275}$. Thermal diffusivity values are nearly independent of T. Mass diffusivity over a wide temperature range is described by a polynomial fit, whereas narrow ranges can be fit to a power law. For benzene, $D_{mass} = 2.6496 \times 10^{-15}T^{4.8211}$. For toluene and pentane, D_{mass} was collected at very low T and is continuous across the freezing point.

water, the fits for ν were simpler: kinematic viscosity follows a simple power law over the temperature range explored.

In addition, the transport properties follow a prescribed order. Both D_{heat} and ν decrease according to the order:

$$\text{benzene} \geq \text{methanol} \sim \text{toluene} > \text{hexane} > \text{pentane} \tag{6.5}$$

The same order holds for $1/D_{mass}$, as expected from its inverse correlation with ν at 298K (Fig. 6.3). This order is controlled by density, except for methanol, which has a much lower molecular mass (Table 6.2). The inverse correlation at 298K does not hold at all temperatures. For example, the product $\nu \times D_{mass}$ equals $1.5\times10^{-6}T^{1.24}$ for hexane or is $2.276\times10^{-7}T^{1.55}$ for benzene, which have simpler fits for D_{mass} than do methanol, pentane, or toluene.

6.3.3.3 *Silicone Oils*

Fig. 6.7 shows mass diffusivity measured for four different silicone oils, which are described by a peak in molecular mass, as listed (Thern and Lüdemann, 1996). Viscosity was inferred from the reported molecular mass, rather than the reported density, which was not entirely consistent with manufacturer's reports. Polydimethylsiloxane fluids are linear-chain polymers with chemical formulas of $(SiO(CH_3)_2)_x$. Because the lowest-viscosity silicone oil (0.65 cSt) has the shortest chain possible, it alone has a fixed composition of $O(Si(CH_3)_3)_2$. More viscous oils contain polymers of various lengths and the range of lengths increases with the peak mass, as is evident in the mass distribution plots of Thern and Lüdemann (1996). Oils from different manufactures may be produced by blending, and so the distributions of mass may differ (Roberts et al., 2017). Therefore, curves for ν in Fig. 6.7 do not correspond exactly to D_{mass} measurements. In addition, evaporation is significant by 400K for the low viscosity oils (Roberts et al., 2017). Because measurements of ν have been made many times for many different oils, ν vs T is well-constrained, leading to a regular trend in the power-law fits. This may not be the case for D_{mass}, as suggested by the less regular trend in the power for T. Because of these differences, we do not plot ratios of the transport properties, and focus on the trends, but not the details.

The temperature response of the three transport properties is regular and similar to that of water and the small molecule hydrocarbons (Figs. 6.5 and 6.6) which have much lower

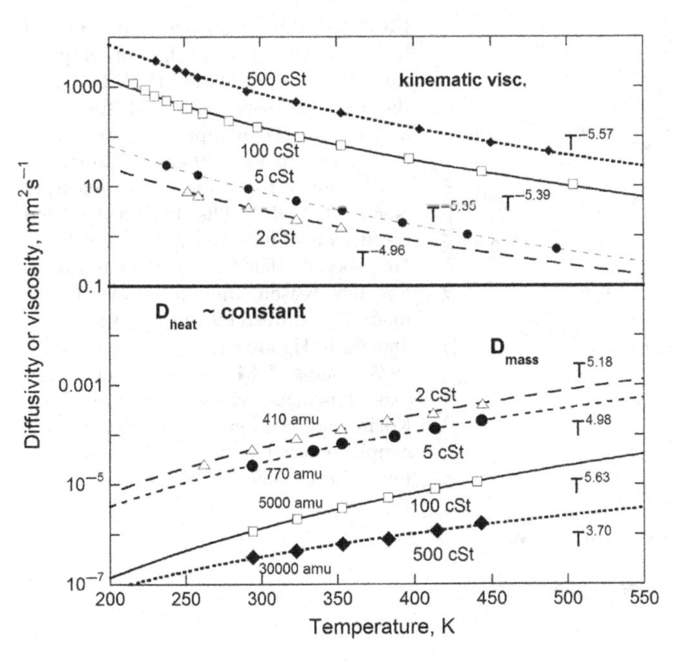

FIGURE 6.7 Transport properties of silicone oils as a function of temperature. Data from Thern and Lüdemann (1996) and the Clearco website, see text. Thermal conductivity data of Roberts et al. (2017) indicate D_{heat} is independent of temperature. The three transport properties were not determined from the exact same substance. Power-law fits are shown, but without the prefactor.

m (18–92 amu). Kinematic viscosity and D_{mass} appear to be inversely correlated whereas ν and D_{heat} appear to converge at high *T*. The temperature dependencies are simpler for the more complex silicone oils. However, departures from power laws are observed at the coldest temperatures accessed for 100 cSt oil (Fig. 6.7).

6.3.3.4 Liquid Metallic Mercury

Large disparities in values and trends with temperature exist among the three transport properties for Hg (Fig. 6.8), as observed for nonmetals (Figs. 6.5–6.7), although transport properties for these two types of matter differ in many respects. Mass diffusivity with weak curvature over a narrow *T* range is like that of nonmetals. The weak curvature for Hg was not fit, given the stated uncertainty and that interlaboratory comparisons of suggest uncertainties up to 20% for the tracer methods that are applied to metals (Galus, 1984). The trend for ν of Hg with *T* is complicated, but similar to a decaying power law, as in nonmetals. Liquid metals offer less resistance to matter flow, which can be attributed to the small size and round shapes of the cations, compared to the larger and less regular shapes of neutral molecules.

Thermal diffusivity of Hg depends moderately and linearly on *T* (Fig. 6.8), rather than being flat and nearly constant, as in the nonmetallic liquids. It is unlikely that atomic motions are involved, due to D_{mass} being 0.05% of D_{heat}. One concern is use of containers in laser flash analysis (LFA) measurements. Schreimpf's (1972) measurements predate three-layer models (Chapter 4). The effect of the supporting walls should be small, and his data were in reasonably agreement with earlier measurements. Thermal conductivity

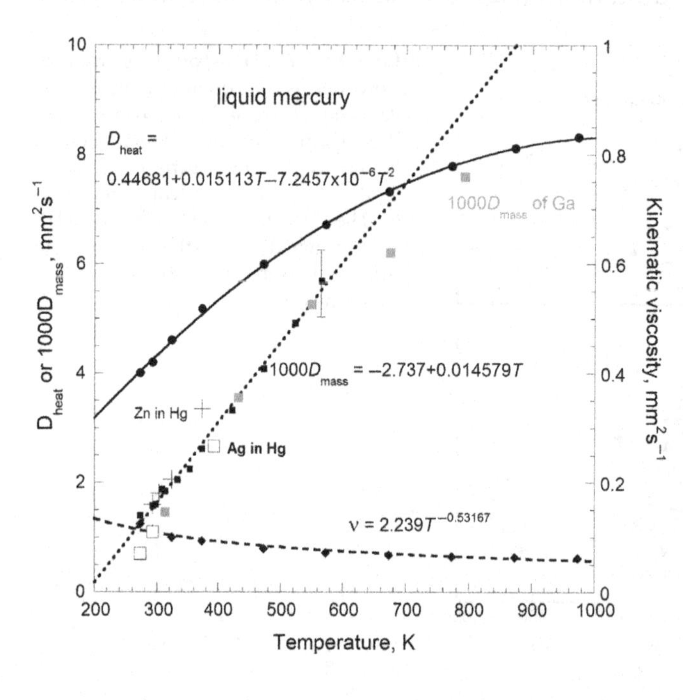

FIGURE 6.8 Transport properties of liquid mercury as a function of temperature over its entire liquid field. Left axis: thermal diffusivity (dots, with the range and fit from Schriempf, 1972) and mass diffusivity (black squares, shown as ×1000 measured values with data from Nachtrieb, 1967). Fits have correlation coefficients of 0.99. Uncertainties in D_{mass} are generally stated to be 5% as indicated. For this reason, only a linear fit was made and diffusivities of Ga, and of Zn and Ag in Hg are also shown (Schriempf, 1973; Galus 1984, respectively). Right axis: kinematic viscosity, compiled by Kozin and Hansen (2013), was fit to a simple power law. For an accurate fit, a fourth-order polynomial is needed.

measurements involve contact losses as well. It is unlikely that the corrections address all the issues in the older data. Hence, additional measurements of D_{heat} for liquid metals are needed.

6.3.4 Effect of Pressure on Transport Properties

As discussed in Chapter 4, measurements of heat transport at pressure are problematic mostly due to contact losses. We therefore only discuss how viscosity and D_{mass} respond to pressure.

Nuclear magnetic resonance methods used to obtain D_{mass} for molecular liquids are accurate (Suárez-Iglesias et al., 2015), and are amenable to studies at pressure. All the compounds listed in Table 6.2 behave similarly. Fig. 6.9A shows isothermal curves for toluene as a function of P. Isobaric data on D_{mass} were also plotted as a function of T: these curves are roughly parallel (not shown). However, as discussed in Chapter 1 and Chapter 5, the fundamental variable is volume, not pressure. Isothermal curves for D_{mass} as a function of volume are parallel at large volumes and essentially linear (Fig. 6.9B). For small volumes, the changes are less rapid, requiring polynomial fits over large ranges of V, whereas small ranges of V and compressed material are exponential. Mass diffusivity for any given liquids is a function of T and V which is affected by the mass of the molecule. At a constant and large volume, the change in D_{mass} with T is roughly linear, from Fig. 6.9B.

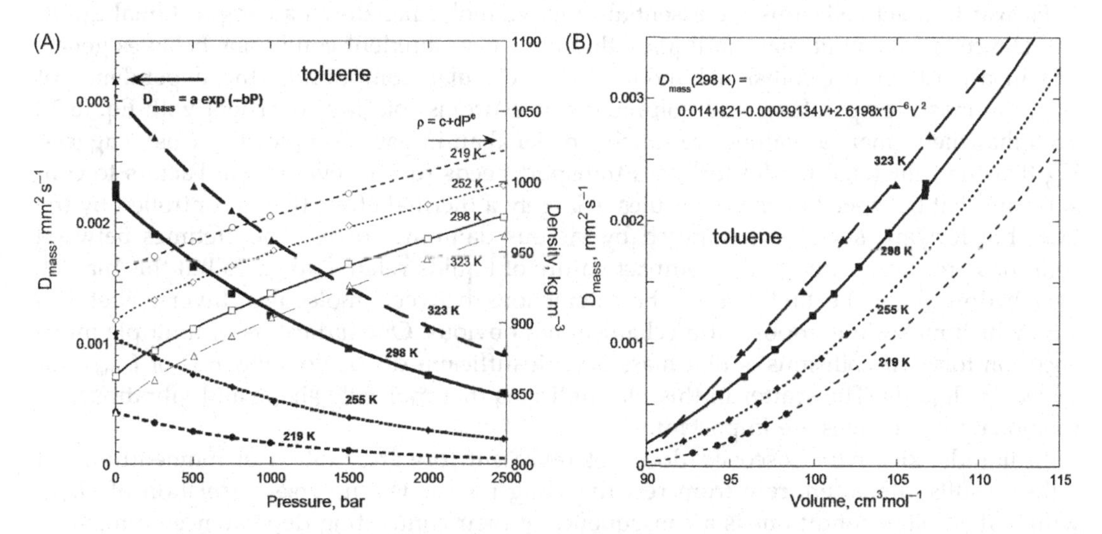

FIGURE 6.9 Mass self-diffusivity and density of liquid toluene collected as a function of pressure and temperature. NMR measurements by Harris et al. (1993), as tabulated by Suárez-Iglesias et al. (2015). Each isothermal data set is shown with a particular symbol and line pattern, as labeled. (A) As P increases, isothermal D_{mass} decays exponentially whereas density increases as a modified power law. (B) As volume increases, isothermal D_{mass} increases. The data above ambient temperature require quadratic fits. Low T data can be fit to exponentials or power laws, due to lower ranges in V (or P).

Although a volume dependence was not specified in Eq. (6.2), rarefication certainly affects speeds, causing u to decrease as V increases. Because D_{mass} increases with V, a strong increase in τ_{mass} is indicated. The trends in D_{mass} are consistent with a collisional mechanism for mass transport, but one that is severely impeded by strong molecular interactions in the liquid state.

It is long known that viscosity increases under compression for pourable liquids except for water below 306K (e.g., Bridgman, 1925). The caused is enhanced frictional drag due to tighter packing of the molecules. The unusual behavior of water stems from its density maximum near 277K, that is, the molar volume of water responds anomalously to pressure around this temperature, and so must viscosity. For measurements of supercooled water and a detailed description of the fascinating changes in viscosity with T and P, see Singh et al. (2017). Hence, viscosity and mass diffusivity for pourable liquids respond in opposite directions to mass, temperature, pressure, and volume, and cannot involve the same mechanism of direct collisions which underlies the kinetic theory of gas.

6.4 MECHANISMS OF TRANSPORT IN LIQUIDS

The contrasting behavior of the three transport properties for liquids with basic physical properties and state variables (molecular mass or size, density, temperature, volume, and pressure) indicates that each property is associated with a distinct mechanism. Yet, some ties exist between the transport properties through the motions and interactions of the molecules and through the equation of state (EOS).

Forward, direct collisions are essential to move molecules down a compositional gradient. Hence, diffusion of mass in liquids that pour near ambient conditions behaves generally in accord with a collisional model. For molecular compounds, the dependence of D_{mass} on mass, temperature, and volume (or pressure) is roughly consistent with Eq. (6.2), such that the numerical values are much smaller than in gas. Complexities exist, suggesting that the collisional model for mass transport needs further evaluation. Factors to consider are that the speed of the molecules is largely a thermal effect that is controlled by the EOS, but forward speed is hampered by viscous damping forces. The lifetimes between collisions are shorter due to the compact nature of liquids relative to gases but this may be overshadowed by the lifetimes of the interactions between molecules. Inverse lifetimes add, which makes the important mechanism not obvious. One indication of multiple influences on forward collisions is that mass is an insufficient descriptor (Fig. 6.1) of D_{mass} for molecular liquids. The scatter in this plot indicates that molecule shape and vibrational or rotational interchanges are important.

In liquids, kinematic viscosity does not represent forward motion of momentum, but rather results almost entirely from resistive drag forces. The inverse correlation of D_{mass} with ν at ambient conditions is a consequence of their contrasting dependence on molecular mass, which was used to represent the size of the molecules. That D_{mass} and ν are affected in opposite directions by mass, T, V, density, and P results from their contrasting mechanisms of propulsion forward by direct collisions and inhibition of motions by frictional glancing collisions. We reiterate that the equations describing mass diffusion and viscosity, derived by Fick and Newton, respectively, are unrelated.

Thermal diffusivity is independent of mass and density for molecular liquids (Figs. 6.1 and 6.2), which together show that volume is unimportant. Because D_{heat} measurements on molecular liquids are subject to ballistic radiative transfer, the temperature dependence, albeit weak, probably is stronger than the intrinsic behavior (Chapter 4). Thus, D_{heat} is largely independent of T, as well as m, density, and volume, which indicates that P has negligible influence. Yet another mechanism is required for this transport property. For liquids, as in gases, forward and glancing collisions as well as interactions of the molecules (generally vibrational) generate photons. Thermal diffusivity is ascertain by sampling photon flux outward.

Metallic liquids similarly show evidence for three different transport mechanisms because their properties also differ greatly. However, transport properties in metallic liquids are governed by density, rather than molecular mass. The mechanisms for flow and mass transport in liquid metals are similar to those in molecular solids, given the results of Section 6.3. However, heat transport in liquid metals differs by being larger and temperature-dependent. The increase should not be due to ballistic transport, since metals are opaque over a wide spectral range. The limited data on liquid metals for D_{heat} suggest that the mechanism operates like that of the solid state (Criss and Hofmeister, 2017; Chapter 9).

The effect of compression on D_{mass} of toluene suggests that the scatter in Fig. 6.1 for the molecular compounds is related to density and volume effects, as well as molecule shape. Tying all the data together in a model requires a deeper analysis than that presented here, and more data than currently available. We have shown that the mechanisms for heat transport in liquids differ from simple, molecular collisions in gas. Yet, liquids, like gases, will have a photon gas intermingled. Plus, the photon gas is sampled during heat transport measurements in all cases (Chapters 1 and 5). To understand this, we next delve into solids with reliable data on D_{heat} as a function of temperature.

Values of transport properties projected to the point mass are consistent with predictions from the kinetic theory of gas, amended to account for glancing collisions affecting ν. The success in applying this model, which was derived for the unrealistic ideal gas, to pourable liquids stems from this unachievable limit being compatible with the model and to combining many parameters in fitting, which can mask imperfections. If point masses could collide, then these collisions would be elastic, but no time evolution would be possible.

References

Andrade, E.Nda C., Dobbs, E.R., 1952. The viscosities of liquid lithium, rubidium and caesium. Proc. R. Soc. A 211, 12–30.

Assael, M.J., Bekou, E., Giakoumakis, D., Frienda, D.G., Killeen, M.A., Millat, J., et al., 2000. Experimental data for the viscosity and thermal conductivity of water and steam. J. Phys. Chem. Red. Data 29, 141–166.

Aurnou, J.M., Olsen, P.L., 2001. Experiments on Rayleigh-Bénard convection, magnetoconvection, and rotating magnetoconvection in liquid gallium. J. Fluid. Mech. 430, 283–307.

Blagoveshchenskii, N., Novikov, A., Puchkov, A., Savostin, V., Sobolev, O., 2015. Self-diffusion in liquid gallium and a hard sphere model. EPJ Web of Conferences 83, No. 02018.

Bleazard, J.G., Sun, T.F., Teja, A.S., 1996. The thermal conductivity and viscosity of acetic acid-water mixtures. Int. J. Thermophys. 17, 111–125.

Bridgman, P.W., 1925. The viscosity of liquids under pressure. Proc. Natl. Acad. Sci. 11, 603–606.

Chaplin, M., 2018. Water structure and science. http://www1.lsbu.ac.uk/water/water_structure_science.html (accessed September 1, 2018).

Coupland, J.N., McClements, D.J., 1997. Physical properties of liquid edible oils. J. Am. Oil. Chem. Soc. 74, 1559–1564.

Criss, E.M., Hofmeister, A.M., 2017. Isolating lattice from electronic contributions in thermal transport measurements of metals and alloys and a new model. Int. J. Mod. Phys. B 31, No. 175020.

de Freitas-Cabral, A.J., de Oliveira, P.C., Moreira, S.G.C., Alcantara Jr., P., 2011. Thermal diffusivity of palm olein and compounds containing β-carotene. Int. J. Thermophys. 32, 1966–1972.

Diamante, L.M., Lam, T., 2014. Absolute viscosities of vegetable oils at different temperatures and shear rate range of 64.5 to 4835 s^{-1}. J. Food Process. No. 234583, 1–6.

Ertl, H., Dullian, A.L., 1973. Self-diffusion and viscosity of some liquids as a function of temperature. AIChE J. 19, 1215–1223.

Filby, J.D., Martin, D.L., 1965. The specific heats below 320°C of potassium, rubidium and caesium. Proc. R. Soc. A 284, 83–107.

Fokin, L.R., Kalashnikov, A.N., 2008. The viscosity and self-diffusion of rarefied steam: Refinement of reference data. High Temp. 46, 614–619.

Fukusako, S., 1990. Thermophysical properties of ice, snow, and sea ice. Int. J. Thermophys. 11, 353–372.

Galus, Z., 1984. Diffusion coefficients of metals in mercury. Pure Appl. Chem. 56, 635–644.

Harris, K.R., Alexander, J.J., Goscinsa, T., Malhotra, R., Woolf, L.A., Dyamond, J.H., 1993. Temperature and density dependence of the selfdiffusion coefficients of liquid *n*-octane and toluene. Molec. Phys. 78, 235–248.

He, X., Fowler, A., Toner, M., 2006. Water activity and mobility in solution of glycerol and small molecular weight sugars: implication for cryo- and lyoperservation. J. Appl. Phys. 100, No. 074702.

Hofmeister, A.M., 2010. Scale aspects of heat transport in the diamond anvil cell, in spectroscopic modeling, and in Earth's mantle: implications for secular cooling. Phys. Earth Planet. Inter. 180, 138–147.

Hofmeister, A.M., Criss, E.M., 2018. How properties that distinguish solids from fluids and constraints of spherical geometry suppress lower mantle convection. J. Earth Sci. 29, 1–20. Available from: https://doi.org/10.1007/s12583-017-0819-4.

Huang, L., Liu, L.S., 2009. Simultaneous determination of thermal conductivity and thermal diffusivity of food and agricultural materials using a transient plane-source method. J. Food Eng. 95, 179–185.

Iida, T., Guthrie, R., Tripathi, N., 2006. A model for accurate predictions of self-diffusivities in liquid metals, semimetals, and semiconductors. Metall. Mater. Trans. B 37, 559–564.

James, D.W., 1968. The thermal diffusivity of ice and water between −40 and +60° C. J. Mater. Sci. 3, 540–543.

Koschmieder, E.L., Pallas, S.G., 1974. Heat transfer through a shallow, horizontal convecting fluid layer. Int. J. Heat Mass Trans. 17, 991–1002.

Kozin, L.F., Hansen, S.C., 2013. Mercury Handbook: Chemistry, Applications and Environmental. Royal Society of Chemistry, Cambridge, UK.

Meyer, R.E., 1961. Self-diffusion of liquid mercury. J. Phys. Chem. 65, 567–568.

Meyers, M.A., Chawla, K.K., 2009. Mechanical Behavior of Materials. Cambridge University Press, Cambridge, UK.

Nachtrieb, N.H., 1967. Self-diffusion in liquid metals. Adv. Phys. 16, 309–323.

Nguyen, L.T., Balasubramaniam, V.M., Sastry, S.K., 2012. Determination of in-situ thermal conductivity, thermal diffusivity, volumetric specific heat and isobaric specific heat of selected foods under pressure. Int. J. Food Prop. 15, 169–187.

Ramires, M.L.V., Nieto de Castro, C.A., 1995. Standard reference data for the thermal conductivity of water. J. Phys. Chem. Ref. Data 24, 1377–1381.

Roberts, C., Graham, A., Nemer, M., Phinney, L., Garcia, R., Stirrup, E., 2017. Physical properties of low-molecular weight polydimethylsiloxane fluids. Sandia Report, SAND2017-1242. https://doi.org/10.2172/1343365 (accessed 20.12.17).

Ross, R.G., Andersson, P., Sundqvist, B., Bäckström, G., 1984. Thermal conductivity of solids and liquids under pressure. Rep. Prog. Phys. 47, 1347–1402.

Sandberg, O., Sundqvist, B., 1982. Thermal properties of two low viscosity silicon oils as functions of temperature and pressure. J. Appl. Phys. 53, 8751–8755.

Schriempf, J.T., 1972. A laser-flash technique for determining thermal diffusivity of liquid metals at elevated temperatures. Rev. Sci. Instrum. 43, 781–786.
Schriempf, J.T., 1973. Thermal diffusivity of liquid gallium. Solid State Commun. 13, 651–653.
Singh, L.P., Issenmanna, B., Caupin, F., 2017. Pressure dependence of viscosity in supercooled water and a unified approach for thermodynamic and dynamic anomalies of water. Proc. Natl. Acad. Sci. USA 114, 4312–4317.
Suárez-Iglesias, O., Medina, I., Sanz, M., Pizarro, C., Bueno, J.L., 2015. Self-diffusion in molecular fluids and noble gases: available data. J. Chem. Eng. Data 60, 2757–2817.
Thern, A., Lüdemann, H.D., 1996. P, T Dependence of the self-diffusion coefficients and densities in liquid silicone oils. Z. Naturforsch. A 51, 192–196.
Touloukian, Y.S., Ho, C.Y., 1981. Properties of Nonmetallic Fluids. McGraw-Hill, New York, NY.
Valeri, D., Meirelles, A.J.A., 1997. Viscosities of fatty acids, triglycerides, and their binary mixtures. J. Am. Oil. Chem. Soc. 74, 1221–1226.
Vargaftik, N.B., Phylippov, L.P., Tarzimanov, A.A., Totskii, E., 1994. Handbook of Thermal Conductivity of Liquids and Gas. CRC Press, Boca Raton, FL (translator: Yu.A. Gorshkov).
Watanabe, H., 1996. Accurate and simultaneous measurement of the thermal conductivity and thermal diffusivity of liquids using the transient hot-wire method. Metrologia 33, 101–115.
Watanabe, H., Kato, H., 2004. Thermal conductivity and thermal diffusivity of twenty-nine liquids: alkenes, cyclic (alkanes, alkenes, alkadienes, aromatics), and deuterated hydrocarbons. J. Chem. Eng. Data 49, 809–825.
Weertman, J., 1983. Creep deformation of ice. Annu. Rev. Earth. Planet. Sci. 11, 215–240.
White, D.B., 1988. The planforms and onset of convection with a temperature-dependent viscosity. J. Fluid. Mech. 191, 247–286.
Xu, Z., Morris, R.H., Bencsik, M., Newton, M.I., 2014. Detection of virgin olive oil adulteration using low field unilateral NMR. Sensors 14, 2028–2035.
Yáñez-Limón, J.M., Mayen-Mondragón, R., Martínez-Flores, O., Flores-Farias, R., 2005. Thermal diffusivity studies in edible commercial oils using thermal lens spectroscopy. Superficies y Vacio 18, 31–37.

Websites

Viscosity of liquids
http://www.engineersedge.com/fluid_flow/kinematic-viscosity-table.htm
http://www.engineeringtoolbox.com/kinematic-viscosity-d_397.html
http://www.viscopedia.com/
Thermal conductivity at 25°C
http://www.engineeringtoolbox.com/thermal-conductivity-d_429.html
Thermal conductivity of various liquids at 300K
http://www.engineeringtoolbox.com/thermal-conductivity-liquids-d_1260.html
http://www.engineersedge.com/heat_transfer/thermal_conductivity_of_liquids_9921.htm
Thermal diffusivity
http://www.engineersedge.com/heat_transfer/thermal_diffusivity_table_13953.htm
Physical properties of the elements
http://periodictable.com/Elements/049/data.html
Steam/water tables
http://thermopedia.com/content/1150Begal (Reprinted, with permission, from NBS/NRC Steam Tables)
https://www.nist.gov/document-12896 (has density as a function of T and P)
Physical properties of silicone fluids from manufacturers
www.gelest.com
www.clearcoproducts.com
Thermophysical properties of methanol https://www.thermalfluidscentral.org/encyclopedia/index.php/Thermophysical_Properties:_Methanol

CHAPTER

7

Thermal Diffusivity Data on Nonmetallic Crystalline Solids from Laser-Flash Analysis

OUTLINE

Truth is stranger than fiction, but it is because fiction is obliged to stick to possibilities; Truth isn't.
Mark Twain (1897).

Measurements, Mechanisms, and Models of Heat Transport
DOI: https://doi.org/10.1016/B978-0-12-809981-0.00007-3

7.1 WHY IS LFA ESSENTIAL TO UNDERSTAND HEAT TRANSPORT?

Because the main goal of this book is to improve microscopic models for heat diffusion, this chapter presents and examines thermal diffusivity data on crystalline solids that are verifiably free from augmentation via ballistic radiative transfer, and from losses through contacts with thermocouples and heaters. Only laser-flash analysis (LFA) meets both requirements, for the following reasons:

Although thermal conductivity data gathered under near steady-state conditions below ~300K are considered to be free of spurious radiative effects, this is commonly not the case, as demonstrated by comparison of LFA measurements of crystalline perovskites to cryogenic, conventional approaches (Hofmeister, 2010). Similar comparisons of silica glass corroborate this finding (Hofmeister and Whittington, 2012). Although blackbody emissions are indeed low for $T < \sim 300$K, any temperature difference between surfaces promotes ballistic radiative transfer at frequencies where the sample is transparent. Therefore, this length-dependent flux affects cryogenic measurements for certain samples at certain wavelengths. Existence of ballistic transport across partially transparent samples was recognized shortly after the LFA method was developed in the 1960s. Circa 1998, LFA data were ameliorated for this effect by use of metal or graphite coatings and improved models (Chapter 4). Comparison of ballistic-free LFA results for the hard, ceramic like materials important to the geological sciences to data from conventional methods (steady-state and Ångström's periodic techniques) quantified contact losses as ~10% per contact (Hofmeister 2007a). Thus, avoiding thermal contacts is also essential to obtain accurate heat transport properties.

Although partially transparent conditions in the visible are obvious from visual inspection, partially transparent conditions in the near-infrared (IR), which is relevant to temperatures of ~100 to ~800K, are not. The degree of transparency depending on sample thickness (Chapter 2), makes the magnitude of this spurious augmentation depend on sample size. For an example, see *T-t* curves for ceramic BN in Ohta and Waseda (1986). Additional factors are temperature, which determines the peak of the blackbody emissions, and the aforementioned material (spectral) characteristics. For an example, see the *T-t* curves for (gray) Si in Milošević (2010). Because conventional measurements utilize a narrow range of sizes of generally 3–5 mm, the amount of ballistic transfer has also been limited to a certain range for any given material and temperature. Consequently, this systematic error mimics a random error, since length has not been considered as an important variable. This behavior, coupled with ballistic augmentation being low below ~300K, explains why spurious radiative effects at low temperatures have gone unrecognized. Uncertainties in any given measurement from the sum of spurious radiative augmentation and compensating contact losses may be sufficiently low to render the data useful for many applications, such as thermal models in geophysics. However, constructing a model of microscopic behavior requires data without such systematic errors.

This chapter is based almost entirely of data collected in the author's laboratory for consistency. Even modern LFA studies use Cowan's model, under the incorrect assumption that ceramics or darkly colored material are opaque, which is not true (see BN and Si examples studied by Ohta and Waseda,1986; Milošević, 2010), so it is not always obvious that spurious effects were eliminated. In some cases, a different model than that of

Mehling et al. (1998) was applied, which adds a small amount of uncertainty in the comparison. Data from other research groups are incorporated when it is certain that either the sample is truly opaque, or ballistic effects were completely removed using the model of Hofmann et al. (1997) and Mehling et al. (1998) or the very similar model of Blumm et al. (1997).

Our database on nonmetals encompasses common rock-forming minerals, closely packed structures expected to be stable inside planets, simple structures, and a wide range of chemical compositions and chemical bonding. Data have been published on 166 single crystals and nearly pure aggregates and unpublished or in review measurements cover 18 more samples. Fewer multicomponent rocks (24 samples in publications) have been explored. This tally does not include the ~100 glasses and melts that are covered in Chapter 10. Exploration of solid-solutions (aka mixed crystals) in the database is somewhat limited, but the series covered are typical of minerals, for example, Al + H ↔ Si or Mg ↔ Fe. The large number of diverse structures and chemical compositions composing our LFA database permit extraction of systematic behavior, and hence allow for definitive evaluation of microscopic models.

A focus on heat transport of planetary materials is scarcely without precedent in the physical sciences: Fourier's (1822) purpose was to understand terrestrial heat flow. Pioneering experiments on the temperature (Eucken, 1911) and the pressure response of thermal conductivity (Bridgeman, 1924) utilized Earth materials. However, data at pressure are not discussed in any detail here because none of the techniques used are sufficiently accurate for the hard solids that are important to geophysics (Chapter 4): more importantly, data presented in this chapter show that tiny thicknesses required to make measurements at very high pressure preclude addressing heat transfer on large scales, meaning greater than ~1 mm.

We have previously shown that diffusion of low-frequency IR light contributes to thermal diffusivity in electrical insulators (Hofmeister et al., 2014). The current, consensus model assumes that the main mechanism is phonon–phonon scattering. Dominance of heat diffusion by a scattering mechanism gives us concern because (1) phonons are pseudoparticles, rather than real entities like the photon, and (2) inelastic scattering is essential for heat transfer, which necessarily produces a photon, and (3) photons are everywhere and clearly are the entity being monitored when ascertaining thermal diffusivity and conductivity in gases (Chapter 5). Therefore, this chapter includes additional information relevant to the behavior of photons in solids when temperatures are changing.

7.1.1 Synopsis of the Growing LFA Database

LFA data prior to 2015 were summarized by Hofmeister et al. (2014), who focused on single crystals and included semi-conductors, and Hofmeister and Branlund (2015), who focused on minerals and polycrystalline substances. Since then, we have published single-crystal studies of micas (Hofmeister and Carpenter, 2015), feldspathoids, and ANA zeolites (Hofmeister and Ke, 2015), all of which contain H-atoms, but vary distinctly in the means of incorporation and the amount. Thus, our thermal diffusivity database is comprised of a variety of different substances, described as follows: (1) elements, although metals are covered in Chapter 9, (2) diatomic compounds, which generally are structurally simple and quite pure, (3) chemically and structurally simple multiatom oxides such as perovskites and spinels with cation disorder, (4) silicates that are structurally complex, with compositions that near end-member (quartz, muscovite, forsterite), (5) silicates with moderate cation disorder (olivines, pyroxenes, micas, ANA zeolites), and (6) silicates with extensive cation disorder (garnets, feldspars, feldspathoids).

Overall, our database is mostly silicates with some cation disorder and nearly end-member oxides and halides, and thus ionically bonded compounds are well-represented. For an example of an extensive solid solution other than that of the garnets, and of another structure with stoichiometric OH^-, new data on chlorite are presented. Because relatively few data exist on semi-conductors in the database, we provide new measurements on PbS (galena) and SiC (moissanite), and further information on Ge and Si. However, because we found no difference in behavior between electrical insulators and semi-conductors, we combine these two categories in Section 7.3. Adding data on PbS and SiC also expand the types of anions and bonding in simple structures. Bonding in galena is a mixture of metallic and covalent in the simple rocksalt structure, whereas the bonding in moissanite is covalent where the structure is complicated by stacking disorder. Thus, measuring SiC allows us to explore the effect of polytypes on D_{heat}. These same samples were studied spectroscopically by Pitman et al. (2008, 2009). Because the database emphasizes the anion O^{2-}, we include data on carbonates with anion $CO_3{}^{2-}$ (Merriman et al., 2018) and provide details on XF_2 phases in the fluorite and rutile structures (in Appendix B). For experimental details, see Chapter 4.

The data are organized according to crystal structure and to how the material interacts with light, that is, whether the material is partially transparent (e.g., silicates, oxides, and alkali halides), or strongly absorbent (e.g., graphite, sulfides, and other semi-conductors). Table 7.1 lists the structures covered examined so far, with examples of the crystal chemistry explored, both of which control both heat transfer and vibrational spectra (Chapter 2). Each crystalline structure is associated with a certain number of vibrational modes of various types, such that three acoustic modes always exist, and furthermore possess the same symmetry as the IR modes. Table 7.1 lists the number of IR active modes expected for each structure. Symmetry analyses of most of these structures exist with varying degrees of details (e.g., Fateley et al., 1972; Ferraro, 1975; Nemanich et al., 1977; Maroni, 1988; McKeown et al., 1999; McKeown 2005). Symmetry analysis and vibrational spectra of the simple structures can be found in many sources. The polytypes of SiC are complicated by resonances, although the main modes are few for the cubic structure and the simplest hexagonal structures (e.g., Pitman et al., 2008).

For petalite, nepheline, and chlorite, the number of IR modes were derived using the method of Fateley et al. (1972). The irreducible representation of the $P2/m$ structure of petalite with two formula units in the unit cell, providing 96 fundamental modes, is:

$$\Gamma = 21A_g(\text{Raman}) + 24B_g(\text{Raman}) + 22A_u(\text{IR}) + 26B_u(\text{IR}) + A_u(\text{acoustic}) + 2B_u(\text{acoustic})$$

For clinochlore with two formula units and a $C2/m$ structure, providing 168 total modes:

$$\Gamma = 39A_g(\text{Raman}) + 39B_g(\text{Raman}) + 42A_u(\text{IR}) + 45B_u(\text{IR}) + A_u(\text{acoustic}) + 2B_u(\text{acoustic})$$

TABLE 7.1 Connection of Structure With the Temperature Dependence of Thermal Diffusivity of Nonmetals

Fit to $D(T)$	Structure	No. IR Modes	Representative Formulae[e]	Cases[f]
FT^G; $G<0$	Diamond	0[a]	C, Si, Ge	xtl, poly
	Zinc blende (3C)	1[b]	SiC-3C	cvd
	Wurtzite (2H)	1 + 1[c]	BeO, AlN	poly
	Polytypes	>1 + 1[c]	SiC-4H, SiC-6H	xtl
	B1	1[b]	MgO, LiF, NaI, KBr, AgCl	xtl, poly, imp
	Brucite	1 + 1[c]	$Mg(OH)_2$	poly
	B2	1[b]	CsCl	xtl, imp
	Fluorite	1[b]	BaF_2, $\sim Zr_{0.90}Y_{0.09}Hf_{0.01}O_2$	xtl, soln
	Li mica	30[d]	$\sim KLi_2Al_2Si_3O_{10}(OH)_2$	xtl, imp
$FT^G + HT$; $G<0$	Diamond	0[a]	Impure Si	xtl, imp
	B1	1[b]	PbS	xtl
	Fluorite	1[b]	CaF_2	xtl
	Graphite	1 + 1[c]	C	poly, glass
	Cubic perovskite	3[b]	$KTaO_3$, $SrTiO_3$, $LaAlO_3$	xtl, imp
	Spinel	4[b]	$MgAl_2O_4$	xtl, imp
	Rutile	1 + 3[c]	TeO_2, MgF_2	xtl
	Corundum	2 + 4[c]	Al_2O_3	ceramic
	Ilmenite	4 + 4[c]	$FeTiO_3$	xtl
	Calcite	3 + 5[c]	$CaCO_3$	xtl, poly
	α-Quartz	4 + 8[c]	SiO_2	xtl, poly, imp, glass
	Zeolite	16[b]	$Na_8Al_6Si_6O_{24}Cl_2$	xtl, imp
	Garnet	17[b]	$\sim Mg_3Al_2Si_3O_{12}$, $Ca_3Fe_2Si_3O_{12}$	xtl, imp, soln
	Cristobalite	20	SiO_2	poly
	Orthoperovskite	25	$YAlO_3$, $NdGaO_3$	xtl, imp
	Orthopyroxene	27	$MgSiO_3$	xtl, imp
	Clinopyroxene	27	$LiAlSi_2O_6$, $CaMgSi_2O_6$	xtl, imp, soln
	Trioct. mica	30	$\sim KMg_3AlSi_3O_{10}(OH, F)_2$	xtl, imp, soln
	Olivine	35	Mg_2SiO_4, Mn_2GeO_4	xtl, poly, imp, soln
	Sanidine	33	$KAlSi_3O_8$	xtl
	Albite	39	$NaAlSi_3O_8$	xtl

(*Continued*)

TABLE 7.1 (Continued)

Fit to $D(T)$	Structure	No. IR Modes	Representative Formulae[e]	Cases[f]
	Plagioclase	39	$CaAl_2Si_3O_6$	xtl, poly, soln
	Petalite	48	$LiAlSi_4O_{10}$	xtl
	Nepheline	20 + 34[c]	$\sim Na_6K_2Al_8Si_8O_{32}$	xtl, soln
	Dioct. mica	57	$\sim KAl_3Si_3O_{10}(OH)_2$	xtl, imp
	Clinochlore	87	$\sim Mg_5(Al, Fe)Si_3O_{10}(OH)_8$	xtl, soln
$\sim aT^{-1} + bT^{-n}$	Corundum	2 + 4[c]	Al_2O_3, Fe_2O_3	xtl, imp

[a]*The fundamental mode is Raman active, but IR overtones are present.*
[b]*Cubic symmetry, peaks are triply degenerate.*
[c]*Doubly degenerate peak associated with tetragonal or hexagonal unit cells.*
[d]*Lowest* D_{heat}.
[e]*Examples are provided that illustrate the range of chemical compositions. For some chemically complex minerals end-member formulae are given. Some structures listed under the non-zero H category have a few compositions where H = 0, but this is generally associated with low temperature data acquisitions.*
[f]*xtl = single crystal; cvd = chemically vapor deposited; poly = polycrystal; cer = fine grained ceramic; soln = extensive solid solutions; imp = minor impurities. Regarding the solid solutions, X and Y are placeholders for cations such as Mg, Fe, or Cr.*
Notes: For most structures, some samples were not measured to sufficiently high temperature to discern departure from a simple exponential. The mica structure is one example.

Vibrations of O–H in hydrous minerals are not included in the total, because these structures can accommodate fluorine in the same site; O–H modes are best represented as additional and controlled by local structural environments. For nepheline in the $P6_3$ structure, a 32 oxygen formula unit is considered, yielding:

$$\Gamma = 21A(\text{IR} + \text{acoustic, Raman}) + 35B + 35E_1(\text{IR} + \text{acoustic, Raman}) + 21E_2$$

7.1.2 Organization of the Chapter, Appendices and Electronic Deposit

Delineating the temperature dependence of D_{heat} is crucial to understand microscopic mechanisms. This is introduced in Section 7.2 and occupies much of this chapter.

Section 7.3 discusses various factors affecting the fits and numerical values of D_{heat}. We focus on behavior near ambient conditions and on data needed to understand the role of photons in heat conduction. Because attenuation of light depends on thickness, Section 7.3 presents an experimental investigation of the dependence of thermal diffusivity on thickness.

Section 7.4 compares thermal diffusivity data on single crystals and ceramics of the same composition to probe the effect of grain boundaries. Data on mixtures (e.g., rocks) are presented, although this material is not yet sufficient in scope to fully understand the behavior.

Section 7.5 summarizes key information that will be used in Chapter 11 to evaluate and improve the microscopic models. However, understanding diffusive radiative transfer is also required to construct a microscopic model: this is covered in Chapter 8 and Chapter 11. The present chapter also makes some comparisons to provide generalities useful for application of the results to the Earth. Regarding applications, data on thermal conductivity are needed. Computing K requires data on D_{heat}, heat capacity, and density as a function of temperature. Appendix C summarizes available information on these properties from the literature.

Appendix B describes the data in the Washington University electronic deposit. This site provides machine-readable text files of thermal diffusivity against temperature for each sample, and other details associated with the NSF supported research program which forms the basis for this book.

7.2 A UNIVERSAL FORMULA FOR THERMAL DIFFUSIVITY FOR NONMETALLIC SAMPLES AS A FUNCTION OF TEMPERATURE

For virtually all nonmetallic samples studied so far, thermal diffusivity decreases as temperature increases above 298K. The slope is highest at ambient temperature but progressively decreases, becoming flat somewhere near 1000K; Fig. 7.1 illustrates typical behavior. In some cases, beyond some high temperature, the slope reverses. The increase in D_{heat} with T at high T reported here is intrinsic, rather than arising from spurious radiative transfer, as in older thermal conductivity studies. For a few cases, D_{heat} near 298K initially increases with T up to ~350K, and then strongly deceases. This behavior is associated with thickness below ~½ mm, and is discussed in detail in Section 7.3.

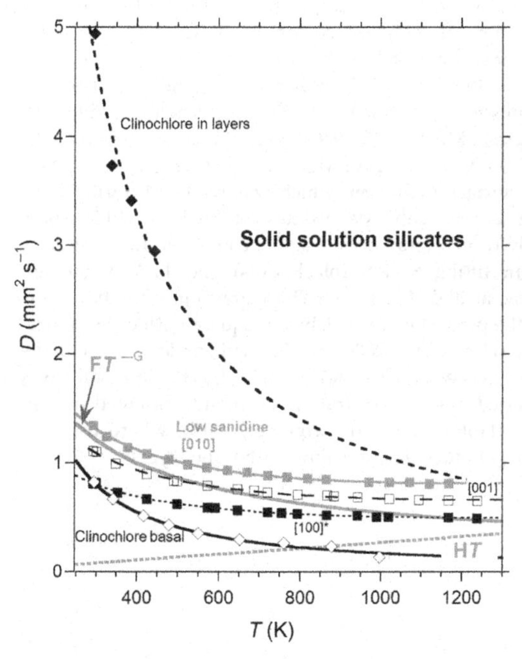

FIGURE 7.1 Examples of structurally complex, anisotropic solid solutions. The formula for clinochlore (variety penninite) of $[Mg_{4.92}Al_{0.69}Fe_{0.43}Cr_{0.02}]Si_{3.25}Al_{0.75}O_{10}(OH)_{7.96}F_{0.06}$ was determined using electron microprobe analysis as described by Hofmeister and Carpenter (2015). This sample has Fe^{3+} and Fe^{2+} in two sites (Appendix B). Two orientations describe its pseudohexagonal symmetry. Data above ~900K was not included in the fit because this was affected by dehydration, consistent with independent measurements of penninite from the Zermatt, Switzerland locality by Brindley and Ali (1950). The basal plate was 0.66 mm thick. To measure flow within the layers required a clamp, large L of 3.01 mm, and termination of the run at low temperatures. The gemmy sample of low sanidine $[K_{0.92}Na_{0.08}Al_{0.99}Fe_{0.01}Si_{2.95}O_8]$ is disordered on both sites, and pseudo-orthorhombic. Values of D_{heat} are only slightly larger than those of glass and melt (Pertermann et al., 2008). The gray solid and dotted lines show the two components of the fit (Eq. (7.1)).

Through examining high-quality LFA measurements above ambient temperature on ~50 nearly gem-quality silicate minerals or optical flats, Hofmeister et al. (2014) showed that diverse nonmetallic single crystals are fit by a simple three parameter formula:

$$D_{heat}(T) = FT^G + HT \tag{7.1}$$

where F, G, and H are constants where H is positive. This book uses the convention that G is negative. This form also fit glasses (exemplified by fused silica and glassy carbon) and polycrystals (exemplified by cristobalite and two graphites). Hofmeister and Branlund (2015) fit 11 additional polycrystalline samples with Eq. (7.1). Fits for end-member chemical compositions in diverse structures are shown in Fig. 7.2, whereas Fig. 7.1 above shows examples of solid solutions. Fits for glasses are given in Chapter 10.

Table 7.2 lists all fits for data solely describing diffusion of heat in single crystals, with some monomineralic and ceramic samples for comparison. The combination of the FT^G term where $G<0$, which describes a strong decrease with T, with the HT term, which describes a strong increase with T, produces the nearly temperature independent high-temperature region of D. The dimensions of F depend on $|G|$, which can be a noninteger. To avoid this oddity, Eq. (7.1) could be recast as

$$D_{heat} = F^* \left(\frac{298}{T}\right)^G + HT \tag{7.2}$$

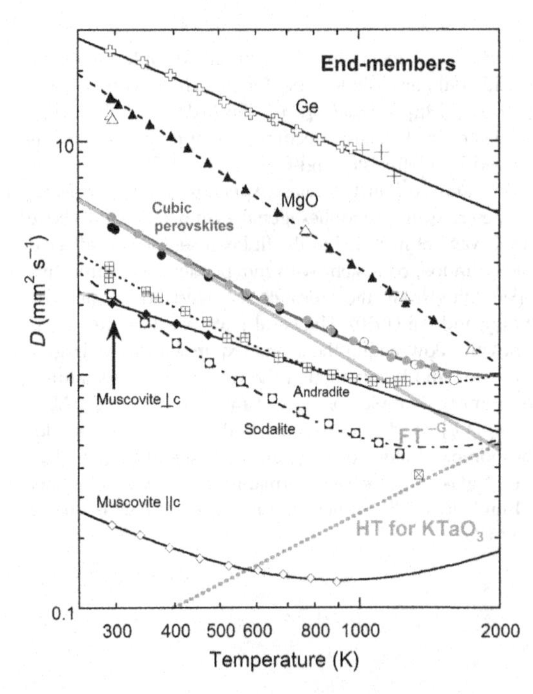

FIGURE 7.2 Log–log plot of thermal diffusivity as a function of temperature for single crystals with end-member (or nearly so) compositions in various structures. Plus signs: Ge, where the highest temperature points were excluded due to incipient melting. Filled triangles: MgO with 0.909 mm thickness (Hofmeister, 2014). Open triangles: MgO with L: 0.51 mm with possible surface hydration, which was removed during heating. Circles: cubic perovskites (Hofmeister, 2010) where only $KTaO_3$ (gray dots) was fit. The result differs slightly from fitting $SrTiO_3$ (black dots) and $LaAlO_3$ (circles, cubic at high T) as well. Thick gray lines: the two parts of the perovskite fit, as labeled. Square with cross: andradite (Hofmeister, 2006). Square with circle or ×: sodalite (Hofmeister and Ke, 2015) where the highest point was omitted due to dehydration. Diamonds: oriented muscovite (Hofmeister and Carpenter, 2015), where use of a clamp limited T_{max} for flow within the layers.

where F^* has the dimensions of thermal diffusivity. Because H is small, $F^* = F \times (298)^{-G}$ nearly equals D_{heat} at 298K. For this reason, Section 7.3.1 explores the connection of $D_{heat, 298}$ with G.

Comparing the forms of Eqs. (7.1) and (7.2) show that trade-offs exist between the coefficients F and G. Trade-offs make the uncertainties in these coefficients frequently substantial. Uncertainties arise in H due to this term describing high-temperature behavior and sufficiently high temperatures were not always reached. Uncertainties in the fits were provided by Hofmeister et al. (2014) and Hofmeister and Branlund (2015). Here, Table 7.2 reports more digits than these works, for better reproducibility of the data. We include χ^2 values to indicate the reasonableness of the fits. If adding the H term provides an insignificant improvement either in the visual fit or in the χ^2 values, we set $H = 0$.

Despite the existence of trade-offs, ranges of parameters are restricted: for the samples studied, $|G|$ ranged from 0.3 to 2, depending on structure, and H was usually near $10^{-4}K^{-1}$. At 298K, D_{heat} varies from 0.1 to 450 $mm^2\,s^{-1}$, which is smaller than the range for F. For typical G and H, the HT term exceeds the FT^{-G} term above about 1300K.

Although the validity of Eqns. (7.1) and (7.2) have been established, the conditions for which H differs from the null value are incompletely understood. To explore these conditions, we need to ensure that the fit only represents the process of heat diffusion. If phase transitions exist or if disordering, dehydration, or oxidation occurs in the temperature range explored, then the fitting reflects the combination of heat diffusion with other processes. Because the original reports attempted to document a wide range of isostructural substitutions, not all the samples yielded high-quality data. Appendix B includes a list of samples excluded from Table 7.2 and why. For example, some small samples had time-temperature curves that were not fit well, consistent with edge-effects perturbing intrinsic values. Several spinels were excluded because disordering occurs with heating: those that are listed in Table 7.2 were runs after the sample was disordered by heating, sometimes multiple times (Hofmeister, 2007).

Our previous reports did not fit materials with reconstructive transitions, but did fit quartz over a wide temperature range by omitted diffusivity data during its displacive α-β phase transition. In this chapter, we only fit data across a phase transition if both D_{heat} and its derivative varied smoothly and continuously with T (e.g., some of the plagioclases). Samples with lambda shapes in their T-t curves are excluded from fitting, unless the transition was at sufficiently high temperature that the low T data provide a reasonably well-constrained fit. This describes petalite: where the data were truncated below its decomposition temperature of 953K (Roy et al., 1950).

Samples that fractured during the runs are not included in Table 7.2 because fracturing creates gaps which add thermal resistance and affect fitting (e.g., Hofmeister and Pertermann, 2008). Likewise, for aggregates, differential expansion can occur between the grains, which has a similar affect (e.g., Pertermann and Hofmeister, 2006). Accordingly, we emphasize the behavior of single crystals in the assessment of Section 7.3, returning to polycrystals in Section 7.4.

The only exception that we have found so far is crystals in the corundum structure (Table 7.3). Yet, the universal fit describes derivatives of the corundum structure and ceramic corundums if the temperature is not too high (Table 7.2). The interesting behavior of the synthetic sapphires is covered in Section 7.3.

TABLE 7.2 Fits to Thermal Diffusivity ($D = FT^G + HT$) in $mm^2\ s^{-1}$ and Maximum Run Temperatures

Sample	Simple Formula	$\nu_{IR,\ max}$ cm^{-1}	L mm	D_{298}	F	G	H	χ^2	T_{max}K	Ref.
Diamond (2 crystals)	C	–	1.355	453	7163600	− 1.7054	0	1000	1030	Hofmeister et al. (2014)
Poly diamond TE	C	–	0.50	139	152190	− 1.1695	0	85	1500	Hofmeister et al. (2014)
Silicon plate	Si 99.999 %	–	2.016	84.5	547080	− 1.5491	0.0022743	10	1600	Hofmeister et al. (2014)
	Si 99.999 %	–	0.525	53.11					350	Hofmeister et al. (2014)
Silicon wafers	Si P-type, B doped, 1–10 ohm-cm	–	0.974	42.4	34441	− 1.1263	0	3.78	400–1570	Hofmeister et al. (2014)
	Si P-type, B doped, 8–12 ohm-cm	–	0.61	63.46	112810	− 1.3154	0.0013263	2	1465	Hofmeister et al. (2014)
	Si N-type, 1900–2800 ohm-cm	–	0.456	52.88	41620	− 1.1597	0	0.29	1465	Hofmeister et al. (2014)
Germanium	Ge N-type, >35ohm-cm	–	0.525	24.8	3169.3	− 0.856	0	0.89	920	Hofmeister et al. (2014)
SiC-3C	SiC	970	0.397	46.9	12556	− 0.97645	0	4.43	1200	This work
SiC-4H	SiC	970	0.33	46.7	36571	− 1.1616	0	9.26	1800	This work
SiC-6H	SiC	970	0.23	24.1	22103	− 1.1136	0	4.45	1600	This work
SiC-6H (5x5mm)	SiC (green)	970	0.335	13.6	21095	− 1.103	0	1.4	500–1600	This work
SiC ceramic black	SiC (6 H + 4 H + 15 R + 3 C)	970	1.073	38 ± 1	18809	− 1.0449	0	0.076	430–1400	This work
BeO ceramic	BeO	1030	0.478	51.35	659290	− 1.6474	0	6.6	1700	Hofmeister (2014)
AlN ceramic	AlN	820	0.555	45.96	32476	− 1.1522	0	5.0	1800	Hofmeister (2014)
Periclase	MgO	738	0.909	16.5					300	Hofmeister (2014)
Alpha/Aesar	MgO	738	0.84	15.2	35764	− 1.364	0	0.05	1100	Hofmeister (2014)
Alpha/Aesar	MgO	738	0.71	15.3					300	Hofmeister (2014)
MTI	MgO	738	0.51	12.5	11884	− 1.2041	0	0.07	1260	Hofmeister (2014)
MTI	MgO	738	0.50	13.06					300	Hofmeister (2014)
MTI	MgO	738	0.465	12.51					300	Hofmeister (2014)
Ceramic periclase	MgO: 2–4% porosity	738	1.1	13.72	16282	− 1.249	0	0.26	2100	Hofmeister (2014)
Poly brucite	~$Mg(OH)_2$~ $\|c$	600	1.205	4.61	331450	− 1.9577	0	0.004	600	Hofmeister (2014)

Poly brucite	~$Mg(OH)_2$~ $\|\|a$	600	1.20	2.87	17782	−1.9456	0	0.001	650	Hofmeister (2014)
LiF-1	LiF	659	1.007	3.462	24502	−1.5653	0	0.051	1090	Yu and Hofmeister (2011)
NaCl-1	NaCl	264	1.989	3.600	38668	−1.633	0	0.0072	1074	Yu and Hofmeister (2011)
NaCl-2	NaCl	264	1.155	3.481	37743	−1.7107	0	0.0018	1074	Yu and Hofmeister (2011)
NaCl-4	NaCl	264	1.085	3.423	37271	−1.6371	0	0.0038	1074	Yu and Hofmeister (2011)
Poly NaCl-7	NaCl	264	1.268	3.000	25911	−1.5951	0	0.0020	1074	Yu and Hofmeister (2011)
NaI-1	NaI	176	1.877	1.137	5611.9	−1.4969	0	0.0004	933	Yu and Hofmeister (2011)
NaI:Tl	NaI (Tl~0.1 wt%)	176	1.851	0.823	4621.5	−1.494	0	0.0005	924	Yu and Hofmeister (2011)
KCl-1	KCl	214	1.100	4.617	90203	−1.7337	0	0.053	1044	Yu and Hofmeister (2011)
KCl-2	KCl	214	2.295	5.221	311550	−1.9381	0	0.016	1044	Yu and Hofmeister (2011)
KBr-1	KBr (Cl~0.8 wt%)	165	1.069	2.260	37743	−1.7107	0	0.0018	1007	Yu and Hofmeister (2011)
KBr-2	KBr (Cl~2.2 wt%)	165	1.812	1.830	10083	−1.5148	0	0.0012	1007	Yu and Hofmeister (2011)
AgCl	AgCl	196	0.677	0.435	155.99	−1.0374	0		730	Yu and Hofmeister (2011)
Galena	PbS	200	1.02	1.60	1302.4	−1.1875	0.00029422	0.003	1150	This work
Poly CsI-3	CsI	85	1.032	1.078	31216	−1.8131	0	0.0001	924	Yu and Hofmeister (2011)
CsI:Tl	CsI (Tl~0.2 wt%)	85	0.999	0.963	9122.7	−1.6132	0	0.0003	894	Yu and Hofmeister (2011)
Fluorite	CaF_2	500	2.04	3.83	92541	−1.783	0.0002164	0.04	1200	Hofmeister et al. (2014)
Frankdicksonite	BaF_2	350	1.12	2.97	19965	−1.55	0	0.003	1050	Hofmeister et al. (2014)
Cubic zirconia	$Zr_{0.92}Y_{0.08}O_2$	650	0.511	0.700	2.3931	−0.26134	0	0.0003	1200	Hofmeister et al. (2014)
Cubic zirconia	~$Zr_{0.87}Y_{0.08}Hf_{0.05}O_2$	650	1.205	0.754	3.316	−0.26134	0	0.0011	1100	Hofmeister et al. (2014)
Graphite AXF	C porosity 20%, grain-size 5 μm, pore size 0.8 μm	1588	3.97		307210	−1.4153	0.0003074	14	1920	See http://poco.com/MaterialsandServices/Graphite.aspx.
Graphite ZXF	C porosity 20%, grain-size 1 μm, pore size 0.3 μm	1588	2.10	50	28570	−1.1196	0.0024098	2.1	1300	Blumm and Lemarchand (2002)
Perovskite-cubic	$KTaO_3$	820	0.547	4.742	3973.9	−1.1882	0.00025285	0.017	1560	Hofmeister (2010)
LSAT-cubic	$La_{0.29}Sr_{0.67}Al_{0.65}Ta_{0.35}O_3$	800	0.516	1.615	34.523	−0.53651	0	0.001	1300	Hofmeister (2010)
Perovskite-1 [100]	$YAlO_3$	750	0.985	3.76	21555	−1.5208	0.00021527	0.08	1800[c]	Hofmeister (2010)[c]

(*Continued*)

TABLE 7.2 (Continued)

Sample	Simple Formula	$\nu_{IR, max}$ cm^{-1}	L mm	D_{298}	F	G	H	χ^2	T_{max}K	Ref.
Perovskite-2 [001]	$YAlO_3$:Tm	750	1.024	2.91	5260.8	−1.3259	0.00026942	0.006	1300	Hofmeister (2010)
Perovskite-3 [100]	$NdGaO_3$	550	0.523	1.93	1742.6	−1.2048	0.00017979	0.003	1500	Hofmeister (2010)
Perovskite-3 [010]	$NdGaO_3$	550	0.945	1.84	3157.9	−1.3164	0.00024113	0.005	1500	Hofmeister (2010)
Perovskite-3 [001]	$NdGaO_3$	550	0.5–0.7	2.02	1894.4	−1.2085	0.0001567	0.005	1500	Hofmeister (2010)
Synthetic spinel	$MgAl_2O_4$	868	0.285	5.28	4849.2	−1.2053	0.00026465	0.18	1650	Hofmeister (2007b)
Spinel (heated)	$Mg_{0.96}Fe_{0.01}Zn_{0.02}Al_2O_4$	868	1.234	5.12	5199.7	−1.2216	0.00029622	0.06	1900	Hofmeister (2007b)
Pink spinel	$MgFe_{0.004}Cr_{0.004}Al_{1.99}O_4$	868	1.075	7.62	13206	−1.3139	0.00018503	0.02	1000	Hofmeister (2007b)
Blue spinel	$Mg_{0.98}Fe_{0.03}Al_{1.99}O_4$	868	0.405	4.70	2231.1	−1.0891	0.00032411	0.08	1000	Hofmeister (2007b)
TeO_2 [110]	TeO_2	770	0.52	1.27	746.81	−1.1242	0.00007389	0.001	800	Yu and Hofmeister (2011)
Sellaite $\parallel c$	MgF_2	620	0.953	8.84	54109	−1.5358	0.00011956	0.073	1350	Yu and Hofmeister (2011)
Sellaite $\perp c$	MgF_2	620	1.023	5.51	46285	−1.5914	0.00017615	0.087	1350	Yu and Hofmeister (2011)
Corundum [0001]	Al_2O_3	870	1.106	12.0	492380	−1.875	0.0006	0.44	1500[a]	Hofmeister (2014)
Corundum [1120]		870	0.993	11.1	571690	−1.915	0.0007	0.44	1500[a]	Hofmeister (2014)
Corundum pink $\perp c$	$Al_{1.998}Cr_{0.0006}Fe_{0.0003}O_3$	870	1.23	10.14					300	Hofmeister (2014)
Corundum red $\perp c$	$Al_{1.996}Cr_{0.033}O_3$	640	1.383	8.57	15172	−1.3191	0	0.05	850	Hofmeister (2014)
Mg-corundum $\parallel c$	$Mg_{0.26}Al_{1.74}O_{2.9}\square_{0.1}$	900	1.208	4.20	4267.1	−1.2328	0.00026221	0.07	1600[a]	Hofmeister (2014)
Mg-corundum $\perp c$		900	0.977	4.02	2746.3	−1.148	b	0.35	1100[a]	Hofmeister (2014)
Ceramic corundum	Al_2O_3 ~0% porosity	640	1.7	11.5	81968	−1.5649	0.0003087	1.6	1950	Hofmeister (2014)
Ceramic corundum	>99.6% Al_2O_3, 1 μm grain size	640	0.499	5.92	2337.7	−1.0489	0	0.07	750	Hofmeister (2014)
Hematite $\parallel c$	Fe_2O_3	662	1.247	2.8	10107	−1.443	0.00020	0.02	1200[a]	Hofmeister (2014)
Hematite $\perp c$		662	0.967	1.9	5947	−1.424	0.00029	0.02	1200[a]	Hofmeister (2014)
Ilmenite $\parallel c$	$Fe_{1.12}Ti_{0.88}O_3$	700	0.78	0.96	11.465	−0.435	0	0.004	1150	Hofmeister (2014)
Calcite $\parallel c$	$CaCO_3$	1550	0.65,1	1.66	3708.2	−1.3602	0.00013001	0.0015	890	Merriman et al. (2018)

Calcite $\perp c$	$CaCO_3$	1550	1.35	1.63	13965	− 1.5987	0.0001945	0.0038	890	Merriman et al. (2018)
Poly. marble	$CaCO_3$ ~4% porosity, ~80 μm grains	1550	0.50	1.288	1341.5	− 1.2184	0	0.0028	880	Merriman et al. (2018)
Quartz [100] HQ	SiO_2	1250	1.414	3.73	176410	− 1.8917	0.00029188	0.011	680	Branlund and Hofmeister (2007)
Quartz [001] HQ	SiO_2	1250	1.044	7.00	210960	− 1.8102	0	0.005	730	Branlund and Hofmeister (2007)
Poly. Quartzite Qmb	SiO_2, 5% porosity, ~40 μm grains	1250	1.145	3.63	43731	− 1.6534	0	0.004	695	Branlund and Hofmeister (2008)
Crypto. agate	SiO_2, <0.2 μm grains	1250	1.68	3.30	44017	− 1.6728	0.00037059	0.008	765	Branlund and Hofmeister (2008)
Poly. cristobalite	SiO_2, ~50 μm grains	1250	0.945	1.85	392370	− 2.1773	0.00052729	0.009	1250	Branlund and Hofmeister (2008)
Fused quartz glass	SiO_2 (36 ppm H_2O)	1250	0.8,1.7	0.85	7.9635	− 0.40912	0.00021884	0.0006	1700	Branlund and Hofmeister (2008)
Xenocryst	$Mg_{2.13}Fe_{0.45}Ca_{0.42}Al_2Si_3O_{12}$	1056	0.687	1.32	51.316	− 0.6715	0.00026876	0.0009	1070	Hofmeister (2006)
Rhodolite	$Mg_{1.68}Fe_{1.14}Mn_{0.02}Ca_{0.16}Al_2Si_3O_{12}$	1050	1.836	1.22	67.526	− 0.71745	0.00024002	0.003	950	Hofmeister (2006)
Py-Al	$Mg_{1.29}Fe_{1.56}Mn_{0.03}Ca_{0.12}Al_2Si_3O_{12}$	1042	1.025	1.43	310.65	− 0.95659	0.00029728	0.02	1350	Hofmeister (2006)
Ternary	$Mg_{1.20}Fe_{1.46}Mn_{0.03}Ca_{0.31}Al_2Si_3O_{12}$	1047	1.24	1.23	133.68	− 0.83305	0.00021042	0.003	1300	Hofmeister (2006)
Ant hill	$Mg_{1.05}Fe_{1.31}Mn_{0.05}Ca_{0.59}Al_2Si_3O_{12}$	1046	1.245	1.25	123.83	− 0.81995	0.00024445	0.0014	1300	Hofmeister (2006)
Al-Gr	$Mg_{0.15}Fe_{2.16}Ca_{0.60}Al_{1.91}Cr_{0.9}Si_3O_{12}$	470	0.622	1.04	85	− 0.78	0	0.003	470	Hofmeister (2006)
Al − Py	$Mg_{0.63}Fe_{2.19}Mn_{0.05}Ca_{0.13}Al_2Si_3O_{12}$	670	1.606	1.25	58.154	− 0.67897	0	0.02	670	Hofmeister (2006)
Al − Sp	$Mg_{0.03}Fe_{1.83}Mn_{1.11}Ca_{0.03}Al_2Si_3O_{12}$	1035	0.836	1.55	583.61	− 1.0507	0.00027028	0.001	1050	Hofmeister (2006)
Spessartine	$Fe_{0.09}Mn_{2.88}Ca_{0.03}Al_2Si_3O_{12}$	1030	1.61	1.99	4080.4	− 1.3574	0.00039257	0.036	1350	Hofmeister (2006)

(Continued)

TABLE 7.2 (Continued)

Sample	Simple Formula	$\nu_{IR, max}$ cm^{-1}	L mm	D_{298}	F	G	H	χ^2	T_{max}K	Ref.
Sp-rich	$Mg_{0.15}Fe_{0.47}Mn_{1.83}Ca_{0.55}Al_2Si_3O_{12}$	1020	1.62	1.18	212.04	−0.92337	0.00026457	0.002	1350	Hofmeister (2006)
Grossular-dry	$Fe_{0.04}Mn_{0.03}Ca_{2.87}Al_{1.91}Fe^{3+}{}_{0.9}Si_3O_{12}$	1008	1.5	2.79	2977.2	−1.2338	0.00037177	0.024	1350	Hofmeister (2006)
Grossular-wet	$Fe_{0.04}Mn_{0.03}Ca_{2.87}Al_{1.91}Fe^{3+}{}_{0.9}Si_3O_{12}$	1008	1.5	2.5	1583.3	−1.1463	0.00031335	0.005	1270	Hofmeister (2006)
Mn-rich Gr	$Mg_{0.09}Ca_{2.91}Al_{1.85}Fe^{3+}{}_{0.05}Mn_{0.10}Si_3O_{12}$	1000	1.2	2.86	629.25	−0.95553	0.00019731	0.028	1250	Hofmeister (2006)
Gr-An 1	$Mg_{0.09}Ca_{2.91}Al_{1.77}Fe^{3+}{}_{0.21}Si_3O_{12}$	1000	1.391	2.82	4579.3	−1.3075	0.0003696	0.04	1270	Hofmeister (2006)
Gr-An 2	$Fe_{0.06}Ca_{2.92}Al_{1.64}Fe^{3+}{}_{0.36}Si_3O_{12}$	1000	1.167	2.77	1278.1	−1.063	0.00036558	0.02	1350	Hofmeister (2006)
Andradite	$Ca_{0.03}Al_{0.03}Fe^{3+}{}_{1.97}Si_3O_{12}$	982	1.85	2.8	2913.8	−1.2331	0.00037354	0.06	1270	Hofmeister (2006)
YAG	$Y_3Al_5O_{12}$	860	1.27	4.13	32078	−1.5768	0.00033312	0.090	1550	Hofmeister (2006)
SmGG	$Sm_3Ga_5O_{12}$	~725	1.598	2.14	286.2	−0.87471	0.00016608	0.005	1120	Hofmeister (2006)
SmGG:Fe	$Sm_3Ga_5O_{12}$ doped with Fe	~725	1.233	2.05	402.12	−0.93488	0.00017648	0.005	1350	Hofmeister (2006)
EuGG:Cr	$Eu_3Ga_5O_{12}$ doped with Cr	~725	1.246	2.64	6963.7	−1.3907	0.00028566	0.003	1250	Hofmeister (2006)
GGG	$Gd_3Ga_5O_{12}$	~725	1.254	1.93	1348.8	−1.1547	0.00020119	0.003	1250	Hofmeister (2006)
En98 [100]	$Ca_{0.01}Mg_{1.96}Fe_{0.02}Al_{0.01}Si_2O_6$	~1150	1.609	2.70	41022	−1.7034	0.00031082	0.077	1250	Hofmeister (2012)
En98 [010]		~1150	1.05	1.93	5612.7	−1.4099	0.00015729	0.017	1350	Hofmeister (2012)
En98 [001]		~1150	1.482	5.12	546540	−2.040	0	0.5	600	Hofmeister (2012)
En96 [210]	$Ca_{0.01}Mg_{1.93}Fe_{0.02}Al_{0.07}Si_{1.97}O_6$	~1150	1.09	2.34	22498	−1.6233	0.00037849	0.01	1000	Hofmeister (2012)
En92 cleavage flake	$Ca_{0.01}Mg_{1.84}Fe_{0.11}Al_{0.03}Si_{1.97}O_6$	1150	1.10	1.98	3752.2	−1.3346	0.00024099	0.006	850	Hofmeister (2012)
En90 [100]	$Ca_{0.02}Mg_{1.8}Fe_{0.15}Cr_{0.01}Al_{0.02}Si_2O_6$	~1150	1.655	1.85	7683.7	−1.4795	0.00033257	0.015	1200	Hofmeister (2012)
En90 [010]		~1150	1.654	1.77	11113	−1.5397	0.00031214	0.006	1200	Hofmeister (2012)
En90 [001]		~1150	1.492	3.10	16833	−1.5308	0.00051125	0.185	1350	Hofmeister (2012)

En89 [100]	$(NaCa)_{0.01}Mg_{1.77}Fe_{0.18}Si_2O_6$	~1150	1.29	1.92	3308	−1.33	0.00023	0.016	1200	Hofmeister (2012)
En89 [010]		~1150	1	1.80	3433.9	−1.3366	0.00016236	0.026	1350	Hofmeister (2012)
En89 [001]		~1150	1.76	3.30					1350^b	Hofmeister (2012)
En75 cleavage flake	$Mg_{1.45}Fe_{0.46}Mn_{0.01}Al_{0.14}Si_{1.92}O_6$	~1150	0.6	1.45	298.2	−0.93789	0.000088529	0.0016	1280	Hofmeister (2012)
Protoenstatite $\|\|c$	$Mg_{1.95}Al_{0.04}Si_{1.98}O_6$, sub-mm lath widths	1070	0.58	2.67	1840.5	−1.1489	0.000050251	0.026	1100	Hofmeister (2012)
Poly Protoenst. ~$\perp c$		1070	0.9	1.04	406.78	−1.0515	0.000043447	0.0015	1100	Hofmeister (2012)
Diopside-Na [100]	$Na_{0.03}Ca_{0.99}Mg_{0.94}Fe_{0.02}Si_2O_6$	1150	1.24	2.54	3894.7	−1. 3056	0.00020033	0.0002	1500	Hofmeister and Pertermann (2008)
Diopside-Na ~$\|\|c$		1150	1.05	3.83	1113.9	−0.99968	0.00017742	0.003	1080	Hofmeister and Pertermann (2008)
Diopside-Fe $\perp c$	$CaMg_{0.92}Fe_{0.031}Si_2O_6$	1150	1.1	2.32	1281.7	−1.1139	0.000042889	0.026	1350	Hofmeister and Pertermann (2008)
Diopside-Fe $\|\|c$		1150	1.23	3.80	2571.7	−1.1483	0.00022473	0.019	1550	Hofmeister and Pertermann (2008)
Augite [010]	$CaMg_{0.77}Fe_{0.08}Al_{0.16}Al_{0.2}Si_{1.8}O_6$	~1150	0.605	1.55	427.76	−0.99331	0.00013529	0.001	1450	Hofmeister and Pertermann (2008)
Augite $\|\|c$		~1150	1.141	2.73	1723.1	−1.1461	0.00032992	0.012	1450	Hofmeister and Pertermann (2008)
Aegirine [110]	$Na_{0.88}Ca_{0.09}Fe_{0.98}Ti_{0.02}Si_2O_6$	~1150	1.279	1.45	3504.1	−1.3768	0.00023586	0.007	1100	Hofmeister and Pertermann (2008)
Aegirine $\|\|$c		~1150	1.28	2.76	1291.7	−1.0908	0.00035625	0.019	1100	Hofmeister and Pertermann (2008)
Spodumene [110]	LiAl $Si_{1.93}O_6$	~1200	1.35	2.10	5186.7	−1.3736	0.00037956	0.079	900	Hofmeister and Pertermann (2008)
Spodumene [001]		~1200	1.10	3.74	4769.8	−1.2759	0.00046607	0.012	900	Hofmeister and Pertermann (2008)
Poly. Jadeite	$Na_{0.98}AlSi_2O_6$, sutured grains	1050	2.745	2.79	2751.6	−1.2203	0.00026142	0.001	900	Hofmeister and Pertermann (2008)

(*Continued*)

TABLE 7.2 (Continued)

Sample	Simple Formula	$\nu_{IR,\ max}$ cm^{-1}	L mm	D_{298}	F	G	H	χ^2	T_{max}K	Ref.
Poly. Jadeite: Fe	$Na_{0.9}Ca_{0.08}Mg_{0.08}Fe_{0.12}Al_{0.80}Si_2O_6$, sutured grains	1050	1.555	1.84	1009.8	−1.1242	0.00031368	0.005	900	Hofmeister and Pertermann (2008)
Fo [001]	Mg_2SiO_4	991	3.5	5.15	12842	−1.3813	0.00027064	0.02	1260	Pertermann and Hofmeister (2006)
Fo:Co [010]	$Mg_{1.99}Co_{0.01}SiO_4$	991	1.102	2.6	34868	−1.6786	0.00017408	0.03	1450	Pertermann and Hofmeister (2006)
Fo:Co [001]		991	1.34	4.17	8853	−1.3482	0.00017684	0.07	1750	Pertermann and Hofmeister (2006)
Olivine [100]	$Mg_{1.84}Fe_{0.16}SiO_4$	990	1.61	3.25	16911	−1.521	0.00040819	0.3	1250	Pertermann and Hofmeister (2006)
Olivine [010]		990	1.88	1.66	3641.8	−1.3649	0.00021739	0.015	1050	Pertermann and Hofmeister (2006)
Olivine [001]		990	1.05	2.69	2245.5	−1.1954	0.0002598	0.015	1150	Pertermann and Hofmeister (2006)
Sinhalite [010]	$Mg_{0.964}Fe_{0.036}AlBO_4$	~1100	1.82	5.3	229260	−1.8883	0.00038205	0.04	1000	Pertermann and Hofmeister (2006)
Chrysoberyl [100]	$BeAl_{1.996}Cr_{0.004}O_4$	1090	1.32	8.99	92903	−1.6299	0.00043952	0.15	1270	Pertermann and Hofmeister (2006)
Chrysoberyl [010]		1090	1.361	6.22	62270	−1.6226	0.00030494	0.013	1270	Pertermann and Hofmeister (2006)
Chrysoberyl [001]		1090	1.215	8.42	60143	−1.5583	0.00032896	0.046	1270	Pertermann and Hofmeister (2006)
Poly. dunite #1	>95% $Mg_{1.8}Fe_{0.2}SiO_4$, up to ~mm grains	990	1.567	2.31	3840.8	−1.3103	0.0001645	0.003	1200	Pertermann and Hofmeister (2006)
Poly. dunite #2	>95% $Mg_{1.8}Fe_{0.2}SiO_4$, up to ~mm grains	990	1.625	2.06	2010.5	−1.2267	0.00019645	036	1370	Pertermann and Hofmeister (2006)
Poly. fayalite	$Fe_{1.93}Mn_{0.07}SiO_4$, sub-mm impurities	950	0.808	1.51	1628.3	−1.2342	0.00014993	0.002	1350	Hofmeister and Branlund (2015)
Poly. knebelite	$Fe_{0.88}Mn_{0.92}Mg_{0.02}SiO_4$, sub-mm impurities	950	1.385	1.245	242	−0.93581	0.000098391	0.003	1300	Hofmeister and Branlund (2015)

Poly. Mn_2GeO_4	Mn_2GeO_4, ~80 μm grains	800	1.08	1.24	2090	−1.3267	0.0002022	0.008	1000	Pertermann and Hofmeister (2006)
Low sanidine [100]	$K_{0.92}Na_{0.08}Al_{0.99}Fe_{0.01}Si_{2.95}O_8$	~1200	0.86	0.8	25.989	−0.62913	0.0001444	0.0006	1200	Pertermann et al. (2008)
Low sanidine [010]		~1200	0.92	1.34	52.178	−0.65754	0.0002668	0.001	1200	Pertermann et al. (2008)
Low sanidine [001]		~1200	1.00	1.11	41.303	−0.64734	0.00020067	0.001	1200	Pertermann et al. (2008)
Albite [100]	$K_{0.02}Na_{0.98}AlSi_3O_8$	~1200	0.8,1.0	1.08	283.74	−0.99622	0.00018615	0.004	1100	Hofmeister et al. (2009)
Albite [010]		~1200	0.86	1.69	223.87	−0.86788	0.00021409	0.001	1300	Hofmeister et al. (2009)
Albite [001]		~1200	1.731	1.35	115.68	−0.7838	0.00011149	0.004	1100	Hofmeister et al. (2009)
FSU 010	$Na_{0.94}Ca_{0.05}Al_{1.05}Si_{2.95}O_8$	~1200	0.741	1.27	51.996	−0.65959	0.00017723	0.0004	1260	Branlund and Hofmeister (2012)
FSU 001		~1200	0.58	1.10	19.411	−0.50745	0.000094654	0.001	1260	Branlund and Hofmeister (2012)
FON 010	$K_{0.03}Na_{0.78}Ca_{0.19}Al_{1.19}Si_{2.81}O_8$	~1200	0.791	0.919	75.451	−0.80058	0.00039441	0.0003	640	Branlund and Hofmeister (2012)
FON 001		~1200	0.885	0.979	135.22	−0.89008	0.00039791	0.0004	640	Branlund and Hofmeister (2012)
FON perp.		~1200	0.919	0.868	493.21	−1.1415	0.00042019	0.0006	640	Branlund and Hofmeister (2012)
FLN 001	$Na_{0.51}Ca_{0.47}Al_{1.47}Si_{2.53}O_8$ has labradorescence	~1200	0.924	0.956	37.746	−0.66129	0.00025614	0.0002	800	Branlund and Hofmeister (2012)
FLN 010		~1200	0.87	0.892	26.129	−0.60957	0.00024322	0.0002	800	Branlund and Hofmeister (2012)
FLN perp		~1200	0.616	0.797	27.195	−0.63303	0.00018773	0.0001	800	Branlund and Hofmeister (2012)
FBM [010]	$Na_{0.39}Ca_{0.59}Al_{1.60}Si_{2.40}O_8$	~1200	0.702	0.794	5.53	−0.351	0.00012	0.0006	1300	Branlund and Hofmeister (2012)
FBM [001]		~1200	0.468	0.776	3.37	−0.262	0.00004	0.0002	1300	Branlund and Hofmeister (2012)
FBM perp.		~1200	0.923	0.632	4.22	−0.343	0.000085	0.0006	1300	Branlund and Hofmeister (2012)

(*Continued*)

TABLE 7.2 (Continued)

Sample	Simple Formula	$\nu_{IR,\ max}$ cm^{-1}	*L* mm	D_{298}	*F*	*G*	*H*	χ^2	T_{max}K	Ref.
FLL 010	$Na_{0.33}Ca_{0.65}Al_{1.65}Si_{2.35}O_8$	~1200	0.877	0.767	7.468	−0.41356	0.00014483	0.0014	1100	Branlund and Hofmeister (2012)
FLL 001		~1200	0.443	0.751	3.4873	−0.27782	0.00010702	0.0001	1100	Branlund and Hofmeister (2012)
FLL perp.		~1200	0.756	0.811	13.736	−0.51964	0.00022756	0.002	1100	Branlund and Hofmeister (2012)
Margarite ‖	$Ca_{0.74}Na_{0.28}Al_2Mg_{0.03}Fe_{0.05}$ $[Si_2Al_2O_{10}][OH]_{1.95}F_{0.05}$	1000	0.265	0.354	46.657	−0.8482	0	0.003	750	Hofmeister and Carpenter (2015)
Muscovite-P ‖	$K_{0.90}Na_{0.10}Al_{1.87}Mg_{0.07}Fe_{0.08}$ $[Si_{3.06}Al_{0.94}O_{10}][OH]_{1.93}F_{0.07}$	1050	0.26	0.227	68.302	−1.0212	0.000072971	0.00003	950	Hofmeister and Carpenter (2015)
Muscovite-P ⊥		1050	3.078	2.03	89.367	−0.66523	0	0.0003	480	Hofmeister and Carpenter (2015)
Muscovite-F ‖	$K_{0.85}Na_{0.12}Al_{1.83}Mg_{0.06}Fe_{0.14}$ $[Si_{3.13}Al_{0.87}O_{10}][OH]_{1.71}F_{0.29}$	1050	0.226	0.181	14.479	−0.76517	0	0.0003	800	Hofmeister and Carpenter (2015)
Muscovite-F ⊥		1050	1.745	1.4	1274.7	−1.1996	0.00050231	0.002	750	Hofmeister and Carpenter (2015)
Phlogopite-A ‖	$K_{0.96}Mg_{2.88}Ti_{0.07}Al_{0.10}$ $[Si_3AlO_{10}][OH]_{1.43}F_{0.57}$	1000	0.20	0.172	69.087	−1.0741	0.000063052	0.00001	980	Hofmeister and Carpenter (2015)
Phlogopite-A ⊥		1000	1.45	2.29	937	−1.06	0	0.01	450	Hofmeister and Carpenter (2015)
Biotite ‖	$K_{0.94}Na_{0.07}Mg_{1.60}Fe_{1.13}Ti_{0.12}$ $[Si_3AlO_{10}][OH]_{1.05}F_{0.94}$	1000	0.261	0.114	2.0041	−0.50054	0	0.00001	800	Hofmeister and Carpenter (2015)
Biotite ⊥		1000	2.80	1.33	46.229	−0.62192	0	0.02	380	Hofmeister and Carpenter (2015)
Zinnwaldite ‖	$K_{0.92}LiAl_{1.09}Fe_{0.71}Mn_{0.07}$ $[Si_{3.33}Al_{0.67}O_{10}]F_2$	1000	0.208	0.220	1.5084	−0.34034	0	0.00002	870	Hofmeister and Carpenter (2015)
Zinnwaldite ⊥		1000	2.76	1.62	7.704	−027351	0	0.0001	470	Hofmeister and Carpenter (2015)

Lepidolite ‖	$K_{0.85}Li_{1.55}[Al_{1.91}Si_{3.27}O_{10}]$ $[OH]_{0.31}F_{1.69}$	1000	0.144	0.126	0.24894	−0.11763	0	0.00001	1010	Hofmeister and Carpenter (2015)
Lepidolite ⊥		1000	2.65	1.16	3.2161	−0.1802	0	0.0002	370	Hofmeister and Carpenter (2015)
Sodalite	$Na_8Al_6Si_6O_{24}Cl_2$	970	1.193	2.22	11445	−1.5125	0.0002036	0.008	1230	Hofmeister and Ke (2015)
Petalite ‖*c*		1210	1.09	1.96	441.76	−0.94713	0.00024307	0.001	930	Hofmeister and Ke (2015)
Petalite ~‖*b*	$LiAlSi_4O_{10}$	1210	1.55	2.07	1076.7	−1.1148	0.00061681	0.0006	930	Hofmeister and Ke (2015)
Petalite (201)		1210	1.18	1.92	1323.9	−1.1613	0.00037817	0.002	860	Hofmeister and Ke (2015)
Nepheline ‖*c*	$\sim Na_6K_2Al_8Si_8O_{32}$	1050	0.985	0.81	10.838	−0.46081	0.000131	0.002	1400	Hofmeister and Ke (2015)
Nepheline ⊥*c*		1050	1.06	0.56	6.67981	−0.44506	0.00012516	0.001	1500	Hofmeister and Ke (2015)
Clinochlore ‖*c*	$\sim Mg_5(Al, Fe)Si_3O_{10}(OH)_8$	1030	0.61	0.826	4948.7	−1.5352	0.000095993	0.0001	1000	This work
Clinochlore ⊥*c*		1030	3.10	5.94	4731.4	−1.215	0	0.13	500	This work

[a] *Fits not ideal: see text.*
[b] *Similar to En90 fit.*
[c] *See Aggarwal et al. (2006) for data below ambient temperature.*
Notes: T_{max} is the upper limit of the data. Horizontal lines separate crystal structures. All fits used starting parameters of $F = 100$, $G = 1$, and $H = 0.001$. The values reported for *F*, *G*, and *H* incorporate the digits needed to reproduce the data.

TABLE 7.3 Fits to Thermal Diffusivity as a Function of Temperature for Single Crystals With the Corundum Structure

Sample	Formula	L mm	D_{298} mm^2 s^{-1}	Equation	χ^2	T_{max} K
Corundum [0001]	Al_2O_3	1.105	12.0	$1927.5\,T^{-1}+4.1579\times10^{10}\,T^{-4}$	0.07	1800
Corundum [1120]		0.993	11.1	$1831.8\,T^{-1}+3.6994\times10^{10}T^{-4}$	0.04	1700
Mg-corundum $\parallel c$	$Mg_{0.26}Al_{1.74}O_{2.9}$[a]	1.205	4.20	Similar to $\perp$c, but fits poorly	–	1075
Mg-corundum $\perp c$		0.977	4.02	$175.23\,T^{-0.72712}+9.6155\times10^{8}\,T^{-3.6008}$	0.14	1600
Hematite $\parallel c$	Fe_2O_3	1.247	2.8	$4965.3\,T^{-1.3154}+2.581\times10^{-11}\,T^{-3.1678}$	0.01	1200
Hematite $\perp c$		0.967	1.9	$1300.2\,T^{-1.1509}+7.1036\times10^{-12}\,T^{-3.4012}$	0.01	1200

[a]*This synthetic material is inferred to have about 0.1 vacancies in the oxygen state, to balance charge. A small amount of hydroxyl is present, but is insufficient to account for the cation charges.*

Notes: Data from Hofmeister (2014). The entire range of measurements was fit, whereas the universal equation can only fit data below 1500K for corundum, see Table 7.2.

Based on the simplicity of the fit (Eqs. (7.1) or (7.2)) and large number of materials it represents, this finding has repercussions for high-temperature models of heat transport. One explanation is that the two terms describing $D_{heat}(T)$ are associated with two distinct microscopic mechanisms: the explanation previously offered was transport of coupled photons and phonons (polaritons: Hofmeister et al., 2014). For a more complete understanding, we examine the database in detail, and fill in some "holes" in the data for reasons discussed in Section 7.1.3. Given the results on heat transport in gas (Chapter 5) and liquids (Chapter 6), our concern is better delineating the role of diffusive radiation in the IR. Variations in optical spectra of crystals (Chapter 2) are germane to radiative type of heat transfer and will be taken into account as much as possible, since we have not yet presented a radiative transfer model.

7.3 SINGLE-CRYSTAL ELECTRICAL INSULATORS AND SEMI-CONDUCTORS

Most of our publications centered on one specific structure (listed in Table 7.1) and focused on end-members. A few publications consider closely related structures constituting mineral families (e.g., perovskites, plagioclase, micas). Table 7.2 includes new results on clinochlore (Fig. 7.1), PbS and SiC (Fig. 7.3), data on calcite and calcite marble (Fig. 7.4) from Merriman et al. (2018), and provides more details and more measurements on C, Ge, Si, and material with the rutile and fluorite structures. Appendix B provides details on these materials.

Despite use of coatings, virtually all measurements had remnants of ballistic transport at high temperature, which required using the model of Mehling et al. (1998). This finding underlies our focus on data collected in the author's laboratory.

Figs. 7.1–7.4 illustrate the essential behavior, wherein temperature, structure, and chemical composition all affect thermal diffusivity. The behavior is summarized as

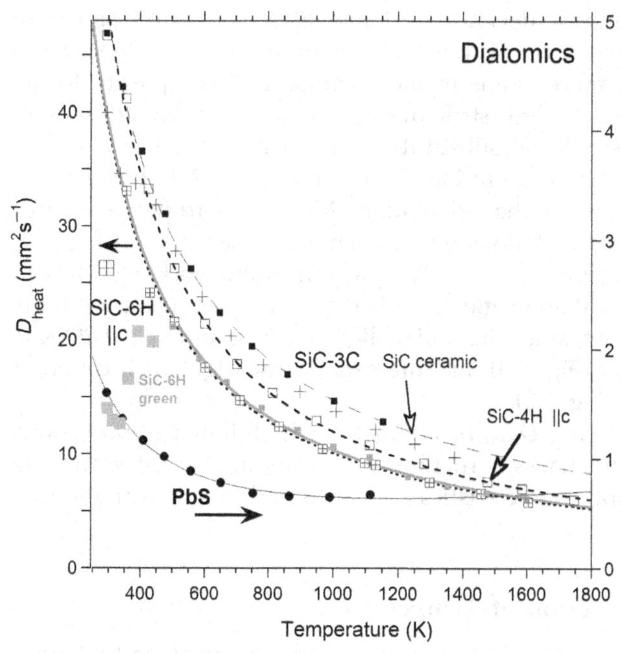

FIGURE 7.3 Thermal diffusivity of simple diatomic compounds with bonding that is covalent (SiC) or mostly so (PbS; $L = 1.02$ mm; right-axis). All samples have low amounts of impurities, see Appendix B. The initial point for SiC-6H (square with cross; $L = 0.23$ mm) was not included in the fit. The 3C sample (filled squares; $L = 0.397$ mm) is light brown and was chemically vapor deposited. The 4H sample (open squares) is 0.333 mm thick. The green SiC-6H sample is less pure than the others, but similar thickness, and shows an initial increase in D_{heat}. Ceramic SiC contains several phases. The need to fit PbS with non-zero H is obvious, whereas the high-T trends for the SiC crystals are all simple exponentials.

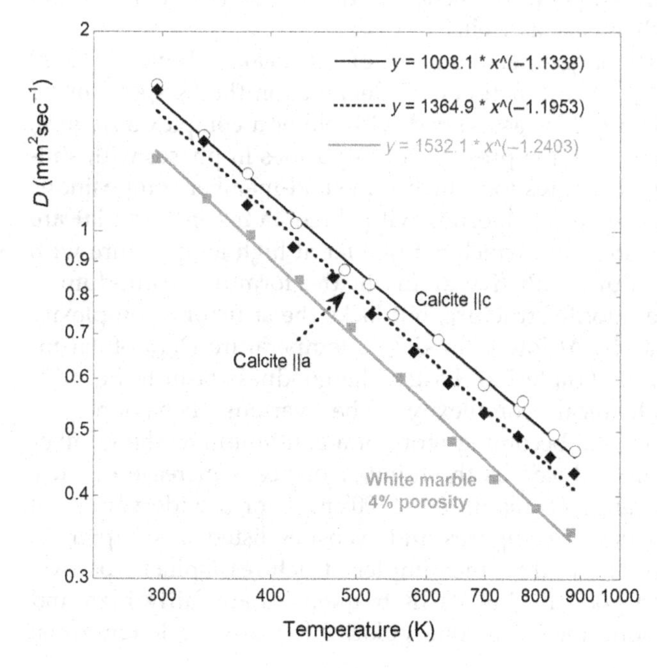

FIGURE 7.4 Thermal diffusivity of calcite crystals and marble, with porosity of 4%. All compositions are very close to $CaCO_3$. These and other carbonate minerals and rocks were studied by Merriman et al. (2018). Power fits are shown on an expanded log–log scale to illustrate the bow-shape associated with a small H term. The porous marble with grain size ~0.08 mm outgassed water in the pores during the run. However, due to scattering we could not determine water content before or after. In grainy material, differential expansion can occur, so that D_{heat} at high temperature is less certain.

follows: temperature causes D_{heat} to mostly decrease as T increases. For simple and pure compounds, thermal diffusivity simply decays exponentially with temperature. More complicated structures, which are tied to more atoms in the formula, add complexity to the dependence of D_{heat} on T (Fig. 7.2). Even slight structural differences have an effect (e.g., the polymorphs of SiC: Fig. 7.3). Isostructural substitutions affect both the initial values and temperature dependence of D_{heat} (cf. MgO in Fig. 7.1 to PbS in Fig. 7.3). If the structure is anisotropic, then D_{heat} depends on the orientation. More anisotropic structures show greater differences between the orientations (cf. clinochlore to sanidine in Fig. 7.1, which respectively have strongly layered and fairly uniform framework structures). Polycrystals generally have lower D_{heat} than single crystals (Fig. 7.4). Also, there are hints in the figures that thickness has an effect, since the initial slight increase in D_{heat} at 298K is only associated with sub-mm samples (Fig. 7.3) and the samples of MgO with different thickness have different D_{heat} at 298K (Fig. 7.2).

To explore the many factors influencing D_{heat} of nonmetals, the following subsections focus on changes with temperature, thickness, structure, polarization, chemistry and the chemical bond, and disorder. IR spectral characteristics appear to underlie the variations.

7.3.1 Thermal Diffusivity as a Function of Temperature

Whether the universal formulae accurately describes any given nonmetal, and whether the HT term in Eq. (7.1) (or its variant, Eq. (7.2)) is needed for a good fit, depend on the structural and chemical complexity of the phase of interest. This statement presumes that some minimum temperature is reached, since H is small and the HT term describes high-temperature behavior. The minimum temperature needed to detect this high-temperature term depends on the structure and chemistry as well.

Either type of complexity creates a greater number of IR peaks (Table 7.1 and Chapter 2), and thus the form for D_{heat} as a function of T depends on the IR spectrum of the phase. From Tables 7.1 and 7.2, finite H is associated with either a complex structure, or a complex chemical composition, or with the presence of impurities in phases with simple chemistry and structure (e.g., Si). Structures for which some end-member compositions have non-zero H and others have $H = 0$ (e.g., fluorites with three atoms in formula) are intermediate in complexity. The only structure which is fit better at high temperature with another formula (Table 7.3) is corundum with five atoms in the formula. Corundum is considerably more complex than the fluorite structure, but lacks the structural complexity of a silicate, which have many IR bands. At low to moderate temperature D_{heat} of corundum can be fit to the universal formula (Table 7.2). Lastly, the goodness-of-fit to Eq. (7.1) improves with structural and chemical complexity. The various behaviors are attributable to the IR spectrum of a crystal becoming more of a continuum as the number of peaks increases. Spikey profiles are reduced as the number of peaks increase because fundamental modes have a limited range of about 100–1200 cm^{-1} for a wide variety of cations-anion pairings, excluding H (see e.g., figures and websites listed in Chapter 2). Regarding corundum, this has more bands than the simpler structures (spinels, perovskites) which need the HT term for a good fit, but its IR frequencies are fairly high and well-separated, and its transverse optic modes in one polarization are the longitudinal optic (LO) models in the other polarization. The power law contributes to D of the corundum phases at all temperatures: it is the high-temperature term that varies from linear.

The HT term is not needed to describe data on virtually all diatomic compounds (Table 7.2). These occupy simple structures and have one strong IR band. Their thermal diffusivity data are fit very well with the simple power law, even to high temperature. The exception is PbS, which differs from the other diatomics by having partially metallic bonding (see below).

A final indication of the importance of the IR spectrum is the behavior of single crystals with the diamond structure. These do not have fundamental IR modes, but do have overtone/combination bands in the IR as well as impurity bands. The universal formulae, with or without the HT term, fits these materials less well than those with IR fundamentals. We did not find a simple formula which matched the data at all temperatures. For Si and Ge, the range of data collection is limited by melting. For diamond, large and thick samples are needed for this extremely thermally conductive material, but such samples are expensive.

7.3.1.1 Behavior at Moderate Temperatures

Because the HT term is nearly negligible at 298K, and only becomes important by ~1000K, the power law describes $D_{\text{heat}}(T)$ at ordinary temperatures. Fitting parameters for the power law can be examined in two approaches: parameters F and G can be compared (Eq. (7.1)) or $D_{\text{heat, 298}} = F^*$ and G can be compared (Eq. (7.2)). These two visualizations reveal links of the strong decrease to material properties.

Fig. 7.5 shows that F exponentially increases with G. In part, this behavior stems from the form of Eq. (7.1), in that the units of F, and thus the size of F depend on the size of G. However, not all solids behave in the same way: different trends of F with G are associated with the various types of chemical bonds in our solids. The different trends are not associated with the presence or size of the high-temperature term, because each trend has phases with both $H = 0$ and with finite H. Our previous comparison of fewer samples (Hofmeister et al., 2014) revealed two trends only, where the less populated trend was associated with monatomic solids (C, Si, and Ge). The greater variety of samples examined here show that covalent compounds have larger F for any given G than do the mostly ionically bonded minerals, whereas the converse is true for heat flow across the layers of micas, which are bonded by van der Waals attraction. Ionically bonded alkali halides have slightly smaller F than the average of the oxides and silicates, which have some covalent character to their bonds. The fit to the alkali halides may not project to low G due to little variation in the fitting parameters (Table 7.2). AgCl and PbS differ slightly from the trend, which is attributed to their bonding not being fully ionic.

The distinct trends in Fig. 7.5 have some connection with structure, but this link is secondary, as follows: most of the covalent compounds occupy the simple diamond structure or its derivative structures, but the graphite structure is layered, and glassy carbon does not have a crystal structure. Although the coordination is similar for all covalent compounds (tetrahedral), the building block of the silicate minerals is the SiO_4^{4-} tetrahedron. Regarding the micas, flow of heat within the layers is described by F and G values as observed for the other nonlayered silicates, and for chlorite, which has layers too, but

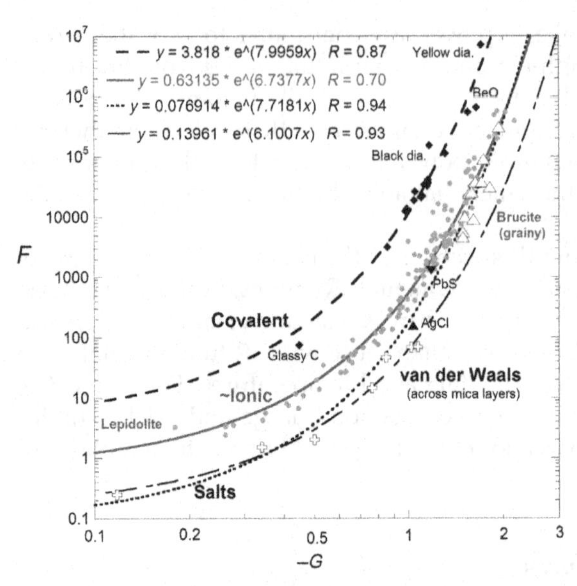

FIGURE 7.5 Comparison of fitting coefficients F and G from Eq. (7.1) for all 173 measurements in Table 7.2. For anisotropic substances, fits involve one to three different sections. Labels in small type indicate certain samples. Exponential fits are listed. Labels in large type indicate bonding. Ceramic BeO and AlN are slightly lower than the trend for covalent bonding due to these being grainy ceramics, as is also the case for a few other samples: of these, brucite is labeled. Salts include CsI, although this has a slightly different (B2) structure.

chlorite layers are stacked and bonded differently than in micas. Hence, the distinct trend across the micas layers can only be due to van der Waals attractive forces.

Visualizing the power law by comparing D_{heat} at 298K to G (Fig. 7.6) provides more information on the effect of structure and IR characteristics. This depiction rests on Eq. (7.2), which shows that the coefficient F^* equals D_{heat} at 298K, if $H=0$. Presenting the data in this manner includes a small contribution to thermal diffusivity at 298K from the HT term, but removes the dependence of F on G that arises through the fitting, and avoids the variable units for F. Because this diagram distinguishes individual structures and most structures tend to have either $H=0$ or finite H, the visualization better isolates the low-temperature behavior inherent to each structure than simply comparing F to G in Fig. 7.5. The results below confirm that the presence or absence of the HT term has a negligible effect on the observed trends.

The size of the power coefficient G depends on the room temperature value for D_{heat} systematically, if the database is sorted into categories with different IR characteristics. The categories in Fig. 7.6 are based on the different structures which are each associated with a certain number of IR modes (Table 7.1). Many categories consider measurements of a single composition, where samples differ in the amounts of impurities, existence of polytypes, or thickness examined. Several categories contain fits with and without the HT term. Samples from various categories were fit to power laws, except for the complex silicates and oxides which cover a broad area on the diagram, but were not subdivided for clarity. The latter categories were not broken down here, but are discussed individually in Section 7.3.1.3.

The frequency of the highest IR fundamental for the compounds (listed in Table 7.2) varies with the categories, as suggested by one of the arrows above Fig. 7.6. Although diamond, Si, and Ge lack IR fundamentals, these materials have overtone/combination bands and impurities which produce IR absorptions at high frequency. Nitrogen impurities in

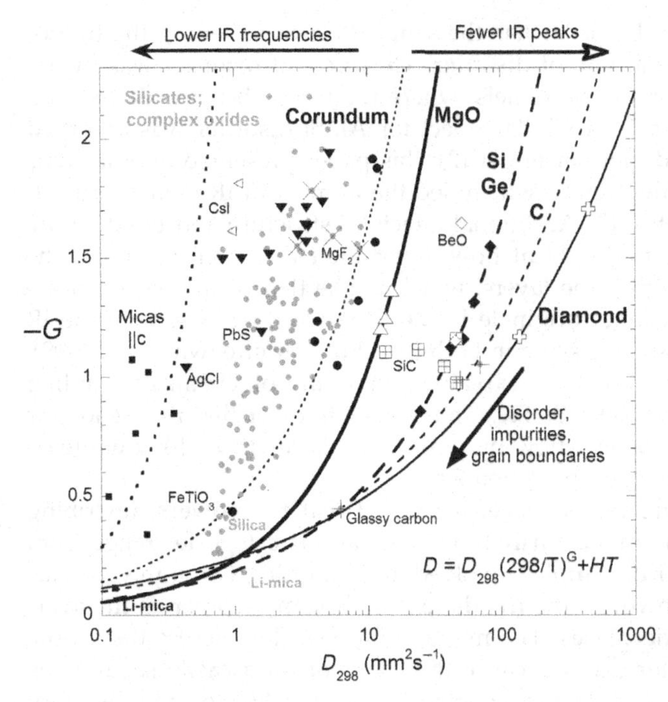

FIGURE 7.6 Comparison of the fitting coefficients of Eq. (7.2) for all 173 measurements in Table 7.2. For anisotropic substances, fits involve one to three different sections. Labels in small type indicate certain samples. Fits are power laws and are close to square roots. The following categories (names in bold type) mostly have $H = 0$: Open plus: diamond, where each point represents two samples; solid diamond: Si or Ge; open diamond: wurtzite ceramics; square with plus: SiC polymorphs; open triangles: MgO; sideways triangle: CsI; filled triangle: rocksalt structure. The following categories mostly have non-zero H: circles: corundum-type structures; X: triatomic fluorides; gray dots: various silicates, perovskites, and spinels. The micas with heat flow $||c$ (squares) have various H values, depending on both temperature and crystal chemistry. Arrows above the graph indicate the competing effects of two key characteristics of an IR spectrum. The arrow within the graph illustrates the effect of other factors.

the yellow diamonds examined here are near 1800 cm^{-1}, which are higher than the IR fundamentals of graphite, which are higher than the overtones in Si bands from ~500 to 1000 cm^{-1} (Stierwalt and Potter, 1962), and higher still than MgO, which in turn has higher frequencies than other compounds in with the rocksalt structure (PbS and alkali halides, except for CsI). The trends for these substances follow the order of the IR frequencies (Fig. 7.6) which shows that $D_{heat,298}$ depends on the IR frequency of these substances. The behavior of the monatomic and diatomic compounds further shows that it is not so important whether the main IR band is a fundamental or an overtone/combination, but where the frequencies lies.

The IR bands in corundum occur at both higher and lower frequencies than the single IR peak of MgO, but are not terribly dissimilar on average. Yet, corundum occupies a different trend than MgO. This difference exists because the number of IR peaks clearly has an effect. Other examples exist. Unlike the other substances, all samples of SiC have similar values for G, but varying $D_{heat,298}$. From Table 7.2, $D_{heat,298}$ decreases as the complexity of the polymorph increases, which produces more IR modes via zone folding (Nakashima and Harima, 1997). A similar relationship exists between diamond and graphite, the latter of which has two IR fundamental modes. Thus, the number of IR modes and the frequencies of the IR modes independently affect the dependence of thermal diffusivity on temperature, as summarized by the arrows above Fig. 7.6.

Lastly, the categories of Fig. 7.6 (and Table 7.2) reveal that several additional factors influence the temperature dependence of D_{heat} at low T. (1) The presence of impurity ions reduces both D_{298} and $|G|$, as exemplified by the Si samples. Specifically, the presence of impurity ions flattens the trends. (2) Disorder has the same effect of flattening the trends, since incorporating impurity ions is a type of disorder. The effect of disorder was distinguished from that of impurities in study of spinels, where the trends became flatter with each successive run (Hofmeister, 2007). A similar effect for Al, Si disorder was observed for feldspar minerals (Branlund and Hofmeister, 2012): this paper presented another type of visualization of the "flatness" of the trends, which tied the number of IR modes and cation disorder to the flatness of silicates. (3) Additional effects of structure and bonding are also evident in the behavior of the micas. Heat flow along the *c*-axis, which crosses the layers, is much slower than flow within the layers, which is like that of the other silicate minerals (Fig. 7.6). This difference cannot originate in the IR spectra, because both the IR frequencies and number of modes are the same in both directions (McKeown et al., 1999). Rather, the special behavior of the micas is explained in terms of chemical bonding (Fig. 7.5). (4) Ceramic samples of diamond, Si, SiC, and MgO fall lower that the trends for their respective crystals. Reduction of heat flow by grain boundaries is well-known (e.g., Smith et al, 2003) and is discussed further in Section 7.4.

Regarding the overall temperature dependence, we found that the powers describing the dependence of G on ambient thermal diffusivity for the various categories range from 0.31 to 0.64. Given the small number of samples involved, to first order, G roughly goes as the square root of D_{heat} at 298K. However, the trends tend to become steeper as the average D_{heat} at 298K for the category decreases. The micas, with heat flow across the layers, show the steepest trend. More scatter exists because the layers of mica easily separate as temperature increases. The steepening as the average D_{heat} at 298K for the category decreases leads to the trends coalescing near the origin. This behavior is consistent with the lowest G and D_{heat} measured, that of the Li-rich mica lepidolite (Hofmeister and Carpenter, 2015). The existence of a coalescence point at $G=0$ is a consequence of the various factors combined (frequency and number of IR modes, impurities and disorder). The coalescence point at $G=0$ describes constant thermal diffusivity, and moreover requires that D_{heat} associated with $G=0$ to be very low, as observed for lepidolite with heat flow across the layers (Figs. 7.5 and 7.6).

7.3.1.2 The Complex Structures

The exponent in Eq. (7.1) for most crystals measured is near unity, that is, the parameter G varies only from 0.26 to 2.2 with a mean value of 1.2. The values of G are clearly tied to structure (Table 7.2). Fairly simple structures are discussed above. Here, we consider the silicates and complex oxides, which were not distinguished in Figs. 7.5 and 7.6.

Of the silicates, feldspars have very low *G* near 0.3–0.85, with *G* decreasing as Al/Si order decreases; garnets and clinopyroxenes have *G* near unity; olivine and orthopyroxenes have *G* near 1.2–1.6. Complex oxides, such as perovskites and yttrium garnets, have *G* near 1.2–1.6, whereas simple oxides, halides, and diamond structure elements have a large exponent of 1.7–2.2. The corundum structure has its own unique fits (Table 7.3), but for fits up to 1500K for the end-members, *G* is intermediate to the simple and complex oxides.

7.3.1.3 The Term *HT* for High-Temperature Behavior

For most samples, the data are bowed from a power law behavior, as shown for calcite in Fig. 7.4. The "bow" increases with temperature, in accord with high-temperature behavior being described by the linear term in Eqs. (7.1) and (7.2). Fits were also attempted with weaker and stronger dependences on T, but no improvement was seen by varying the power of the HT term (Hofmeister et al., 2014). Studies of a wider variety of materials to higher temperature are needed to fully constrain the high-temperature behavior.

Some cases exist where H cannot be resolved. Frequently, these cases involve low temperatures. Examples are the two Fe-rich garnets which were measured below 600K, and many of the micas for heat flow inside the layers. Whether H can be resolved also depends on its magnitude, which varies by a factor of 50 (Table 7.2).

To better understand the conditions needed to observe H, we consider diatomic compounds with the B1 structure. Most could be fit simply by a power law (Table 7.2). Many halides have low IR frequencies and become unstable at high T, so the temperature range is too small to extract H (with magnitude observed so far) for many of these materials. For PbS, over a somewhat larger temperature range (to 1000K), the HT term is essential (Fig. 7.3). In contrast, MgO was fit by FT^G, even though almost 2000K was reached (Fig. 7.2). Is H simply zero for ionic diatomics? We think not because other factors contribute to detecting finite H. Fig. 7.7 illustrates two limiting factors in resolving H. First, a minimum temperature is needed, and on average H increases with temperature. This is unexpected because with higher T_{max}, one should be able to resolve smaller H. But, accessing higher T requires more refractory materials, that is, oxides or carbon, rather than alkali halides or hydrous materials. Thus, Fig. 7.7A mixes experimental conditions with material properties. Second, the parameter H also increases as G increases (Fig. 7.7B). This is expected because as G increases, the power law itself becomes "flatter" at high T, and thus a larger value of H is required to observe the "bowing" shown in Fig. 7.4. Thus, for the diatomics, which all have high G-values, the value of H must be high to resolve this fitting coefficient. However, as shown in Fig. 7.8A, the value of H depends on the IR

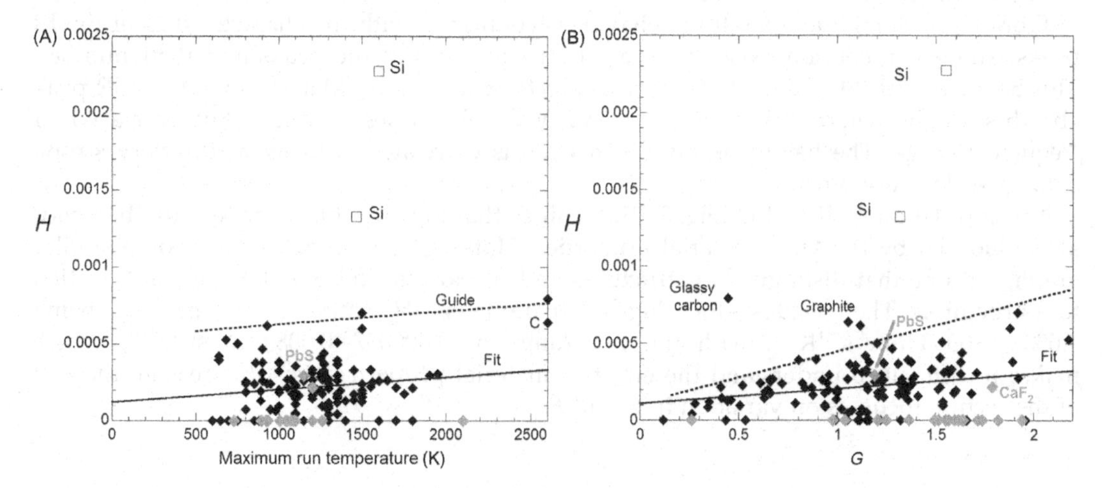

FIGURE 7.7 Dependence of the fitting coefficient H of Eq. (7.1) on (A) the maximum run temperature and (B) the power coefficient G. Diamonds indicate fundamentals for one peak (gray) or many (black, with linear fit). Squares represent overtone frequencies for materials that have finite H, but no fundamental IR modes. The dotted line is a guide to the maximum value of H, excepting the Si samples.

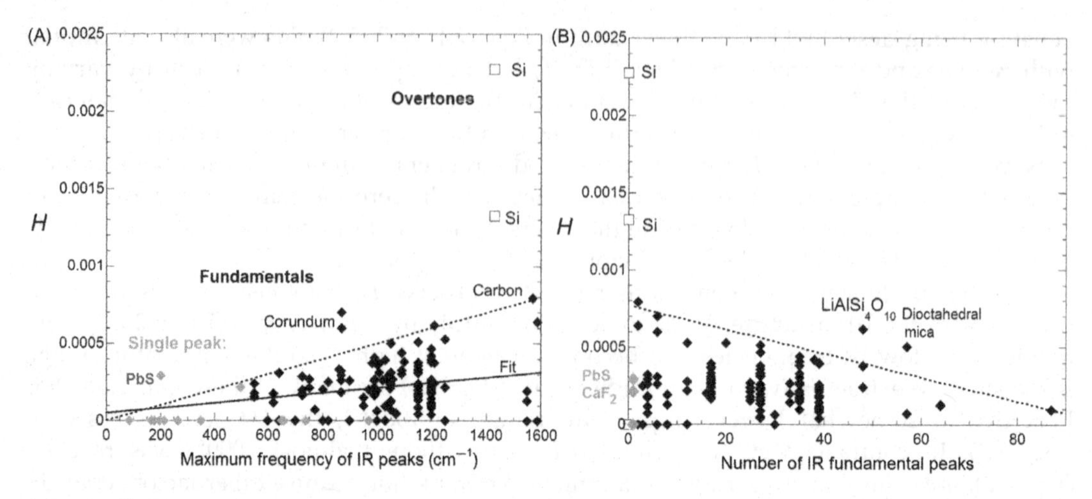

FIGURE 7.8 Dependence of the fitting coefficient H of Eq. (7.1) on (A) the maximum frequency of the IR peaks and (B) the number of IR fundamental modes. Diamonds indicate materials with one IR peak (gray) or many (black, with linear fit). Squares represent overtone frequencies for materials that have finite H, but no fundamental IR modes. The dotted line is a guide to the maximum value of H, excepting the Si samples.

frequency. The combination of low IR frequencies, low maximum run temperatures, and high G for the diatomics, make it difficult to resolve the high-temperature term, which should be very small. Only for PbS with low G and high T_{max}, could H be quantified. However, galena has a metallic character to its bonding, which is more covalent than ionic. The general decrease of H with the number of IR modes (Fig. 7.8B) should help in the observation of H for diatomics, but apparently is insufficient. If the HT term exists for the ionic diatomics, it is very, very small.

From Fig. 7.8, high-temperature behavior is connected with the characteristics of the IR peaks, and a competition exists between the frequencies of the peaks and their number. This behavior indicates that IR absorption affects heat transfer. Materials with one IR peak absorb strongly, whereas those with several peaks absorb less strongly, but over a wider frequency range. The height of any given peak is correlated with its width: very strong peaks are also very broad.

The correlation of H and G (Fig. 7.7B) signifies that low and high-temperature behavior are influenced by the same material properties. Hofmeister et al. (2014) provide a similar graph, but one that distinguishes structures and found that lower H for a given G is tied to structure. The values for high-T parameter H vary little, mostly being 0.0002–0.0003 mm^2 s^{-1}K^{-1}, with the full range of 0.000007–0.008 mm^2 s^{-1}K^{-1}, which makes it difficult to understand the effect of material properties. For this reason, the rest of the section focusses on variation in G and F.

7.3.2 Polarization in Heat Transfer and IR Spectroscopy

Another observation suggests that IR modes play an important role in heat transfer. Differences between thermal diffusivity of the three crystallographic orientations of olivine are greatest in measurements made with LFA from thin disks, but are smaller when measured with methods that use long cylindrical shapes (see Fig. 8 in Hofmeister and Branlund, 2015). Light propagating along **Z** in a solid has its amplitude perpendicular to **Z** (dipoles vibrating along *X* and *Y* directions are directly coupled to the electromagnetic wave, stimulating the transverse optic (TO) modes of the lattice: Fig. 2.6). For thin disks, interactions with LO modes are minimized, as occurs in IR experiments. As thickness increases, LO modes, which are indirectly coupled to an applied electromagnetic wave, are increasingly excited. Increased participation of LO modes mixes polarizations during heat flow across tall cylinders because LO modes related to the TO modes in the perpendicular orientation.

During physical scattering, a thin disk geometry would creat a large cone of forward propotation, whereas a tall cylinder would create a narrow cone (Fig. 2.6D and E). For a scattering mechanism, greater anisotropy would be observed for the tall cylinders, whereas for an absorption and re-emission mechanism, greater anisotropy is expected for the thin disks, as is observed.

7.3.3 Thermal Diffusivity as a Function of Sample Thickness

Because thermal diffusivity (and conductivity) are considered to be material properties, few studies have investigated the length (L = sample thickness) dependence. Indeed, Fourier's law (Eqs. (3.2) and (3.3)) presumes length independence. Generally, these studies are on ceramics, which have grain-size effects (Section 7.4) or concern thermal barrier coatings, where thickness is an important variable (e.g., Campbell, 2010). These studies considered the length dependence to arise from certain experimental conditions, rather than being intrinsic. Yet, for single crystals and silica glass, over lengths that are conducive to collecting good data and with reasonable length to diameter ratios, thermal diffusivity at 298K varies with L (Fig. 7.9). This behavior implicates radiative transfer as the mechanism, as explained below and explored further in Chapter 9 and Chapter 11. Because this finding has ramifications in many fields of physical science, experimental evidence for an intrinsic length dependence of D_{heat} is discussed in detail.

Our previous studies focused on a narrow range of thickness where D_{heat} is fairly constant (e.g., quartz in Fig. 7.9), thereby permitting understanding of effects of temperature, structure, and IR characteristics on D_{heat} (Section 7.3.1) and effects of impurities (Section 7.3.4). As shown below, the variation of thermal diffusivity with thickness being largely independent of temperature and material characteristics underlies the systematic behavior observed in LFA experiments, as long as ballistic radiative effects are removed.

7.3.3.1 *Effect of Thickness at Ambient Temperature*

Much of the data in Fig. 7.9 comes from our published studies, although sample length was not always listed. Examples are alkali halides, quartz, and muscovite mica. These data include samples with low amounts of impurities. In detail: (1) The quartz data are averages over ~10 data collections at 298K, so are well-constrained, but variable amounts of impurities and inclusions are present (Branlund and Hofmeister, 2007). A few samples were colored: pink quartz has hematite-like color centers and is higher than the average

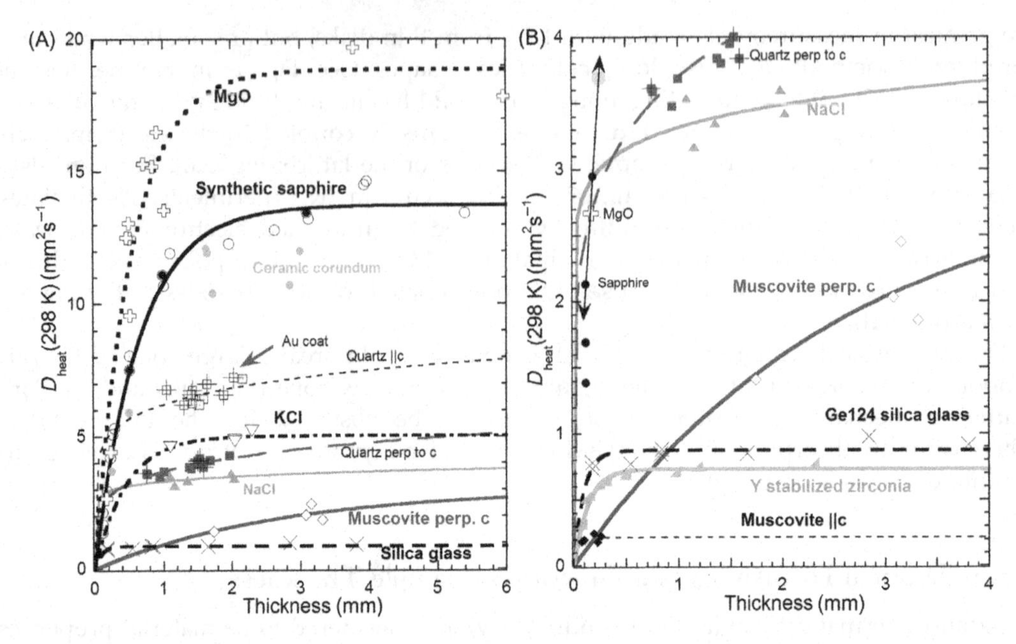

FIGURE 7.9 Dependence of D_{heat} at 298K on thickness. (A) All samples. (B) Expanded view, focusing on insulating samples and thin samples. Fits are least squares. Heavy lines and bold labels indicate samples fit with the exponential, Eq. (7.3). For narrow ranges of L, a power law fit was used (see Table 7.4). Open plus: MgO. Circles: corundum, where black dots indicate (1120) orientations. Gray dots: Al_2O_3 ceramic. Squares: quartz in two orientations, where +: gold coated; open triangles: KCl; dark gray triangles: NaCl; X: Ge124 silica glass; right light gray triangles: yttrium stabilized cubic zirconia; diamonds: muscovite, for each direction.

trend. All data were collected over lengths appropriate for accurate measurements. The trends for the two orientations are similar to each other, and to trends for other samples. Therefore, impurities (or lack thereof) contribute divergence from the trend with L. Use of gold sputter coatings did not affect the results on average, as shown. (2) Several different samples of halides were studied by Yu and Hofmeister (2011). Minor impurities exist in some samples, so Fig. 7.9 only includes the purest samples from this study (NaCl and KCl). (3) All muscovite samples are impure. The variation is random and small except for fluorine substituting for OH^-. Measurements within the layers have a greater uncertainty, as is evident in Fig. 7.7B, so the steep trend is uncertain. Across the layers, which involve more measurements, a thickness dependence like that of quartz is indicated.

Some of our published studies on synthetic and nearly pure substances suggested a thickness dependence, but the data were on relative few sections. Therefore, we supplemented our published data with new measurements of MgO, Al_2O_3, and silica glass. (1) To avoid variation in water content for silica glass, Fig. 7.9 only shows data on dry (24 ppm H_2O) Ge124, which is a fused quartz. Samples with $L \sim 1$ mm were part of the Hofmeister and Whittington (2012) study. We added data on thicker sections, all cut from the same rod, which show that the dependence on L becomes weaker as L becomes larger. At large L, D_{heat} of fused quartz asymptotes to a constant. (2) Cubic MgO samples are subject to surface hydration, so a suite of samples was measured immediately after opening the sealed packages from the manufacturers. The odd point is a sample from Hofmeister (2014) which was suggested to have surface hydration. At high T, the measurements matched fresh samples of similar thickness. Below L of 1 mm, we found a strong dependence of D_{heat} at 298K on L, with some scatter. A larger cube provided data at high L, for which flat behavior is seen. For $L > 4$ mm (with is substantial compared to a 10 mm widths of these sections), we have concerns about two-dimensional cooling to the surroundings, yet similar trends are observed for all samples on average. (3) Our study of synthetic sapphire, which is dry and pure, but anisotropic, utilized L near 1 mm. A thickness dependence existed in the earlier data (Hofmeister, 2014), but was attributed to orientational differences. Data collected from thinner sections, all from a single manufacturer, indicate that D_{heat} at 298K depends on length, rather than whether heat flows along the c- or a-axis for the anisotropic corundum structure. Scatter is seen at high L, and we are again concerned about aspect ratio and two-dimensional heat losses, as well as orientation for these 12.7 mm diameter sections. The thicker sections were from various manufacturers and were mostly along the c-axis. Data for the $L = 4$ mm sample were collected under several conditions (variable cap aperture, laser power, and graphite coating thickness), where results converged to $D_{heat} = 14.7 \pm 0.3\ mm^2\ s^{-1}$ at 21.1°C. This particular sample is slightly wedged, and so the orientation is possibly the cause. However, the 3.10 mm sample is considerably off the c-axis orientation, but this did not affect thermal diffusivity. Examining the ~4 mm sample under crossed polarizers showed that it did not extinct: this anomalous birefringence suggests that strain is present, which is known to affect D of glasses (Chapter 10). In any case, at high L, the thickness dependence is weak, and the data scatter about a constant value (Fig. 7.9A).

Because the muscovite data for flow $||c$ (across the layers) is not terribly convincing and the other weakly diffusing sample is amorphous, not crystalline, we provide new data on (cubic) yttrium stabilized zirconia, which is technologically important as thermal barrier coatings and readily available, due to use as diamond substitutes. Two measurements are described by Hofmeister et al. (2014). Thin samples were purchased from MTI Corporation and thick samples from Morion Inc. and Pretty Rock Minerals. Micro X-ray fluorescence indicates that all compositions are essentially the same. Our YSZ samples all have about 8 wt% yttrium oxide and about 1 wt% hafnium oxide (which is difficult to separate from Zr metal or oxides) as is typical. As observed for the high D samples, different trends exist at low and high L.

A single, simple formula did not exactly describe the behavior of thermal diffusivity at 298K overall thicknesses studied for all samples. However, we could fit the data on several samples closely, and could reasonably approximate most curves with the function:

$$D_{heat}(L) = D_{\infty}\left[1 - \exp(-bL)\right] \quad (7.3)$$

The best fits include the cubic crystals and isotropic glass. The parameters in Eq. (7.3) represent thermal diffusivity for a very long sample (D_{∞}) and attenuation of a signal (b or its inverse). Values are listed in Table 7.4. Samples measured over narrow ranges mostly could not be reliably fit using Eq. (7.3). To reproduce all data, Table 7.4 also provides fits using two other simple formulae that apply to either large or small thickness (but not both), and indicates the starting and stopping L for the fits. The power law approach does

TABLE 7.4 Dependence of Thermal Diffusivity at 298K on Thickness

Sample	D_∞ $mm^2\ s^{-1}$	b mm^{-1}	D_{heat} (298K, thin) $mm^2\ s^{-1}$	Max. L mm	D_{heat} (298K, thick) $mm^2\ s^{-1}$	Min. L mm	R
MgO	18.867	1.8394	$31.26L-15.969L^2$	1.0	$16.044\ L^{0.094619}$	0.60	0.83
Al_2O_3	13.738	1.546	$18.937L-7.7369L^2$	1.2	$11.199\ L^{0.15347}$	0.80	0.90
Quartz $\parallel c$					$6.2318\ L^{0.13383}$	1.0	0.50
Quartz $\perp c$					$3.6959\ L^{0.18288}$	0.75	0.88
KCl	5.0269	2.2283			$4.5303\ L^{0.12629}$	1.0	0.82
NaCl					$3.3186\ L^{0.072277}$	1.0	0.43
Ge124 glass	0.87352	11.111			$0.85214\ L^{0.072983}$	0.15	0.88
YSZ	0.73397	7.5016	$4.8101L-9.2481L^2$	0.30	$0.72745\ L^{0.095981}$	0.30	0.95
Muscovite $\parallel c$	0.2176	25.436			$0.25472\ L^{0.11409}$	0.18	0.47
Muscovite $\perp c$	2.9905	0.3805			[a]	[a]	0.78

[a]*Power law fits are not reported because these do not provide a flat region, like the other samples.*
Notes: The linear correlation coefficient R applies to the power law fit for large L, except for muscovite $\perp c$.

not represent the steep trend near $L=0$. The power law exponents being similar is compatible with the power law being a reasonable substitution when the spread in L is insufficient to utilize Eq. (7.3). For the steep trend associated with thin sections, we provide a polynomial fit for comparison. With polynomial fits, the limit of $L \to 0$ provides $D_{heat} \to 0$: hence, the intercept are very small. Table 7.4 omits the unnecessary intercept.

The power law for the thick samples provides D_{heat} at 298K for 1 mm samples (the prefactor), rather than D_∞, and an average power of 0.11 ± 0.04. The power does not appear to vary much, and the prefactor is consistently smaller than D_∞, except for muscovite. Notably, the position of the transition from "thin" to "thick" behavior increases slightly with D_{heat} for $L=1$ mm and $T=298$K. An asymptote seems to describe high thickness. This is best constrained for silica glass and YSZ, which are flat beyond $L \sim 1$ mm, and trend to $D=1.0$ and 0.80 $mm^2\ s^{-1}$, respectively. A power law fit for muscovite with heat flow within the layers is not reported because Eq. (7.3) describes all data. Interestingly, the temperature dependence for muscovite $\perp c$ with $L=1.74$ mm was like very thin samples of other materials (see below). More data are needed to constrain the thickness dependence for micas because impurity variation in the suite examined likely affects the values of D.

For thin samples, the fits point to the origin. If no medium exists, then heat cannot be diffused. The observed dependence stems from heat transfer being radiative, and thus not being exactly a material property, as discussed below. Lastly, the existence of limiting values for D_{heat} for both small and large thickness is revealing: Eq. (7.3) describes attenuation, which is definitely compatible with diffusion of radiation. Admittedly, intermediate diffusivity samples were not sufficiently tested, but the trends in Fig. 7.9 indicate that Eq. (7.3) is generally true at ambient temperatures.

7.3.3.2 *Effect of Thickness and Temperature Combined on Thermal Diffusivity*

Fig. 7.3 shows that for some thin samples of SiC, D_{heat} increases with T for a small range of temperatures above ambient. The polycrystalline diamond shown in Fig. 2f of Hofmeister et al. (2014) is another example. We also observed upturns for Si, as indicated in Table 7.2. The upturn also existed in data on mica for heat flow within the layer. For this reason, D_{heat} at 298K was extrapolated for muscovite-F, to provide the point at $L = 1.74$ mm in Fig. 7.9B. The extrapolation is only slightly lower than data on other muscovite samples and would only change the fit by a small amount.

This curious upturn near ambient temperature (Fig. 7.3) is associated with many of the thin samples used to construct Fig. 7.9. This section probes the cause. Due to purity considerations and availability of material, the effect of T and L together on D_{heat} is quantified here using synthetic sapphire. Even though this phase differs from the universal formula at high temperature, corundum-structure solids follow similar behavior to the simpler oxides and more complicated silicate minerals in Figs. 7.5 and 7.6, which describe $D_{heat}(T)$. Quantifying the thickness dependence for this "exception" is relevant to the microscopic mechanism of heat transport.

For Al_2O_3, the upturn in D_{heat} slightly above 298K exists in all data collected from thicknesses below 0.3 mm, and becomes more pronounced as L decreases. Examples are shown in Fig. 7.10A. The upturn contributes greatly to the rapid decline as $L \rightarrow 0$ to the limit of $D_{heat} \rightarrow 0$ at 298K (Fig. 7.9). If we exclude the upturn during the fitting, then the

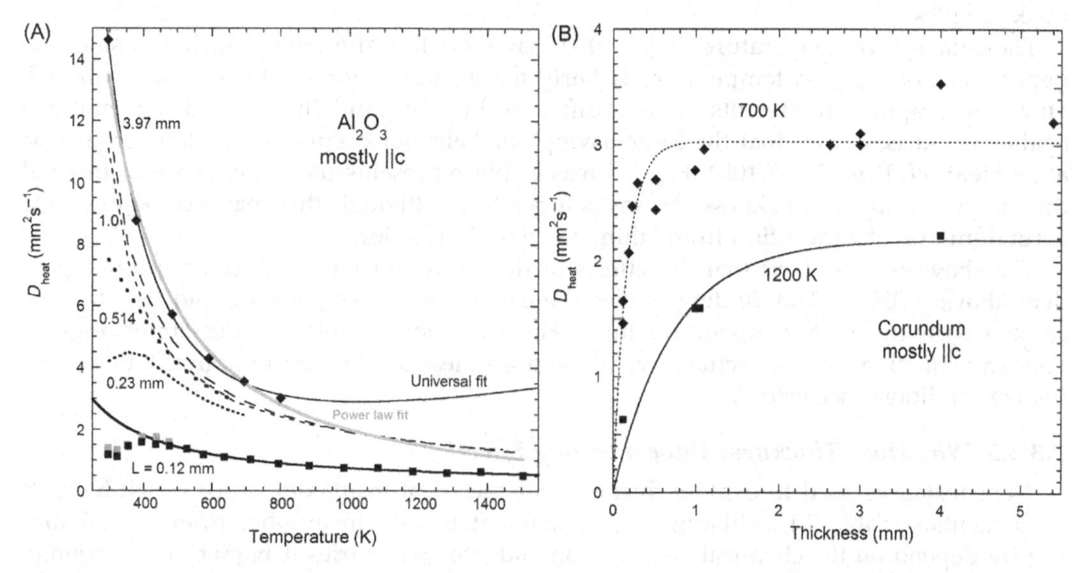

FIGURE 7.10 Thermal diffusivity of synthetic sapphire (actually corundum, Al_2O_3). (A) Dependence of D_{heat} on temperature for selected, representative thicknesses. Solid lines are least squares fits to thin and thick sections, where individual data points are shown as various symbols. Patterned lines link data points. Sections with $L = 0.514$ mm (heavy dots) and 0.993 mm (short dashes) are 1120 faces. All others are 0001 faces. The $L = 3.97$ mm section is best represented by averaging the two fits. (B) Dependence of D_{heat} on thickness for 700K, which involves many samples, but some data were extrapolated from 500–600K. Results for 1200K are from very few samples. For the 4 mm sample, we averaged the two fits shown in part (A) for the 1200K point.

temperature dependence of all sections can be fit to the universal Eq. (7.1). Data were not collected to sufficiently high temperature to accurately constrain the H for the 4 mm thick sample, but the two fits of Fig. 7.10A suggest $H = 0.001\ mm^2\ s^{-1}K^{-1}$, which is similar to values in Table 7.2. The thick (~4 mm) section has large $D_{heat,298}$, F, and $|G|$, whereas for the thin (~0.1 mm) section, these parameters and also H are quite small. The response of D_{heat} to temperature for any given value of L are consistent with the trends in Figs. 7.5 and 7.6. Thus, excluding the initial upturn, all sections of corundum follow the temperature behavior described in Section 7.2. This finding demonstrates that our results for the effect of thicknesses on D_{heat} of corundum and other materials (Table 7.4) represent intrinsic behavior. The results of Figs. 7.9 and 7.10 combined indicate that sections of corundum with L below ~1 mm represent "thin" behavior. Thus, for this thickness of corundum, fitting over a wide range of temperatures is unlikely to be accurately describe by the universal formula: although an upturn is not obvious in the data, it exists. Thermal diffusivity data from a 2- or 3-mm sample would be better for geophysical applications for this thermally conductive material.

We have not amended our empirical Eqs. (7.1) and (7.2) to address the upturns. Lower temperature data are needed on a variety of substances to establish a more complicated formula for the dependence of D on both L and T near 298K and below. However, the upturn in $D_{heat}(T)$ for the thin samples points toward the origin, so a simple, linear increase is compatible with the available data. This behavior indicates radiative transfer at low T in the thin samples, just as the HT term indicates radiative behavior at high T in the thick samples.

The data at high temperature (Fig. 7.10A) may seem less affected by thickness, since the dependence of D_{heat} on temperature is fairly flat at high T for the thicknesses explored. However, graphing the results for two different, but substantially elevated temperatures against thickness shows that the same asymptotic behavior occurs at high temperature as at ambient (cf. Figs. 7.9–7.10B). Fig. 7.3 reasonably represents the dependence of thermal diffusivity on sample thickness for any temperature, although this may not be exact for corundum over the transition from "thin" to "thick" behavior.

The above results show that the effects of thickness and temperature on are independent above ~350 K. This finding allows the equations for temperature and length to be combined in the high temperature limit. However, for samples in the "thin" regime, Fourier's equations are not actually valid, because these are limited to optically thick (diffusive) conditions, see below.

7.3.3.3 *Why Does Thickness Have a Strong Effect?*

The substances used to explore thickness are electrical insulators, all of which have IR fundamentals (Table 7.1), although the fundamentals vary in number from 1 to 57 and spectra depend on the chemical composition and site symmetries (Chapter 2). All nonmetals have overtone/combination bands at higher frequency, which are considerably weaker than fundamental modes, as shown in Fig. 2.10A and B. This difference underlies the dependence of thermal diffusivity on length, as follows:

Sections of corundum which are 0.12 mm thick are optically thick for the fundamentals but optically thin in the overtone region, whereas sections which are 9 mm thick are optically thick for the entire overtone region. As shown in Fig. 2.10D, a section of ~1 mm thick is optically thin for the weakest, highest frequency overtones only. The frequency where corundum is optically thick gradually increases with length. The shape of the curve of Fig 2.10 is remarkably similar to that of D_{heat} against thickness at 298K (Fig. 7.9). Fig. 2.10D listed a logarithmic fit. The frequency where these materials are optically thick is also described by power laws with L:

$$\begin{aligned} \nu_{corundum} &= 1346.6L^{0.11243}; \quad R_s = 0.97 \\ \nu_{silica} &= 1941.9L^{0.10547}; \quad R_s = 0.95 \\ \nu_{calcite} &= 1834.6L^{0.15313}; \quad R_s = 0.99 \end{aligned} \tag{7.4}$$

The powers are remarkably similar to the powers describing D_{heat} vs thickness (Table 7.4).

Fused silica and calcite behave similarly to corundum even though their spectra differ (Fig. 2.10A–C). Thus, nearly constant D_{heat} exists for several mm thicknesses because the sample is optically thick to all IR modes, which diffuse the heat. For samples well below ~1 mm, only the fundamental modes diffuse heat, whereas the weak, near-IR modes allow ballistic radiative transport. For intermediate lengths, some but not all of the near-IR absorptions can diffuse heat. The change in D_{heat} is rapid below ~1 mm because the overtones close to the fundamentals are much more intense and also fall on the shoulders of the fundamentals.

As temperature increases, bands broaden and weaken. However, the great disparity in strength between the IR fundamentals and overtones remains. For this reason, D_{heat} at moderately high temperature depends similarly on L as at ambient T (Fig. 7.10).

The value of the asymptote D_{∞} depends on the material, as does the attenuation coefficient b. Both of these are tied to the spectra. The trends in Figs. 7.5, 7.6, and 7.9 are clearly connected with the number of IR peaks and their frequencies. From Fig. 7.10, thermal diffusivity will diverge from the universal formula with temperature and from the exponential dependence on length depending on the nuances of the spectra. Calcite, with its many IR sharp bands at high frequency, would provide an interesting test case. Chapters 8 and 11 provide further discussion of the radiative mechanism.

7.3.3.4 Thickness and Layered Models

Available models for layered geometries assume that Fourier's law holds. Once steady-state is reached, this assumption should be reasonable, presuming that a thermal conductivity appropriate to the thickness of the layers is used. The long-standing Eq. (3.29) should be valid: for heat flow in series, conserving energy during steady-state yields:

$$\frac{L}{K} = \sum \frac{L_i}{K_i} \tag{7.5}$$

However, the time-dependent behavior of samples with layers is a concern. Fourier's laws cannot be valid when D depends on the spatial coordinate in the direction of flow, so a radiative transfer model warrants consideration. The problem can be stated in a different way: with D depending on distance, the linear approximation utilized by Fourier is invalid. Nonlinear equations pertain. Chapter 11 returns to this issue.

7.3.4 Effects of Isostructural Substitutions, Impurities, Disorder, and Structure

In this section, comparisons are made between samples, and thus the thickness varies, which adds uncertainty. Much of our mineral data were collected from thicknesses of 0.5–2.5 mm, where D_{heat} depends weakly on thickness. Minerals tend to be thermally insulating, which, fortunately yield similar values for D_{heat} at 298K over thicknesses of 0.4–2 mm (Fig. 7.9B). Samples with $D_{heat} > 2\ mm^2\ s^{-1}$ at 298K yield similar values at large thicknesses of 2–4 mm (Fig. 7.9A), which are unfortunately larger than values typically used during data collection. For this reason, thickness must be considered when comparing high D_{heat} samples at 298K. At higher temperatures, the variation of D_{heat} with thickness is reduced (Fig. 7.10). Therefore, effects of chemical substitution and so forth are best determined by comparing samples at high T.

Common substitutions in rock-forming minerals are the focus. Substitution of Fe^{2+} for Mg^{2+} is uncomplicated. However, multiple charge states of iron complicate matters, because ferric iron can substitute for Al on tetrahedral sites or for Mg in octahedral sites. Therefore, we begin by comparing isostructural end-members to probe how ionic masses alter thermal diffusivity. After this, effects of impurities on end-member compositions and solid solutions are discussed, under the restriction that the crystallographic symmetry is unchanged. For this case, the cation sites are disordered. If the vibrational peaks of the end-members in a binary solid solution overlap, then the spectra changes continuously in frequency from one end-member compound to the other, which is known as one-mode behavior. But if the spectral peaks do not overlap, both end-member peaks exist such that their intensities vary continuously across the binary in accord with proportions, which is known as two-mode behavior (Chang and Mitra, 1968). Complicated structures with many peaks typically exhibit one-mode behavior for the high frequency peaks associated with tightly bonded units but two-mode behavior for low frequency peaks associated with the divalent cations in larger sites. Thus, solid solutions common in geophysically important minerals (e.g., heavy Fe for light Mg in olivine) have more peaks than as listed in Table 7.1. Two- and one-mode behavior affect thermodynamic properties in different ways (e.g., Hofmeister and Pitman, 2007), so impurities should impact thermal diffusivity in accord with the spectral subtleties.

An important substitution in minerals (Al^{3+} for Si^{4+}) requires a coupled substitution either involving the low valence cations or addition of protons to oxygen anions to balance charge. Isostructural substitutions are not common: more typically, this exchange includes some structural variation, as exemplified by feldspars. For this reason, the effect of structure is revisited here. Earlier, Section 7.2 focused on structure as manifest in vastly different numbers of IR modes, and how few vs many modes affects the temperature dependence of thermal diffusivity. Here, we focus on silicate minerals, which all have large numbers of IR modes, but differ in arrangements of the atoms.

7.3.4.1 *Mass Effects Without Disorder*

To avoid effects of impurities, this section covers synthetics. Although many of our silicate samples are nearly end-member (e.g., the olivine family), these still have some impurities, which contribute cation disorder.

Data collected on many alkali halides in the B1 structure (Yu and Hofmeister, 2011) and on some synthetic garnets (Hofmeister, 2006) reveal the effect of exchanging one ion in various sites. Larger cation mass lowers thermal diffusivity for the garnets (except that $Eu_3Ga_5O_{12}$ does not fit the trend). Larger anion mass for the alkali halides lowers D_{heat} (e.g., KCl > KBr > KI), but larger cation mass raises D_{heat} (e.g., KI > NaI > LiI). Bond length is germane to these comparisons. Thermal diffusivity more consistently increases as the IR frequency increases, which represents a combination of mass, bond length and bond strength. The strongest correlation is inverse between $D_{heat, 298}$ and density of all halides, including those with silver instead of the alkalis and with Cs in the B2 structure. Density reflects the combination of bond length and mass. For the garnets, mass and density increase together, but this is not the case for the alkali halides, where small cations are situated in the "holes" in the anion sublattice.

Synthetic oxide perovskites involve simultaneous substitution of two cations. For the perovskites, density and mass increase together, and for the same structure, these increases drive D_{heat} down (Hofmeister, 2010).

7.3.4.2 *Effects of Cation Disorder Without Structural Changes*

Cation disorder accompanies incorporation of impurities and behavior of thermal diffusivity across solid solutions. Small amounts of impurities generally do not alter the structure of a mineral, which permits isolating the effect of ionic substitution on heat transport. That impurity ions lower thermal diffusivity and thermal conductivity is well-known (e.g., Slack, 1964), and is evident in Table 7.2.

Pyroxene is the only silicate mineral where the effect of an extensive solid solutions without structural changes has been quantified using LFA. However, this substitution only involves half of the binary (Fig. 7.11). Irregularities exist in the trends, which are

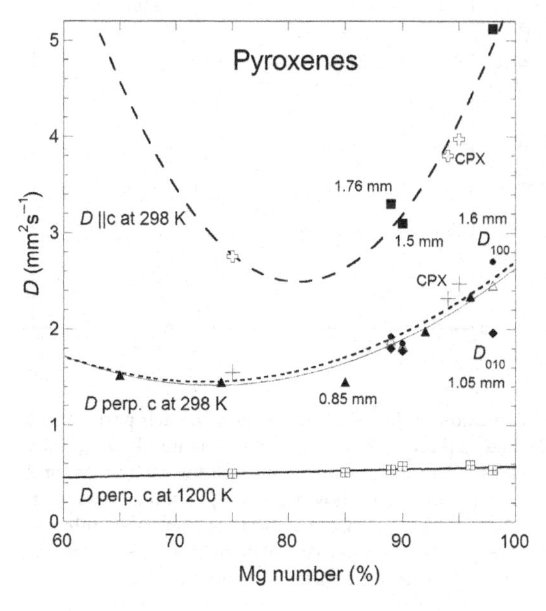

FIGURE 7.11 Effect of cation disorder on thermal diffusivity of pyroxenes at ambient and elevated temperature. Small amounts of Co, Ca, and Mn are not distinguished from Fe^{2+}. Plus symbols: diopside and augite clinopyroxenes, which are included in the fits shown as broken lines. Others are orthopyroxenes where the triangles: D_{210}; circles: D_{100}; squares: D_{001}. Thickness (extremes are listed) likely contributes to the scatter for $D > 2\ mm^2\ s^{-1}$. Source: *From: Hofmeister, A. M., 2012. Thermal diffusivity of orthopyroxenes at elevated temperature. Eur. J. Mineral. 24, 669–681. Reprinted with permission.*

attributable to thickness differences as indicated. The steep decrease observed in D from the Mg end-member pyroxenes (Fig. 7.11) was also observed for Fe substitution in the spinel structure (Hofmeister, 2007). The steep descent in D_{heat} in response to impurity incorporation has permitted detection of this effect in other materials in previous studies using various techniques.

The following findings are independent of thickness variations. For solid solutions, the lowest values of D or k_{lat} are measured near the middle of each binary, with steep decreases from end-member values, as shown for pyroxenes (Fig. 7.11). The type of ion or the structure appears to be unimportant. For example, thermal diffusivity of various crystals and glasses decreases during small amounts of anion substitution (OH^- for O^{2-}: Hofmeister et al., 2006). Because impurities reduce D from the end-member values, binary compositions have D lower than either end-member: hence, the bow in the curve of Fig. 7.11.

The behavior is consistent with a radiative transfer mechanism because impurities create additional absorption bands and because more IR bands leads to lower thermal diffusivity. Impurities also broaden existing bands, which has a similar effect.

7.3.4.3 Structural Effects and Solid Solutions

If the substitution of an impurity is not random in a crystal structure, then the symmetry is reduced. Thermal diffusivity thus decreases in response to increases in both structural and chemical complexity. Combined effects are exemplified by plagioclase feldspars (Fig. 7.12A). Most of the samples studied by Branlund and Hofmeister (2012) are volcanic

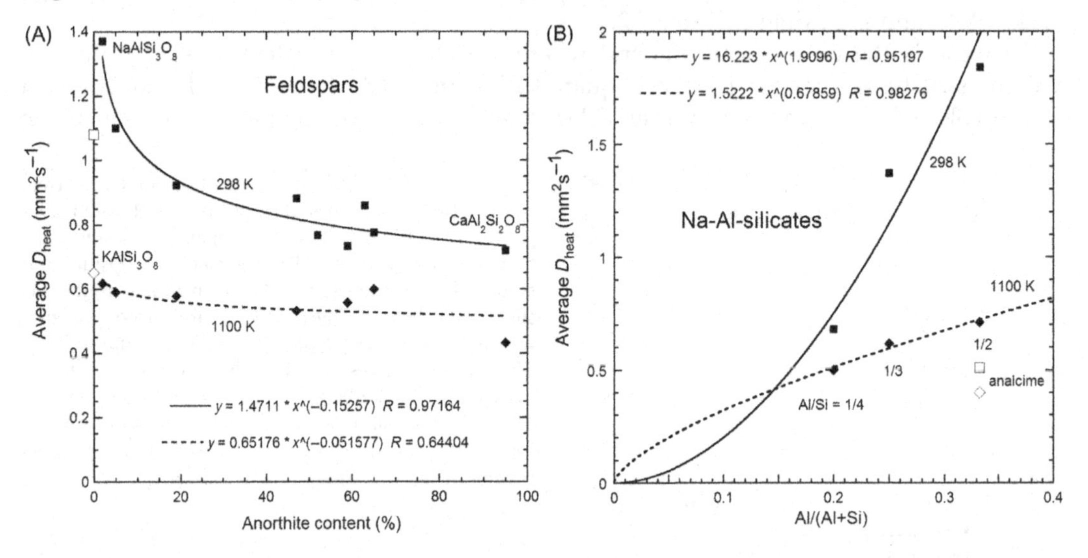

FIGURE 7.12 Dependence of thermal diffusivity on substitution of Al for Si in silicates. In each part, sample thickness is similar and D-values are orientationally averaged. Squares: D_{heat} near 298K. Diamonds: D_{heat} near 1100K. Lines are least squares power law fits, with fits listed. (A) Coupled substitution of Ca for Na and Al for Si in plagioclase feldspars, where thickness is slightly below 1 mm. Sanidine is shown for comparison as open symbols. Note that anorthite has a lower symmetry structure than the plagioclase. (B) Near end-member minerals with Na nearly equal to Al in various structures, where thickness is 1 mm or above. Analcime (open symbols) has a much larger unit cell than jadeite and the others, and is hydrated. Data from various papers in Table 7.2.

and disordered. The anorthite structure differs substantially, so the shape of the curve differs from that of isostructural binaries. Scatter in the plot is associated with structural variations and existence of multiple, intergrown phases in many feldspars.

Isolating disorder from structural effects during Al/Si exchange is difficult due to partitioning among sites and the requirement of a charge-coupled substitution. Fig. 7.12B tries to remove the latter by considering minerals with Na:Al ratios of unity, and variable Al/Si ratios. As Al increases, the thermal diffusivity increases, which is opposite to a disordering effect. The trend is entirely structural, since D_{heat} of analcime, which has Al/Si = ½ as in jadeite, is much lower than observed for this pyroxene Table 7.2.

7.4 GRAINY INSULATORS: EFFECTS OF PORES, THEIR FILLINGS AND GRAIN SIZE

Conventional methods have established that grainy material (e.g., ceramics or rocks) generally has lower values of thermal transport properties than do single crystals. This behavior is shown in Fig. 7.3 for SiC, Fig. 7.4 for calcite, and Fig. 7.9 for corundum. The reduction could involve the many factors which characterize a grainy material. The most obvious parameter is the size and size distribution of the grains; average grain-size has been the focus of investigations of ceramics (Smith et al., 2003). Grain cohesion plays a part as poorly cemented materials will have gaps between grains. Although cohesion should be strong in high-temperature refractory material, and therefore a reduction in heat transport is unexpected, differential expansion during experiments at ambient pressure can reduce internal thermal contact, particularly with large grains. Cohesion is likely a factor for rocks formed in low-pressure, low-temperature environments or altered by low-temperature metamorphism (retrogression) or hydrothermally. Porosity, which also has a size and distribution, is important, as evidenced by marble which as ~4% porosity and a much larger decrease in D from crystal values than the ceramics with negligible porosity (cf. Figs. 7.3, 7.4, and 7.9). Permeability may be relevant, since this pertains to flow. The contents of the pores are germane (Branlund and Hofmeister, 2008). Because pores cannot be empty, a grainy substance can never be one material, although substances containing air-filled pores are described as monomineralic (or homogenous ceramics). The presence of air is indicated by opacity of the sample, as is evident in production of completely transparent material, but very fine-grained material in spark-fusion syntheses (Chaim et al., 2004).

Efforts made to quantify heat transfer in ceramics and quartzites using LFA data (Smith et al., 2003; Branlund and Hofmeister, 2008) predate recognition that thickness of the sample affects thermal diffusivity. Similar thicknesses were used in these two studies. Smith et al. (2003) mention thickness of samples prior to pressing. Branlund and Hofmeister (2008) considered porosity and grain boundaries, whereas Smith et al. (2003) focused on resistance at grain boundaries. The next subsection examines on the effect of porosity on samples of similar lengths.

7.4.1 Single-Phase Ceramics and Monomineralic Rocks

Crystalline aggregates of several rock-forming minerals (graphite, various olivines, protoenstatite, quartzite, cristobalite, pure calcite marble) and various ceramics (MgO, SiC BeO, AlN, and Al_2O_3) are included in Table 7.2. The parameters describing D_{heat} as a function of T are consistent with single-crystal behavior as evidenced in Figs. 7.4 and 7.5, but the ambient temperature thermal diffusivity is lower than single crystals with similar compositions and thicknesses. However, for many cases with a large decrease in D_{heat} near 298K,the ceramic is relatively thin. The ceramics being generally thinner than the crystals (Table. 7.2) causes thermal diffusivity to be lower for a different reason than grain-size.

For monomineralic rocks with thickness similar to those of single crystals, reduction of D_{heat} is not substantial. Fig. 7.13 shows that cryptocrystalline agate with small grains and well-cemented nearly pure quartzite with large grains have diffusivity similar to less conductive orientation, which dominates the bulk property (2:1 ratios are needed for the average). For the thicknesses utilized for silica, the change in D_{heat} with L is low (Fig. 7.9). Results for forsteritic olivine are similar. All comparisons involve grainy material that is strong (welded). In these examples, the temperature response of the grainy material differs slightly from that of the single crystal. An obvious explanation is differential expansion of the quartzites and dunites with large grains, because extremely fine-grained agate has D_{heat} which is similar to the (100) direction of quartz at all temperatures. Porosity is also relevant, since the departure increases with the porosity, which is negligible in agate, very low in dunite, but several per cent in quartzite.

For coarse-grained material with low porosity, the series formulation (Eq. 7.5) is unlikely to be valid because paths exist for the heat to flow around the pores. Also, with

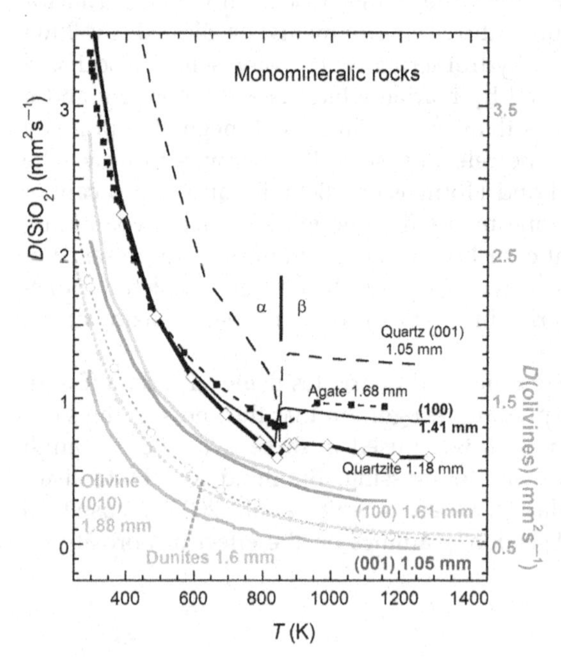

FIGURE 7.13 Comparison of essentially monomineralic rocks to oriented single crystals of olivine and quartz, with thicknesses as labeled. Fits in Table 7.2 were limited to temperatures below the displacive $\alpha-\beta$ transition in SiO_2 near 850K. Solid gray: the three orientations of orthorhombic olivine. Broken gray indicates coarse-grained dunites. Data from Pertermann and Hofmeister (2006). Data on SiO_2 from Branlund and Hofmeister (2007, 2008). Coarse-grained dunites and quartzites are similar to the average D or lowest orientation, respectively, at low temperature, but decrease more rapidly at high T, likely due to cracks forming during differential expansion. Cryptocrystalline agate remains near the less conducting orientation at all temperatures. Source: *From: Figure 13 in Hofmeister, A.M., Branlund, J.M., 2015. Thermal conductivity of the Earth. Treatise in Geophysics, second ed. In: Schubert, G. (Ed.) (In Chief), vol. 2, Mineral Physics (In: Price, G.D. (Ed.)). Elsevier, The Netherlands, pp. 584–608; reprinted with permission.*

air in the pores for which thermal conductivity is miniscule, Eq. (7.5) indicates that the conductivity (diffusivity times storativity) will be very low, rather than similar to the single-crystal values at the moderate porosities of ~10% measured for quartzites. We propose that when the pores are small compared to the mineral grains, which describes the rocks explored here, the existence of pores serves to expand the material, while not substantially affecting heat flow. That is, the length measured includes both the crystal and pores, yet only the crystal transmits the heat. This correction can be substantial, because dimensionally $D \sim L^2/\tau$ (Chapter 3). For a small amount of pores, with fillings that contribute negligibly to heat diffusion:

$$D_{\text{aggregate}} = D_{\text{crystal}}(1 - 2\varpi) \tag{7.6}$$

where ϖ is the fraction of the pores. Eq. (7.6) presumes that the grains are well-adhered. This fraction should be on a length basis, but an isotropic variation permits substitution of the volume fraction. In practice, the area fraction is commonly measured (microscopic images are two-dimensional). Quartzites follow the relationship of Eq. (7.6) quite closely (Fig. 7.14A). We omitted sample Qmb for which density was greater than quartz. Because porosity is difficult to measure accurately and grain plucking during polishing likely creates additional pores, this relationship needs to be tested against a greater range of materials.

Eq. (7.6) was also applied to opals, which differ from quartzites by being assembled of spheres of silica (cristobalite) with diameters near visible wavelengths (~650 nm for the orange samples). Our opals with 8 wt % water have quite low values relative to D_{298} of

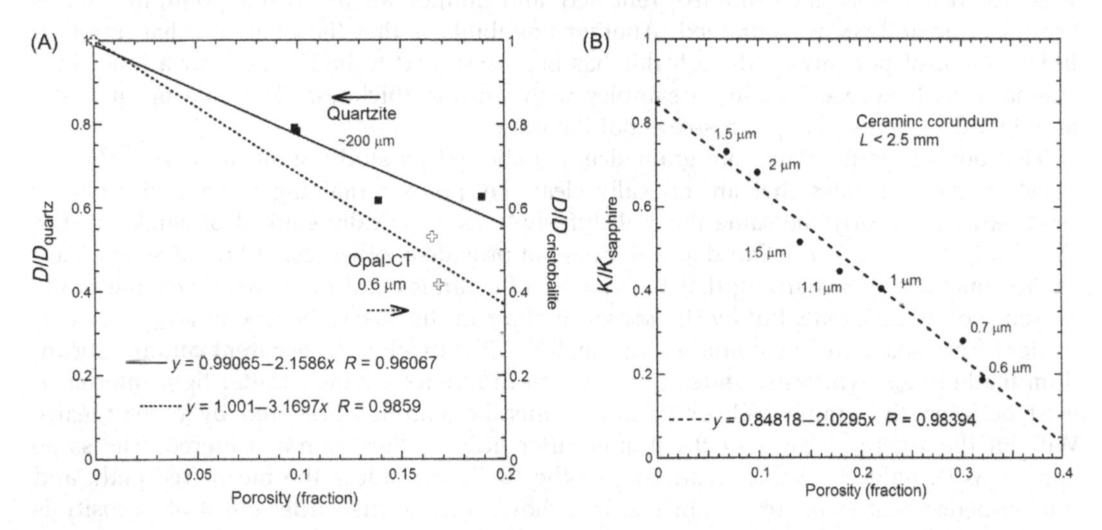

FIGURE 7.14 Dependence of ratios of transport properties at 298K to measured porosity. (A) Thermal diffusivity ratios of SiO_2 samples at 298K. The quartzites are coarse grained (0.12–0.42 mm), extremely tough (welded), with densities near that of quartz. Similar thicknesses were used in LFA measurements. Opals were orange, indicating similar size SiO_2 spheres, and have cristobalite as the phase. Data from Branlund and Hofmeister (2008), except that porosity for the opals was calculated from water contents, using an on-line converter for two phase systems from weight to volume percent (see website list). (B) Thermal conductivity ratios for ceramic corundum referenced to the $L = 2.2$ mm single-crystal value from Fig. 7.9. All other data from Smith et al. (2003). The fit indicates a 0% porosity value lower than the single crystal, see text for discussion.

cristobalite. The trend of Fig. 7.14A shows that thermal diffusivity of the opals is reduced more than a simple expansion of the sample can explain. For opals, water fills the pores. Although the diffusivity of water is low (0.143 $mm^2 s^{-1}$, see Chapter 6), this is certainly not negligible, and unlike air, the heat capacity of water is high. Therefore, under prediction of opal D from Eq. (7.6) (the steep slopes) indicates that water filling the pores participates in diffusing heat across an opal. This deduction is based on Eq. (7.5) describing a series of stacked layers, which shows that the lower diffusivity material has a strong influence. Using the storativities of water and cristobalite (4.18 and 1.56 $J cm^{-3}K^{-1}$, respectively), assuming that the heat capacity of opal is represented by a weighted average, and treating L in Eq. (7.5) as proportional to the volume occupied by each phase, gave $D_{heat,298} = 0.887$ and 0.880 $mm^2 s^{-1}$ for the opals, which is within 1% of the average of the two measurements (0.98 and 0.77 $mm^2 s^{-1}$, respectively). No free parameter exists in this approach.

As a further test of Eq. (7.5), Fig. 7.14B shows LFA data on corundum ceramics (Smith et al., 2003). Thermal conductivity and porosity values were obtained from their Fig. 6, because the tabulated values appear to be modeled. The slope of 2 indicates that pores alter the effective lengths of monomineralic ceramics. The intercept is not the single-crystal value of 38 $mm^2 s^{-1}$, which is appropriate to a 2 mm sample of sapphire. Several possibilities exist, such as a very small amount of impurities existing in their samples. Table 7.2 indicates the high sensitivity to Cr impurities, which sometimes exist in aluminous chemicals. More likely, binder was not completely removed during sintering these 2.5 mm samples. Sintering times were short, 2 hours. The low porosity Al-23 sample shown in Fig. 3.9 was heated to 1800K, then thinned, reheated, and thinned again. At this point, the single-crystal value at 298K was attained. Another possibility is that the grain size has an effect independent of porosity. Although this has been assumed to be the case for a long time, comparisons have been made on samples with varying thickness. The data on quartzite and opal-CT suggest that grain size is not the issue.

That porosity is the issue, not grain size, is indicated by study of alkali halides. Pressed pellets of alkali halides that are optically clear (i.e. pores containing water and air were removed) have nearly the same thermal diffusivity as chemically equivalent single crystals (Yu and Hofmeister, 2011). Our database does not provide another test of Eqs. (7.5) and (7.6).

We conclude by remarking that the opacity of ceramics and minerals is not due to the presence of grains alone, but by the presence of air in the spaces between the grains, as is evident from spark-fusion syntheses and material like lucolox (transparent ceramic corundum made in gel synthesis). Index of refraction differences, which scatter light internally, exist between the grains and air in monomineralic samples prepared by other means. Without the air, the indices of the grains differ little so light is not scattered. The same applies to IR light as visible. Scattering of the IR light reduces the mean free path, and thus impedes heat transport. From Fig. 7.14, however, the first-order effect of porosity is less material in a set length.

7.4.2 Multicomponent Rocks

The database contains measurement of 24 different silicate rocks, with some duplicate measurements to check for effects of inhomogeneity and some measurements of samples oriented with respect to foliation or lineation. Table 7.5 provides fits to the universal

TABLE 7.5 Thermal Diffusivity Fits for Rock Samples ($D = FT^G + HT$, in $mm^2\ s^{-1}$)

Sample	Mineralogy	L mm	D_{298} $mm^2 s^{-1}$	F	G	H	χ^2	T_{max} K	Ref.
Aubrite meteorite	93% En99, 5% diopside, 2% albite	1.135	1.81	17380	−1.626	0.00032351	0.015	1000	Hofmeister (2012)
En65 [210] gem	Layers: 95% En65, 5% $FeTiO_3$	0.53	1.52	318.15	−0.94101	0.000090025	0.0007	920	Hofmeister (2012)
Fayalite rock	70% Fa, 15% iron oxide, ~15% Ca, Fe-rich silicate	1.632	1.27	336	−0.986	0.00009	0.005	1200	Pertermann and Hofmeister (2006)
Fayalite slag #1	~85% Fa, ~10% Pb-bearing flux, <5% FeS	1.12	0.8	151	−0.933	0.000129	0.003	1200	Pertermann and Hofmeister (2006)
Fayalite slag #2		1.87	1.81	490.74	−1.0635	0.00014965	0.001	1300	Pertermann and Hofmeister (2006)
Monticellite	60% $CaMgSiO_4$, 40% epidote	1.485	1.17	130	−0.837	0.000161	0.007	1200	Pertermann and Hofmeister (2006)
Hortonolite	50% $Mg_{1.2}Fe_{0.8}SiO_4$, 50% $FeTiO_3$	1.112	1.47	250	−0.910	0.00014	0.001	1000	Pertermann and Hofmeister (2006)
Rhyolite #1	Fine-grained quartz and sanidine	0.552	1.86	370.23	−0.93078		0.0005	570	Whittington et al. (2009)
Rhyolite #2		0.998	2.18	1112.7	−1.10		0.004	770	Whittington et al. (2009)
Granite #1	Quartz, ksp, albite	0.964	2.11	4063.6	−1.3247		0.006	770	Whittington et al. (2009)
Granite #2		0.98	1.527	454.69	−0.99854		0.0003	770	Whittington et al. (2009)
Schist ⊥#1	Quartz, biotite, garnet, starolite	0.854	1.525	389.85	−0.98172	0.0001612	0.0002	770	Whittington et al. (2009)
Schist ‖ #1		0.83	1.78	184.5	−0.81871		0.002	770	Whittington et al. (2009)
Schist ‖ #2[a]		0.86	1.66	[a]					Whittington et al. (2009)
Schist ‖ #3		0.97	1.456	1874.1	−1.2742	0.00035736	0.0007	670	Whittington et al. (2009)
Schist ‖ #4		0.73	1.28	131.73	−0.81627		0.0002	670	Whittington et al. (2009)
Felsic granulite A[b]	qtz, k-spar, <5% garnet, <2% mica, <1% kyanite	1.08	1.483	1026.6	−1.1573		0.005	770	Nabelek et al. (2010)
Felsic granulite B		1.08	1.477	652.81	−1.077		0.005	740	Nabelek et al. (2010)
Felsic granulite C		1.066	1.31	762.31	−1.1221		0.0011	750	Nabelek et al. (2010)
Mafic gneiss A[b]	qtz, k-spar, plag, 5–20% biotite, 5–10% garnet, trace spinel	1.035	1.487	359.25	−0.9661		0.0007	745	Nabelek et al. (2010)

(*Continued*)

TABLE 7.5 (Continued)

Sample	Mineralogy	L mm	D_{298} mm^2s^{-1}	F	G	H	χ^2	T_{max} K	Ref.
Mafic gneiss B		0.54	1.46	221.6	−0.88531		0.0009	650	Nabelek et al. (2010)
Mafic gneiss C		1.08	1.178	280.27	−0.96017		0.0008	730	Nabelek et al. (2010)
Garnet tonalite	18% Qtz, 57% Plg, 15 % amph, 6% biot, 2% gar	1.217	1.378	207.22	−0.88594		0.003	770	Merriman et al. (2013)
Tonalite	39% qtz, 59% plag, 4% biot, 4% amphib	0.787	2.716	2493.5	−1.1943		0.01	820	Merriman et al. (2013)
Tonalite gneiss	9% qtz, 52% plag, 20% ampb, 7% biot, 7% cpx	1.12	1.615	349.63	−0.94574		0.0004	772	Merriman et al. (2013)
Trondhjemite	10% qtz, 45% plag, 9% biot, 35% amph	1.22	1.99	6229.5	−1.414	0.00015619	0.002	730	Merriman et al. (2013)
Monzonite	36%plag, 19% biot, 38%k-spar, 5% ser	1.21	0.822	12.828	−0.48094		0.0003	770	Merriman et al. (2013)
Komatiite AW04	58%plag, 27% amph, 13% chlorite	1.19	1.25	6110.1	−1.5017	0.00028011	0.002	700	Merriman et al. (2013)
Komatiite TF07#1	28% olv, 11% oxide, 60% chlorite	1.36	1.25	58.38	−0.6797	0.00012424	0.0005	870	Merriman et al. (2013)
Komatiite TF07#2		0.983	1.329	171.43	−0.86547	0.0002881	0.00008	770	Merriman et al. (2013)
Tholeite basalt#1	48% plag, 4% biot, 14% opx, 29% cpx, 5% oxide	1.09	1.405	48.745	−0.6521		0.003	1170	Merriman et al. (2013)
Tholeite basalt#2		0.67	1.04	50.153	−0.68868	0.00014144	0.0001	760	Merriman et al. (2013)
Amphibolite #1	4% qtz, 11% plag, 71% amph, 2% oxide,12 % ser	1.21	1.23	93.96	−0.76426		0.001	670	Merriman et al. (2013)
Amphibolite #2		0.905	1.49	357.19	−0.97071	0.0002006	0.001	670	Merriman et al. (2013)
Banded iron	45% qtz, 55% oxide	1.03	3.63	18416	−1.4928		0.007	700	Merriman et al. (2013)
Granulite KB10	4% qtx, 28% plag, 7% amph, 30% gar, 31% cpx	1.14	1.06	116.14	−0.83211	0.00010458	0.0004	1263	Merriman et al. (2013)
Granulite KB13	35% plag, 8% biot, 12%amph, 3%kspar, 13% opx, 26% cpx	1.12	1.09	476.94	−1.0868	0.00034838	0.0002	710	Merriman et al. (2013)
Granulite KB14	13% plag, 37% biot, 16% amph, 33% opx, 1% oxide	1.16	1.09	240.32	−0.96347	0.00031541	0.0006	700	Merriman et al. (2013)

[a] *Data on schist #2 were affected by dehydration: the data were scattered, precluding a reasonable fit.*
[b] *The felsic granulite (with grainsize of 30 μm) and mafic gneiss (with grainsize of ~40 μm) were oriented. For Sections A, heat flow was parallel to the lineation; for Sections B, heat flow was parallel to the folication and perpendicular to the lineation; for Sections C, heat flow was perpendicular to the foliation.*
Notes: T_{max} is the upper limit of the fit, not the data. All fits used starting parameters of $F = 100$, $G = 1$, and $H = 0.001$. The values reported for F, G, and H incorporate the digits needed to reproduce the data. Numbers #1, #2 etc. indicated different slices of the same rock. For rocks with quartz, the fits were limited to temperatures below the displacive transition at 850K. With higher proportions of quartz, the data are affected at lower temperatures. The cutoff was determined from the time-temperature curves: see Branlund and Hofmeister (2007) for a description of the thermal effects of the transition. For more detailed descriptions of the samples and data.

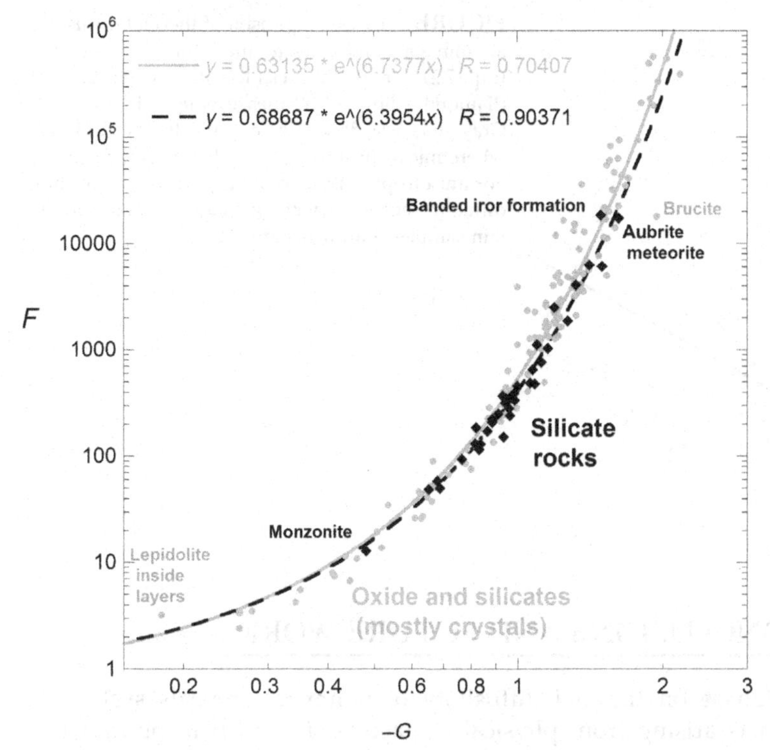

FIGURE 7.15 Comparison of fitting coefficients F and G from Eq. (7.1) for silicate rocks to oxide and silicate crystals. Black diamonds show all 37 measurements in Table 7.5. Gray dots are data in Table 7.2, excluding elements, SiC, alkali halides and mica for flow $||c$. For anisotropic substances, fits involve one to three different sections. Labels in small type indicate certain samples. Exponential fits to the trends are listed. Labels in large type indicate material type.

equation of these diverse igneous and metamorphic samples. Many of our samples have quartz, which necessitated that the fitting be terminated at 660–770K, before the displace transformation occurs. The temperature is higher for samples with little quartz. Due to the small range of temperature, the coefficient H was mostly below the limit of detection. The power law describes grainy mixtures quit well, as indicated by the low χ^2 values.

The F parameter for the silicate rocks depends on G just as defined by the oxide and silicate crystals (Fig. 7.15). Smaller ranges of F and G exist for the rocks than the minerals, which is attributable to averaging and that the rocks have phases with relatively low thermal diffusivity. The rocks with the highest F and G values are the banded iron formation (mostly quartz) and aubrite (mostly $MgSiO_3$), which have a simple mineralogies and phases with high D. Rocks with low F and G values (basalts) lack the high D mineral quartz, but have plagioclase in abundance, with very low D. Most of the rocks contain a substantial proportion of chemically and structurally complex phases, such as chlorite and amphibole (Table 7.5). However, neither the complexity of the phases, nor the grain size, nor the existence of mixtures in a wide range of proportions, nor the presence of pores seem to alter the temperature dependence of the thermal diffusivity from the universal formula (Eq. (7.1)).

The dependence of D_{heat} at 298K on $|G|$ for the rocks is also similar to that of silicate crystals (Fig. 7.16). The crystals have higher thermal diffusivity at ambient temperature because many are nearly pure. Also, the crystals generally have light cations such as Mg. This comparison corroborates that heat transport in mixtures depends on the phases composing the aggregate.

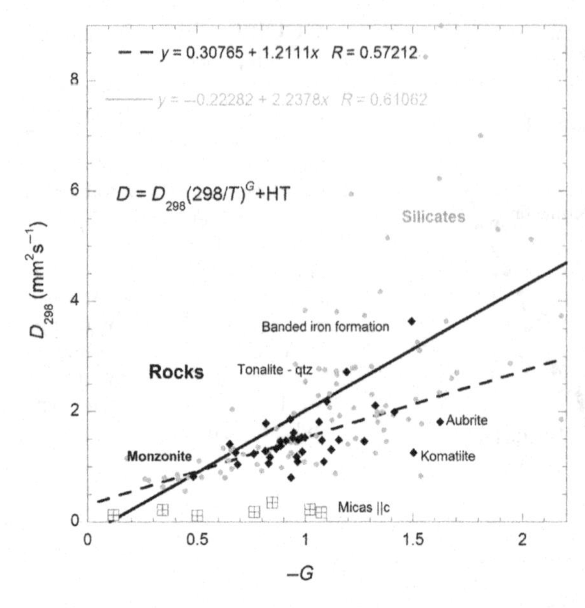

FIGURE 7.16 Comparison of thermal diffusivity at ambient conditions to the fitting coefficient G (Eq. (7.2)) for silicate rocks to silicate crystals. Black diamonds show all 37 measurements in Table 7.5. Gray dots are data only on silicates in Table 7.2, where micas for flow $||c$ are shown as the squares. For anisotropic substances, fits involve one to three different sections. Labels in small type indicate certain samples. Linear fits are shown.

7.5 CONCLUSIONS AND FUTURE WORK

Exploring the large database on thermal diffusivity of nonmetals reveals systematic behavior because the artifacts arising from physical contacts and from ballistic radiative transfer were removed. The following behavior describes solids that are not metallic:

- The temperature dependence of thermal diffusivity follows a simple formula.
- Parameters describing $D(T)$ near ambient temperature are strongly affected by bonding.
- Thermal diffusivity depends on thickness, such that $D \to 0$ as $L \to 0$ and D is nearly constant at large L.
- Thermal diffusivity of monomineralic rocks and ceramics is governed by the material property and porosity in a simple equation, if the pores are air-filled, up to porosities of $\sim 40\%$.
- If the pores are filled with water, with its high heat capacity (storativity), steady-state behavior is described using adiabatic formulae for flow in series.
- Mixtures of phases also follow the universal equation.
- Impurities reduce thermal diffusivity strongly, as was known some time ago.
- The unrecognized dependence of thermal diffusivity on length has led to intrinsic behavior being overlooked, and misattributions of porosity affects to grain boundary resistance.

Although our database is mostly crystals, the glasses investigated behave in the same manner.

All findings point to diffusive radiative transfer being the mechanism for heat flow inside nonmetallic solids. In particular, the dependence of D on length (Figs. 7.9–7.10) and optical thickness on length (Fig. 2.10) have the same shape and both change rapidly near the same value of L. This behavior exists due to great differences in absorption strengths of the fundamental IR modes and the overtones. Other clues are the relationship of thermal diffusivity near 298K with the number of IR modes, and the influence of bonding, which is related to intensity of the absorptions. Very high D_{298} materials have no IR fundamentals, but very weak overtone/combinations are present.

Thermal diffusivity depending on thickness has both theoretical and practical ramifications. The thickness dependence in LFA data suggests that problems exist in the 3-omega and thermoreflectance studies, which sample very small penetration depth, but provide values similar to conventional measurements of mm samples, see Chapter 4. Regarding the high pressure measurements of very small samples, these almost always assume a value at ambient pressure based on previous measurements of thick samples. Measurements are not made at ambient pressure, but are extrapolated. For these extremely small samples in the diamond anvil cell, D should be zero. Although it may seem puzzling that a "reasonable" pressure derivative is obtained, one must recognize that the geometry changes in the experiments, and so the investigations are probably monitoring changes in the configuration, not in heat transport. The importance of geometry to heat flow measurements is well-known and evident in Fourier's equations. Changes in the configuration are tied first and foremost to the bulk modulus, which theoretically is related to heat transport, via the carrier speed (see Chapter 9). This connection (D = speed × mean free path) explains obtaining pressure derivatives that are compatible with theory.

This chapter did not focus on high-temperature behavior as much as on low due to the limitations of the database, and because a model for diffusive radiative transfer is needed to make full use of the information presented here. The nature of the *HT* term, which comes into play at high T requires diffusion of radiation. As discussed above, thin crystals also yield D increasing with T, where this increase is observed near 298K. A model for diffusion of heat must take this behavior into account.

The database is not complete: further study is needed of (1) the thickness dependence and its connection with spectra, and (2) the effect of variables thought to be important at constant thickness. In particular, changing the contents of pores needs to be explored. Water residing in pores is the likely cause of behavior previously ascribed to grain boundary resistance. Effects of impurities need revising because purer samples tend to be larger, but may not be large enough that the regime of constant thermal diffusivity is attained.

The length dependence has technological implications. Thermal diffusivity of coatings is germane to many applications, but is very difficult to measure, particularly as ballistic radiation often exists, but layered models do not address this effect (Chapter 4). Our results show how to infer D of a coating from a series of measurements. A radiative transfer model is needed for accurate extrapolation, however.

References

Aggarwal, R.L., Ripin, D.J., Ochoa, J.R., Fan, T.Y., 2005. Measurement of thermo-optic properties of $Y_3Al_5O_{12}$, $Lu_3Al_5O_{12}$, $YAlO_3$, $LiYF_4$, $LiLuF_4$, BaY_2F_8, $KGd(WO_4)_2$, and $KY(WO_4)_2$ laser crystals in the 80-300 K temperature range. J. Appl. Phys. 98, 103514.

Blumm, J., Lemarchand, S., 2002. Influence of test conditions on the accuracy of laser flash measurements. High Temp. High Pres. 34, 523–528.

Blumm, J., Henderson, J.B., Nilson, O., Fricke, J., 1997. Laser flash measurement of the phononic thermal diffusivity of glasses in the presence of ballistic radiative transfer. High Temp.- High Pres. 34, 555–560.

Branlund, J.M., Hofmeister, A.M., 2007. Thermal diffusivity of quartz to 1000 degrees C: effects of impurities and the α-β phase transition. Phys Chem. Minerals. 34, 581–595.

Branlund, J.M., Hofmeister, A.M., 2008. Factors affecting heat transfer in SiO_2 solids. Amer. Mineral. 93, 1620–1629.

Branlund, J.M., Hofmeister, A.M., 2012. Heat transfer in plagioclase feldspars. Amer. Mineral. 97, 1145–1154.

Bridgeman, P.W., 1924. The thermal conductivity and compressibility of several rocks under high pressures. Am. J. Sci. 7, 81–102.

Brindley, G.W., Ali, S.Z., 1950. X-ray study of thermal transformations in some magnesian chlorite minerals. Acta Cryst. 3, 25.

Chaim, R., Shen, Z., Nygren, M., 2004. Transparent nanocrystalline MgO by rapid and low-temperature spark plasma sintering. J. Mater. Sci. 19, 2527–2531.

Campbell, R.C., 2010. Approximations in the use of two and three layer analysis models in flash diffusivity measurements. Thermal Conduct. 30, 328–338.

Chang, I.F., Mitra, S.S., 1968. Application of a modified random-element-isodisplacement model to long-wavelength optic phonons of mixed crystals. Phys. Rev. 172, 924–933.

Eucken, A., 1911. Über die temperaturabhängigkeit der wärmeleitfähigkeit fester nichtmetalle. Ann. Phys. 34, 186–221.

Fateley, W.G., Dollish, F.R., McDevitt, N.T., Bentley, F.F., 1972. Infrared and Raman Selection Rules for Molecular and Lattice Vibrations: The Correlation Method. Wiley Interscience, New York, NY.

Ferraro, J.R., 1975. Factor group analysis for some common minerals. Appl. Spectrosc. 29, 418.

Fourier, J.B.J., 1822. Théorie Analytique de la Chaleur. Chez Firmin Didot, Paris. (translated in 1955 as The Analytic Theory of Heat. Dover Publications Inc, New York, NY (A. Freeman, Trans).

Hofmann, R., Hahn, O., Raether, F., Mehling, H., Fricke, J., 1997. Determination of thermal diffusivity in diathermic materials by laser-flash technique. High Temp.-High Press. 29, 703–710.

Hofmeister, A.M., 2006. Thermal diffusivity of garnets at high temperature. Phys. Chem. Miner. 33, 45–62.

Hofmeister, A.M., 2007a. Pressure dependence of thermal transport properties. Proc. National Academy Sci. 104, 9192–9197.

Hofmeister, A.M., 2007b. Thermal diffusivity of aluminous spinels and magnetite at elevated temperature with implications for heat transport in Earth's transition zone. Am. Mineral. 92, 1899–1911.

Hofmeister, A.M., 2010. Thermal diffusivity of perovskite-type compounds at elevated temperature. J. Appl. Phys. 107 (103532).

Hofmeister, A.M., 2012. Thermal diffusivity of orthopyroxenes at elevated temperature. Eur. J. Mineral. 24, 669–681.

Hofmeister, A.M., 2013. Heat transport properties of cristobalite and discussion of "snowflake" formation. Can. Mineral. 51, 705–714.

Hofmeister, A.M., 2014. Thermal diffusivity and thermal conductivity of single-crystal MgO and Al_2O_3 as a function of temperature. Phys. Chem. Miner. 41, 361–371.

Hofmeister, A.M., Whittington, A.G., 2012. Effects of hydration, annealing, and melting on heat transport properties of fused quartz and fused silica from laser-flash analysis. J. Non-Crys. Solids. 358, 1072–1082.

Hofmeister, A.M., Branlund, J.M., 2015. Thermal conductivity of the Earth. Treatise in Geophysics. In: Price, G.D. (Ed.), vol. 2, Mineral Physics, second ed. Elsevier, The Netherlands, pp. 584–608. (In: Schubert, G. (Ed.), In Chief).

Hofmeister, A.M., Carpenter, P., 2015. Heat transport of micas. Can. Mineral. 53, 557–570.

Hofmeister, A.M., Ke, R., 2015. Thermal diffusivity of feldspathoids and zeolites as a function of temperature. Phys. Chem. Mineral. 42, 693–706.

Hofmeister, A.M., Pertermann, M., 2008. Thermal diffusivity of clinopyroxenes at elevated temperature. Eur. J. Mineral. 20, 537–549.

Hofmeister, A.M., Pitman, K.M., 2007. Evidence for kinks in structural and thermodynamic properties across the forsterite-fayalite binary from thin-film IR spectra. Phys. Chem. Mineral. 34, 319–333.

Hofmeister, A.M., Yuen, D.A., 2007. Critical phenomena in thermal conductivity: implications for lower mantle dynamics. J. Geodyn. 44, 186–199.

Hofmeister, A.M., Dong, J.J., Branlund, J.M., 2014. Thermal diffusivity of electrical insulators at high temperatures: evidence for diffusion of phonon-polaritons at infrared frequencies augmenting phonon heat conduction. J. Appl. Phys. 115 (163517). Available from: https://doi.org/10.1063/1.4873295.

Hofmeister, A.M., Pertermann, M., Branlund, J., Whittington, A.G., 2006. Geophysical implications of reduction in thermal conductivity due to hydration. Geophys. Res. Lett. 33, L11310.

Hofmeister, A.M., Pertermann, M., Branlund, J.M., 2007. Thermal conductivity of the Earth. In: Schubert, G., Price, G.D. (Eds.), Treatise in Geophysics, vol. 2 Mineral Physics. Elsevier, The Netherlands, pp. 543–578.

Hofmeister, A.M., Whittington, A.G., Pertermann, M., 2009. Transport properties of high albite crystals and near-endmember feldspar and pyroxene glasses and melts to high temperature. Contrib. Mineral. Petrol. 158, 381–400.

Maroni, V.A., 1988. An analysis of the vibrational characteristics of zeolites using factor group methods. Appl. Spectrosc. 42, 487–493.

McKeown, D.A., 2005. Raman spectroscopy and vibrational analyses of albite: From 25 °C through the melting temperature. Am. Mineral. 90, 1506–1517.

Mehling, H., Hautzinger, G., Nilsson, O., Fricke, J., Hofmann, R., Hahn, O., 1998. Thermal diffusivity of semitransparent materials determined by the laser-flash method applying a new mathematical model. Inter. J. Thermophys. 19, 941–949.

Merriman, J.D., Alan, G.W., Hofmeister, A.M., Nabelek, P.I., Benn, K., 2013. Thermal transport properties of major Archean rock types to high temperature and implications for cratonic geotherms. Precambrian. Res. 233, 358–372.

Merriman, J.M., Hofmeister, A.M., Whittington, A.G., Roy, D.J., 2018. Temperature-dependent thermal transport properties of carbonate minerals and rocks. Geosphere 14, 1961–1987.

McKeown, D.A., Bell, M.I., Etz, E.S., 1999. Vibrational analysis of the dioctahedral mica: $2M_1$ muscovite. Am. Mineral. 84, 1041–1048.

Milošević, N., 2010. Application of the laser pulse method of measuring thermal diffusivity to thin alumina and silicon samples in a wide temperature range. J. Therm. Sci. 14, 417–423.

Nabelek, P.I., Whittington, A.G., Hofmeister, A.M., 2010. Strain heating as a mechanism for partial melting and ultrahigh temperature metamorphism in convergent orogens: Implications of temperature-dependent thermal diffusivity and rheology. J. Geophys. Res. 115, B12417.

Nakashima, S., Harima, H., 1997. Raman investigation of SiC polytypes. Phys. Status Solidi A 162, 39–64.

Nemanich, R.J., Lucovsky, G., Solin, S.A., 1977. Infrared active optical vibrations of graphite. Solid State Commun. 23, 117–120.

Ohta, H., Waseda, Y., 1986. Measurement of thermal diffusivity of inorganic materials at elevated temperature by the laser flash method. High Temp. Mater. Proc. 7, 179–184.

Pertermann, M., Hofmeister, A.M., 2006. Thermal diffusivity of olivine-group minerals. Am. Mineral. 91, 1747–1760.

Pertermann, M., Whittington, A.G., Hofmeister, A.M., Spera, F.J., Zayak, J., 2008. Thermal diffusivity of low-sanidine single-crystals, glasses and melts at high temperatures. Contrib. Mineral. Petrol. 155, 689–702. Available from: https://doi.org/10.1007/s00410-007-0265-x.

Pitman, K.M., Hofmeister, A.M., Corman, A.B., Speck, A.K., 2008. Optical properties of silicon carbide for astrophysical environments I. New laboratory infrared reflectance spectra and optical constants. Astron. Astrophys. 483, 661–672.

Roy, R., Roy, D.M., Osborn, E.F., 1950. Compositional and stability relationships among the lithium aluminosilicates: eucryptite, spodumene, and petalite. J. Am. Ceram. Soc. 33, 152–159.

Slack, G.A., 1964. Thermal conductivity of pure and impure silicon, silicon carbide, and diamond. J. Appl. Phys. 35, 3560–3465.

Smith, D.S., Fayette, S., Grandjean, S., Martin, C., Telle, R., Tonnessen, T., 2003. Thermal resistance of grain boundaries in alumina ceramics and refractories. J. Amer. Ceram. Soc. 86, 105–111.

Stierwalt, D.L., Potter, R.F., 1962. Lattice absorption bands observed in silicon by means of spectral emissivity measurements. J. Phys. Chem. Solids. 23, 99–102.

Twain, M., 1897. Following the Equator: A Journey Around the World. American Publishing Co, Hartford, CT.

Whittington, A.G., Hofmeister, A.M., Nabelek, P.I., 2009. Temperature-dependent thermal diffusivity of Earth's crust: implications for crustal anatexis. Nature 458, 319–321.

Yu, X., Hofmeister, A.M., 2011. Thermal diffusivity of alkali and silver halides. J. Appl. Phys. 109, 033516. Available from: https://doi.org/10.1063/1.3544444.

Websites

Poco graphite (accessed 02.02.18)
poco.com/Portals/0/Literature/Semiconductor/IND-109441-0115.pdf
http://poco.com/MaterialsandServices/Graphite.aspx
volume to weight fraction and vice versa calculator (accessed 10.02.18)
https://netcomposites.com/guide-tools/tools/calculators/volume-weight-fraction/
Data from the Washington University laboratory are available at http://epsc.wustl.edu/~hofmeist/thermal_data/

CHAPTER

8

A Macroscopic Model of Blackbody Emissions With Implications

Anne M. Hofmeister, Everett M. Criss and Robert E. Criss*

OUTLINE

* E.M. Criss is an employee of Panasonic Avionics Corporation, but prepared this article independent of his employment and without use of information, resources, or other support from Panasonic Avionics Corporation.

Measurements, Mechanisms, and Models of Heat Transport
DOI: https://doi.org/10.1016/B978-0-12-809981-0.00008-5

Scientific writers, not less than others, write to please, as well as to instruct, and even unconsciously to themselves, (sometimes) sacrifice what is true to what is popular. ***Frederick Douglass, Commencement Address at Western Reserve College, July 12, 1854.***

Measurements of heat transfer in gas, pourable liquids, and nonmetallic solids point to diffusion of radiation being an essential component of heat transport (Chapter 5, Chapter 6 and Chapter 7). Diffusive radiation of heat is a direct outcome of a kinetic theory of gas that incorporates the inelastic nature of collisions (Chapter 5). Long ago, Lommel (1878) associated blackbody (BB) emissions from solids with inelastic losses, and specifically discussed damping of atomic vibrations (Kangro, 1976), but was unable to provide a formula for thermal emissions. His ideas were set aside in the early 1900s. Yet, strong theoretical support for inelasticity exists in the laws of thermodynamics, when these are framed in terms of the macroscopic behavior of light (Chapter 1 and Chapter 5). Fourier's laws, which describe heat transfer as energy flowing through a medium (Chapter 3), are also consistent with diffusive of radiation explaining conduction in solids.

Solids also transport heat by ballistic radiation in the near-infrared (IR) when transparent. This added complication hindered accurate time-dependent measurements of thermal diffusivity of partially transparent insulators until the 1990s, when techniques were developed to suppress ballistic effects and models were derived to remove its remnants in laser flash experiments (e.g., Blumm et al., 1997). Thus, modern methods accurately quantify diffusion of heat (Chapter 7). But, before we can model the microscopic interactions of light with condensed matter that cause D_{heat} to vary with structure, composition, and temperature in Chapter 11, we need to understand the macroscopic process of diffusive radiation. Because radiative transfer can be ballistic, diffusive, or both (Chapter 1), this process is inherently difficult to understand, yet the many manifestations of radiative transfer make this process important on scales from the microscopic to astronomical. Understanding how matter responds to light is a prelude to understanding radiative transfer because all measurements involve material, including the "vacuum" of space. Data on the interactions are ascertained using spectroscopy (Chapter 2). The present chapter makes use of these results.

Models for both types of radiative transfer center on the temperature and frequency (or wavelength) dependence of BB emissions. However, real materials differ from a perfectly absorbing BB (Chapter 1 and Chapter 2), and so measured BB spectra invariably depart from this ideal to some degree. Comparing the experiments with the model requires making some assumptions and applying instrumental corrections. In addition, measurements of absolute intensity are difficult, so historical comparisons did not constrain the prefactors in Planck's formula (Eq. 2.3). Acceptance of Planck's function was motivated by the need to explain the photoelectric effect, which was interpreted as quantization of light. Because the BB function and its meaning are incredibly important to physics, this chapter probes theory and experiment thoroughly, from several angles.

Section 8.1 shows that Planck's function does not require quantization of frequency, and is inconsistent with the independently measured Stefan–Boltzmann constant. A key issue is differences between energy, flux, and intensity. Because the spin of light is not part of any model of BB emission, Section 8.2 develops a macroscopic model that includes this motion, based on thermodynamics and Maxwell's equations. Our model accounts for the adiabatic and isothermal conditions describing BB emissions, explains how the second law is met through independent analyses of energy and entropy, and shows that energy in the spin and translation of electromagnetic (EM) waves are linked through the zeroth law (the same temperature pertains). The origins of two erroneous factors of 2 in Planck's formula are explained. Section 8.3 covers experimental evidence for the continuous nature of the frequency of light, and how certain behaviors and properties of matter impose restrictions on interactions and uptake of light energy. The photoelectric effect is explained. Historical measurements of BB spectra are shown to be inadequate to the task of confirming Planck's law. Section 8.4 places Planck's formula in historical context, emphasizing incompatibility with thermodynamic requirements and electrodynamics. Section 8.5 summarizes, and explains how partial solutions, experimental ambiguities, out-of-synch developments, and consensus opinion has stymied progress towards understanding the continuous and simple nature of BB emissions.

8.1 MATHEMATICAL IMPLICATIONS OF PLANCK'S FORMULA

Eqs. (2.3abc) list Planck's well-known formula for the "intensity" of a BB (I_{BB}) for various spectral units. First, "intensity" is a misnomer, as the units of I_{BB} are different than those of intensity (W m^{-2}). Actually, "I_{BB}" is a distribution function, see below and Section 8.2. Nonetheless, this section uses the most familiar version, where I_{BB} is cast as a function of frequency in hertz ($\underset{\sim}{\nu}$, cycles per second):

$$I_{BB}(\nu, T) = \frac{2h\underset{\sim}{\nu}^3}{c^2}\frac{1}{\exp(h\underset{\sim}{\nu}/k_B T) - 1} \tag{8.1}$$

which has units of W m^{-2} sr^{-1} Hz^{-1} (=J (m^2-sr)$^{-1}$). By definition, the sphere contains 4π steradians (sr). Henceforth in this chapter, the tilde is dropped, because converting from Hertz to wavenumbers is trivial. Planck's formula is connected with emissions in a cone

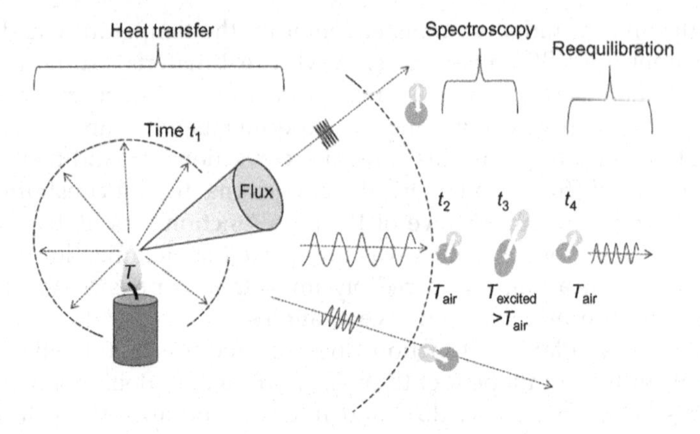

FIGURE 8.1 Schematic of how blackbody emissions are produced and measured, with a comparison to spectroscopic measurements. Left, thermal and visible emissions from a source (candle) travel outwards linearly (arrows). For a point source, the emissions are radial, defining a spherical front at some time t_1, but for realistic shapes, the sphere is incomplete (dashed arc). The cone illustrates that for any source, the radiation defines part of a sphere, and so the heat emitted is defined as energy per area per steradian (medium gray end of cone, labeled flux). Stefan's experiment measures the heat being transferred at some given distance over a defined area. Right, interaction of light from some source with molecules in the air. At some time t_2, the light crosses the gas, interacting with some molecules when certain conditions are met. Some molecules uptake heat, for example, in a vibrational transition, which can be described as the excited molecules having a higher temperature. Spectroscopic measurements record the uptake of the energy as a loss from the impinging flux. At some later time, the excited molecule releases the excess and returns to the equilibrium state of the surrounding air.

from a point source of light (Fig. 8.1, LHS), while assuming that all light produced has the same speed (c, in dilute gas).

Planck's first derivation in 1900 of Eq. (8.1) did not specify the constants, did not invoke energy quantization, and did not use statistical mechanics, as he was a critic of Boltzmann, which he much later regretted (~1942, see Kangro, 1976). Planck portrayed his second attempt in 1901, which included the constants, as a mathematical construction (Kragh, 2000). The quantization interpretation came later. Historic data motivating his efforts are shown in Fig. 8.2. Irrespective of the assumptions made and the mathematical manipulations used, the form of Eq. (8.1) has various consequences, as follows:

8.1.1 How Many Photons?

According to Marr and Wilkin (2012), the number of photons in the quantum description associated with the flux per steradian is defined by:

$$N_{\text{photons}} \text{ per steradian} = \int_0^\infty \frac{I_{\text{BB}}}{h\nu} d\nu = 4\zeta(3)\frac{{k_B}^3}{h^3c^2}T^3 \tag{8.2}$$

where $\zeta(3)$ is the Riemann Zeta function of 3, which is irrational with a value near 1.20206, and is known as Apéry's constant. The total number of photons crossing a spherical shell

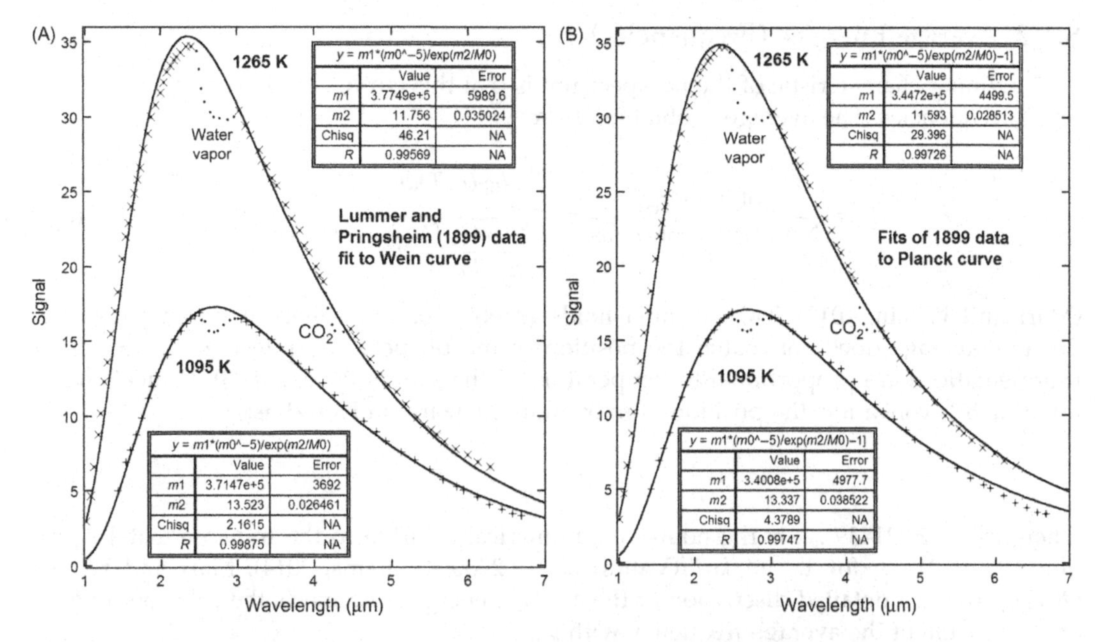

FIGURE 8.2 Spectral basis of the blackbody distribution function. Measurements are those Lummer and Pringsheim (1899); data over the greatest wavelength range are shown. The Y-axis is an electronic measurement that is scaled (corrected), see Section 2.3. Dots: data points for $T = 1265$ and 1095 K. $\times$ and $+$: data used in the fits shown, which excluded absorptions of water vapor and CO_2. The experimentalists attributed these absorptions to the atmosphere, but as shown below, these are absorptions by the prism. If these artifacts are included in the fitting, our algorithm matches the calculated values shown by the authors. (A) Least square fits to the distribution function of Wien (1897), with coefficients and uncertainties listed in the boxes. (B) Fits to the distribution function of Planck (1900, 1901). Neither Wien's nor Planck's model is clearly better than the other, due to experimental uncertainties.

is 4π times the result, from the definition of a steradian, and because the speed of light is independent of frequency.

Comparing the left and middle terms of Eq. (8.2) shows that the function I_{BB} describes the distribution of frequencies in BB emissions. The physical constant h is needed because the formula actually pertains to the distribution of energies. The RHS shows that N cannot possibly be an integer. First and foremost, temperature is involved, which is a continuous variable that can take on any positive, finite value. In addition, π and $\zeta(3)$ are irrational numbers, so no matter how many decimal places are used, or how precisely the physical constants are determined, the number of photons composing BB emissions cannot be an integer, unless temperature itself is quantized in a manner that somehow offsets the irrational numbers. Contrary to the standard view, Planck's formula is not consistent with the discretization of light in terms of its frequency, because an integer number of photons (N_j) should then exist for each ν_j, and the sum of these integers ($N = \Sigma N_j$) should also be an integer.

8.1.2 Average Energies (Frequencies)

A defining characteristic of the BB spectrum is that this curve is represented by two different frequencies. The average is obtained from:

$$\langle E \rangle = \frac{\text{Total energy}}{\text{Number of photons}} = \frac{\int_0^\infty I_{\text{BB}}(\nu, T) d\nu}{\int_0^\infty \frac{I_{\text{BB}}(\nu, T)}{h\nu} d\nu} = \frac{\pi^4}{30\zeta(3)} k_B T, \tag{8.3}$$

(Marr and Wilkin, 2012). This formula holds whether or not spherical symmetry exists. But this average does not match the position of the BB peak, despite the fact that both representations are proportional to temperature of the source (Wien, 1893). Planck's function (Eq. 8.1) constrains the position of maximum intensity (in Hertz) as:

$$\nu_{\text{max}} = \frac{w_3 k_B}{h} T \tag{8.4}$$

where $w_3 \cong 2.821439 \ldots$ is the nontrivial, numerical solution to the transcendental equation $3\ (1-e^{-x}) = x$, for $x = h\nu/k_B T$ (Valluri et al., 2000; Williams, 2014). Marr and Wilkin (2012) provide a detailed discussion of the average energy, as regards the color of our Sun and confusion of the average frequency with ν_{max}.

Two important frequencies, representing the most probable (v_{max}) and average (v_{avg}) frequencies, are associated with the BB spectrum due to its asymmetry. This behavior contrasts with the symmetric shapes of weak peaks in many absorption spectra (Chapter 2). The differences between BB and absorption spectra are related to conditions being, respectively, optically thick and optically thin (Section 8.3). Temperature being a continuous variable implicates that frequency is continuous also.

8.1.3 Comparison to Flux Measurements

The formula for total radiant energy per area per time, $\mathfrak{R} = \sigma T^4$ (Eq. 1.24), was first ascertained by Stefan (1879). Due to Boltzmann's subsequent efforts, the proportionality constant bears both names. Its numerical value of $\sigma = 5.670 \times 10^{-8}\ \text{W m}^{-2}\ \text{K}^{-4}$ is experimentally determined. An important example is provided by the Sun. The measured Solar luminosity of $\sim 4 \times 10^{26}$ W is compatible with Eq. (1.24), computed using a temperature of 6000 K from applying Wien's displacement law (Eq. 1.25) to the Solar spectrum, and using the measured radius (s) to compute the surface area ($4\pi s^2$) of the nearly spherical Sun. Other types of stars behave similarly: this is a hallmark of "main sequence" stars, for which luminosity, mass, radius, and temperature are all linked by power laws (e.g., Zombeck, 2007). The Solar example verifies that to provide the flux per steradian, $\mathfrak{R}$ must be divided by 4π.

The flux per steradian at any given frequency (in Hertz) is also allegedly provided by I_{BB}. Integration provides the flux at all frequencies per steradian. From Marr and Wilkin (2012):

$$\text{flux}(\text{in Wm}^{-2}\text{sr}^{-1}) = \int_0^\infty I_{BB} d\nu = \frac{2\pi^4}{15} \frac{{k_B}^4}{h^3 c^2} T^4 \tag{8.5}$$

However, equating Eqs. (8.5) and (1.24) yields a flawed result, namely:

$$\frac{\sigma}{4\pi}T^4 = \frac{2\pi^4}{15}\frac{k_B{}^4}{h^3c^2}T^4\ ? \tag{8.6}$$

The flaw lies in the theoretical numerical value of the RHS being fourfold larger than the experimental determination of the LHS. The factor of 4 is an integer within experimental uncertainties. This discrepancy is known, but has not been logically explained in any source that we could find. Commonly no explanation is given, or the Stefan–Boltzmann law is provided without the bothersome factor of 4, or integration is computed over one hemisphere only, illogically yielding π, not 2π.

The integral of Eq. (8.5) is nontrivial, and probably first appeared in the English literature in the comprehensive mathematical handbook by Gradshteyn and Ryzhik (1965), who cite Fikhtengol'ts (1948) for a similar solution. Earlier efforts tried to quantify σ from flux experiments and independently to ascertain whether the Planck or the Wein function was correct from fitting data from spectral experiments (e.g., Coblentz, 1916: Section 8.3).

8.1.4 Summary

The above mathematical findings provide counterevidence to light consisting of quantized, discrete frequencies, and reveal a fourfold inconsistency between Planck's formula and the experimentally determined physical constant describing the total flux (σ_{SB}). These shortcomings are serious and have implications beyond the scope of this book.

Although emission spectra have confirmed the basic shape of the BB curve proposed by Planck, these data *do not provide absolute intensities* because surface reflections are ubiquitously present and have been confused with emissivity varying from unity (Chapter 2, more in Section 8.3). Many problems exist with the underlying assumptions made by Planck and his predecessors (Section 8.4). In particular, because vibrations are anharmonic and damped (e.g., Lommel, 1878), vibrational energies differ from transitional energies, which are probed in spectroscopy. A macroscopic model is developed first (Section 8.2), in order to elucidate the thermodynamic implications of light waves spinning, and to incorporate other concepts that were incompletely developed in the 1900s, but pertain to thermal emissions.

8.2 MACROSCOPIC ANALYSIS OF THERMAL EMISSIONS

All matter is warm ($T > 0$ K) and emits light, which underlies the thesis of the universal nature of BB emissions. After covering general constraints of macroscopic thermodynamics and electromagnetism, we develop a model through an analogy with the classical (elastic) ideal gas. Just as the macroscopic kinetic theory of matter gas did not require detailed knowledge of the nature of atoms, a macroscopic kinetic theory of BB emissions need not address the microscopic character of light. Accounting for the spin of EM waves (Poynting, 1909) is a crucial part of our macroscopic model, and allows us to explain the photoelectric effect (Section 8.3.1) without calling on quantization of frequency.

8.2.1 Requirements of the Zeroth Law

Measuring BB emissions at some distance from a source (Fig. 8.1) presumes that the emissions are not perturbed between the source and detector, that is, that the expansion of light is adiabatic and isothermal. These experimental conditions are that of equilibrium, where the thermal emissions (heat) are linked to the source via equal temperatures (Fig. 8.1). From the zeroth law (Fowler and Guggenheim, 1939), if the BB emissions retain the characteristic temperature of the source, then they are in thermal equilibrium with that source. The process of equilibration plays no role in the zeroth law. Consequently, how thermal radiation is produced is irrelevant, and so exploring emissions through their link to vibrations in an absorbing material is misplaced. This section presents an alternative.

8.2.2 Macroscopic Properties of Light: Energy, Wavelength, Frequency, Momentum, and Spin

Early work provided information regarding the energy associated with light. That energy increases as wavelength decreases predates Herschel (1800), but ascertaining the precise inverse relationship required accurate gratings, and therefore involved efforts well into the 1800s. Table 8.1 summarizes: see Chapter 2 for details and references. Importantly, what is measured is actually a flux = energy per period of a cycle.

Of great importance was the work of Thomas Young (1804), who demonstrated that light has properties associated with transverse waves. Young's finding connected wavelength to inverse frequency via the speed, which had been measured in the 1700s, and appears to be a constant in air and more dilute media. This connection links the energy associated with the translational motion of light to its frequency:

$$\nu = c/\lambda \text{ and thus } E \propto \nu \tag{8.7}$$

and wavelength, although the proportionality constant was not constrained by Young (Table 8.1). The physical constant ($h = 6.62609923 \times 10^{-34}$ J s) describing the linear dependence of energy on frequency was provided later by Planck (1901). Moreover, this motion has an associated momentum:

$$\text{linear momentum} = E/c \tag{8.8}$$

from Maxwell's macroscopic equations (Table 8.1). Measurements of radiation pressure on surfaces confirmed that light has linear momentum (Nichols and Hull, 1901; Lebedev, 1901).

Light pushes things in its way, despite being wave-like and having no detectable mass. Because light is wave-like, it has a spatial dimension (which is commonly referred to as amplitude, via a mechanical analogy), yet it occupies no volume, so there is no limit to the packing of waves in space. Infinite packing stems from light waves negligibly interacting with other light waves. Classically, infinite packing is expressed in the principle of superposition, which is observed for waves along strings or in water. Furthermore, light rays can be highly focused. On laboratory scales, the interaction of light with itself has only been detected when high-intensity, high-frequency, very short pulse laser beams cross in air or when similar, but lower frequency, laser beams cross in glass (Naudeau et al., 2006). Note that a dense medium is essential for low frequency interactions. The presence of

TABLE 8.1 Key Discoveries and Their Mathematical Representation Important to Thermal Emissions, Focusing on Frequency, With Some Modern Physical Constants

Discovery	Who (When)	Mathematical Equivalent	Eq. or Fig.	Notes
Thermal peak; infrared region	Herschel (1800)	Provided a qualitative spectrum with a peak in λ	n.a.[a]	Showed luminosity differs from the effect of heating
Energy of light	Over centuries	$\text{Energy} \propto \frac{1}{\lambda}$	n.a.[a]	Evident in Herschel's spectrum, but wavelength was not yet quantified; see Barr (1960)
Light has a frequency	Young (1804)	$\nu = \frac{c}{\lambda}$; thus $E \propto \nu$	8.7	Constant (h) specified later
EM theory	Maxwell (1865)	$R = \frac{(n-n_{\text{med}})^2+k^2}{(n+n_{\text{med}})^2+k^2}$	2.14	One result is given; describes reflection, refraction, absorption; see Chapter 2 or Wooten (1972)
Light has momentum	Maxwell (1865)	$\text{Linear momentum} = \frac{E}{c}$	8.8	Maxwell's predicted radiation pressure verified by Nichols and Hull, 1901; Lebedev, 1901
Light has spin[b]	Poynting (1909)	Evident in polarization[b]	Fig. 8.3	Orbital angular momentum also exists: Allen et al. (1992; 1999)
Heat is light proved	Melloni (1850)	$E = f(T)$	8.9	Equation also describes ideal gas
Kinetic theory of gas	Clausius (1857)	$\text{KE}_z = \frac{1}{2}m\langle u_z^2\rangle = \frac{1}{2}k_B T$	5.13	Describes one degree of freedom per particle in the ensemble
Location of BB peak	Wien (1893)	$\nu_{\text{maximum}} = b_\nu T$	1.25[c]	Supports linking the above equations
Total flux measured	Stefan (1879)	$\Re = \frac{E_{\text{tot}}}{\text{area time}} = \sigma T^4$	1.24	Key thermodynamic constraint
Shape of BB spectra	Wien (1897)	For example, $I_{BB} \propto \frac{\nu^3}{e^{h\nu/k_B T}}$	2.4[c]	Equivocal ~1900s experiments; Several competing models

[a] *Early measurements of light (see Barr, 1960) involved prisms and assignments of wavelengths using dispersion equations, for example, of Cauchy. Herschel (1800) labeled the X-axis of his plot of intensity as categories of red, yellow, etc., and clearly indicated that heat was much "redder" than the solar peak. Quantification followed the invention and use of gratings by Fraunhofer (1821).*

[b] *Spin angular momentum being associated with circular polarization of light was discovered by Poynting (1909) and is schematically illustrated in Fig. 8.3. Orbital angular momentum was deduced and latter demonstrated in laser experiments by Allen et al. (1992; 1999). See Padgett et al. (2004).*

[c] *The equations of Wien (1893, 1897), presented in Chapter 1 were cast in terms of wavelength, which is how light was measured, prior to the invention of interferometers and Fourier transform instruments.*

ionization suggests that the medium plays an important role in the high-frequency laser beam interactions.

8.2.2.1 Spin Is Crucial, But Was Neglected in Previous Models

When the new quantum physics was developing, Poynting (1909) proposed that circularly polarized light has spin angular momentum, based on considering the mechanics of a rotating shaft. Light also has orbital angular momentum, but experimental confirmation of this distinct property requires lasers, and hence is a recent discovery (Allen et al., 1992, 1999). Padgett et al. (2004) summarize both phenomena, discuss experiments, and provide some history.

Spin angular momentum of light was discovered after the development of the various models for BB emissions, and was not given much consideration until after quantization of light was accepted. Yet, both types of angular momentum can be discussed using Maxwell's macroscopic equations, which do not require quantization. Furthermore, both motions can be explained using classical mechanics. Because spin is associated with a single EM wave, this property is relevant to BB emissions, and is explored in detail.

Rather than explaining spin by presenting the mathematics underlying propagating EM waves, as done in many textbooks (e.g., Lorrain and Corson, 1970; Rojansky, 1971; Jackson, 1999), we will *visualize* the spin of a propagating light wave, relying on verbalizations of Maxwell's macroscopic theory in these sources and Poynting's (1909) discussion. The starting point in textbooks is a highly simplified picture of a forward moving, linearly (plane) polarized EM wave with a certain frequency (Fig. 8.3, top). Light has perpendicular components arising from the electric (∃) and magnetic (Ħ) field intensities, although only one wave actually exists. For simplicity and clarity, and because only the electric field is probed via optical spectroscopy, the lower sections of Fig. 8.3 and this subsection focus on just ∃.

The standard view of plane-polarized light (Fig. 8.3, top) is not conducive to understanding either spin or BB emissions, which are unpolarized, and embodies the erroneous idea that the total energy will periodically disappear and reappear (see below). One example of BB emissions is the electrically heated SiC source commonly used in IR spectrometers. Passing this light through a wire-grid polarizer, which filters out all directions of ∃ but

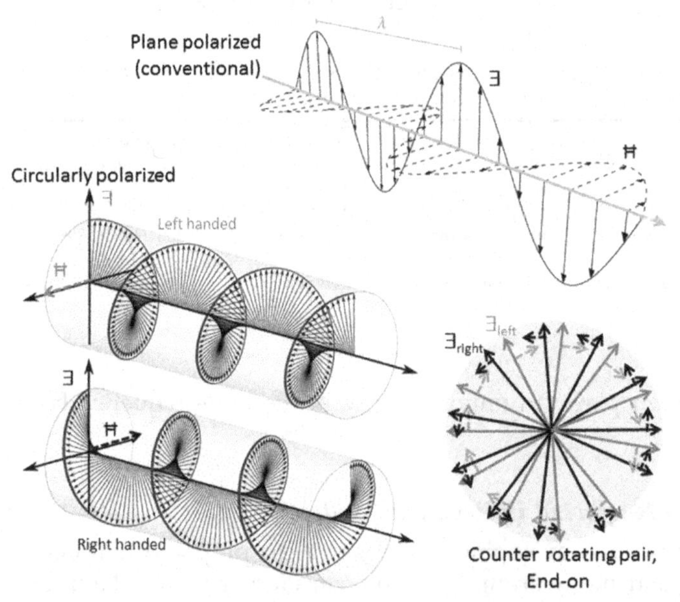

FIGURE 8.3 Schematics of plane and circularly polarized of EM waves. Top, the classical "snapshot" of a plane-polarized wave, showing the relationship of the electric ∃ and magnetic Ħ fields, which are perpendicular to each other and to the direction of motion, see text for discussion. Bottom left, circularly polarized light focusing on the electric field, which spins while light moves forward. Both right- and left-handed configurations exist. The accompanying magnetic field for each handedness is shown only at the origin. In actuality, Ħ has zero magnitude where ∃ is maximal, see text. Equal proportions yield unpolarized light. Bottom right, illustration of how synchronized, counterrotating circularly polarized light might sum to produce a planar electric field. Wave images from Creative Commons, slightly modified.

one, forces the light to be linearly (plane) polarized, and allows the experimentalist to probe anisotropy in matter. With other equipment, circular polarization can be explored (see Fig. 2 in Padgett et al., 2004). Circular polarization is relevant to BB emissions, as follows.

Circular polarization is best described in terms of one entity, where the direction of the field ∃ rotates progressively around the direction of propagation as the wave travels forward (Fig. 8.3, bottom left). This representation originated with Poynting (1909), who proposed that light has spin by comparing its behavior with the turning of a solid shaft. Like bolts that can be "normally" or "reverse" threaded, right- and left-hand circular polarizations both exist (Fig. 8.3, left). As shown, ∃ has the same size everywhere for any given handedness.

With Fig. 8.3 (bottom half) in mind, and still only considering ∃, plane-polarized light is mathematically equivalent to superimposing (summing) circularly polarized waves that are rotating in opposite directions. One possibility is that the wave components are synchronized while the size of ∃ for each component is the same, allowing the handedness to cancel. Because any number of waves can exist in some volume of space, the components rotate through each other, as illustrated in an end-on view (Fig. 8.3, bottom right). Other (elliptical) polarizations can be produced through various combinations. Because thermal emissions are unpolarized, we do not discuss these variations in detail. The simplest picture possible of BB radiation, is that this "gas of light" consists of an equal amount of right- and left-handed emissions. These might be superimposed in various ways, but the amounts (beams, rays, photons, or corpuscles) must be equal. Consequences are:

- Spin angular momentum existing in perfect balance means that there is rotational energy, but no net spin angular momentum.
- Rather than unpolarized light being a wave with a fixed direction and varying amplitude, as in Fig. 8.3, top, light produced thermally spins with a fixed field strength, but the direction of ∃ varies with time, as in Fig. 8.3, bottom left.
- BB emissions contain equal amounts of right- and left-handed elements because angular momentum must be conserved during atomic collisions (discussed further below).

8.2.2.2 Paired, Phase Shifted, Electric and Magnetic Fields of Unpolarized Light

The picture in the previous section is incomplete because electric and magnetic fields are intimately linked. With spin of one-handedness, ∃ is constant but changes direction. To satisfy Maxwell's equation, Ħ spins just the same, and consequently, the components of the ∃ and Ħ field vectors are 90 degrees out of phase, as shown at the origin for circularly polarized light in Fig. 8.3, but this does not define the magnitude of Ħ. Two visualizations are presented of the fields coupling. Various representations are needed to understand light, because light is not actually a wave, but is a thing that travels in space, and spins while producing paired EM fields.

A 90-degree phase shift between ∃ and Ħ is required because the power of the EM wave is constant (Poynting, 1884). The size of the field strengths are related through the permeability and permittivity of the medium, which together define the speed of light. Constant power is then obtained through this 90-degree phase shift because the cosine-squared and sine-squared sum to unity (Fig. 8.4A). Constant power was the starting point of Poynting's (1909) analysis, where he proposed light spins. Utilizing Poynting's (1884)

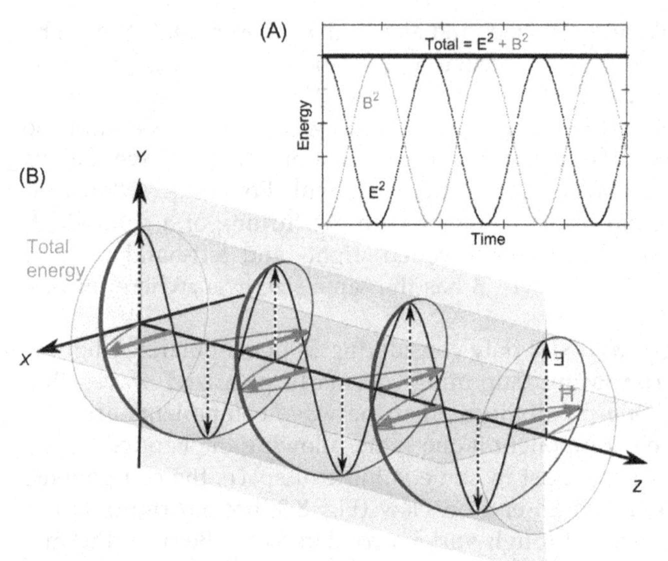

FIGURE 8.4 New schematics of circularly polarized light. (A) Graph of squared-sine and squared-cosine functions, which sum to unity. This phase shift meets the requirements of Poynting's theorem. (B) Visualization of the phase shift of the electric and magnetic fields, where the field vectors are indicated only along y and z axes. Connecting their maxima traces a spiral along the surface of a cylinder, as discussed by Poynting (1909). This visualization shows that the period of the spin is identical to the period of the forward motion.

theorem and Maxwell's equations provides the view in Fig. 8.4B of light approaching an observer to the right. Here, the spin is not shown: instead, the magnitudes of the field components are shown in the two transverse directions. This visualization is relevant to plane-polarized light, and further more lacks the disappearing and reappearing energy in the conventional and incorrect representation of Fig. 8.3, top.

The coupled fields of ∃ and Ħ provide a pattern remarkably similar to that of the well-known double helix chain of DNA molecules (Fig. 8.5A). This deduction can be obtained by imagining the pattern in Fig. 8.3 (left) to be rotating and advancing, which leads to the recognition that the magnetic field follows the electric field around. For a mathematical description, see Poynting (1909), who unfortunately did not provide any illustrations. The double helix is offered here as an analogy because it defines the surface of Poynting's rotating shaft at constant power (Fig. 8.5B) and suggests the spinning motion of circularly polarized light of a certain handedness. Another mechanical analogy is that of a spinning bullet (Fig. 8.5E). As is well known, the spin of a bullet stabilizes its straight line trajectory. Also, each bullet only spins in one direction. Based on the above:

- The 90-degree phase shift locks the fields together, so that the beam is continuous and not described as wave packets, but is better visualized as a double helix.
- The electric and magnetic fields never disappear together simultaneously, but their individual magnitudes in given, orthogonal directions vary sinusoidally.
- Periods of spin and of the transverse amplitude variation are identical.
- Spin and the straight line motion of light are connected.

8.2.2.3 *Classical Physics of Spinning, Unpolarized Eelectromagnetic Waves*

An *unpolarized* EM wave is characterized by frequency and spin (Figs. 8.3–8.5). Frequency describes forward motion, linear momentum, translational energy, and when

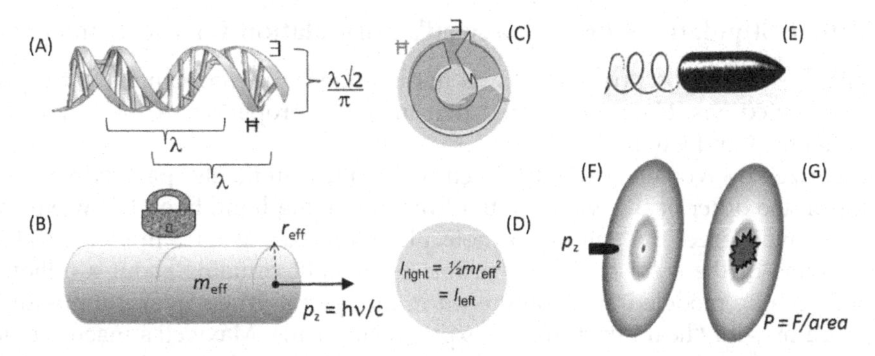

FIGURE 8.5 Analogies and links between amplitude and various momentums of light. (A) Side view, showing linear momentum of circularly polarized light, for which Ǝ and Ħ spin as interlocked helixes. Although the energy, momentum and effective mass are associated with one wavelength or cycle, these are inseparable, leading us to represent light as (B) a train of locked cylinders. (C) End view of the helix, showing phase shifted rotation of the field vectors. (D) End view, showing that the moment of inertia of a disk or cylinder is identical, so the length is immaterial: light could consist of very thin disks or extremely long cylinders. (E) Analogy with a spinning bullet, which moves in a straight line, like light. Without spin, bullets wobble. (F) Momentum of a bullet is constant before impact. Similarly, light has a momentum in space, without reflection from any surface. (G) Pressure exists when the bullet hits the target. Destruction of the bullet is analogous to absorption of light by a medium, whereby the light is converted into a vibration or moves an electron to a higher energy state. Both the light and the medium are changed during absorption, just as the bullet and target are changed during impact, dramatically and irreversibly for the latter.

the light strikes a surface, radiation pressure (Fig. 8.5G). Spin describes motions in the two, transverse directions, providing a lateral size, which has been described as amplitude (intensity) via analogy to plane waves of classical mechanics, but this is not correct considering the mathematical description of spin of Poynting (1909). Rather than having energy in a sinusoidal amplitude as in the conventional picture of a plane-polarized wave (Fig. 8.3, top), unpolarized light of BB emissions has a rotational energy analogous to that of a spinning shaft. The period of the spin is inseparable from the period associated with forward motion (Fig. 8.4B). We consider the simplest case of equivalence.

As in familiar circular motions, a larger radius means more angular momentum, and thus more energy. Hence, light has two types of energy: one type is visualized as translation of the wave and is very familiar. The other type arises from the change in the direction of the fields as light moves forward (Figs. 8.3, left and 8.5A), and is unfamiliar, being more difficult to represent mathematically than translations. The difficulty in understanding spin is evident in the extended debate on Earth's shape (e.g., Todhunter, 1873).

Angular momentum is introduced in physics with reference to linear momentum (e.g., Halliday and Resnick, 1966). Translational and rotational motions differ fundamentally, because the latter has both velocity and size (Fig. 8.5). Classically, the amplitude (size) of a wave depends on the manner in which it is produced. BB emissions have a thermal origin. So far, from the above:

- Thermal production of EM waves sets some restrictions on amplitude.
- These restrictions are governed by macroscopic thermodynamics since T is key.

8.2.3 Further Stipulations From Maxwell's Formulation for Electromagnetism

Maxwell's (1865) macroscopic model was not intended to describe light, yet it predicted its speed in rarified gas. Hertz confirmed that light has properties of EM waves (see, for example, Chapter 1 and Chapter 2).

As summarized by Wooten (1972), IR spectroscopy measurements spatially resolve behavior of solids on the order of the wavelength of the impinging light. Even UV wavelengths of ~100 nm are much larger than atoms or molecules with sizes of ~0.3 nm (e.g., water molecules). A volume of the solid with dimensions of IR light contains about a billion atoms. Hence, solids can be modeled as a continuum in interpreting their interaction with light in optical spectra or with "heat" in diffusion experiments. Thus, Maxwell's macroscopic equations are relevant to spectroscopy and to heat transfer. These equations lead to representing the medium (e.g., air, the "vacuum" of space, or dense metal) either in terms of a complex dielectric function, or of a complex index of refraction, or as the pair of reflectivity with absorptivity (Wooten, 1972). Much information can be extracted (Chapter 2). Equations important to spectroscopy and BB emissions are the absorption coefficient ($A = 2\pi\nu k$) and reflectivity (Eq. 2.14, see Table 8.1), which also contains the absorption index k and the real index of refraction n.

A completely absorbing BB ($A = k = \infty$) has $R = 0$ (Eq. 2.14), in contrast, strongly absorbing materials strongly reflect (Chapter 2). This contradiction shows that the BB end member is unachievable. In contrast, whitebodies ($R = 1$) can actually exist, being described by the permissible values of $n > 1$ and $k \cong 0$. Metals asymptote to $R = 1$ in the far-IR (e.g., iron: Fig. 2.13). For insulators, reflectivity and n are smaller, with reflectivity somewhat less than unity for very low ν, whereas $k \to 0$ (e.g., Fig. 2.12). In the far-IR, electrical insulators reflect weakly but absorb little.

The limit at $\nu \to 0$ of $k \to 0$, while n remains finite, is not only as observed (Chapter 2; Wooten, 1972), and is implicit in Maxwell's equations, but moreover is compatible with thermodynamic law. As the energy dwindles to nothing, it is increasingly difficult to heat a sample because it does not absorb, and increasingly difficult to cool the sample because it cannot emit (Kirchhoff's law, absorptivity = emissivity, see also Chapter 2). The combination makes $T = 0$ K unattainable. The sample can reflect in the limit of $\nu \to 0$, because the momentum also dwindles to nothing in the limit, so reversal of the direction of zero energy light does not introduce energy into the material, should this entity exist.

The opposite limit of $\nu \to \infty$ requires that $n \to 1$ and $k \to 0$, in which case, the sample neither reflects nor absorbs. This behavior is required because light with infinite frequency has infinite momentum, and so its reversal at a surface contributes infinite energy to the solid. Regarding the data, most measurements stop in the UV (Figs. 2.13 and 8.4) where the strong, charge transfer interactions occur, making R and A high for a substantial interval. Beyond the UV region, X-ray scattering and other processes occur, which complicates optical spectroscopic measurements, but confirming data do exist, see Palik (1998).

Over the region of moderate frequency that is connected with ordinary temperatures (Fig. 2.2), finite n and k (or R and A) are observed. Most substances show peaks (Chapter 2), whereas spectra of graphite (Fig. 8.6) show almost imperceptible absorption peaks. This behavior, and fairly constant high A (or k), make graphite a reasonable proxy for a BB. Yet, graphite strongly reflects, unlike a BB.

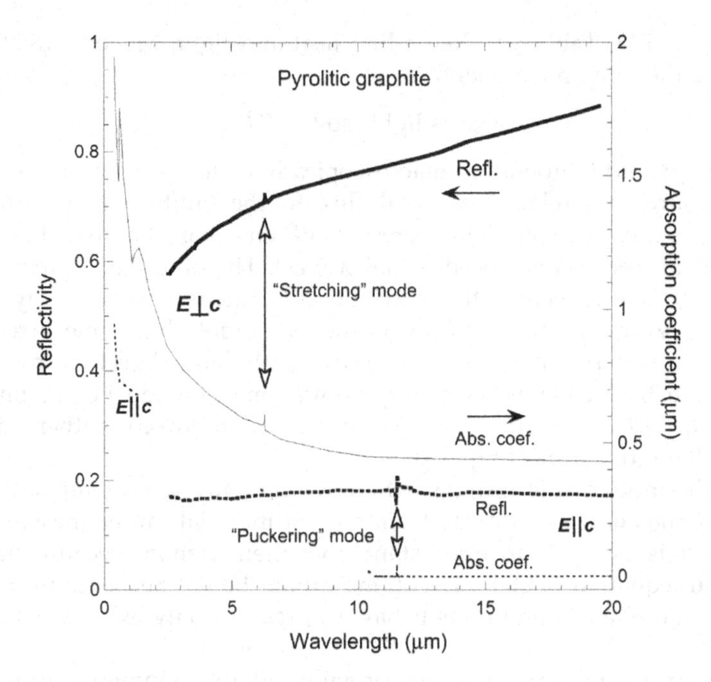

FIGURE 8.6 Reflection and absorption spectra of pyrolytic graphite. This material is formed at high temperatures and is nearly single-crystal, unlike polycrystalline graphite. However, pyrolytic carbon has some covalent bonds between the layers. Due to the material being soft, surfaces do not polish well and reflectivity is approximate. Heavy lines and left axis: reflectivity for the two orientations, as labeled, after Nemanich and Lucovsky (1977). The derivative shape of the peak in $E \| c$ is commonly observed for high-frequency peaks, and is termed a Christiansen feature. Light lines and right axis: the absorption coefficient computed from Eq. (2.9) ($A = 2\pi\nu k$) using the compilation of Borghesi and Guizzetti (1991). Data in the near-IR region are missing for one polarization, because these measurements commonly cannot distinguish k from zero, which is not possible at finite frequency (Section 8.1.3). Double arrows connect the fundamental vibrational modes in the two types of measurements.

In more detail, the macroscopic model used to extract values of n and k from data on R, involves the dispersion relations of a Kramers–Kronig analysis (Eq. 2.15). This approach, based on complex variables, describes the slow changes, the limits discussed above, and the local, rapid changes near a vibrational mode or an electronic transition (see Wooten, 1972). Utilizing the formulation requires estimating high and low frequency behavior, which fortunately are constrained as discussed above, see Wooten (1972), who discusses causality.

Regarding details of the vibrational modes, and also of electronic transitions, classical dispersion analysis is applied (Wooten, 1972). This approach stems from the damped harmonic oscillator model of Lorentz (Eqs. 2.16 and 2.17). Because damping describes continuous energy loss, this depiction of solids combines quantization on the local microscopic scale of atoms with continuum behavior over a larger scale.

8.2.4 Heat Energy, Temperature, and the State Function Describing Ideal Blackbody Emissions

Heat is not temperature (e.g., Chapter 1), but these two quantities are closely related. Development of the kinetic theory of gas provided the link for the simplest conceptual state of matter (Eq. 5.13, Table 8.1). Regarding heat and light, Melloni (1850) was the first to show these are the same phenomenon:

$$\text{heat is light, so } E = f(T). \tag{8.9}$$

Linking light energy and temperature macroscopically is the focus of this section.

One key measurement relates the total flux to the emitter temperature (Eq. 1.24). Measuring flux per area per steradian represents BB emissions because these propagate as a growing spherical front (same speed for all waves). The other key observation is the BB spectrum (Fig. 8.2), which records the distribution of intensity (not energy) as a function of frequency: however, spectroscopic experiments cannot determine absolute intensity because the source is a real material and optical paths and detectors involve other real materials (Section 8.3). The mathematical representation of the ideal emissions is the distribution function, $I_{BB}(\nu,T)$. These two key determinations are linked mathematically through integrating (i.e., the numerator of Eq. 8.3).

Unlike most thermodynamic systems, where we can only sense temperature or temperature differences, the energy flux away from the hot material can be measured. This additional information is essential to understand the phenomenon, because light cannot be represented by an equation of state. A volume cannot be defined because rays interpenetrate and superimpose, and due to this behavior, pressure only exists when light strikes a surface.

Thermal emissions from a point source provide radially symmetric growth of the heat front (Fig. 8.1). The heat front (but not light itself) has a radius and thus volume is the independent variable in addition to temperature. Because the relevant independent variables are T and V, the Helmholtz free energy is the appropriate state function:

$$F = E - TS \tag{8.10}$$

Casting the dependent variables (entropy S, pressure P) and internal energy E as functions of T and V completely addresses behavior (e.g., Chapter 1; Pippard, 1974). However, time playing a key role in BB emissions creates problems in applying the equation of state in addition to superposition. Fortunately, propagation of BB emissions (Fig. 8.1) has features in common with diffusion, plus the known dependence of BB emissions on T and wavelength (or frequency) is highly restricting, especially when coupled with the zeroth law (Section 8.2.1). The nature of EM waves (Section 8.2.2) imposes another set of restrictions. Further constraints exist.

8.2.5 Requirements of the First and Second Laws on Blackbody Emissions

Laboratory measurements show that total flux of thermal emissions is represented by a single temperature (Eq. 1.24, Table 8.1). Hence, conditions are isothermal. The first law further requires adiabatic conditions, since no heat is added to the light before we measure its flux. The fanning out of the rays (Fig. 8.1) and constant light speed means that the density of light energy in the volume of space decreases away from the source (Fig. 8.7A). Specifically, each shell of equal volume of space in a sphere surrounding the point source has the same amount of energy, but the energy density of all of the rays drops rapidly.

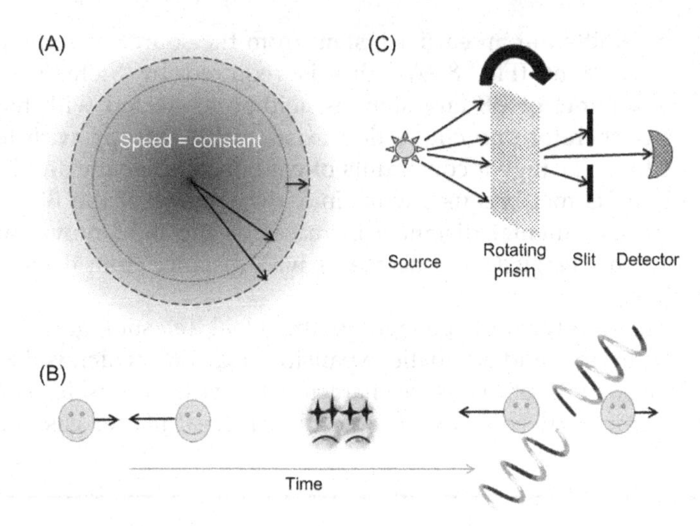

FIGURE 8.7 Schematics of BB emissions and experiments. (A) Flux diffusing in Stefan's experiment for a continuously producing source. Because speed is constant, spherical shells of flux propagate outward (dashed circles). Because T = constant for all, $E = E(T)$ is constant (adiabatic) and so energy density in space decreases with radius. As in the isothermal, free expansion of ideal gas, S increases with radius, so $S = S(V)$; this process is diffusion, and the EOS is therefore irrelevant. (B) Schematic of conservation laws operating during an inelastic collision. Because light is all energy with little momentum, and atoms are the opposite, linear momentum of the light pair is conserved, and so must the angular momentum be conserved, via opposite spins. (C) Spectral measurement of BB flux, showing essential elements of the instrument. Sources (Sun), prisms (dotted parallelogram), slits (heavy lines), detectors (cone), lenses (not shown), and mirrors (not shown) are real materials which absorb and reflect as a function of frequency. Lenses and mirrors can be avoided, but not easily, because the prism cannot be near the hot source. All elements distort the spectra from the ideal, so absolute intensity is not determined.

The density of light in space is independent of density of individual light beams, as illustrated by the shading of Fig. 8.7A differing from the arrows therein, which represent individual rays. The density in space is controlled by *diffusion*, whereas the energy density of the rays is a physical property (quantified below). If light is produced in a pulse, a propagating, growing front describes the emissions (dashed circles in Fig. 8.7A). If light is produced continuously (steady-state), the emissions describe decreasing energy density with radius (shading in Fig. 8.7A).

The second law requires entropy of a BB wave front to increase with time. Setting $dT = 0$ in the definition of entropy (Eq. 1.8, repeated here):

$$dS = \left.\frac{\partial S}{\partial T}\right|_V dT + \left.\frac{\partial S}{\partial V}\right|_T dV \tag{8.11}$$

requires that S of BB gas depends on the volume (of space, not of the light beam). The temperature dependence of entropy is irrelevant to this thermodynamic problem. As BB emissions move radially away from the filament (Figs. 8.1, 8.7), its entropy increases with radius, time, and volume, which are all related to the speed. Because speed is constant, all light emitted at any instant advances together, so the flux over a surface at any given time represents the energy emitted at an earlier instant from the source at some particular T. If T is constant, E is constant (Fig. 8.7A): this is required by Stefan's equation (1.24, Table 8.1). Thus, the volume of BB emissions is neither connected with temperature nor with total energy, and therefore no connection exists of BB entropy with temperature or internal energy, due to the stringent constraints of combined adiabatic and isothermal conditions. The second law is met because, with time, the volume of the BB wave front (the surface area times an incremental distance) increases as the light moves away from the source. Consequently, the *energy density* decreases with distance from the source. This process is radial diffusion.

The paragon of thermodynamic pedagogy, the ideal (elastic) gas, behaves in this manner during an isothermal and adiabatic expansion (Fig. 1.6), which is likewise irreversible (Chapter 1). Text Box 8.1 compares its characteristics with that of BB emissions. Many parallels exist, for example, the velocity of molecules in ideal matter gas does not depend

TEXT BOX 8.1

CHARACTERISTICS OF GASES CONSTITUTED OF MATTER AND OF LIGHT.

Classical Ideal Monatomic Gas	*Blackbody Emissions*
Tiny *atoms* of identical mass translate in all directions with various velocities and collide. Angular momentum also exists inside these tiny spheres.	Propagating *waves* move unidirectionally at constant speed with variable frequency, while spinning at that frequency with variable amplitude.
Elastic collisions	*Noninteracting* → no collisions
Packing is limited Atoms have finite mass and volume.	*Packing is unlimited* Waves can interpenetrate.
Static A representative volume is defined by equation of state.	*Dynamic* Volume is defined by advance of a wave front with time.

First and second laws[a]

Under adiabatic and isothermal *irreversible* expansion of either gas, the total translational energy is a function of T only whereas the entropy is a function of V only.

Average velocity $\leftrightarrow T$	*Average frequency* $\leftrightarrow T$

Virial theorem[a]

Linear momentum balancing in all directions describes a bound state of a certain length scale. Energy is not exchanged between different bound states. Bound states do not evolve without external input.

TEXT BOX 8.1 (*cont'd*)

Angular momentum	*Spin momentum*
Electrons orbiting the tiny dense nucleus define atom size and have energy, constituting a bound state that cannot contribute to the thermal average of large-scale translating motions.	The counterrotating paired electromagnetic fields define lateral wave size and has energy, but this energy does not contribute to T because the restricted circulation constitutes a small scale bound state. Forward travel is unbounded.
Zeroth law[a]	
Bodies or systems with the same temperature are in thermal equilibrium (communicability).	
Rotations or vibrations are described by the same T as the translations	*Energy of spin is described by the same T as that associated with advance of the BB emission front*
Molecule behavior is usually discussed in terms of these thermal reservoirs (Chapter 5), but communicability holds for any scale, including electrons orbiting a nucleus	The frequency of the wave and its spin are interrelated in a cloud of light that is produced thermally.

[a] *These laws or theorems describe rules or conditions that apply to both types of gas.*

on volume, nor does the frequency of the EM waves depend on the volume containing the light. Rather, known frequency changes are connected with the speed of the emitter, and are known as Doppler shifts (e.g., Halliday and Resnick, 1966). For both light and ideal matter gases, internal energy depends on temperature only and entropy depends on volume only.

The linked behavior of the ideal matter gas and BB gas is largely a consequence of the thermodynamic relations listed in Table 1.1. The equations for the dependent variables (S, P) in terms of the independent variables (T, V) show that the restriction of energy to a function of T only requires S to be only a function of V.

Differences also exist between these gases. Importantly, for the matter gas, only the initial and final states are considered, and so the process can be treated as static and time independent. In contrast, the expansion of thermal emissions at immense speed is a dynamic process, where time is the independent variable. The equation of state and other thermodynamic constructions, such as energy–entropy connections are irrelevant. Yet, the latter inappropriately underlie the formulae for BB intensity of Wien and Planck (Section 8.4).

8.2.6 Constraints on Blackbody Emissions From the (Elastic) Kinetic Theory of Gas

Point masses can neither collide nor interact, which is why finite atomic size was an early amendment to the classical kinetic theory (Chapter 5). The tiny sphere amendment explains the existence of reversals and introduces the concept of collisional cross section, but has no effect on the energy computed in the classical model, since elasticity was not relinquished. Hence, no collisions and no interactions have the same mathematical consequence on the internal energy of an ideal gas as does the postulate of reversible, elastic collisions. This behavior, and the other parallelisms in Text Box 8.1, indicates that some of the known results for the (elastic) ideal matter gas can be utilized in analyzing BB emissions, which constitutes a special type of light with a characteristic temperature.

Kinetic theory (Eqs. 5.12 and 5.13, see Table 8.1) links translational energy to temperature for an average particle in an ideal gas. For light, energy is proportional to its frequency (color), as discovered in the 1800s (Barr, 1960). Combining these constraints (Eqs. 5.13, 8.7–8.9 in Table 8.1), yields:

$$\langle E \rangle = \frac{1}{2} h^* \langle \nu \rangle = \frac{1}{2} k_B T, \tag{8.12}$$

where the constant h^* has the units of Planck's constant. Below, we will be show that $h^* = h$. The factor of ½ arises because translations of light are in a straight line in one direction only, and because each degree of freedom yields $\frac{1}{2}k_B T$ in kinetic theory. The factor of ½ in the KE term is consistent with the momentum of light as currently defined, even though the energy of Eq. 8.12 differs from that of the conventional treatments, see below.

The peak of BB emissions does not equal the average frequency $<\nu>$, yet ν_{max} depends linearly on T as well (e.g., Table 8.1). Consequently, all frequencies linearly depend on temperature, although the constant of proportionality is not necessarily identical for all frequencies. This varying proportionality is the essence of I_{BB}, which is a type of distribution function, describing the contribution of any given frequency to the translational energy at some specific temperature. Thus, anywhere temperature enters into the distribution I_{BB} of Eq. (8.4), it must be accompanied by a frequency (or wavelength). From Eq. (8.12), temperature in I_{BB} enters as the dimensionless ratio:

$$x = \frac{h^* \nu}{k_B T} \tag{8.13}$$

Eq. (8.13) suggests that $h^* = h$, which was independently determined long ago by fitting BB spectra (Fig. 8.2).

The constant k_B, which refers to energy per particle, is related to the gas constant R_{gc}, which refers to energy per mole of gas, and was determined experimentally in the 1700s (e.g., Fegley, 2015). The ratio x also relates to the ability of light of a certain frequency to do work. The capability is evidenced by light exerting radiation pressure when it strikes a surface (Fig. 8.5F,G), and stems from light having linear momentum whether or not a surface is impacted.

8.2.7 Constraints of the Virial Theorem and Zeroth Law on the Spinning Blackbody Emissions

Light has two different types of energy: (1) high-speed linear translation and (2) spin, which features prominently in Maxwell's laws due to the curl being mathematically equivalent to spin, which was recognized 44 years later by Poynting.

In a classical presentation, light of a certain frequency has an effective mass (m_{eff}). Spin is a much different type of motion than translation, leading to a different representation. Circulation gives spin a magnitude, which has been connected to amplitude through projections of the fields in the $x-y$ planes (Fig. 8.4B), but the power is actually constant and defines a surface. The classical picture of spin also involves an effective mass. For this motion, a disk or cylinder shape pertains (Fig. 8.5B,D).

The energy reservoirs of translation and spin of light are distinct because the scales and types of motion are different (Fig. 8.5). The spinning motion is restricted in space, and moreover involves a balance of linear momentum in all directions (in the stationary frame of reference of the moving wave). This describes a bound state, and therefore the Virial theorem applies (Hofmeister and Criss, 2016). The spin period does not change as the wave moves forward. These two reservoirs do not exchange energy because the translational motion of the wave is unbounded: not only do the scales differ, but the forward motion depends on time. The circular motion does not depend on time because there is no unique point on the circle, just as in a perfectly circular orbit of a celestial object. As a consequence, the energy of light that we infer is that computed over a cycle.

Furthermore, when light traverses from one medium to another, these two energy reservoirs behave differently at the interface. The importance of spin is evident in the mathematics of refraction and reflection at a surface (e.g., Lorrain and Corson, 1970). In contrast, linear momentum is connected only with the translatory motion. With light moving at constant speed, radiation pressure (force/area) only exists when light impacts a surface and changes direction or loses energy. These differences are illustrated in Fig. 8.5, where light is compared to a spinning bullet.

The rotational and translational reservoirs act independently, yet are linked in BB emissions, owing to the thermal origin. Under equilibrium conditions, these two reservoirs must have the same temperature, from the zeroth law. Extending Eq. (8.12) gives:

$$\mathrm{KE}_z = \frac{1}{2}m_{\text{eff}}\langle u_z^2\rangle = \frac{1}{2}h^*\langle\nu\rangle = \frac{1}{2}k_BT = \mathrm{RE}_{\text{spin,left}} = \mathrm{RE}_{\text{spin,right}} \tag{8.14}$$

such that the total energy is kinetic and equals $\mathrm{KE}_z + \mathrm{RE}_{\text{spin, left}} + \mathrm{RE}_{\text{spinright}}$. The effective mass m_{eff} does not imply quantization, but rather is a classical description of the kinetic energy over one cycle of forward translations.

From Fig. 8.5, the translating and spinning waves have an affinity to a bullet. Rearranging the second and third terms from the left of Eq. (8.14), recognizing that light speed pertains and that the relationship ($\mathrm{KE} = {}^1\!/\!{}_2 m_{\text{eff}}c^2$) holds for a wave of any frequency in the BB thermal field yields:

$$m_{\text{eff}} = \frac{h^*\nu}{c^2} \tag{8.15}$$

The associated classical momentum is $p_z = m_{\text{eff}}c = h\nu/c$, which can be derived from Maxwell's equations using the KE of Eq. (8.14) and total reflection (e.g., Halliday and Resnick, 1966). Total reflection pertains because this depicts no energy losses when an interface is encountered. Total absorption is depicted in Fig. 8.5G, where the consequence is disappearance of the bullet during destruction of the target. Light is not altered when absorption does not exist. Thus, the momentum of unperturbed light is computed from

Maxwell's equations assuming total transmission (no absorption) at an imaginary interface, which provides the following requirement:

- Light traversing space isothermally (adiabatically) with a momentum of $p_z = h\nu/c$ over a cycle must have a translational energy $KE_Z = \frac{1}{2}h\nu$ over this cycle.

Classically, spin possesses rotational energy of $\frac{1}{2}I(2\pi\nu)^2$. This spin pertains to each handedness (Eq. 8.14, which are analogous to coordinated motions in x and y at any given instant). The length of the light beam is immaterial because an infinitely thin disk has the same moment of inertia as a very long cylinder: $I = \frac{1}{2}m_{eff}r_{eff}^2$, where r_{eff} is the radius corresponding to the amplitude (Fig. 8.5D).

The same effective mass must pertain to both translational and spin motions: combining the above gives the average effective radius as:

$$\langle r_{eff,BB} \rangle = \frac{\langle \lambda \rangle}{\pi\sqrt{2}} = \frac{c}{\pi\sqrt{2}\langle \nu \rangle} \tag{8.16}$$

- Thus, the beams with a diameter $\sim 0.2\lambda$ are bullet shaped over a cycle of either field.

The energy density in the average BB beam over a cycle is:

$$\left\langle \frac{K.E.}{\lambda\pi r_{eff,BB}^2} \right\rangle = \frac{\pi h^* \langle \nu \rangle}{\langle \lambda \rangle^3} = \frac{\pi h^*}{c^3}\langle \nu \rangle^4 = \frac{\pi k_B^4}{c^3 h^{*3}} T^4 \tag{8.17}$$

Thus, the average beam in thermally produced light has an effective diameter proportional to its wavelength, and moreover, the energy density strongly increases with frequency to the fourth power. Visible light packs a punch! We tentatively suggest that Eqs. (8.16) and (8.17) describe all light, such that nonthermal production packs many such beams in the identical space, thereby increasing intensity. Our reasoning is that all light is electromagnetic, although our macroscopic model does not elucidate formation. Examining the spirals of Fig. 8.4B indicates that energy density is related to the tightness of the coils: in matter, the analogy is that closer packing of atoms creates a higher mass density.

In support of the numerical values for the effective radius, we note that the wavelength in the visible ~ 500 nm and with an associated diameter of ~ 100 nm is far larger than interatomic distances in condensed matter of ~ 1 nm. This finding is consistent with use of X-rays to sample interatomic spacing. Because the latter process involves scattering, energy losses are unimportant and so it is the size of X-ray light that is important, both longitudinally and traverse.

Note that the diminishing energy density in space (Fig. 8.7A) is unconnected with the constant beam energy density (Figs. 8.4 and 8.5). The former diffusional aspect arises because the beams fan out (Fig. 8.1) and is a consequence of the second law.

8.2.8 Constraining Flux From Our Macroscopic Model by Comparison to Measured Flux

Flux (energy per area per time) for the average BB beam, associated with its forward translation, is derived from the energy arriving over the time it takes one wavelength to pass through an imaginary plane in space, that is, over one cycle (Fig. 8.7):

$$\Im_{\text{beam}} = \frac{\frac{1}{2}h^*\langle\nu\rangle}{\left(\pi r^2_{\text{eff,BB}}\right)(1/\langle\nu\rangle)} = \frac{\pi h^*\langle\nu\rangle^4}{c^2} = \frac{\pi k_B^4 T^4}{h^{*3}c^2} \tag{8.18}$$

To conserve linear momentum, as required in any classical physics approach, two such beams must exist (one in each hemisphere). To conserve angular momentum, these spin with opposite handedness. Paired beams result from a single collision between a pair of molecules in a gas (Fig. 8.7B), as well as in bulk material. Therefore, the flux per steradian contributed by the average frequency in each beam is:

$$\Im(\langle\nu\rangle) = \frac{\Im_{\text{beam}}}{2\pi} = \frac{k_B^4}{2c^2h^{*3}}T^4 = \frac{h^*\nu^4}{2c^2}. \tag{8.19}$$

The factor in Eq. (8.16) is 2π, not 4π because two beams exist. Eqs. (8.18) and (8.19) yield Stefan's law, but not the constant σ, because it describes flux contributed by the *average* frequency. This flux per steradian does not represent the sum of *all* beams in the thermal assembly, nor the *average of the assembly,* simply because beams can superimpose. Superposition prevents us from relating the area of the surface in space (Fig. 8.1) to area of any one beam. Nonetheless, Eq. (8.18, RHS) requires that:

$$\sigma \propto \frac{k_B^4}{h^{*3}c^2} \propto \frac{k_B^4}{h^3c^2} \tag{8.20}$$

The above picture of BB emissions is incomplete because we have not (yet) determined how the waves of different frequencies contribute to the average flux at any given temperature, and so did not provide the total flux. Nonetheless, we have shown that combining Maxwell's equations with thermodynamics in a continuum model leads to Stefan's law. In more detail, we have shown that the counterrotating, oppositely directed pairs of spinning waves generated during inelastic collisions creates a BB flux that is proportional to T^4, and have ascertained which physical constants are involved, and in what ratio. One collision may generate one pair or multiple pairs. This is immaterial, as we can only "count" the emissions, not the collisions. In either case, an additional numerical factor pertains.

Here, we note that BB radiation can be produced by collisions of translating atoms in a real gas of atoms or molecules, or by collisions of vibrating atoms in a metal or a nonmetallic solid or in a disordered molecular liquid, or electronic d–d transitions (when things get really hot, Fig. 2.2). No macroscopic model can address such diverse behaviors or their origins on an atomic scale. Instead, macroscopic models rely on observations. Experimental determinations of flux (e.g., Coblentz, 1916) provide the numerical constraint:

$$\sigma = \frac{2\pi^5}{15}\frac{k_B^4}{h^3c^2} \tag{8.21}$$

But because the parameters in Eqs. (8.18)–(8.21) are lumped (see Transtrum et al., 2015), we cannot determine the proportionality between h and h^* from flux measurements. Spectral studies are needed.

8.2.9 Derivation of I_{BB} via a Macroscopic Model

Before proceeding, the obvious needs restating. (1) I_{BB} must depend on frequency and temperature. (2) Any factor of temperature entering into this distribution function needs to be accompanied by a frequency as specified by Eq. (8.13), which links discoveries in the 1800s. (3) All finite frequencies are populated, and so the distribution function must be positive. (4) For an ideal BB, which represents equilibration of all process to a single temperature, a smooth distribution of frequencies exists between the limits of $\nu \to 0$ and $\nu \to \infty$. (5) The distribution must be in accord with Eq. (8.18), which describes the flux associated with the average frequency, but is not the average flux. (6) Total flux is defined by:

$$\Im_{\text{total}} = 4\pi \int_0^\infty g[e(T)]de = 4\pi \int_0^\infty g(\nu, T) d\left(\frac{1}{2}h^* \nu\right) \equiv 4\pi \int_0^\infty I_{\text{BB}}(\nu, T) d\nu \qquad (8.22)$$

Lastly, (7), the final form for the I_{BB} function must match spectra (Fig. 8.2).

8.2.9.1 *Importance of Limiting Frequencies*

Weber (1888) recognized that the limiting values at $\nu \to 0$ and $\nu \to \infty$ need to be considered to devise an appropriate function for BB emissions. The distribution function at each limit is constrained by Maxwell's laws, specifically, Eq. (2.14) (Table 8.1), but thermodynamics pertains to these thermal emissions, so we extend the discussion of Section 8.2.3, which concerned light, but not necessarily thermal emissions:

Light with null frequency requires temperatures of 0 K, and is therefore impossible based on thermodynamic law. This finding provides a strong constraint on the distribution function. Eq. (8.13) connects temperature of BB emissions to some average frequency. As $\nu \to 0$, the absorption of any medium (gas, "vacuum," metal) approaches zero, but reflection is finite, and moreover nearly perfect for a metal. If a large amount of light existed near this limit, it would shed momentum and energy to the surface, thereby heating it (i.e., by being absorbed). This contradiction does not exist if as $\nu \to 0$, then $I_{\text{BB}} \to 0$ at a faster rate. This decline is needed to avoid more light being associated with colder bodies than with warmer: if the converse existed a cold body could warm a hotter body. Mathematically, I_{BB} must depend directly on ν at this limit, and the dependence must be stronger than linear. For the opposite limit of $\nu \to \infty$, Maxwell's equations require that both absorptivity and reflectivity approach 0. If any light exists with extremely high frequency it would not be reflected, and so would penetrate the interface, and therefore would create heat (be absorbed). This contradiction is resolved if light does not exist as $\nu \to \infty$ or does not interact with the material. The intensity should asymptote to zero (similar to the kinetic theory of gas). The slopes required for I_{BB} at the limits show that I_{BB} must have a maximum in frequency, as observed.

To construct I_{BB}, we start with one of the limits to determine a trial form, and then ascertain how this trial mathematical form must be modified to satisfy the other limit. Iterating may be necessary. The low frequency limit is considered first, on account of the third law, which states that T can approach absolute zero, but can never reach 0 K.

8.2.9.2 *Blackbody Intensity Function Near Absolute Zero*

Obviously, as $T \to 0$, so must $\nu \to 0$. Moreover, when frequency is infinitesimally small, the differential dν is no different than ν. At infinitesimal frequency, the integral of Eq. (8.22) is trivial to evaluate, giving the flux per steradian as νI_{BB}. The average is already

TABLE 8.2 Derivation of the Dependence of the Blackbody Distribution Function (I_{BB}) on Frequency and Temperature From Limiting Values Imposed by Nernst's Third Law, Maxwell's Equations for Electromagnetism, and the Virial Theorem

Step	Limit	Conditions	I_{BB} Trial Function	$\Re$ Yielded	Comments
1	$T \to 0$	$\nu \cong T$; $I_{BB} \to 0$	$\frac{h^* \nu^3}{2c^2}$	$\propto \nu^4 \propto T^4 \to 0$	No peak: so modify
2	$\nu \to \infty$; $T \neq 0$	$I_{BB} \to 0$	$\frac{h\nu^3}{2c^2 e^{h\nu/k_BT}}$	$\propto \nu^4(1 - h\nu/k_BT)$ for $\nu \to 0$	I_{BB} now has a peak, and $h = h^*$ from spectral data, but problems now exist at low ν; so modify
3	$\nu \to 0$; $T \neq 0$	$I_{BB} \to 0$	$\frac{h\nu^3}{2c^2\left[e^{h\nu/k_BT} - 1\right]}$	σT^4	All limits satisfied; spectra matched; total flux matched

known (Eq. 8.19), and moreover $<\nu> = \nu$ at infinitesimally small ν and T. Combining Eqs. (8.22) and (8.19) times 4π at the lower limit gives:

$$I_{BB}(T \sim 0) = \frac{h^* \nu^3}{2c^2}. \tag{8.23}$$

Temperature is not involved, and only one frequency of light exists at this limit.

Eq. (8.23) provides our initial trial function, step 1, in Table 8.2. Clearly this equation is insufficient, although the requirements near 0 K (Section 8.2.9.1) are satisfied. Namely, everything goes to zero and rather quickly.

8.2.9.3 Constraints Imposed by the High-Frequency Limit

To improve upon the trial function of step 1, the opposite limit of $\nu \to \infty$ is next considered. The trial function of step 1 blows up at high frequency, and thus needs to be divided by some factor which is very large for large frequency and finite temperature. However, this factor must also go to a constant for very small frequency (which requires very small T), to not make a complete mess of the trial function which describes flux at the limit of 0 K. Not many simple functions behave in this manner. But the exponential does, and furthermore the exponential function describes attenuation and enters into Newton's law of radiative cooling. Attenuation is obviously involved in inelastic interactions, where some portion of the energy is lost in each collision.

For all x, the exponential function is defined by:

$$e^x = 1 + x + \frac{x^2}{2} + \frac{x^3}{6} + \frac{x^4}{24} + \ldots = \sum_{j=1}^{\infty} \frac{x^i}{j!} \tag{8.24}$$

Furthermore, the dimensionless variable x is already defined by Eq. (8.13) (also see Text Box 8.1).

The resulting second attempt to construct I_{BB} (line 2 of Table 8.2) resembles that derived by Wien (1897), albeit from a different series of steps and without constraining the physical constants (Section 8.4). A similar equation was independently fit to the early data of Paschen and Lummer and Pringsheim (Fig. 8.2), which were collected at fairly high frequency. The BB curves constrain the physical constant which relates frequency to energy, because the other physical constants (k_B and c) were previously determined ($R_{gc} = N_A k_B$ from gas experiments, and lightspeed from astronomical measurements). Therefore:

- Because the exponential function fits spectral data (Fig. 8.2) towards high frequencies with $x = h\nu/k_BT$, then the constant $h^* = h$.

For consistency with spectral measurements, we set $h^* = h$ in our second trial function (step 2). This link was also implicit in the forms for momentum and energy of light, over one cycle.

8.2.9.4 Cross-Checking the Low T *Constraint*

The simple exponential of step 2 does not cover all conditions because it provides an incorrect limit for flux at small frequency and temperature (but not exactly zero) because $e^{-x} = 1 - x$ for small x. To improve the second trial function:

- Unity is subtracted from the exponential to provide correct limits for small T as $\nu \to 0$.

This step leads to a result proportional to that of Planck (Eq. 8.1) in step 3.

No further steps are required. This function fits observed spectra, for which absolute intensities are not determined (Section 8.3). As a cross-check, we integrated the function in Table 8.2 (see, for example, the step by step process in BB Calculator or other websites). For I_{BB} to yield Stefan–Boltzmann's coefficient, the factor of 2 must be in the denominator, as in Table 8.2, not in the numerator, as in Planck's formula of Eq. (8.1) (discussed further below).

Other solutions may exist, but would be more complicated. Based on parsimony we stop here.

8.2.9.5 Brief Comparison With Efforts Circa 1900

One may argue that our method of limits depends on knowing the answer, that is, the form of the BB curve was experimentally known prior to our analysis. This situation is usually the case for any macroscopic model, because the goal of a macroscopic model is to explain the observations. Kangro (1976) lists many experimenters (Herschel, Fraunhofer, Tyndall, Crova, Paschen, Weber, and Langley) who provided curves in advance of the analysis of Wien (1893). The need for an exponential function was voiced by Lommel (1878), and this function was combined with power laws to represent the earliest mathematical representations of BB spectra (Michelson, 1887; Weber, 1888). The powers in these early theories were derived by comparison to Stefan's T^4 result for the total flux. Kangro (1970, pp. 106–107) describes in detail the steps that Wien took to meet this constraint, which involved no less than postulating five radiation theorems. Planck (1900) built his model on the efforts of Wien and of Michelson, and on early data. Thus, as is usually the case, theory follows experiment. Section 8.4 discusses model development further, but for completeness, this following section describes where the extra factor of 4 in Planck's formula (Eq. 8.1) arises.

8.2.9.6 The Distribution Function

The form for I_{BB} in Table 8.2, reproduces experimental determinations of σ, and spectral curves, but differs from Planck's formula by a factor of 4. In view of the energy representation of the integral (middle terms in Eq. 8.23), the factor $2h \times d\nu$ in Planck's formula (Eq. 8.1) represents the energy contribution of each frequency, whereas in our approach (Table 8.2), energy is represented by the factor $\frac{1}{2}h \times d\nu$. The latter form stems from the analogy of the kinetic energy and momentum of light to one-dimensional translations of some entity with an effective mass at speed c. Planck's method was subsequently interpreted as assuming energy of photons $= h\nu$. This assignment accounts for one of the factors of 2.

The other factor of 2 exists because Planck (1900, 1901) double counted. Planck summed the energy of a collection of many oscillators (e.g., Kangro, 1970). From the Virial theorem, <KE> = <PE> (potential energy) for a simple harmonic oscillator. Energy conservation is independent of the Virial theorem, which instead describes balancing linear momentum over space (Hofmeister and Criss, 2016). Hence, the total energy is 2<KE> for either a single simple harmonic oscillator or for a collection. Planck (1900, 1901) considered total energy, see, for example, Kangro (1970). However, light only has kinetic energy and no potential energy, and thus modeling light by analogy to oscillators, which do have <PE>, amounts to double counting. Again, the underlying flaw is that the microscopic nature of light cannot be determined from BB emissions.

That our result provides the correct numerical value for the Stefan–Boltzmann constant supports the energy of light being $\frac{1}{2}hv$ over a cycle, which is expected from Maxwell's equations for unabsorbed light. Our result:

$$I_B(\nu, T) = \frac{h\nu^3}{2c^2\left[e^{h\nu/k_B T} - 1\right]} \tag{8.25}$$

does not alter the extracted value of the peak (Eq. 8.4), which matches long-standing experimental observation, nor does it alter the *average* frequency, but it does provide half the average energy than that obtained by applying Eq. (8.3) to Planck's formulae. This could be tested with difficult, absolute measurements of emission from bare metals, for which reflectivity data can be accurately determined, are pinned at low frequency, and seem not to depend strongly on T (Fig. 2.13). Again, the energy considered for light is per cycle, but the existence of a cycle does not imply quantization.

8.3 EVIDENCE FOR CONTINUUM BEHAVIOR FOR LIGHT BUT DISCRETE BEHAVIOR FOR MATTER

The theoretical arguments of Section 8.2 indicate that light can have any frequency, because ν is a continuous function that is directly linked to temperature, which is clearly a continuous function. Experimental evidence discussed here shows that energy quantization actually arises from the properties of matter, not of light. Crucially, spectroscopic methods used to characterize BB emissions themselves differ in fundamental ways from those used to probe how light interacts with matter. The two types of spectroscopic experiments (Fig. 8.1) record different processes. The photoelectric effect is an absorption experiment, which probes the behavior of electrons rather than the intrinsic properties of light. Our discussion below considers not only energy, but the two momentums as well. The latter properties are equally important in interactions, but the focus heretofore has been on the scalar variable of energy. Flux is also important, which does not equal energy as discussed above.

8.3.1 Light Having Momentum Explains the Photoelectric Effect in a Continuum Model

High-frequency light literally kicks electrons out of a metal in photoelectric experiments. Low frequency light, even at high intensity, seems unable to accomplish this feat (e.g., Hertz, 1887), which led to the hypothesis that light was quantized (Einstein, 1905). At that time, it was accepted that heat is light and that heat in metals is carried by electrons (the Wiedemann–Franz law, as explained by Drude, 1900). More recent models actually consider that the electron trades heat back and forth with the lattice during its journey (e.g., Ashcroft and Mermin, 1976). Thus, in historic and recent models, low frequency IR light moves electrons. So, what does the minimum energy in the photoelectric effect actually signify?

Light having both linear and spin angular momentum and depositing energy and momentum over a finite time interval explains the photoelectric effect in a continuum model, if we account for the electrons in the metal already being in motion, and that the amount of light reflected or absorbed by any given material depends on frequency:

Foremost, linear momentum of light is connected with its forward motion and is quantified by $h\nu/c$. This amount is associated with a time interval, specifically a cycle of spin. Transfer of linear momentum from light to an electron during an interaction displaces it from an equilibrium position while adding kinetic energy to the electron. To eject an electron from the metal requires applying sufficient force that its momentum augmentation allows the electron to break through the surface, despite its strong coulombic attraction to the cations near the surface of the metal. Strong forces require both absorption of light by the electron and large linear momentum of the light.

Also relevant is the effective mass of light in the visible, which is $\sim 10^{-6}$ times the mass of the electron (Eq. 8.15). The fastest electrons in a metal are those near the Fermi level with velocities of $\sim 10^6$ m s^{-1} (e.g., Burns, 1990), so the *momentum* of a stable, but fast, electron is $\sim 10^4 \times$ larger than light can provide over one cycle, whereas the *energy* of the fast electrons are only $\sim 100 \times$ the energy of light delivered in one cycle. To provide an electron with the additional energy and momentum that it needs to escape requires absorption of light over some interval of time (~ 10 cycles in the visible, just to double the electron's velocity). The frequency of the light gives us a clue as to how much more is needed, for a given metal.

Fig. 8.5 compares light beams to bullets. The translational energy, but not the spin energy, of a bullet is damaging. Similarly, the translation of light, but not its spin, has the greatest effect on perturbing the electrons of a metal. The electron, like the bullet, must damage the surface to escape, which requires force and an increase in its momentum, over that in the stable configuration.

Some trepidation must accompany comparing light to a bullet. Nonetheless, visible light is dense energetically and bullets are dense as regards mass. IR light is not dense energetically, and is quite ineffective at displacing electrons from their equilibrium positions. Using IR light in a photoelectric experiment may be compared to shooting ping-pong balls at a target on a rifle range: whereas the lead bullets make a hole (are absorbed), the ping-pong balls bounce back (are reflected), even if they move very fast, like a bullet. A high flux of the ping-pong balls can help, but this would need to be immense to compensate for their low momentum.

Returning to light and electrons, low momentum IR light can perturb the momentum of the electrons, but the delivery rate must be immense in order to provide an individual electron with the extra momentum needed to breach the surface. IR light is reflected from metals whereas visible light is absorbed (Fig. 2.15), so momentum delivery of light to the electrons also depends on both the frequency and the specific metal (e.g., gold, silver, and copper have their distinct colors due to spectral differences in the visible). Regarding flux, lasers were not available circa 1900. Regarding heat, this low frequency energy actually does jolt the electron sufficiently that a small heat current can be established for a short time (Criss and Hofmeister, 2017; Chapter 9). *So, it is not that light is quantized in frequency that causes the photoelectric effect, but that the light must have sufficient momentum and be delivered at a sufficient rate to eject electrons.* Here, we note that the observed threshold in the ~1900's experiment is equipment controlled, and the minimum frequency was obtained via linear extrapolation. A minimum might exist as regards the product of flux and frequency, but given current models of electron behavior, an exponential decline from the visible to the IR is expected (see figures in Chapter 9).

The energy associated with the spin of light beam is connected with the transverse fields, creating an effective size far larger than electrons and their orbits about nuclei (Eq. 8.15). Lower energy light not only has lower linear momentum, but the energy density is lower. Although the energy density can be increased by superimposing light beams, this has no effect on linear momentum, according to Maxwell's equations. Energy density is not flux. Intensity is tied to flux, but amplitude is not. Instead, amplitude is related to size, and to energy density.

It is well known that only considering energy is insufficient to describe interactions of matter with matter, and so only considering energy is thus insufficient to describe the interactions of light with matter. Forces and momentum must be also considered (Section 8.5).

8.3.2 How Quantization Is Manifest in Spectra

Spectroscopic absorption measurements (Fig. 8.1, RHS) are based on incomplete transmission of the light impinging on a material, that is, certain frequencies are attenuated to some amount. The measurements reveal distinct absorption peaks for virtually all matter, where the results depend on the material probed and the spectral range being studied. These spectra are unlike BB spectra. As detailed below, quantization is evident in peak area being proportional to the number of atoms involved in producing the peak, and the number of peaks being connected with the number of transitions expected from symmetry.

To conduct a quantitative absorption experiment, conditions must be optically thin (Chapter 2), so that some amount of the light at every frequency reaches the detector. These experiments are ballistic and nonequilibrium, in contrast to Stefan's emission experiments, which involve optically thick conditions and equilibrium between the source and its emissions.

Fig. 8.1 (RHS) depicts an absorption measurement conducted after Stefan's emissions have traveled through air in the laboratory. From t_2 to t_3, light absorbed by some atmospheric constituent (e.g., CO_2) is recorded in an absorption experiment. Sometime after the measurement (e.g., at t_4), the excited molecules reequilibrate with the temperature of their surroundings by releasing their extra energy. Generally, studying reequilibration is not part of spectroscopic measurements. An important exception is ultrafast (fs) spectroscopy experiments on metal films and surfaces (e.g., Aeschlimann et al., 2000; Kruglyak et al., 2005; Patz et al., 2013), whereby electrons are briefly excited with an intense visible or UV laser pulse, and then the changes in the material after stimulation are measured using another fast, pulsed laser and a state-of-the-art electronic detection system. These measurements demonstrate that the temperature of the excited electron gas is ~500 K higher than the sample's bulk temperature, but only for a brief time, fs to ps. The variation in equilibration times of the electrons is associated with the presence and type of magnetic interactions, and how the latter are coupled with vibrations.

Conventional IR–visible spectroscopy also probes nonequilibrium conditions of excited states, as illustrated in Fig. 8.1 and also in a generic energy level diagram (Fig. 8.8). The spectra do not record the energies of vibrations or of electronic states, but rather describe differences between states. Harmonic oscillators, with evenly spaced energy levels, have peaks with energies similar to, or multiples of, the ground state, which has caused confusion about what exactly is being measured.

For gases, absorptions in response to added energy of a light beam consist of narrow lines (Fig. 2.7) suggesting discretization. However, the lines associated with water vapor are clustered, forming a broad group, bifurcated at the center. These clusters correspond to broad featureless bands in condensed matter of the same composition (Fig. 2.7: cf. vapor to water to ice). Superposition of narrow lines and clustering demonstrate that rotational transitions are coupled to vibrational transitions. Without this coupling, rotational transitions cannot be detected in optical spectra. However, neither the vibrations

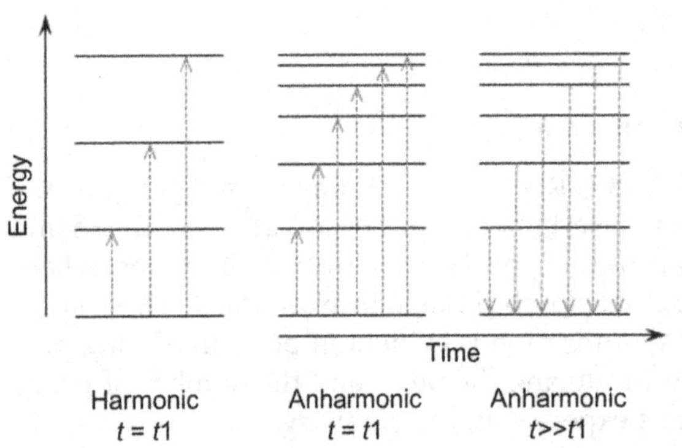

FIGURE 8.8 Energy level diagrams for a quantized harmonic oscillator with evenly spaced energy levels (left), and a quantized anharmonic oscillator, with unevenly spaced excited states (middle and right side). The lowest state is connected with the temperature of the experiment. Incoming light excites a transition, rather than describing a stable state.

nor the rotations of homogenous diatomic molecules (e.g., H_2) are sampled by IR spectra, although both are present, unless some additional process operates (Chapter 2). Therefore, the manifestation of rotations as discrete transitions in gas spectra does not indicate discretization of every rotation, but rather reveals that certain portions of the rotational continuum are sampled. For a transition to be observed, matching the energy (frequency) is required, but insufficient.

Neither are translations of the gas molecules sampled using optical spectroscopy. These exist, as demonstrated by heat capacity data (Chapter 5). In contrast, certain translations of atoms in a solid can be excited by IR light. Translational motions of atoms in a solid are manifest in absorption spectra because these are restricted by symmetry and hindered, which transforms the long excursions of molecules in a gas into a back-and-forth localization in condensed matter, that is, a vibration. Atoms translating in concert and in opposition are termed acoustic and optic modes; these are excited by sound and light, respectively. Similarly, three-dimensional rotations of molecules in the gas phase become twisting or rocking motions in condensed matter. Consequently, absorption spectra of gas and condensed matter of the same chemical composition differ in detail (Fig. 2.7).

Importantly, if the vibrations in a solid are sampled by a different means, specifically via inelastic scattering of neutrons, the experimenter can map out changes in the vibrational frequencies across the Brillouin zone, which is closely related to the unit cell (e.g., Burns, 1990). The representation of the data in terms of spacing and orientation of planes of atoms, rather than of repeated positions of individual atoms, is more convenient in depicting scattering (e.g., the familiar technique of X-ray crystallography). These changes in energy (frequency) are continuous. However, the range of frequencies across the Brillouin zone is limited to below roughly 500 cm^{-1}, through the relationship of optic and acoustic modes during zone-folding, which describes symmetry differences in closely related structures (e.g., Burns, 1990). The distinct character of vibrational modes (e.g., bending vs. stretching) evidenced in IR spectra of solids is blurred inside the Brillouin zone due to dispersion and crossing of the branches.

In summary, optical spectroscopy describes quantization inherent to the material explored, but in a manner that is restricted by the technique, and yields an incomplete view of the microscopic processes. For optical sperrtroscopy, nonequilibrium, transient behavior is measured under optically thin conditions, is limited to interactions at the center of the Brillouin zone, and depicts transitions, not energy configurations (Fig. 8.8).

8.3.2.1 The Beer–Lambert Law

Strong evidence for quantization in absorption spectra is provided by the Beer–Lambert law. This law states that peak area in optical spectroscopy is proportional to the number of transitions producing the peak (Lambert, 1760, who refers to an earlier essay by Pierre Bouguer; Beer, 1852). Peak height can be used to ascertain concentration if the peak shape does not change. Because the different transitions have various physical causes, absorption peaks for any given material have different heights, as shown in the many examples of Chapter 2. Fig. 8.6 shows the two tiny peaks of graphite, superimposed on continuous, but not constant, strong reflection or absorption curves.

Calibrating peak area against concentration allows inference of the number of participating atoms. This aspect makes Beer's law useful, particularly for impurity ions, and has led to the inference of Earth's mantle being slightly hydrated (Bell and Rossman, 1992). The concentration of OH^- in a crystal is proportional to the number of hydrogen atoms present, which is an integer, no matter how large the crystal or how well hydrated the sample is. Quantization is connected with the distinct atoms composing matter.

8.3.2.2 *Reflection and Emission Spectra*

The same information in absorption spectra is obtained in reflection data (Fig. 8.8) which require optically thick conditions, but in actuality probe a depth consistent with optically thin conditions (Chapter 2). Both measurements of the interaction of light with a material are appropriately analyzed with macroscopic models that are equivalent to the equations of Maxwell (1865). This is uncontestable.

Emissivity is tied to absorptivity via Kirchhoff's law; these quantities are equal for matter in quasiequilibrium with its surroundings. As in absorption spectra, optically thin conditions are required. However, experimental conditions are generally optically thick, and also involve reflection at an interface, and thermal gradients at the surface (Chapter 2). Misinterpretations of data are common. Due to the complexity and importance to BB studies, Section 8.3.4 provides a detailed analysis.

8.3.2.3 *Energies From Spectra*

Discrete peaks record transitions and thus energies of transitions from some ground state to an excited state (Fig. 8.8). Optical spectra of any type do not quantify the energy associated with the ground state. Hence, these data do not provide the energy of a vibration. The transitions do provide the frequency associated with the transition, which necessarily involves the energy of light helixes over a cycle. Due to the transitions being governed by more than energy, and that vibrations are both damped and anharmonic, as evidenced in peak widths and spacing of peaks in overtone-combination spectra, respectively (e.g., Mitra, 1969), a microscopic model is needed to connect energies of light with the transitions. We will return to this in Chapter 11, along with the factor of $^1/_2$.

8.3.3 The Blackbody Continuum

BB emissions constitute continuum behavior because inelastic interactions are required for temperature to evolve. Chapter 5 discusses behavior of gas, showing that tiny losses of energy during the multitude of collisions between translating atoms underpins the transport properties of monatomic and more complex gases. These losses altogether provide the thermal emissions, which leave the gas unopposed, but are in equilibrium with the interactions. Energies of the translations define a continuum, and so do the inelastic losses. Neither can the energies reach the limits of 0 or ∞.

In condensed matter, translations (and rotations) of molecules become vibrational motions inside the crystal lattice, which are restricted by symmetry (Table 7.1). The positions of the vibrational peaks can be described as discrete harmonic oscillations, but the measured peaks have finite widths that can be quite broad (Chapter 2, several figures). The damped harmonic oscillator model describes spectra of the vibrational modes, and also of electronic transitions in condensed matter quite well (Wooten, 1972). Damping involves gradual loss of energy, is well understood in mechanics (e.g., Symon, 1971), and was first connected to production of BB emissions by Lommel (1878). Atoms in a solid collide repeatedly with their neighbors; each interaction involves an infinitesimal loss of energy, which cannot be quantized, even if the vibrations and transitions between states of the vibrations are discrete. For gases, some damping occurs, since the peaks are not infinitely tall delta functions (Fig. 2.7), confirming inelasticity of the interactions.

Thermal emissions are both ubiquitous and continuous. To maintain equilibrium, a substance must have a source of thermal energy. This may be via physical contact with the environment (e.g., coffee in a pot sitting on a hotplate), or via radiation (e.g., Earth receiving Solar input), or via electricity (a tungsten filament in an incandescent lightbulb). Without a direct heat source, a body cools by radiating its heat, until it reaches the temperature of the surroundings. Even then the emissions do not stop, rather, flux received is flux radiated. With a heat source that maintains the body at some constant temperature, the emissions are connected with that particular temperature. For the candle of Fig. 8.1, the heat source is combustion.

The sum of all losses provides the thermal emissions. Due to reflection at interfaces, and the presence of thermal gradients, the emissions of a large solid are modified from those of an ideal BB, as described in Chapter 2 and below. The bottom line is that absolute intensities have not been determined for BB emissions.

8.3.4 Effect of Experimental Conditions on Emission Spectra and Blackbody Curves

Before we examine BB spectra from ~1900, modern examples are provided to illustrate the large and variable effects that material properties have on emission spectra. These examples include materials used historically. This analysis assumes that the spectra are gathered under nearly isothermal conditions, and that processes such as fluorescence are not germane.

8.3.4.1 Effect of the Source Material on Emission Spectra

Emission spectra of large crystals of calcite ($CaCO_3$) in Fig. 2.15 shows that after removal of the BB curve, the results closely match $1 - R$, where R is reflectivity. However, this result is not the emissivity, because the measurements of Christensen et al. (2000) involved a thermal gradient. Rather, the spectra represent the product $\varepsilon(1 - R)$, where $\varepsilon \cong 1$ under the optically thick conditions of the experiments, as derived by McMahon (1950) and Gardon (1956), who used the thermodynamic considerations. Bates (1978) provided equations for all conditions, by considering forward and backwards scattering at an interface (Chapter 2, Section 2.6). Ascertaining emissivity differing from unity in the strongly absorbing mid-IR region requires submicron thicknesses, not sub-mm (for examples of minerals, see McAloon and Hofmeister, 1993; Hofmeister and Bowey, 2006).

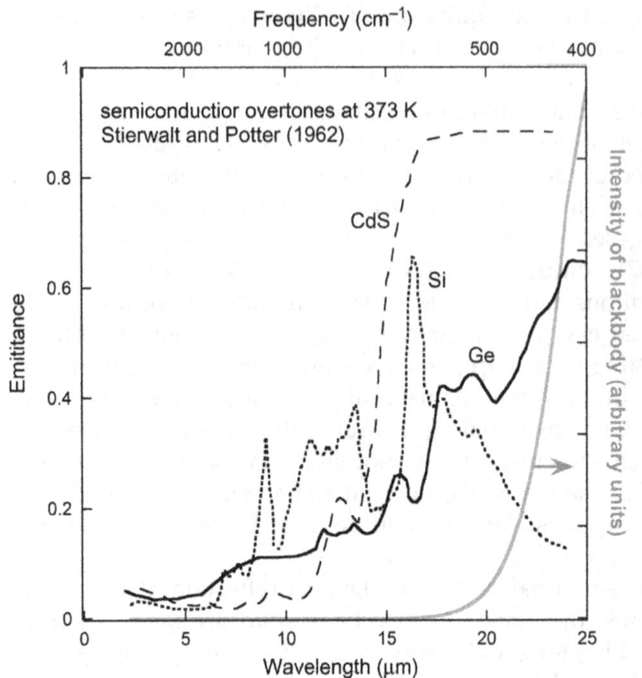

FIGURE 8.9 Emission spectra of semiconductors at 373 K, over near-IR to mid-IR wavelengths. Black curves are data from Stierwalt and Potter (1963). For single-crystal Si, thickness = 1.68 mm and resistivity = 30 ohm-cm. For polycrystalline Ge, thickness = 2.0 m and resistivity = 40 ohm-cm. For CdS, thickness = 5.1 mm. Spectra at slightly higher and lower temperature are similar, but do change with *T*. Gray curve shows ideal BB emissions at 373 K, on an arbitrary scale; the peak is at much lower wavelength.

Fig. 2.7 shows changes in absorption spectra with thickness. When the conditions become optically thick, the peaks are flat topped, and relief between peaks and valleys is reduced. Likewise, emission spectra approach $\varepsilon \cong 1$ as sample thickness increases.

Emission spectra in the mid- and near-IR from mm-size semiconductors (Fig. 8.9) contrast greatly with that of partially transparent insulators, by strongly resemble *absorption* spectra. The contrast extends to visible wavelengths (~0.7–0.4 μm), where semiconductors appear gray-to-black and opaque. Stierwalt and Potter (1963) explored the overtone-combination vibrational bands, which are very weak. Thus, ~mm-sized samples of Si and Ge are optically thin conditions in mid-IR experiments (Fig. 8.9). This circumstance is due to the lack of IR fundamentals for these diamond structure materials (Table 7.1). In contrast, CdS changes from being optically thin near 1 μm to opaque and highly reflecting near 15 μm. Unlike Si or Ge, CdS has strong absorption at longer wavelengths than shown, for either of its polymorphs (zinc blende or wurtzite structure, see Table 7.1). The sample of CdS is optically thick below wavelengths of ~15 μm and provides an emissivity $\cong 1 - R$. The experimental configuration of Stierwalt and Potter (1963) did not avoid the thermal gradient problem in the Christensen et al. (2000) database, but since the overtones are weak, reflectiionvity is not evident in emission spectra for Si and Ge. For silicates, reflectivity varying in the mid-IR has caused misinterpretations, as discussed by Hofmeister (2014).

Source materials used in emissions experiments around 1900 have much simpler spectra. Spectra of graphite (Fig. 8.6) differ between the two orientations. But because $E \perp c$ describes two orientations and graphite flakes tend to deposit in this orientation, the reflectivity is rather constant in the wavelength range explored by Lummer and Pringsheim (1899) and many others (Kangro, 1976). Of equal importance is that the

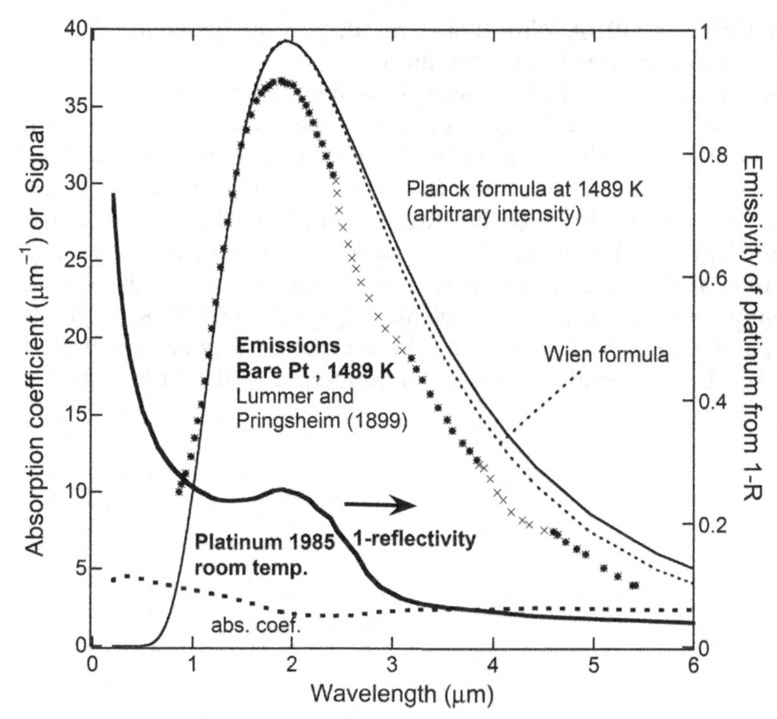

FIGURE 8.10 Comparison of emittance spectra of bare platinum at 1489 K from Lummer and Pringsheim (1899) to modern reflectivity data at 293 K, from the compilation of Lynch and Hunter (1985) and to proposed BB spectra. Although the reflectivity will change with temperature, this should be weak, as shown for Fe in Fig. 2.13.

absorption peaks of graphite are very weak and superimposed on a strong, uniform absorbance. This combination renders graphite or graphitic coatings quite suitable to explore the underlying BB curve. A graphitic source would provide the BB curve, but with intensity that is reduced by a constant reflectivity. But this curve is unconnected with quantization, since the tiny vibrational peaks of graphite (Fig. 8.6) are superimposed on the continuum, and would show as reflection features, if at all. Temperature broadening of the peaks and low resolution of historic dispersive instruments (see below) prevented detection of graphite's quantized behavior.

Emissions of bare platinum metal (Lummer and Pringsheim, 1899) do not match the formula of either Wien or Planck at short wavelengths (Fig. 8.10). This mismatch is due to the reflectivity of Pt changing with frequency over the near-IR to visible, as shown, which similar to Fe where emissivity is little affected by temperature or phase (Fig. 2.13). Thus:

- The thermodynamic correction of $1 - R$ describes departure of the source from ideal I_{BB} curves in both historic and modern data, regardless of whether the material is a partially transparent insulator (Fig. 2.15), a semiconductor (Fig. 8.9) or a metal (Figs. 2.13 and 8.10).

8.3.4.2 *Effect of Prism Absorptions on Historic Spectra*

Fig. 8.2 shows digitized data of graphitic emissions by Lummer and Pringsheim (1899). As reported by these authors, the spectra for 1065 K and lower temperatures are fit by the formula of Wien (1897) whereas Planck's formula (Eq. 8.1) provides a slightly better match above 1065 K. Note that intensities are not absolute, and so the prefactor is not part of the fitting, as clearly stated by Coblentz (1916). With this in mind, was the better match with Planck's formula actually outside experimental uncertainty?

Concerns indeed existed at this time. In response, Lummer and Pringsheim (1900) extended their measurements to longer wavelengths, but provide a short table, rather than spectral data. These authors state that the formula of Wien represents the observations down to 18 μm, within uncertainty, but felt that neither form was correct (Kangro, 1976). To address the criticisms of Nernst and Wulf (1919), Rubens and Michel (1921) provided "proof of Planck's equation" also in the form of brief tables comparing a few points. Coblentz (1916) provided data at 1600 K that agreed with the Planck's formula (his Fig. 2, which is reproduced in many sources). However, Coblentz's spectrum at 1370 K, which is reproduced in the upper panel of Fig. 8.11, reveals a severe problem: Fits are predicated on correctly applying a large instrumental correction for refraction of the fluorite prism

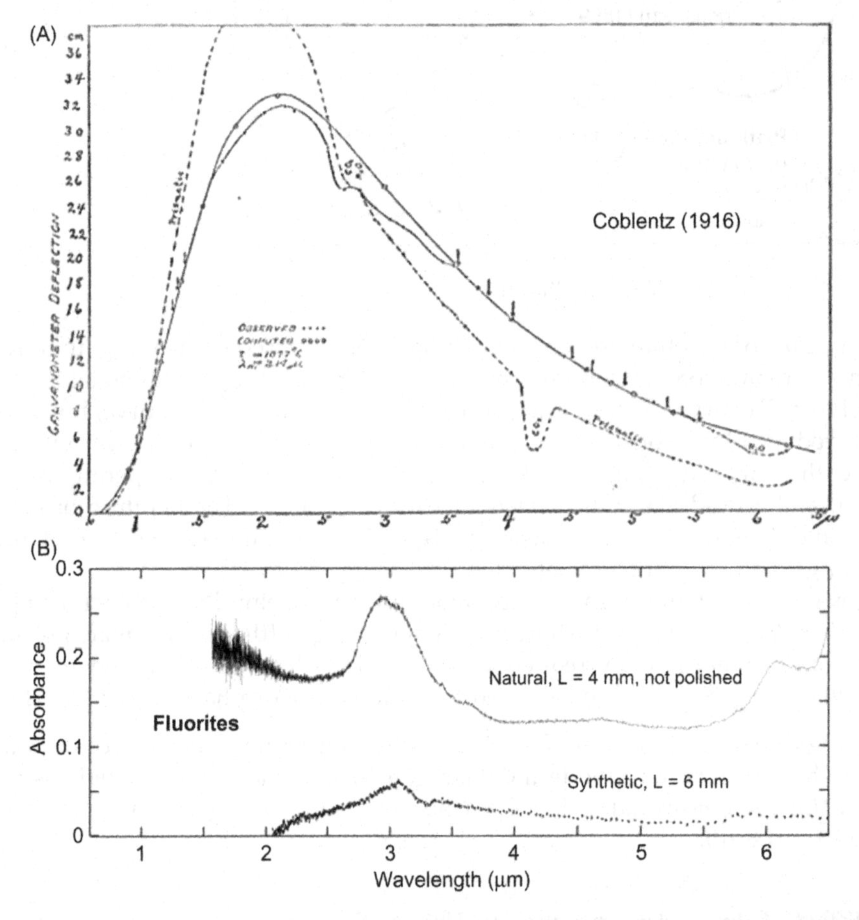

FIGURE 8.11 Blackbody spectral curves from Coblentz (1916) compared to absorption spectra of two fluorites. Top panel, *Y*-axis is signal from the bolometric detector. *X*-axis is the wavelength in μm. Dashed line: raw data, labeled "prismatic." Solid curve with dots: corrected data. Solid line with circles: the ideal Planck curve. Tiny arrows indicate points used by Coblentz to determine the maximum. His incorrect assignments to atmospheric absorptions are indicated. Bottom panel, spectra of thick slabs of unpolished natural fluorite and polished pure synthetic. The data were not corrected for reflection, which should be rather constant in this region.

used to disperse the wavelengths. Fig. 8.7E provides a schematic of the apparatus: light from the source is refracted by the prism, forming a "rainbow." Part of the rainbow passes through a slit and on to a detector. By either turning the prism or moving the slit the experimenter can isolate small wavelength ranges.

The situation is worse than as described in the historic papers. The bands attributed to water vapor at 3.3 and 6 μm in Fig. 8.11 are not only too broad for this assignment, but do not have the known relative intensities, nor the expected bifurcated peak shape (Fig. 2.7). The band attributed to CO_2 is likewise inconsistent with vapor spectra. The water features are better matched by impurity bands in a natural fluorite from the Washington University mineral collection (Fig. 8.11, bottom panel). Absorbance of their prisms were not accounted for, and moreover this behavior violates the assumption of the experiments. Coblentz mentioned that one of his fluorite prisms was green. Other materials used at that time for prisms and windows (natural quartz and calcite, salts, crown glass, mica) have water impurities as well. Comparing the panels in Fig. 8.11 also indicate that resolution for the historic experiments was quite low. This actually may have helped provide fits, via smoothing. Other experimental uncertainties existed. The most important is probably in ascertaining wavelength, which requires standards in dispersive instruments.

In summary, the baseline absorbance being fairly constant, except for the impurity bands (Fig. 8.11), and the strong control of the BB curve by the exponential function at low wavelength and by the power function at short wavelength (Eq. 8.25) enabled the fits in Fig. 8.2. The measurements of the early 1900s were incapable of ascertaining which distribution function best described the BB curve shape. Our deduction is supported by recent papers, which evaluate limitations of similar source materials as blackbodies and the fairly strong influence of the geometry of the cavity in producing BB radiation (DeWitt and Nutter, 1988).

Proof of the frequency and temperature functionality is evident in subsequent data. However, absolute intensity has not yet been determined to the best of our knowledge, for several reasons: (1) the thermal gradient provides a $1-R$ correction to I_{BB} (McMahon, 1950; Gardon, 1956; Bates, 1978), which is clear from data on diverse substances (Figs. 8.9 and 8.10, Eqs. 2.13, 2.15); but (2) the need for the $1-R$ correction has been overlooked (e.g., Christensen et al., 2000) due to confusion of emissivity with Stewart's result (see discussion by Hofmeister, 2014). Kirchhoff's result pertains to optically thin conditions and nearly equilibrium conditions, which are rarely met in emissions measurements. Absolute intensity cannot be determined for sources considered most appropriate, since graphite does not polish well, and the spectra (Fig. 8.6) are far from the constant R ideal.

Determining absolute intensities is possible for a bare metal source, heated in laboratory vacuum, but requires measurements of absolute reflectivity at similarly high temperatures, which is difficult. Due to belief in Planck's formula (8.1), this formula is used instead of data, which are indeed difficult to measure and analyze.

8.4 HISTORICAL DEVELOPMENT OF MODELS OF BLACKBODY EMISSIONS

That light waves have spin was not known when models of BB emissions were developed. To the best of our knowledge, a classical analysis that includes spin has not heretofore been presented. The incompleteness of this picture has led to an inaccurate description of BB intensity, particularly insofar as intensity is associated with motion perpendicular to the observed translations of light rays (Figs. 8.3 and 8.4). Additional errors in theoretical assessments stem from entropy being misunderstood, the incompleteness of thermodynamic laws circa 1900, and ignoring the crucial role of time. Below, some details are provided, based on Kangro's (1976) summary in English of results from key papers in the German literature. Kangro's annotations of translated theoretical papers (e.g., Planck, 1972) are illuminating.

8.4.1 Early Efforts

BB emissions were once presumed to originate only from condensed matter. Early investigators focused on explaining their own data (Kangro, 1976). The first to connect vibrations of atoms in solids with emissions was Lommel (1878), who argued that damping of vibrations was the process creating heat, which involved exponential decay, but he did not derive any formulae for spectral characteristics. Vladimir Michelson (1887; 1888) also considered that vibrations of atoms were involved. He derived:

$$I(\lambda, T) = c_1 \frac{T^{3/2}}{\lambda^6} \exp\left(-\frac{c_2}{T\lambda^2}\right) \tag{8.26}$$

by assuming that the intensity of the light of any wavelength should be proportional to the velocity of atoms producing that light. This function provides an asymmetric peak that is too narrow compared to the data. From fitting data, Weber (1888) deduced:

$$I(\lambda, T) = c_1 \frac{\pi}{\lambda^2} \exp\left(c_2 T - \frac{1}{c_3^2 T^2 \lambda^2}\right) \tag{8.27}$$

Although the predicted peak is too broad, Weber's equation describes the BB maximum, with a formula and coefficient similar to that of Wien (1893). Independently, Kövesligethy (1890) also deduced that the maximum wavelength should be inversely dependent on temperature, and provided Stefan's T^4 law. Kangro, who is fluent in German, found his arguments difficult to follow.

These early analyses arrived at incorrect forms for the BB curve because early data had more problems than the 1898–1921 data covered in Section 8.3. But, clearly, all participants were aware that both an exponential function and some powers of wavelength were needed to describe BB curves, and that Stefan's law must also be satisfied. Schirrmacher (2003) refers to this approach in science as "experimenting theory."

Wien (1893) began by focusing on equilibrium thermodynamics. The basis is flawed because the volume of space occupied by light depends on time, not the characteristics of light. At that time, the zeroth law of communicability was not in place and entropy was viewed as relating to heat only, either real or uncompensated, with the latter being a vague and mysterious concept. The complete differentials were not in place (Bridgman, 1914), isothermal expansion of ideal matter gas was misunderstood, and reversible was considered as realistic. The approach of Wien also considers formation of the emissions in the cavity, and considers interactions as elastic, yet does not conserve energy. Wien (1897) modified his early theory to consider gas particles as centers of radiation. Wien did his best to incorporate EM theory, including the energy in the intensity (Poynting, 1884), but the spin of light was not yet recognized, and thermodynamics was incomplete at that time.

8.4.2 Efforts Circa 1900

The year 1900 was quite busy, but the focus was not on thermodynamics. In February, Thiesen (1900) modified the power in the formula of Wein from λ^{-5} to $\lambda^{-4.5}$. In June, Lord Rayleigh (1900) provided:

$$I(\lambda, T) = c_1 \frac{T}{\lambda^4} \exp\left(-\frac{c_2}{\lambda T}\right) \tag{8.28}$$

His analysis combines Stephan's T^4 law with the displacement law. This was followed by Planck's (1900) paper giving Eq. (2.3), where the exponent is in the denominator and accompanied by the factor of unity, but with unknown constants. Unlike Rayleigh's version, temperature appeared only in the exponential factor. The logic centered on modifying the energy–entropy connection from Wien, which is in error, as discussed above.

According to Kangro (1976), Planck stated in a letter of 1897 "I, too, am of the opinion that it is in principle quite hopeless to deduce the velocity of irreversible processes, e.g., friction or heat conduction in gases, strictly from the present gas theory." For this reason, Planck's model describes light as a conservative problem, which is contrary to observations (Fig. 1.1: light cannot be put back into the flashlight). Planck was aware that vibrating atoms were damped, but viewed the connection of intensity of light with the energy of the resonator to be independent of the damping. Classically this is impossible: amplitude decreases with time during damping (e.g., Symon, 1971). Fig. 8.7 provides additional evidence that this hypothesis is unsupportable. The energy of light produced by a resonator is connected with *transitions* between energy levels, not with the values of each energy level. Thermal emissions differ from stimulated absorption only in the direction: thermal emissions involve release of heat to access a lower energy state, whereas absorptions describe uptake of heat to access a higher energy state. These transitions involve damping through time-*dependent* losses: see Hofmeister and Criss (2016) for comparison of various classical oscillators. Many have confused the energies of vibrational peaks with vibrational states: this is connected with the even spacing of levels in harmonic oscillations (Fig. 8.7).

Experimentalists at this time (e.g., Rubens and Kurlbaum, 1900) were unhappy with all available formulae, which motivated further theoretical efforts. Rayleigh (1900) obtained a limiting value for long wavelengths by assuming that only so many light waves can be packed into some given volume. Light does not behave in this manner, as known from geometrical optics. Rayleigh's (1900) formula was accepted, despite the lack of a peak, the flawed assumption, and not conserving energy, because another well-respected scientist (Kayser, 1902) showed that Planck's equation reduces to Rayleigh's.

Planck (1900, 1901) modified his original theory to incorporate Rayleigh's work and also the statistical work of Boltzmann, while retaining the assumption of perfect and elastic oscillations. Entropy depending on energy and thus on temperature underpins his analysis, which requires that energy must then also depend on volume. This is untrue.

Quantization of light was inferred independently by Einstein (1905), who interpreted Hertz's experimental data on the photoelectric effect (Section 3.1). Also, additional measurements of the BB curves (Coblentz, 1916; Rubens and Michel, 1921) seemingly agreed with Planck's revised formulation, so his formula (Eqs. 2.3 and 8.1), which included physical constants, was accepted. His formula was not compared with experimental determinations of the Stefan–Boltzmann constant, but should have been. This omission underlies acceptance of a flawed result (Section 8.1).

8.4.3 Spin of Light

None of the models of I_{BB} incorporate spin. This property of light is equally as important as its forward motion, because it provides light with an effective lateral size that is proportional to its wavelength (Section 8.2). Linking of electric and magnetic fields in a double helix addresses cross sections and intensity, while emphasizing the importance of angular momentum and linear momentum to interactions of light with matter. All light appears to be the same, except that its energy density differs. Matter is different. It is obviously discrete and comes in many varieties (e.g., neutrinos, electrons, protons, molecules, raindrops, geese, moons, and stars). Importantly, discrete entities differ from quantization.

8.5 SUMMARY AND FUTURE WORK

The essence of quantization is that transitions have a certain energy that is characteristic and limited to some exceedingly narrow range, and therefore light associated with any given transition must have this particular energy or narrow range of energies. In this quantum picture, the frequencies of the photons interacting with matter are limited, but their number is not, since the transition can be alternately be stimulated and relaxed. The BB curve represents the opposite case, where the frequencies of the light can take on any value, but distribution of light over frequency is specified. The connection of the total energy of emitted BB radiation with temperature underscores this finding, since temperature is not only a macroscopic parameter, it is one of the two fundamental state variables which are both continuous, that is, not quantized. The other continuous state variable is volume, which is unrelated to the BB emissions, but is of great importance in describing matter.

The experiments circa 1900 were heroic, as much was at stake regarding advancement of science (and scientific reputations). Excitement and confidence in regards to the new physics of quantization seems to have impacted objectivity. The fits produced support Planck's equation, yet the material used (bare Pt metal or platinum black and some sort of graphite coating) should not have produced the ideal curve in these early experiments. The matches largely rest on the substantial corrections made to account for the index of refraction of the prisms, while neglecting significant prism absorptions. Eagerness to match the emissions to the latest idea reveals the unfortunate influence of "fashion" in science, as pointed out by Frederick Douglass (1854), in advance of these debates.

Scientists are generally conservative, so the rapid change of the early 1900s is somewhat puzzling. This too should be placed in context: in the very late 1800s many considered everything about (classical) physics was known. Actually, central concepts of thermodynamics at that time were incomplete or flawed (Chapter 1). These unrecognized problems have greatly impeded understanding conductive and radiative heat transfer, and their relationships to each other and to thermodynamics.

To understand the microscopic nature of light and of heat transfer as well, classical experiments need repeating with improved fast electronics and high-intensity, tunable lasers. Heat conduction in metals has been misunderstood because the time dependence in nonequilibrium experiments was only recently explored (Criss and Hofmeister, 2017). Similarly, laser pulses can now be delivered in fs. These have been applied to spectroscopic study of electrons in metals (e.g., Aeschlimann et al., 2000; Patz et al., 2013), but vibrations in insulators have not been similarly probed. Similarly, emission spectra could be revisited, for example, of cold metals with their nearly constant reflectivity. The tendency to infer the fundamental nature of light from special interactions of visible light with matter is far too limiting to reveal its true nature. Experiments probing the spin of light at low frequency are particularly needed.

References

Aeschlimann, M., Bauer, M., Pawlik, S., Knorren, R., Bouzerar, G., Bennemann, K.H., 2000. Transport and dynamics of optically excited electrons in metals. Appl. Phys. A 71, 485–491.

Allen, L., Beijersbergen, M.W., Spreeuw, R.J.C., Woerdman, J.P., 1992. Orbital angular momentum of light and the transformation of Laguerre–Gaussian laser modes. Phys. Rev. A. 45, 8185–8189.

Allen, L., Padgett, M.J., Babiker, M., 1999. The orbital angular momentum of light. Prog. Opt. 39, 291–372.

Ashcroft, N.W., Mermin, N.D., 1976. Solid State Physics. Holt, Rinehart and Winston, New York.

Barr, E.S., 1960. Historical survey of the early development of the infrared spectral region. Am. J. Phys. 28, 42–54.

Bates, J.B., 1978. Infrared emission spectroscopy. Fourier Transform IR Spect. 1, 99–142.

Beer, A., 1852. Bestimmung der absorption des rothen lichts in farbigen flüssigkeiten. Ann. Phys. 162, 78–88.

Bell, D.R., Rossman, G.R., 1992. Water in the Earth's mantle: the role of nominally anhydrous minerals. Science 255, 1391–1397.

Blumm, J., Henderson, J.B., Nilsson, O., Fricke, J., 1997. Laser flash measurement of the phononic thermal diffusivity of glasses in the presence of ballistic radiative transfer. High Temp. High Press. 29, 555–560.

Borghesi, A., Guizzetti, G., 1991. Graphite (C). In: Palik, E.D. (Ed.), Handbook of Optical Constants of Solids, Vol. 2. Academic Press, San Diego, CA.

Bridgman, P.W., 1914. A complete collection of thermodynamic formulas. Phys. Rev. 3, 273–281.

Burns, G., 1990. Solid State Physics. Academic Press, San Diego, CA.

Christensen, P.R., Bandfield, J.L., Hamilton, V.E., Howard, D.A., Lane, M.D., Piatek, J.L., et al., 2000. A thermal emission spectral library of rock-forming minerals. J. Geophys. Res. 105, 9735–9739.

Clausius, R., 1857. The nature of the motion we call heat. Phil. Mag. 14, 108–127.

Clausius, R., 1870. On a mechanical theorem applicable to heat. Philos. Mag. Ser. 4 40, 122–127 (translated from Sitz. Nidd. Ges. Bonn, p. 114 1870).

Coblentz, W.W., 1916. Constants of spectral radiation of a uniformly heated enclosure or so-called blackbody. II. Bull. Bureau Stand. 13, 459–477.

Criss, E.M., Hofmeister, A.M., 2017. Isolating lattice from electronic contributions in thermal transport measurements of metals and alloys and a new model. Int. J. Mod. Phys. B 31 (No. 175020). Available from: https://doi.org/10.1142/S0217979217502058.

DeWitt, D.P., Nutter, G.D., 1988. Theory and Pratice of Radiation Thermometry. John Wiley & Sons, New York.

Douglass, F., 1854. The Claims of the Negro. Lee, Mann, and Co, Rochester NY.

Drude, P., 1900. Zur elektronentheorie der metalle. Ann. Phys. 306, 369–402.

Einstein, A., 1905. Über einen die Erzeugung und Verwandlung des Lichtes betreffenden heuristischen Gesichtspunkt. Annalen der Physik. 17 (6), 132–148 (translation by Dirk ter Haar in 1967): http://users.physik.fu-berlin.de/~kleinert/files/eins_lq.pdf.

Fegley Jr., B., 2015. Practical Chemical Thermodynamics for Geoscientists. Academic Press\Elsevier, Waltham Massachusetts.

Fikhtengol'ts, G.M., 1948. Kurs Differentsialnogo I Integral'nogo Ischisleniya (Course in Differential and Integral Calculus), Vol. 2. Gostekhizdat, Moscow.

Fowler, R., Guggenheim, E.A., 1939. Statistical Thermodynamics. Cambridge University Press, Cambridge, UK.

Fraunhofer, J., 1821. Formation of spectrum upon diffraction from a framework of wire. Denkschr. König Akad. Wiss., München 8, 1–76.

Gardon, R., 1956. The emissivity of transparent materials. J. Am. Ceram. Soc. 39, 278–287.

Gradshteyn, I.S., Ryzhik, I.M., 1965. Table of Integrals, Series, and Products. Academic Press, New York.

Halliday, D., Resnick, R., 1966. Physics. John Wiley & Sons Inc, New York.

Herschel, W., 1800. On the power of penetrating into space by telescopes; with a comparative determination of the extent of that power in natural vision, and in telescopes of various sizes and constructions; illustrated by select observations. Phil. Trans. R. Soc. Lond. 90, 49–85.

Hertz, H., 1887. Ueber den Einfluss des ultravioletten Lichtes auf die electrische Entladung [On an effect of ultraviolet light upon the electrical discharge]. Annalen der Physik. 267, 983–1000. Available from: https://doi.org/10.1002/andp.18872670827.

Hofmeister, A.M., 2014. Carryover of sampling errors and other problems in far-infrared to far-ultraviolet spectra to associated applications. Rev. Mineral. Geochem. 78, 481–508.

Hofmeister, A.M., Bowey, J.E., 2006. Quantitative IR spectra of hydrosilicates and related minerals. Mon. Not. R. Astron. Soc. 367, 577–591.

Hofmeister, A.M., Criss, R.E., 2016. Spatial and symmetry constraints as the basis of the Virial theorem and astrophysical implications. Can. J. Phys. 94, 380–388.

Jackson, J.D., 1999. Classical Electrodynamics, 3rd Edition John Wiley & Sons, New York.

Kangro, H., 1976, Early History of Planck's Radiation Law (R.E.W. Maddison, trans.). Taylor and Francis, London, UK.

Kayser, H., 1902. Handbuch der Spectroscopie. Verlag Von S, Hirzel, Leipzig, Germany.

Kövesligethy, 1890. Wladimir Michelson's spektraltheorie. Mathematische und Naturwissenschaftliche Berichte aus Ungarn 7, 24–35.

Kragh, H., 2000. Max Planck: the reluctant revolutionary. Phys. World 13, 31–35.

Kruglyak, V.V., Hicken, R.J., Ali, M., Hickey, B.J., Pym, A.T.G., Tanner, B.K., 2005. Measurement of hot electron momentum relaxation times in metals by femtosecond ellipsometry. Phys. Rev. B 71, 233104.

Lambert, J.H., 1760. Photometrie: Photometria sive de mensura et gradibus luminis, colorum et umbrae (Photometry, or, On the measure and gradations of light, colors, and shade). Verlag Von Wilhelm Engelmann, Leipzig, Germany.

Lebedev, P.N., 1901. Experimental examination of light pressure. Ann. der Physik 6, 433–459.

Lommel, E., 1878. Theorie der Absorption und Fluorescenz. Ann. Phys. 238, 251–283.

Lorrain, P., Corson, D., 1970. Electromagnetic Field and Waves, second ed. W.H. Freeman and Company, San Francisco, CA.

Lummer, O., Pringsheim, E., 1899. Die Vertheilung der Energie im Spectrum des schwartzen Körpers und des blancken Platins. Verhandl. Deut. Physik. Ges. 1, 215–235.

Lummer, O., Pringsheim, E., 1900. Über die Strahlung des schwarzen Körpers fur Lange Wellen. Verh. d. Deut. Phys. Ges 2, 163–180.

Lynch, D.W., Hunter, W.R., 1985. Comments on the optical constants of metals and an introduction to the data for several metals. In: Palik, E.D. (Ed.), Handbook of Optical Constants of Solids. Academic Press, San Diego, CA, pp. 275–376.

Marr, J.M., Wilkin, F.P., 2012. A better presentation of Planck's radiation law. Am. J. Phys. 80, 339–405.

Maxwell, J.C., 1865. A dynamical theory of the electromagnetic field. Proc. R. Soc. Lond. 13, 531–636.

McAloon, B.P., Hofmeister, A.M., 1993. Symmetry of birefringent garnets from infrared spectroscopy. Am. Mineral. 78, 957–967.

McMahon, H.O., 1950. Thermal radiation from partially transparent reflecting bodies. J. Opt. Soc. Am. 40, 376–380.

Melloni, M., 1850. La Thermochrôse ou la Coloration Calorifique. Joseph Baron, Naples. <https://play.google.com/books/reader?id = 5NgPAAAAQAAJ&printsec = frontcover&output = reader&hl = en&pg = GBS.PR3> (accessed 21.2.17.).

Michelson, W., 1887. Essai théorique sur la distribution de l'énergie dans les spectres des solides. Journal de Physique Théorique et Appliquée 6, 467–479.

Michelson, W., 1888. Theoretical essay on the distribution of energy in the spectra of solids. Phil. Mag. 25, 425–435.

Mitra, S.S., 1969. Infrared and Raman spectra due to lattice vibrations. In: Nudelman, S., Mitra, S.S. (Eds.), Optical Properties of Solids. Plenum Press, New York, pp. 333–451.

Naudeau, M.L., Law, R.J., Luk, T.S., Nelson, T.R., Cameron, S.M., Rudd, J.V., 2006. Observation of nonlinear optical phenomena in air and fused silica using a 100 GW, 1.54 μm source. Opt. Express 14, 6194–6200.

Nemanich, R.J., Lucovsky, G., 1977. Infrared active optical vibration of graphite. Solid State Commun. 23, 117–120.

Nernst, W., Wulf, T., 1919. Über eine Modifikation der Planckschen Strahlungsformel auf experimenteller Grundlage. Verhandl. Deut. Physik. Ges. 21, 294–337.

Nichols, E.F., Hull, G.F., 1901. A preliminary communication on the pressure of heat and light radiation. Phys. Rev. 13, 307–320.

Padgett, M., Courtial, J., Allen, L., 2004. Light's orbital angular momentum. Phys. Today 57, 35–40.

Palik, E.D., 1998. Handbook of Optical Constants of Solids. Academic Press, San Diego, CA.

Patz, A., Li, T., Ran, S., Fernandes, R.M., Schmalian, J., Bud'ko, S.L., et al., 2013. Ultrafast observation of critical nematic fluctuations and giant magnetoelastic coupling in iron pnictides. Nature Commun. 5 (3229).

Pippard, A.B., 1974. The Elements of Classical Thermodynamics. Cambridge University Press, London, UK.

Planck, M., 1972. Original Papers in Quantum Physics: German and English edition. Taylor and Francis, London, UK, Haar, D.T., Brush, S.G. (Trans).

Planck, M., 1900. Zue Theorie des Gesetzes der Energievertheilung im Normalspectrum. Vh. DPG 2, 237–245.

Planck, M., 1901. Über das Gesetz der Energievertilung im Normalspektrum. Ann. Physik (Ser. 4) 4, 553–563.

Poynting, J.H., 1884. On the transfer of energy in the electromagnetic field. Phil. Trans. R. Soc. Lond. 175, 343–361.

Poynting, J.H., 1909. The wave motion of a revolving shaft, and a suggestion as to the angular momentum in a beam of circularly polarised light. Proc. R. Soc. A 82, 560–567.

Rayleigh, L., 1900. Remarks on the law of complete radiation. Phil. Mag. Ser. 49, 539–540.

Rojansky, V., 1971. Electromagnetic Fields and Waves. Prentice-Hall, Englewood Cliff, NJ.

Rubens, H., Kurlbaum, R., 1900. Über die Emission langwelliger Wärmestrahlen durch den schwarzen Körper bei verschiedenen Temperaturen. Sitzungsberichte der Königlich Preussischen Akademie der Wissenschaften zu Berlin 2, 929–941.

Rubens, H., Michel, G., 1921. Prüfung der Planckschen Strahlungsformel. Phys. Z. 22, 569–577.

Schirrmacher, A., 2003. Experimenting theory: The proofs of Kirchhoff's radiation law before and after Planck. Hist. Stud. Phys. Biol. Sci. 22, 299–355.

Stefan, J., 1879. Über die beziehung zwischen der warmestrahlung und der temperatur. Wiener Ber. II 79, 391–428.

Stierwalt, D.L., Potter, R.F., 1963. Infra-red spectral emittance of Si, Ge, and CdS. In: Proc. International Conference on the Physics of Semiconductors, Exeter, 1962. (Institute of Physics and Physical Society, London), pp. 513–520.

Symon, K.R., 1971. Mechanics. Addison-Wesley Publishing Company, Reading, MA.

Thiesen, M., 1900. Über das Gesetz der schwarzen Strahlung. Verhandlungen der Deutschen Physikalische Gesellschaft 2, 65–70.

Todhunter, I., 1873. A History of the mathematical theories of attraction and the figure of the Earth, Volume I. MacMillan and Co, London, p. 474.

Transtrum, M.K., Machta, B.B., Brown, K.S., Daniels, B.C., Myers, C.R., Sethna, J.P., 2015. Perspective: sloppiness and emergent theories in physics, biology, and beyond. J. Chem. Phys. 143. Available from: https://doi.org/10.1063/1.4923066.

Valluri, S.R., Corless, R.M., Jeffrey, D.J., 2000. Some applications of the Lambert W function to physics. Can. J. Phys. 78, 823–831.

Weber, H.F., 1888. Untersuchungen über die Strahlung fester Körper. Sitzungsberichte der Königlich Preussischen Akademie der Wissenschaften zu Berlin 2, 933–957. Available from: https://archive.org/stream/sitzungsberichte1888deutsch#page/932/mode/2up.

Wien, W. 1893. Eine neue Beziehung den Strahlen schwarze Körper zum zweiten Haupsatz den Wärmetheorie. In: Sitzungsbereichte der Koniglich Preussicchen Akademie der Wissenschaften, Berlin, pp. 55–62.
Wien, W., 1897. On the division of energy on the emission-spectrum of a black body. Phil. Mag. S. 43, 214–220.
Williams, B.W., 2014. A specific mathematical form for Wien's displacement law as ν_{max}/T = constant. J. Chem. Educ. 91, . Available from: https://doi.org/10.1021/ed400827f623.
Wooten, F., 1972. Optical Properties of Solids. Academic Press, Inc, San Diego, CA.
Young, T., 1804. Bakerian Lecture: experiments and calculations relative to physical optics. Phil. Trans. Roy. Soc. 94, 1–16.
Zombeck, M.V., 2007. Handbook of Space Astronomy and Astrophysics. Cambridge University Press, Cambridge, UK.

Websites

Blackbody integrator
http://www.spectralcalc.com/blackbody/integrate_planck.html.
Riemann sums
https://en.wikipedia.org/wiki/Riemann_sum#Right_Riemann_sum.

CHAPTER

9

Transport Properties of Metals, Alloys and Their Melts From LFA Measurements

E.M. Criss* and A.M. Hofmeister

OUTLINE

* E. M. Criss is an employee of Panasonic Avionics Corporation, but prepared this article independent of his employment and without use of information, resources, or other support from Panasonic Avionics Corporation.

DOI: https://doi.org/10.1016/B978-0-12-809981-0.00009-7

"I had an immense advantage over many others dealing with the problem inasmuch as I had no fixed ideas derived from long-established practice to control and bias my mind, and did not suffer from the general belief that whatever is, is right." ***Henry Bessemer, 1905.***

Motivation behind this chapter includes understanding planetary cores, which are iron alloys, although the specific composition is debated. For simplicity, this chapter uses the term "metals" to denote both pure elements and their alloys in the metallic state, and adds the qualifiers "elemental" or "pure" to "metal" where needed. The term "alloy" is used to denote any metal with a complex atomic composition. Alloys can have a variety of structures, ranging from homogenous systems with an abundance of elemental defects, to systems with clearly defined domains. Critically, these domains can be alloys themselves. Thus, differences between elements and metal alloys are much greater than between end-member minerals and their solid solutions, where the crystal lattice is often little perturbed by cation substitutions. In an ionically bonded substance, the lattice is defined by the anions, with cations occupying the "holes." But for metals and alloys, the cations define the lattice, with a small number of electrons orbiting loosely. This type of bonding gives metals ductility and a forgiving lattice, resulting in the domains of alloys.

Heat transport in metals has been investigated over a long time, and similar to gases, the model in place has historical roots and was based on an incomplete understanding of atoms. Therefore, as in Chapter 5, both data and models are covered here, with a focus on electron behavior. This chapter is based on available data and especially on our recent publication (Criss and Hofmeister, 2017), where diverse metals were measured in a manner that distinguishes the rapid electronic from the slow vibronic transport of heat. Prior to our study, proportions of the lattice and electronic contributions to thermal conductivity measurements were derived from models (e.g., Klemens and Williams, 1986) rather than by experimentally isolating either of these components. We also developed a new model: some of this is included in Chapters 3 and 5. The present chapter summarizes essential results of this study.

9.1 AVAILABLE DATA CONTRADICTS A MOSTLY ELECTRONIC MECHANISM IN METALS

Metals were among the earliest substances probed in heat transport studies due to availability, importance to physics, and use in technical applications from antiquity. Early measurements of Wiedemann and Franz (1853) near ambient temperature and under approximately steady-state conditions revealed a connection between electrical resistivity (the inverse of electrical conductivity, σ) and thermal conductivity (K). The conclusion was that electrons carry heat in metals in a thermal current. Lorenz (1872) deduced the proportionality constant. The nearly free electron model (Drude, 1900) provided further theoretical justification. Currently, a quantum model (Sommerfeld, 1928), which uses Fermi–Dirac statistics for electron speed and heat capacity, defines the Lorenz constant (L_S) in the Wiedemann–Franz (WF) law:

$$K_{ele} = \frac{\pi^2}{3}\frac{k_B^2}{e^2}\sigma T = L_S \sigma T \qquad (9.1)$$

where k_B is Boltzmann's constant, e is electron charge, and lifetimes (τ) are obtained from the DC conductivity (σ). This is modeled as:

$$\sigma = \frac{N}{V} e^2 \frac{\tau_{\text{ele}}}{m_e} \tag{9.2}$$

where ele indicates electrons, m_e is electron mass, N is the number of electrons in the volume, V.

Many researchers have pointed out the failures of the WF law (e.g., Hust and Sparks, 1973; Zheng et al., 2007; Wakeham et al., 2011; Lee et al., 2017). Although the WF law is believed to apply to pure elements, the data from Gale and Totemeier (2004) countermand its validity (Fig. 9.1). We use this reputable compilation because other sources often report K calculated from Eq. (9.1), since resistivity measurements are easier to preform and more accurate. Data on iron, nickel, and steels from this compilation do not match the WFL law, except for Ni at a single temperature: 470K (Fig. 1 in Criss and Hofmeister, 2017). For these geologically important materials, pressure derivatives of their thermal and electrical conductivities differ, which contradicts assumptions of the model (Sundqvist, 1982). For certain materials, the WF law is reasonably accurate near 298K, but a temperature-dependent fitting function is needed otherwise. This adaptation is in violation of the underlying assumptions.

Acceptance of the physics and geophysics community has not waivered, probably because a reasonable explanation for the phenomenon, accompanied by unequivocal counterevidence involving a wide variety of metals and/or alloys, was not previously available. Next, we summarize previous data that led us to closely examine heat transport in metals. Almost all of the data depicted in Section 1 were obtained prior to 1973 and are elements with small, but non-negligible, impurities.

Thermal conductivity experiments convolute thermal diffusivity with heat capacity (Chapter 3), so how fast heat moves is inseparable from how much heat is being moved. If electrons dominate heat transport of the elements, then the measured (meas) value should be:

$$K_{\text{meas}} \cong K_{\text{ele}} = \rho c_{\text{ele}} D_{\text{ele}} \tag{9.3}$$

where ρ is metal density and c is specific heat on a per mass basis. Strong Coulombic interactions make electron–electron scattering the most probably event (e.g., Ashcroft and Mermin, 1976) as is confirmed by fs-spectroscopy (see below). Combining the above equations allows us to recast WFL law to portray thermal diffusivity near ambient conditions:

$$D_{\text{ele}} = \frac{L_S \sigma T}{\rho c_{\text{ele}}} = \frac{L_S \sigma M}{\rho \gamma_S} \tag{9.4}$$

where M is molar weight and γ_S is Sommerfeld's constant, defined by considering heat capacity on a per mole basis:

$$C_{\text{ele}} = \gamma_S T = \frac{\pi^2}{2} \frac{R_{\text{GC}} Z}{T_F} \tag{9.5}$$

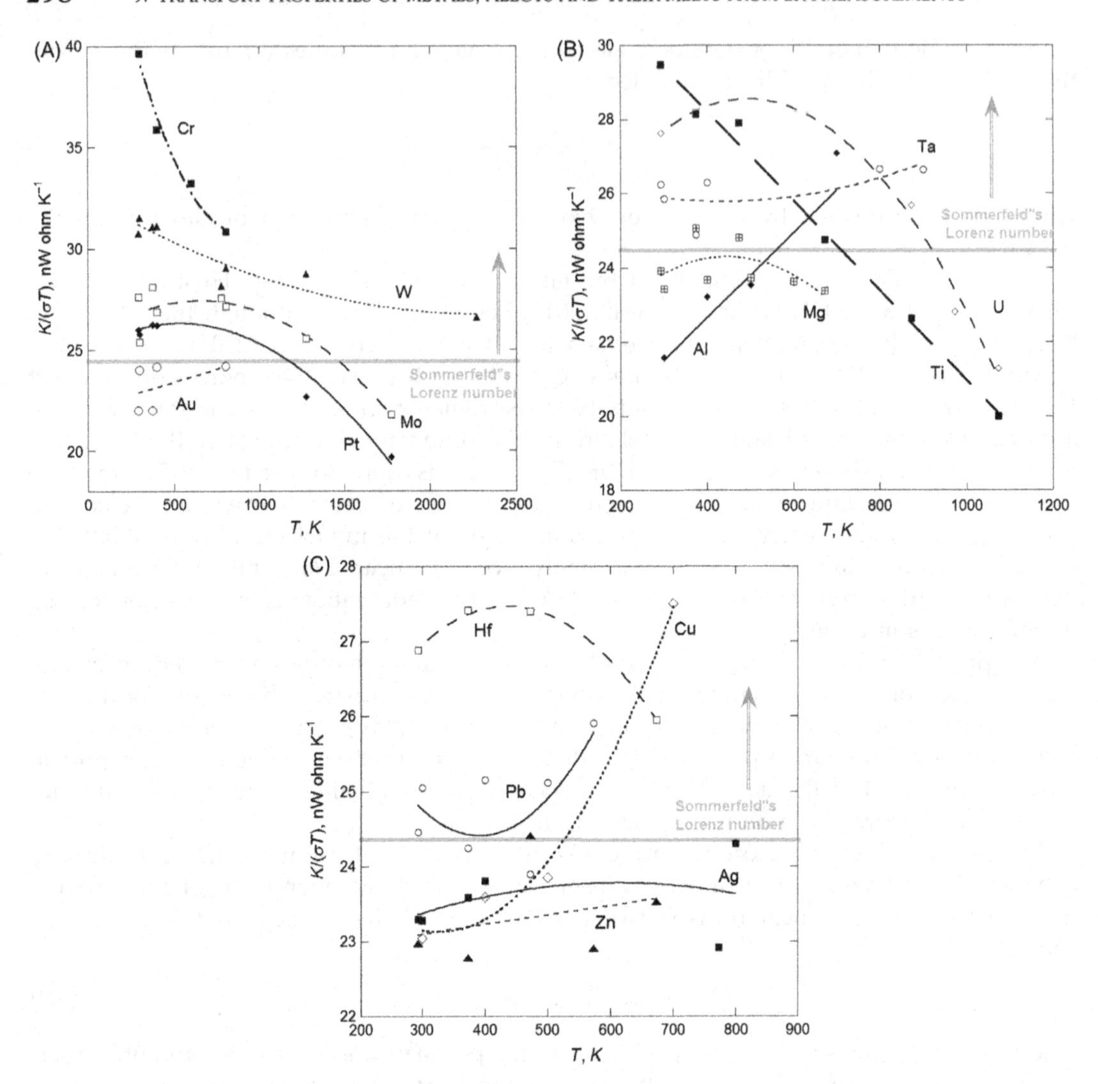

FIGURE 9.1 Mismatch of conductivity data of pure metals with the Wiedemann–Franz law. Gray: Sommerfeld's Lorenz number for the ratio of thermal conductivity to electrical conductivity × temperature (Eq. (9.1)). Gray arrow indicates the direction of the factor of 3 error, which stems from kinetic theory describing fluctuations, not flow (Chapter 5). Black symbols: data on elements from Table 14.2 of Gale and Totemeier (2004). We selected elements that had four or more temperatures for which measurements of both K and σ existed.

where R_{GC} is the gas constant, Z is the valence, and T_F is the Fermi temperature of the nearly free electron gas (e.g., Kittel, 1971). Available data show that an electronic mechanism clearly misrepresents thermal diffusivity of nearly pure elements (Fig. 9.2).

The other end-member possibility is that lattice (lat) vibrations control heat transport in metals. Should this be the case:

$$K_{meas} \cong K_{lat} = \rho c_{lat} D_{lat} \cong \rho c_{meas} D_{meas} \tag{9.6}$$

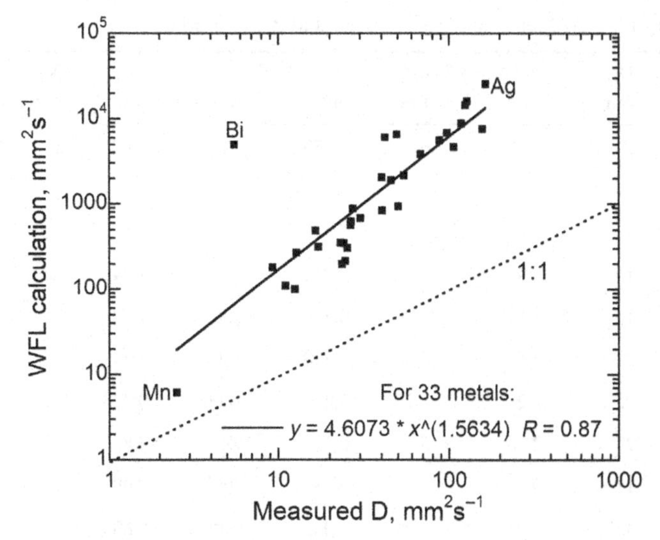

FIGURE 9.2 Discrepancy between thermal diffusivity measured for 33 metals near equilibrium with Sommerfeld's model (Eq. (9.4)). Many of the measurements are from Touloukian et al. (1973): data are listed in Table 9.1. Electrical resistivity is from various websites listed at the end of the chapter. Source: *Reproduction of Fig. 1b in Criss, E.M., Hofmeister, A.M., 2017. Isolating lattice from electronic contributions in thermal transport measurements of metals and alloys and a new model. Int. J. Mod. Phys. B 31, No. 175020. doi:10.1142/S0217979217502058.*

The RHS utilized the fact that the lattice stores virtually all the heat of metals (e.g., Ashcroft and Mermin, 1976). Excellent agreement exists between Eq. (9.6) and measurement at 298K (Fig. 9.3), and at other temperatures, as is known (e.g., Touloukian et al., 1970; 1973). If D had been calculated using c_{ele}, which is required if electrons are the dominant carrier, then the computed values would be 100 × those shown.

Note that measured diffusivity of metallic and non-metallic elements cover similar ranges. Semiconductors (Ge, graphite, Si) have $D = 25–90\ \mathrm{mm^2\,s^{-1}}$, as in moderately efficient metallic elements.

Agreement between measured K and that calculated from D and heat capacity was previously rationalized in terms of the electron interchanging heat with the lattice. However, the electron has negligible heat to give to the lattice, and if it did lose its excess in this manner, the electron could not continue to transport heat. In addition, it is very difficult for electrons to uptake heat from the lattice because only electrons with high energies, near the Fermi level, can be promoted. Thus, electrons which can be excited are "hotter" than the lattice on average (Section 9.2 provides further discussion).

Crucially, measured thermal diffusivity of metals closely matches that of nonmetallic elements (Fig. 9.3). Semi-conductors and diamond lie above the 1:1 line, because ballistic radiative transfer affects K but was removed from D in the laser-flash analysis (LFA) studies shown. Diamond was not included in this fit due to its extreme ballistic transfer and location beyond any metal. Fits for the metals having scatter is attributed to differences in impurities, because purification techniques have improved since much of the data were collected (c. 1970).

The existence of vibrational transport is exemplified by behavior across the Fe–Ni binary (Fig. 9.4). A minimum is associated with disorder. Electrical conductivity behaves similarly near the end-members, but is much more complicated near the middle of the binary. The domains of alloys seem to affect the electrical property more.

TABLE 9.1 Thermal Diffusivity Data From LFA on Metals and Alloys at Ambient Conditions

Name	Composition %	D_{lat} $mm^2\ s^{-1}$	D from K[e] $mm^2\ s^{-1}$	u_p[e] $km\ s^{-1}$	$(V_{atom})^{1/3}$[e] Nm	$1/\sigma$ nm-ohm	λ_{lat}[e] nm	λ_{IR}[k] nm
Al	99.999 Al	102		6.40	0.255	26	15.0	15
Al-1100	99Al + 0.95 Si, Fe	99	96.3			29		
Al-6061	96–99 Al	69				40		
Ti	99.995 Ti	10	9.4	6.01	0.260	400	1.54	52
V	99.8 V	13	10.4	5.98	0.240	200	1.72	21
Cr	99.95 Cr	27	29.4	6.63	0.230	130	4.44	47
Mn	99.9 Mn	2.51	2.2	5.15	0.230	1600	0.423	45
Fe	99.99 Fe	23	22.6[f]	5.96	0.228	97	3.80	28
Hexahedrite	94 Fe + 6 Ni	11.8	9.6[g]	5.67[i]	~Fe	200[j]	2.08	
Invar	64 Fe + 36 Ni	4.0	2.9[g]	4.79[i]	~Fe	820[j]	0.835	
Permalloy	55Fe + 45Ni	5.28				390[j]		
Steel-SRM	62Fe + 20Ni + 16Cr	3.66	3.92[f]	6.08	~Fe	593	0.602	
Steel-310	52Fe + 20Ni + 25Cr	3.35[b]				780		
AlNiCo-8	34Fe + 35Co + 15Ni + 7Al	6.2	3–13[h]		~Fe	530		
Co	99.95 Co	26.2	26.7	5.61	0.222	60	4.76	31
$Co_{94}Mo_6$	93.7 Co + 6.3 Mo	8.0	n.m.	n.m.	~Co	330	~Co/3	
Ni (pure)	99.995%Ni	25.4	22.9	5.62	0.222	70	4.08	29
Ni	99.5% Ni	17.1	22.9					
Inconel-600	74Ni + 16Cr + 8Fe	3.43[c]				1030		
Ni-400	66.5Ni + 31.5Cu	~6.8				450		
Cu-OFHC	>99.999 Cu	151	117	4.78	0.228	17.1	31.6	14
Cu 110	99.9 Cu	108–124	117					
Cu-rich	$Cu_{99}Cr_{0.8}Zr_{0.08}$	100[d]						
Brass	62Cu + 35Zn + 3Pb	34.3	35.6	4.15	~Cu	66	8.26	
Bronze	63Cu + 25Zn + 6Al + 3Fe + 3Mn	16.7	13.9	4.56	~Cu	172	3.66	
$Zr_{97}Hf_3$	>97Zr + <3Hf	9.7	12.7	2.6	0.285	420	4.89	
Mo	~Mo	56[a]	53.9	5.0	0.249	50	10.8	
Pd	99.9% Pd	25.4	24.5	4.46	0.245	100	7.04	48
Ag	99.9 Ag (hard)	175.7	174	3.65	0.258	16	47.8	12
Sn	99.9% Sn	55	42	3.35	0.300	110	12.6	18

(Continued)

TABLE 9.1 (Continued)

Name	Composition %	D_{lat} mm^2 s^{-1}	D from K^e mm^2 s^{-1}	u_p[e] km s^{-1}	$(V_{atom})^{1/3e}$ Nm	$1/\sigma$ nm-ohm	λ_{lat}[e] nm	λ_{IR}[k] nm
Ta	99.95% Ta	~30	24.4	4.19	0.262	130	5.84	29
W	99.9% W	92	67	5.22	0.251	50	17.6	27
Pb	99.9998% Pb	24.73	24.3	2.17	0.312	200	11.2	50
Bi	99.99% Bi	~10	6.7	2.19	0.329	1300	3.06	

[a] *From Rudkin et al. (1962)*
[b] *Blumm et al. (2007)*
[c] *Blumm et al. (2003)*
[d] *Rhode et al. (2013)*
[e] *Unless noted otherwise, for the elements, http://www.periodictable.com/Properties/A/ThermalConductivity.v.html was used to provide resistivity, and compute thermal diffusivity from* k/C *and to compute the structural properties from the data provided on bulk modulus, lattice constants, and structure. For alloys, we likewise used http://www.matweb.com/, where steel 316 was used as a proxy, and http://eddy-current.com/conductivity-of-metals-sorted-by-resistivity/ or ASTM standard metals.*
[f] *Thermal diffusivity and conductivity from Henderson et al. (1998a,b).*
[g] *Calculated from Ho et al. (1978)*
[h] *From http://www.advancedmagnetsource.com/Alnico.pdf; our D-value indicates* $K = 22\ W\,m^{-1}K^{-1}$.
[i] *From Ledbetter and Reed (1973) and http://www.nickel-alloys.net/invar_nickel_iron_alloy.html#_Physical_properties.*
[j] *Hust and Lankford (1984)*
[k] *Penetration distance of infrared photons in elements calculated from data in Ordal et al. (1985) or Lynch and Hunter (1991).*
Notes: After Tables 1 and 2 from Criss, E.M., Hofmeister, A.M., 2017. Isolating lattice from electronic contributions in thermal transport measurements of metals and alloys and a new model. Int. J. Mod. Phys. B 31, No. 175020. doi:10.1142/S0217979217502058., unless noted otherwise. Previous measurements on *K* are linked to *D* of the less pure samples.

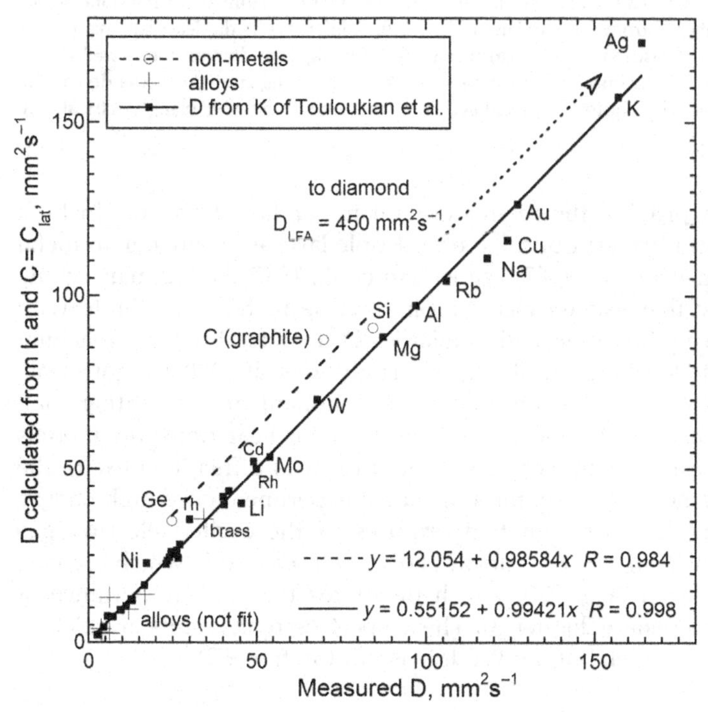

FIGURE 9.3 Comparison of *D* calculated from measured *K* using measured heat capacity (e.g., Touloukian and Buyco, 1970), which describes the lattice, to direct measurements of. Data on *K* from Touloukian et al. (1970) and various websites, most of which trace to this source. Direct measurements of *D* from Touloukian et al. (1973) and Hofmeister et al. (2014) describing 39 metallic elements and 4 nonmetals, respectively. See Table 9.1. Source: *From Fig. 4 of Criss, E.M., Hofmeister, A.M., 2017. Isolating lattice from electronic contributions in thermal transport measurements of metals and alloys and a new model. Int. J. Mod. Phys. B 31, No. 175020. doi:10.1142/S0217979217502058, who provide* D *of alloys.*

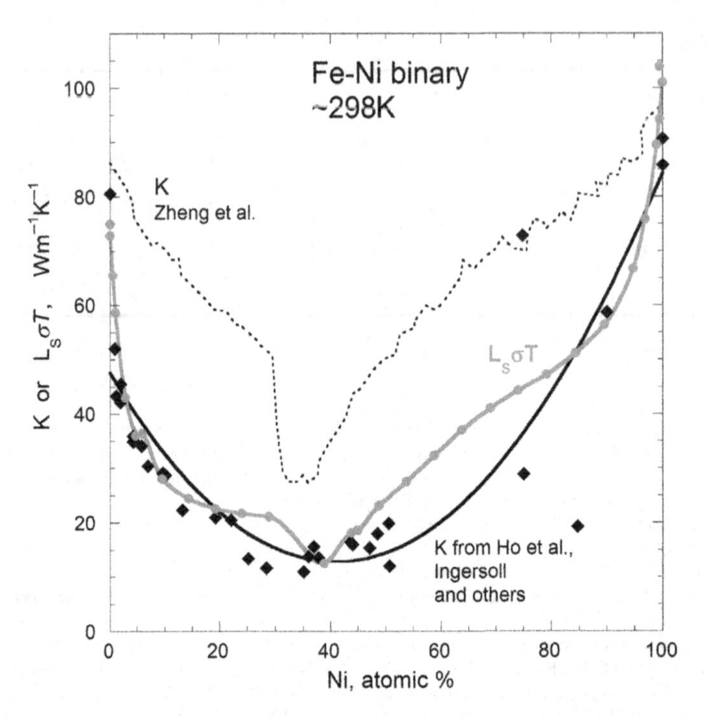

FIGURE 9.4 Dependence of thermal and electrical conductivities on nickel content for Fe–Ni alloys. Diamonds: *K* measured mostly by Ingersoll (1920), with additional data from Silverman (1953), Watson and Robinson (1961), Bungardt and Spyra (1965), and Moore et al. (1971). Ho et al. (1978) summarize the data, which have uncertainties of 5–10%. End-member *K* from Touloukian et al. (1970). Solid curve is the least squares fit to a second-order polynomial. Dotted line: *K* measured across a diffusion couple using time-domain thermoreflectance (Zheng et al., 2010). The different values are attributed to thermoreflectance results being strongly model dependent (Chapter 4). Gray circles: thermal conductivity equivalent calculated from the Lorentz number using electrical conductivity (Ho et al., 1983), and 300K.

The inability of electrons to provide the main mechanism for heat transport in bulk metals is implicit in results from ultrafast optical spectroscopic laser experiments on metal films and surfaces (e.g., Elsayed-Ali et al., 1987; Schoenlein et al., 1987; Aeschlimann et al., 2000). These experiments show that excited electrons in diverse metals equilibrate with other electrons over ~fs, while equilibration with the lattice is considerably later, reaching ~ps (e.g., Beaurepaire et al., 1996; Bauer et al., 2015; Kruglyak et al., 2005; Bigot et al., 2009; Patz et al., 2013). Electron–electron scattering is the dominant process, rather than electron–phonon scattering, which is the mechanism assumed in heat transport models (e.g., Ziman, 1962), based on results such as Fig. 9.2. Because it is difficult to excite the electrons due to their miniscule heat capacity, these ultrafast experiments use high energy light and high fluxes, resulting in very high temperatures of the excited electron gas (~500K above ambient: Corkum et al., 1988; Brorson et al, 1990). Electrons which are easier to excite also reequilibrate faster (Fig. 9.5), which illustrates the great importance of heat capacity to transport phenomena (Chapter 3). These spectroscopic results are consistent with the nearly free electron model, but the WF law is not (Section 9.2).

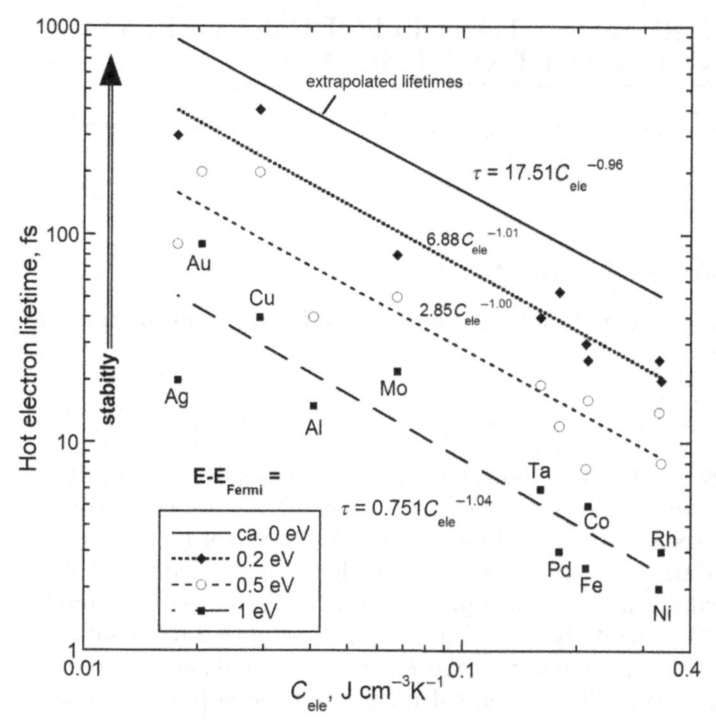

FIGURE 9.5 Dependence of excited electron lifetimes from ultrafast spectroscopy on the electronic heat capacity. Data on lifetimes represent multiple studies of each substance, and were taken from graphs in Bauer et al. (2015). Lifetimes for very small energy excesses above the Fermi level were obtained by extrapolating the experimental values. Data on C_{ele} from Aven et al. (1956) and Gopal (1966). Inverse fits exist whether the basis for electron heat capacity is per mass, per mole, or per volume. In all cases, the correlation coefficients vary from 0.73 to 0.99. Source: *From Fig. 2 of Criss, E.M., Hofmeister, A.M., 2017. Isolating lattice from electronic contributions in thermal transport measurements of metals and alloys and a new model. Int. J. Mod. Phys. B 31, No. 175020. doi:10.1142/S0217979217502058.*

Fast recovery (Fig. 9.5) is consistent with fast speeds of the electron, which naturally leads to the question: is electronic heat transport in metals transient? Because heat transport is a time-dependent process, transient measurements are essential, as discussed in the preceding chapters. Electron speeds are $\sim 100\times$ faster than acoustic speeds associated with lattice perturbations and specifically with heat conduction in graphite, Ge, and Si, which have similar diffusivity as metals (Fig. 9.4). Therefore, almost all measurements of thermal diffusivity (e.g., Parker et al., 1961), which have focussed on characterizing heat transport in metals as equilibrium is being approached (i.e., these have measured D compatible with K from steady-state methods), are unlikely to resolve the possible transient signal of the electrons. This signal should be weak, from the connection of the rise in temperature to the heat capacity of the electrons: this finding is based on Parker et al. (1961), see Chapters 3 and 4.

To address mechanisms for heat transport near room temperature and above, Criss and Hofmeister (2017) measured the response of metals and alloys immediately after the pulse and found a signal akin to transient signals in insulators that arise from high-frequency light but slower, that is, consistent with electron speeds (Chapter 4). Section 9.3 covers the transient behavior that immediately follows the laser pulse showing consistency with properties and expected behavior of electrons. Section 9.4 describes the longer term approach to equilibrium via vibrational interactions. Section 9.4 includes data of others on metallic melts, due to importance to planetary interiors, and the few reliable, low pressure data on solid metals (i.e., not diamond anvil cell studies, see Chapters 4 and 7). Section 9.5 covers implications.

9.2 THEORETICAL REASONS THAT ELECTRONS CARRY INSIGNIFICANT HEAT IN METALS

The numerical agreement of the quantum electronic model for K is unimpressive (Fig. 9.1). Theoretical problems are summarized below.

9.2.1 Equilibrium Conditions are Described

Sommerfeld's (1928) model of thermal conductivity is based on the mean free gas description of electron transport:

$$K_{ele} \cong \frac{1}{3}u_{ele}^2 \tau_{ele} \frac{N}{V} C^*_{V,ele} = \frac{1}{3}u_{ele}\lambda_{ele}\frac{N}{V}C^*_{V,ele} \quad \text{(historic)} \tag{9.7}$$

where u = velocity, λ = mean free path, and * indicates a per mole basis. Eq. (9.7) is equivalent to $K_{ele} = \frac{1}{3} < K_{ele} >$, where the brackets specify that average values are being considered. For a detailed derivation, see Criss and Hofmeister (2017). The factor of ⅓ describes local fluctuations in the three Cartesian directions under isothermal conditions, which is inconsistent with Fourier's equations and invalid during heat conduction from thermodynamic law (Chapters 3 and 5). Consequently, L_S in Eq. (9.1) is a factor of 3 too small, and the thermodynamically correct value does not describe K for the elements at over moderate temperatures (Fig. 9.1). The mismatch between calculated and measured D shown in Fig. 9.2 is made substantially worse by removing the erroneous factor of ⅓.

9.2.2 The Model Contains Trade-Offs

Ambiguous behavior is characteristic of models which contain lumped (multiplied) factors, whereby data can be fit if even if the physics is incorrect (Transtrum et al., 2015). Fig. 3.5 provides a visualization of tradeoffs of combined variables. The WF law embodies several ambiguities inherent to "sloppy models."

One reason is the manner in which parameters combine. The simplicity of the WF law ($K \propto \sigma$) results from factors in the product $u_{ele}C_{ele}$ canceling. These factors cancel regardless of whether the quantum product ($u_{ele}C_{ele}$) or the classical product ($u_{gas}C_{gas}$) is used. This ambiguity is demonstrated by the classical paper by Drude (1900) who provided nearly the same formula as Eq. (9.1) (e.g., Ashcroft and Mermin, 1976; Kittel, 1971; Burns, 1990).

Second, the WF law describes vibrational transport equally well as electronic. The $u_{gas}C_{gas}$ term could represent $u_{lat}C_{lat}$ (this was used in Drude's calculation) instead of $u_{ele}C_{ele}$. Importantly, oscillations of the vibrating cations set up an AC current among the conduction electrons. The latter is a consequence of electron speeds being much faster than those of massive cations. Under thermal equilibrium, no electrical resistance is associated with this AC microcurrent, since the fluctuation is local. But, during time-dependent heat flow, thermal disequilibrium between electrons and phonons propagates down the temperature gradient, and so does the AC microcurrent (Criss and Hofmeister, 2017).

Third, the product of u and C for electrons equals the product associated with the lattice. Specifically, electrons move at speeds $\sim$100 to 1000$\times$ those of sound, but have heat capacities about 0.01 times that of the lattice, as is well-known.

9.2.3 Use of Incorrect Sum Rules

The sum of $K_{meas} = K_{lat} + K_{ele}$ was proposed by Meissner (1915), based on fitting a few experimental data points to $K = A + BT^{-1}$, where A and B are constants presumed to represent different mechanisms. Other forms, $K_{meas} = A + BT$ (Smith and Palmer, 1935) or $K_{meas} = AT + BT^2 + F/T$ (Ewing et al., 1957), better represent subsequent data. The simple sum probably was accepted based on analogy to parallel electrical currents.

Parallel flow that conserves energy requires considering heat capacity. This need is obvious because the rise in temperature of a material in response to added heat is governed by volumetric heat capacity, as in the model of Parker et al. (1961) for LFA. General sum rules for parallel flow are derived in Section 3.5 of Chapter 3, after Criss and Hofmeister (2017). For metals, the electronic heat capacity is much lower than the lattice heat capacity, so Eq. (3.25) holds and

$$K_{meas} \cong K_{lat} \tag{9.8}$$

This result is irrespective of the rate at which the electrons carry heat, and is holds whether or not electron and lattice mechanisms are interdependent. Eq. (9.8) mainly results from conservation of energy.

9.2.4 Relying on Thermal Conductivity, Not Diffusivity

Fourier (1822) presented two equations, one of which pertains to steady-state and is governed by K, while the other concerns time-dependent behavior and is governed by D, presuming that the temperature changes are small. Microscopic models for thermal conductivity are compared to steady-state behavior for validation, whereas models for thermal diffusivity are verified via comparison to transient measurements. Because heat capacity combines with carrier speed in microscopic models of K (e.g., Eq. (9.7)), modeling K has an unavoidable ambiguity. By considering thermal diffusivity, different carriers can be distinguished, on the basis of their characteristic speeds.

9.2.5 Electrons are Excited by Energies in the Visible, not in the Infrared

The WF law is a steady-state model, which does not address how electrons are excited. Yet, this process must be explained when discussing the feasibility of the model, because excitation of the electrons is a prerequisite for electronic conduction of heat.

Experiments by Hertz (1887) on the photoelectric effect showed that high energy (visible) light is needed to give electrons in the conduction band sufficient energy to breech the surface. Modern, fs-spectroscopy quantifies how visible light excites electrons from the valence into the conduction band (e.g., Bauer et al., 2015). Infrared light is ineffective in exciting electrons in either experiment. Yet, when a metal receives an IR pulse, heat is amply diffused across the sample (e.g., Parker et al., 1961). So, some process other than electron collisions must be capable of absorbing and transmitting low energy heat.

Spectroscopic measurements of reflectivity show that interaction of metals with light are strongly frequency dependent. Results for iron (Fig. 2.15) and platinum (Fig. 8.6) are representative. As frequency increases, reflectivity declines from near perfection in the far-IR, while the absorptivity increases. The deeper penetration of visible light, plus its higher energy, permits stimulation of electrons in sufficient number to be detected in the photoelectric effect and ultrafast spectroscopy experiments. Shallow penetration of the IR light, with its low energy stimulates very few electrons, and thus electrons cannot transport of large amounts of heat during thermal conductivity experiments.

Additionally, thermally excitable phonons greatly outnumber thermally excitable electrons (those with near the Fermi level), and hence phonons uptake the vast majority of heat (IR light) entering the solid. Because the heat content of the metal is in its lattice vibrations, only interactions of incoming energy with vibrations can propagate significant amounts of heat.

9.2.6 Thermal Changes in Electron States Have Been Misrepresented

Electron transitions and interactions with vibrations have been wrongly described in order to reconcile the data with the WF law. One example is Fig. 9.3, which shows that the relevant heat capacity is that of the lattice.

Electron–electron scattering should have a high probability due to Coulombic forces being strong (e.g., Ashcroft and Mermin, 1976). This is indeed the mechanism during electrical conduction, where the unified flow of the electrons is impeded by the vibrating cations with their cloud of valance electrons. The latter process is the origin of electrical resistivity.

For electron–phonon scattering to instead be the dominant mechanism in heat transfer, the electron–electron collisions must somehow be prohibited by another process. The consensus picture is that electron–electron collisions do not exchange heat energy because a thermally excited electron lacks a state to fall into during electron–electron collisions (Ashcroft and Mermin, 1976; Burns 1990). This picture is incorrect regarding heat transfer because it treats electron population statistics as independent of temperature (Criss and Hofmeister, 2017). Clearly, the statistical distribution of states in the "hot" region differs from that of the "cold." The energy levels do not exactly match, including near the Fermi level (Fig. 9.6A), so the transition is much more complicated than the textbook picture of two electrons simply trading places. In essence, when a "hot" conduction electron travels down the thermal gradient, its electric field distorts the fields of the thermally stable conduction electrons as it enters a colder region of the metal. During a "collision," the interacting electrons form a metastable state (Fig. 9.6B) that has some finite lifetime. After this event, electronic temperature is higher in the region of interest (Fig. 9.6C).

Actually, electron–phonon scattering is the process which is improbable due to low availability of destination states for the electron. When the carrier electron interacts with valence electrons, it sheds its extra energy to the lattice. Thus, the surrounding field of conduction electrons remains essentially unchanged, and so additional destination sites

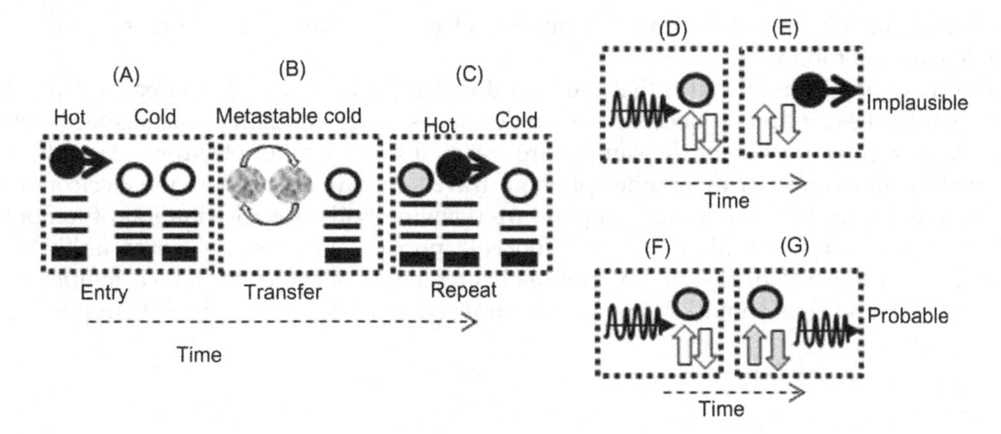

FIGURE 9.6 Schematics of electron and phonon behavior in metals. In all panels: dotted outline: a small volume of the metal in a temperature field; dashed arrow indicates time; various small circles: individual electrons in different states; horizontal lines: energy levels associated with the region of space where the electron of interest resides; large shaded circles: atoms; pairs of open arrows indicate atomic motions; thick black arrow: electron motion down the thermal gradient and termination of motion; thin arrow with squiggle: phonon motion. Right panel: Sequence of events during electron–electron scattering. (A) Initial state. At the hot end (left), a conduction electron (black dot) is excited above the top of the Fermi levels (horizontal bars) and travels towards the cold end (right), where a small proportion of the electrons (circles) are near the top of their Fermi levels, which differ from the levels in the hot region. (B) Thermalization of conduction electrons. In order for the "hot" and "cold" electrons to equilibrate, they interact (stippled circles and curved white arrows). Energy levels are not shown during this brief, metastable state. (C) Progression of heat across the volume. Through the interaction, the incoming excited electron has lost some of its energy, and is now part of the equilibrium, but hot, state (gray circle). This energy was transferred to the middle electron (black dot) which then travels into the colder region. Repetition of these steps (i.e., thermal diffusion) brings the electron population across the metal into the hot state. Right panel: Movement of heat by phonons and electrons. (D,F) The initial state where a phonon enters some volume in the metal. Due to the relative speeds, the electron population has already transformed to the hot state (gray circle), but the lattice is in the cold state (white arrows). Two different exchanges are possible: (E) Excitation of an electron, which increases the temperature of the electron gas, which is already hot, while leaving the cold lattice in the cold state. (G) Excitation of the lattice to the hot state. The electron giving energy to a phonon is irreversible, because it advances the thermal fields toward equilibration. Source: *Modified after Fig. 3 of Criss, E.M., Hofmeister, A.M., 2017. Isolating lattice from electronic contributions in thermal transport measurements of metals and alloys and a new model. Int. J. Mod. Phys. B 31, No. 175020. doi:10.1142/S0217979217502058.*

are ordinarily unavailable for the cooled, conduction electron. Exceptions are defects, grain boundaries, or impurity atoms because these provide additional energy levels. However, impurities do not enhance conduction but instead impede (Fig. 9.4). Furthermore, ultrafast optical spectroscopy demonstrates that electron–phonon interactions are far less probable than electron–electron scattering.

The behavior of a hot electron after it undergoes a "collision" is very important. If the collision is with an electron near the Fermi level, the heat can be propagated electronically (Fig. 9.6A–C). If instead, any given hot electron interacts with the lattice, then this electron's thermal energy is lost to the lattice, becoming part of lattice conduction at slow acoustic speeds. If the diffusional sequence for electron transport in a metal is electron–phonon collisions followed by a phonon–electron event (Fig. 9.6D,E), then the associated speed cannot be u_{Fermi} but instead lies between electronic and sound speeds, depending on event probabilities. However, electrons taking heat from the lattice is problematic, as follows:

Thermodynamic law dictates that the heat donor must be at a higher effective temperature than the heat recipient. Because electronic speeds are faster than acoustic speeds, electrons in any given portion of the sample are warmed first by other electrons. Any phonon (excited by an electron or by another phonon) travels into an area where the electron temperature is higher than the lattice temperature (viewing both as a distribution of temperature-energies). These stipulations make inelastic phonon–electron collisions unlikely to contribute to thermal diffusivity. Vibrational interactions elevate the lattice temperature (Fig. 9.6F,G), and thus should dominate heat transfer in bulk metals. Fig. 9.3 supports this contention.

9.3 ELECTRON TRANSPORT IN METALS AND ALLOYS

Disequilibrium of electrons and phonons during transient heating events, as demonstrated by ultrafast spectroscopy, is a consequence of electrons and phonons obeying different statistics, of Coulombic interactions between electrons being strong, and of electrons moving up to 1000 × faster than phonons. Disequilibrium during heating is thus compatible with the nearly free electron model. Because the transient technique of LFA records the evolution of temperature with time, Criss and Hofmeister (2017) were able to distinguish signals from electronic and vibronic transport. This section summarizes results of that study. Due to electronic transport being transient, it is not relevant to the slow cooling or heating of large planetary bodies. However, the process may be relevant to impacts involving core materials, or perhaps certain highly metalliferous ore deposits, so the results of this paper on electronic transport of heat are summarized here.

9.3.1 Raw Time–Temperature Curves Show a Rapid, Weak Rise Preceding the Main, Slow Rise

Mechanisms with different speeds are distinguishable in transient experiments because the speed of the heating front is directly related to carrier speed (Hofmeister, 2010). In insulators, ballistic transport in the visible precedes the main, slow diffusive warming, as is evident in the time–temperature curves of LFA data. In metals, diffusion of electronic heat precedes diffusion of the lattice heat. Fig. 9.7 shows raw data for steel. Manganese T–t curves are in Fig. 4.5. For both, a small rapid rise was observed following the laser pulse, and then a large and slow rise followed. The large and slow rise provided thermal diffusivity consistent with previous measurements.

Of the 31 different alloys and metals examined by Criss and Hofmeister (2017), the rapid rise could be analyzed for 20 of the samples. Whether or not the signal could be detected depends on various physical properties of that metal (Section 9.3.2). Importantly, long lengths are needed, generally >5 mm, rather than the 1–4 mm lengths typically probed. For alloys (steel) and disordered metals (Mn), shorter lengths could be used. In

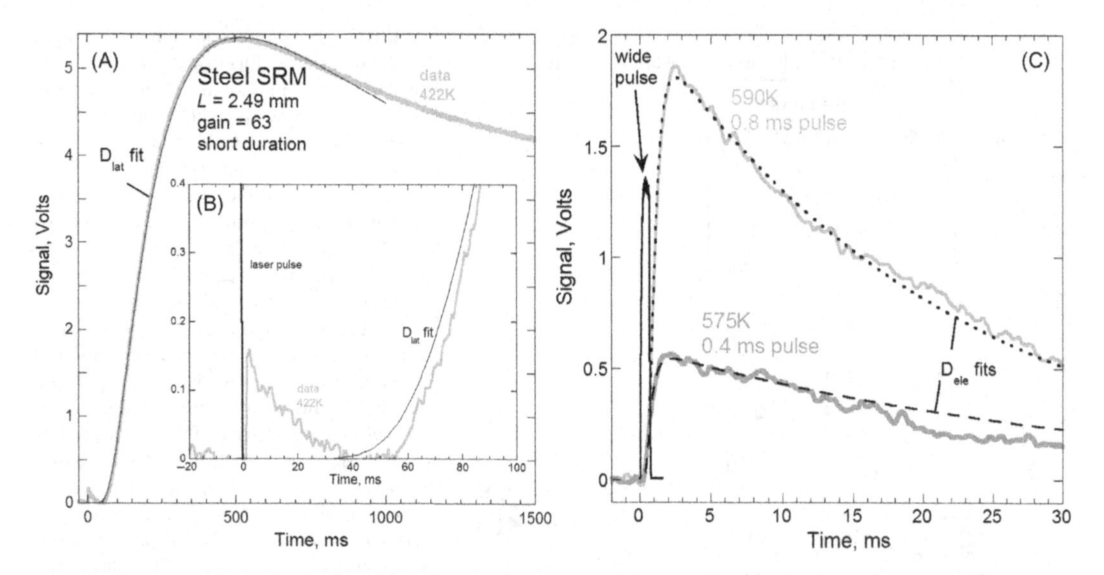

FIGURE 9.7 Temperature–time curves of nonmagnetic steel. Gray: baseline corrected data. Black: fits. One data collection is shown over long time in panel (A) with an expanded view of the rapid rise in inset (B). Panel (C) shows that the intensity of the rapid rise correlates directly with the energy input. The large voltage for these two shots is associated with high gain. Source: *Modified after Fig. 8 of Criss, E.M., Hofmeister, A.M., 2017. Isolating lattice from electronic contributions in thermal transport measurements of metals and alloys and a new model. Int. J. Mod. Phys. B 31, No. 175020. doi:10.1142/S0217979217502058.*

addition, short collection times are essential for the best signal-to-noise ratio. Both experimental conditions help to probe fast electronic diffusion. Some runs were collected so that both signals were present in a single data collection (Fig. 9.7A). This allowed Criss and Hofmeister (2017) to ratio the peak heights, which were consistently low for the rapid rise, and as expected for c_{ele} being $\sim 1\%$ of c_{lat}.

The rapid rise is not an artifact. This signal is distinguished from light leakage based on the different shape of the rise, which was confirmed by purposefully allowing light leakage in some experiments (Fig. 4.4). The electron signal is differentiated from ballistic radiation based on intensity of the latter increasing with temperature as T^3. The shape of the rapid rise (Figs. 9.7 and 4.5) shows the process is diffusional. Also, characteristics of the rapid rise are not tied to those of the slow rise, although each depends on composition and temperature, and responds to phase transitions, including magnetic.

9.3.2 Conditions for Observing the Rapid Rise

For the electrons to flow in response to addition of heat, they must both absorb the heat and carry the heat. The first property is associated with the penetration of light, whereas the second is described by C_{ele}. The map of Fig. 9.8 shows that the rapid rise can be detected and quantified when the combination of both properties are large (compensating trade-offs). Sample size also has an effect because this increases the flux of intensity to the detector, thereby enhancing the signal-to-noise ratio. For a weak signal, this is important.

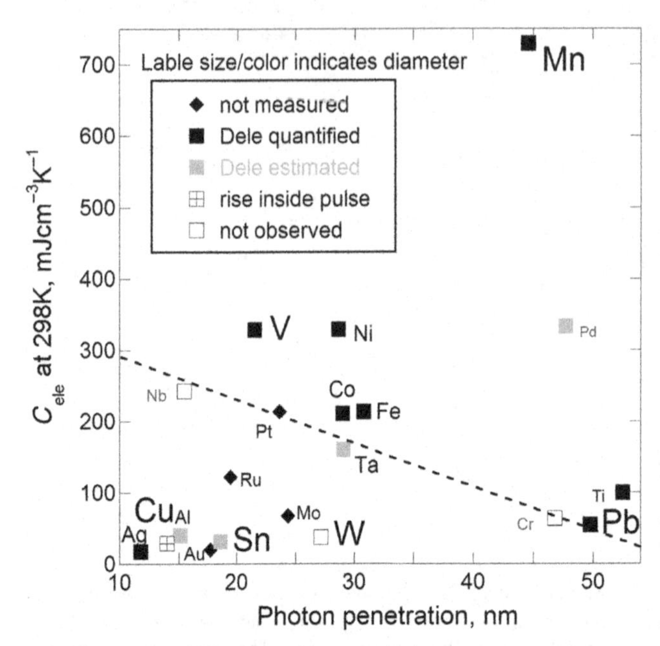

FIGURE 9.8 Conditions where the rapid rise is observed. Electronic heat capacity from the data in Gopal (1966). Diameter pertains, due to noise level. Photon penetration depths were calculated from data in Ordal et al (1985) and Lynch and Hunter (1991). Map of rapid rise characteristics associated with C_{ele} and photon penetration. Inset lists the symbols. Large font size for labels designates diameters >13 mm. Medium font size designates typical $d = 12-13$ mm. Small font size $= d \sim 9.5$ mm. Gray and small font size $= d \sim 6.5$ mm. The dotted line divides metals with rapid signals that could be quantified from metals with weak or nonexistent rises. Diameters below ~ 6.5 mm inhibit detection. Source: *Reproduction of Fig. 12b of Criss, E.M., Hofmeister, A.M., 2017. Isolating lattice from electronic contributions in thermal transport measurements of metals and alloys and a new model. Int. J. Mod. Phys. B 31, No. 175020. doi:10.1142/S0217979217502058.*

The rapid rise is detected for a few metals with low C_{ele} if their electrical resistivity is also very low (Al, Ag). The latter property permits the electrons to travel further, which makes the current larger (more intense) and thus easier to detect. Increasing the intensity of the laser pulse has a similar effect. Detection of the rapid rise depends consistently on the physical properties of the metals and on the experimental conditions (Criss and Hofmeister, 2017).

9.3.3 Behavior of Thermally Excited Electrons

Characterization of the rapid rise is inherently limited by penetration of heat (IR light), the low carrying capacity of the electrons (C_{ele}), their fast speeds, and resistive or magnetic degradation of the electronic heat current. Criss and Hofmeister (2017) were able to characterized D_{ele} in 20 different metals and alloys. The temperature dependence for elemental Ni is shown here (Fig. 9.9) because its Curie transition at low temperature allowed study of both magnetic and nonmagnetic states over reasonably large temperature ranges.

Generally, D_{ele} increases with T at the lowest temperatures accessed, reaches a peak at moderate temperature, and then decreases. The curves, which are represented by a second-order polynomials (Table 3 in Criss and Hofmeister, 2017), are disrupted by phase transitions. The magnetic transition has a strong effect on D_{ele}, in a manner which is consistent with fs-spectroscopy, where magnetic interactions demonstrably shorten the lifetimes of excited electrons. This is clear from our data on nickel (Fig. 9.9). Magnetic interactions also affect D_{lat}, but weakly, as occurs in insulators such as garnets and spinels (e.g., Hofmeister, 2006, 2007). Thus, transport by the electrons and by the lattice differ greatly (Fig. 9.9). In particular, thermal diffusivity of the electrons is large (~ 1000 mm^2 s^{-1}

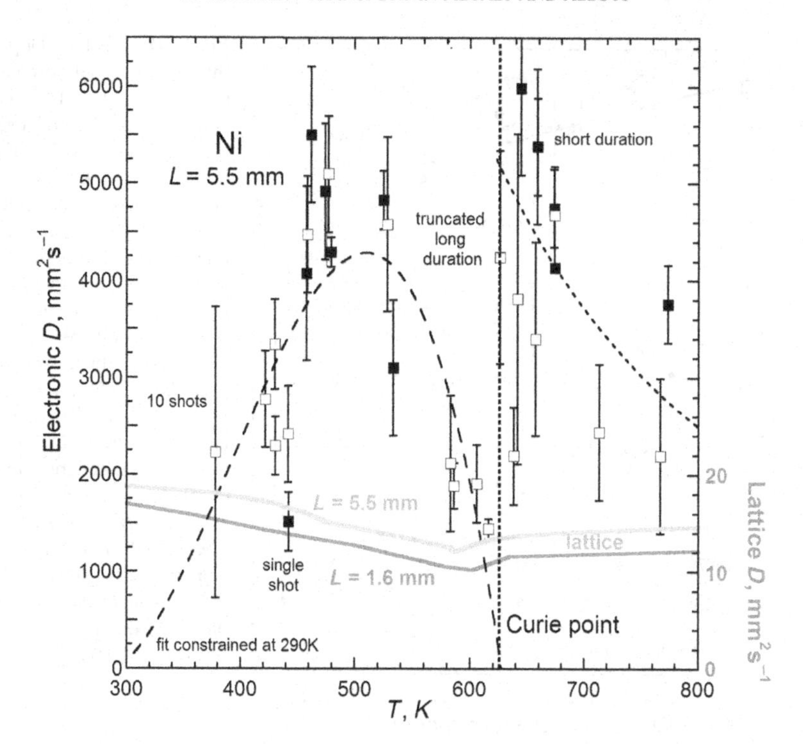

FIGURE 9.9 Temperature dependence of D_{ele} and D_{lat} obtained from the same rod of Ni, in many different runs. The electronic contribution (left axis) is shown as symbols with error bars and a least squares fit (listed in Table 9.3). The lattice contribution (right axis) is shown as heavy gray lines without symbols. Filled squares: short data collection times. Open squares: truncations of long collection times, where both the rapid and slow rises were recorded. The symbol at 373K averages six datapoints from two runs: most of the other symbols average three datapoints. Differences in D_{ele} between short and long collection times above the Curie point are attributed to the very high values not being well-constrained for the lower resolution and higher noise accompanying high T. At the Curie point, while the magnetic transition was occurring, the results vary widely (1870–7000 $mm^2 s^{-1}$) and were not included in the fits, which include short and long collection times. Above T_{Curie}, a power-law fit best describes the data. The magnetic transition has a much smaller effect on D_{lat}, than on D_{ele}. Source: *Reproduction of Fig. 13c of Criss, E.M., Hofmeister, A.M., 2017. Isolating lattice from electronic contributions in thermal transport measurements of metals and alloys and a new model. Int. J. Mod. Phys. B 31, No. 175020. doi:10.1142/S0217979217502058.*

for alloys, ~4000 $mm^2 s^{-1}$ for most elements) compared to the lattice contribution, which ranges from about 2 $mm^2 s^{-1}$ for Mn and alloys to 175 $mm^2 s^{-1}$ for Ag near room temperature (Fig. 9.2). High values of D_{ele} result from high electron speeds. Values of K_{ele} are much smaller, and similar to K_{lat}, due to C_{ele} being small: this trade-off underlies the ambiguity of the WF model.

Various physical properties effect D_{ele}. At room temperature, D_{ele} decreases as resistivity increases (Fig. 9.10). Alloying and ferromagnetism do not seem to have an effect on D_{ele}, except for strongly magnetic AlNiCo alloys. In contrast, D_{lat} is strongly affected by both properties.

Criss and Hofmeister (2017) found an inverse correlation of D_{ele} with C_{ele}, which is consistent with Fig. 9.5, and attributable to interactions increasing as the number of excited

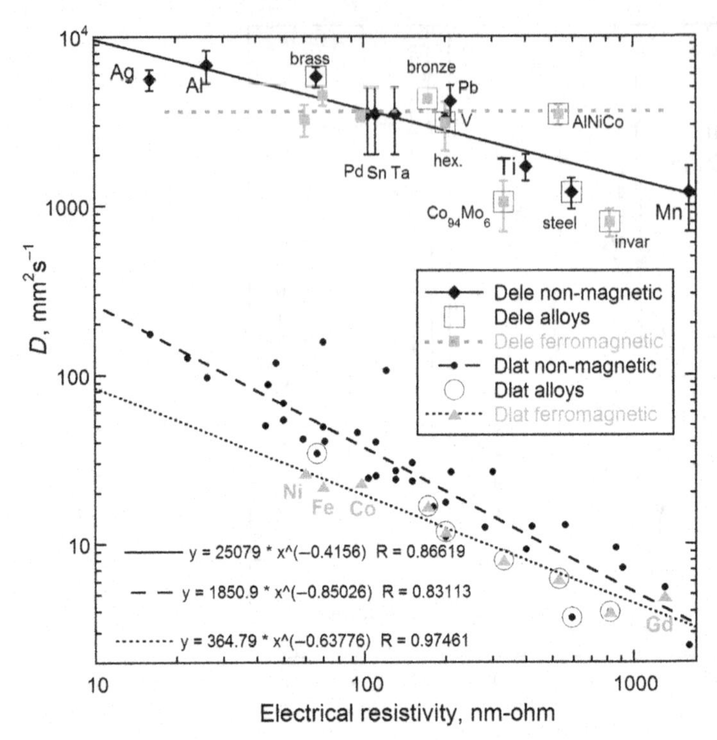

FIGURE 9.10 Dependence of thermal diffusivity on electrical resisitivity. For D_{ele}, the most accurate values at 473K are used to avoid uncertainties associated with projection. Measurements of Ni (see Fig. 9.9 at the Curie point) and invar (see Criss and Hofmeister, 2017) indicate that the non-magnetic state has higher D_{ele} than the magnetic state. For D_{lat}, our data (Table 9.1) and measurements of Touloukian et al. (1973) are used, which are uncertain by ~2% and ~5%, respectively. Least squares fits are listed. Electronic and lattice heat transport are affected differently by magnetism and cation disorder. Source: *Reproduction of Fig. 17 of Criss, E.M., Hofmeister, A.M., 2017. Isolating lattice from electronic contributions in thermal transport measurements of metals and alloys and a new model. Int. J. Mod. Phys. B 31, No. 175020. doi:10.1142/S0217979217502058.*

electrons increase. That electrons impede their own flow explains the maximum in D_{ele} versus temperature (Fig. 9.9). At low T, D_{ele} increases with T as the population capable of transmitting heat grows. However, once some sufficiently large electron population is excited, the carriers interfere with each other, shortening their mean free paths and lowering D_{ele}. A connection with electron speeds was not found in this study, but Fermi speeds vary only by a factor of 2 for the metals examined. Regarding K_{ele}, its size at low temperature depends on the number of electrons in the outer shells: see Criss and Hofmeister (2017) for this correlation, fits, and other details.

9.4 LATTICE BEHAVIOR

Copious data exist on thermal conductivity for metals (e.g., Touloukian et al., 1970). This can be converted to thermal diffusivity because $C_{lat} = C_{meas}$ is well-known (e.g., Touloukian and Buyco, 1970). However, problems exist with conversion near the Curie point for elements because the transition is steep and may not occur at exactly the same position for the various measurements. For this reason, thermal diffusivity data are presented (Table 9.1), with a few exceptions.

At ordinary temperatures, the lattice uptakes the majority of heat as it enters the metal and must transport this heat, as is evident from Fig. 9.3, because the electronic carrying capacity is almost negligible. The electrons contribute transiently, due to their fast speed, and thus the two mechanisms are independent, as assumed in the nearly free electron model. This is helpful for geophysical applications, which concern the lattice only. This section emphasizes materials relevant to Earth's core.

9.4.1 Thermal Diffusivity of the Elements

Most LFA data concerns fairly pure elements (Table 9.1). Like insulators, thermal diffusivity generally decreases from ambient conditions and is flat at high temperature, but variations exist (Fig. 9.11). Exceptions are Mn and Pd. One concern is the presence of impurities (discussed below). The increase in D_{lat} with T for Mn could be due to ordering, however, this sample was slightly porous, which could affect its apparent diffusivity. The origin of the increase in D_{lat} versus T for Pd is unclear. It does not seem that weak magnetism affects the temperature response since Mn, Pd, Al, Ti, V, Mo, and Ta are paramagnetic, whereas Ag and Pb are diamagnetic. In contrast, ferromagnetic transitions strongly affect D_{lat}, providing a λ-shaped curve as in garnets or spinels. Nonetheless, over short ranges in temperature, the trends are close to linear. Accurate fits over an extended range of temperature (for other than ferromagnetic metals) may require second- or third-order

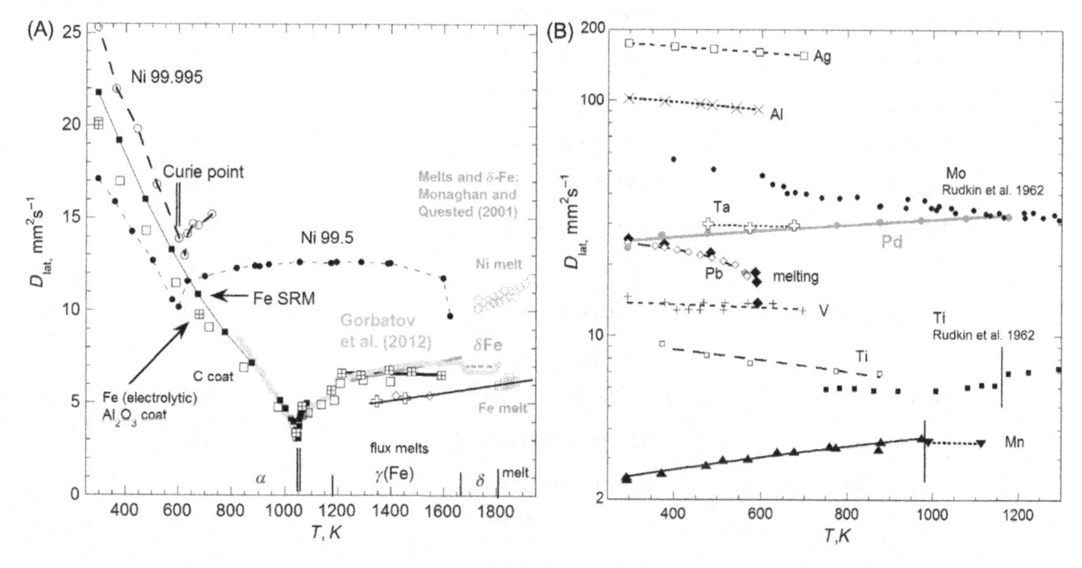

FIGURE 9.11 Lattice thermal diffusivity of nearly end-member elements (A) Ferromagnetic Fe and Ni, mostly from LFA (black symbols from our laboratory). Circles: high purity Ni. Dots: Ni. Filled squares: SRM iron. Open squares: high purity iron. Square with cross: high purity iron coated with Al_2O_3 with a linear fit shown. Open diamonds: permalloy melt (Ni 37%). Open cross: Glorietta Mountain meteorite melt (Ni: 11%). Light gray dots and line: plane-wave method of Gorbatov et al. (2012) on nearly pure Fe. Medium gray symbols: Nishi et al. (2003) on Fe and Ni melts, as labeled. Dark gray lines and symbols: γ-, δ- and melted Fe phases measured by Monaghan and Quested (2001). Fine line: trend of melting promoted by contact with graphite and of the end-member Fe. Vertical lines on the X-axis mark phase transitions of Fe. Ni only occurs in the γ-phase. (B) A variety of nonmagnetic elements, as labeled. The Y-axis is logarithmic. Fits are linear, except for Pb, which melted at low temperature. Vertical line denotes a structural transition in Mn. Data on Mo and on Ti at high T from Rudkin et al. (1962).

polynomials. However, we did not attain high temperatures because the focus was on transient behavior of the electrons, and the earlier studies on the elements only used the simple model of Parker et al. (1961), which does not account for cooling to the surroundings. In addition, the samples of Rudkin et al. (1961) were not described and are unlikely to be very pure. For these reasons, Table 9.2 reports linear fits from recent measurements.

TABLE 9.2 Fits to Thermal Diffusivity Data Above 290K as a Function of Temperature for Metals and Alloys

Name	Fit to D in mm^2 s^{-1} for T in K	T_{max}, K
Al	$112.86-0.036667T$	650
$Al_{99}Si_1$ [f]	$102.32-0.044028T$	800
$Al_{90.4}Si_{9.4}$ [f]	$98.002-0.048992T$	800
Ti	$10.482-0.0043376T$	900
V	$14.381-0.0020433T$	700
Mn	$2.004+0.001855T$	970
AlNiCo magnet	$4.8061+0.0044766T$	700
Invar	$2.8858+0.0026858T$	1400
SRM 1460	$2.8957+0.002614T$	800
Steel 310[c]	$2.7121+0.0024508T$	1300
Inconel[d]	$2.9435+0.002336T$	1300
Cu[e]	$126.0-0.035864T$	1360
$Cu_{99}Cr_{0.8}Zr_{0.08}$ [a]	$106.06-0.020325T$	800
Brass	$0+0.15961T-0.0001464T^2$	650
Bronze	$18.686+0.0016521T$	910
$Sn_{99}AgCu_{0.5}$ [b]	1.33 (nearly constant)	430
$Sn_{96}Ag_{3.5}Cu_{0.5}$ [b]	1.25 (nearly constant)	430
Pd	$22.645+0.0085463T$	1200
Ag	$189.49-0.049311T$	700
Ta	$30.394-0.001855T$	700

[a] *Average value from Rohde et al. (2013).*
[b] *Kim et al. (2009)*
[c] *Blumm et al. (2007)*
[d] *Blumm et al. (2003)*
[e] *From Neztsch website, similar to stock copper.*
[f] *Kim et al. (2016)*
Notes: *Data from Criss, E.M., Hofmeister, A.M., 2017. Isolating lattice from electronic contributions in thermal transport measurements of metals and alloys and a new model. Int. J. Mod. Phys. B 31, No. 175020. doi:10.1142/S0217979217502058, unless noted otherwise.*

Regarding lead, incipient melting likely affects its thermal diffusivity. The data (in $mm^2 s^{-1}$ for T in K) from ~290 to 600K are represented by:

$$D = 46.908 - 0.16205T + 0.00040304T^2 - 3.6495 \times 10^{-7}T^3 \quad (9.9)$$

Small amounts of impurities strongly affect D_{lat} (Fig. 9.11A) and K_{lat} (Fig. 9.4), by introducing disorder into the lattice. For small amounts, domains are not formed, as commonly occurs in alloys; rather, the microscopic behavior follows that of electrical insulators (Chapter 7). This observation explains why our data on oxygen-free copper and on tungsten are much higher than previous measurements (Table 9.1) which were made in the 1970s, from less pure material. Silver and lead being refined since antiquity can be made pure, so our data agrees with data from the websites.

The structure at 298K clearly has an effect. The next section incorporates data on K for a clear picture of the variations at 298K.

Structural transitions affect D_{lat} in various ways. For nearly pure Fe, the transition from BCC to FCC causes a small increase in D_{lat}, whereas the transition to another BCC structure (δ) causes D_{lat} to decrease. The flat trend we observed for FCC Fe is the same as that measured by Parker et al. (1961), also for the SRM. The trends for the FCC (γ) and BCC (δ) phases vary between studies (Fig. 9.11A), which examined different materials. Overall, these trends are flat, as shown by Monaghan and Quested (2001) who depict earlier data as well. Upon melting, D_{lat} drops abruptly (Fig. 9.11A). The highly disordered element manganese shows only a break in slope at the transition, whereas data on Ti from Rudkin et al. (1962) show a jump upon transition from HCP to BCC near 1155K, similar to Fe (Fig. 9.11B). The increase in D_{lat} from FCC to BCC is compatible with data at 298K of all metals (below).

Lattice transport being hindered by magnetic interactions and cation disorder, is observed in electrical insulators (e.g., yttrium iron garnet and magnetite: Chapter 7). Substantial hindrance is evident in Fig. 9.10, where the ferromagnetic elements define one D_{lat} trend with electrical resistivity, even though several different crystal structures and electronic configurations are represented.

9.4.1.1 Mean Free Paths and Lifetimes During Lattice Transport

Lifetimes far smaller than measurement times can be inferred because (repetitive) diffusional processes are probed. From dimensional analysis (Chapters 3 and 5):

$$D_{lat} = u_p \Lambda_{lat} = u_p^2 \tau_{lat} \quad (9.10)$$

where u_p is the velocity of the compression wave (i.e., the longitudinal acoustic mode) and Λ is the mean free path. The longitudinal acoustic velocity pertains because this alone changes the volume of the unit cell during motions (Criss and Hofmeister, 2017). The bcc and fcp structures have only acoustic modes, whereas the hcp structure has an additional Raman mode, and the more complicated structures have many modes. The contribution of optic modes is small, because their velocity at zone center is much smaller than the acoustic modes, and can reasonably be neglected in a simple dimensional analysis. Comparing results from Eq. (9.10) for Λ_{lat} to the average interatomic distances (computed from molar volumes and density obtained from website databases) in Fig. 9.12 shows that the

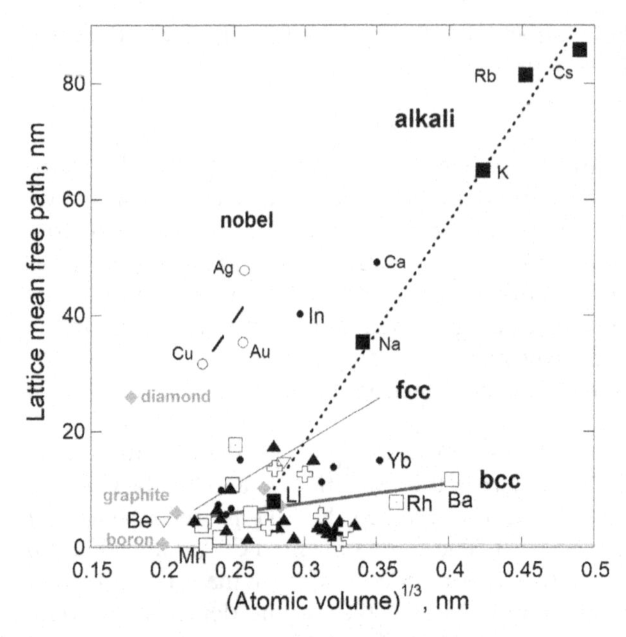

FIGURE 9.12 Comparison of mean free paths calculated from thermal diffusivity (or K/C_{meas}) at 298K to the mean interatomic distance. Physical properties from Table 9.1, except D for C, Si and Ge nonmetals (gray diamonds) are given in Chapter 7. Circles and dashed line fit: fcc noble metals. Dots: fcc and slightly distorted fcc. Filled square and dotted line fit: bcc alkali metals. Open square: other bcc. Various triangles: hcp. Open plus: distorted structures. Thick gray line: 1:1 correspondence. Source: *After Fig. 22a in Criss, E.M., Hofmeister, A.M., 2017. Isolating lattice from electronic contributions in thermal transport measurements of metals and alloys and a new model. Int. J. Mod. Phys. B 31, No. 175020. doi:10.1142/S0217979217502058.*

mechanism is compatible with vibrations for three reasons: (1) Λ_{lat} exceeds the average interatomic distances (computed from molar volumes and density obtained from website databases). For highly distorted Mn and the metalloid Te, the excess is small. (2) Linear trends exist for metals with similar structures (discussed further below). (3) Nonmetals (B, C, Si, and Ge), for which heat conduction is vibrational, have similar trends to metals.

For each of the alkali and noble metal series, which both have a single s-type electron above closed shells leading to simple structures, stretching the lattice increases the distance phonons travel between collisions, as expected. The bcc series occurring to the right of the fcc series in Fig. 9.12 is consistent with bcc being more damped and less compact (since bcc has 14 nearest neighbors whereas fcc has 12 nearest neighbors). Overall, bcc structures have shorter λ_{lat} values than do fcc (cf. dots to open squares in Fig. 9.12). This explains the difference in D during isothermal transitions of Fe (Fig. 9.11A). HCP structures generally have low λ_{lat}, which is attributed to the presence of Raman modes which could interact with acoustic modes, thereby shortening λ. For this reason, D of Ti increases from hcp to bcc (Fig. 9.11B). Similar behavior has been observed for insulators, wherein a strong inverse correlation exists between D values and the number of infrared modes (Chapter 7). Notably, semiconducting elements (B, graphite, Si, and Ge) cluster with most metals, whereas diamond, which has solely phononic transport, clusters with the noble metals, which most closely approach independence of electrons from phonons.

Vibrational lifetimes associated with heat flow are ~0.1 to 100 ps from Eq. (9.10), which is compatible with independent measurements of fs to ps for many materials (e.g., Bron, 2012). The trends in lifetimes (Fig. 9.13) are largely structural, consistent with a lattice mechanism. The inverse of phonon lifetimes is shown because this effectively a damping

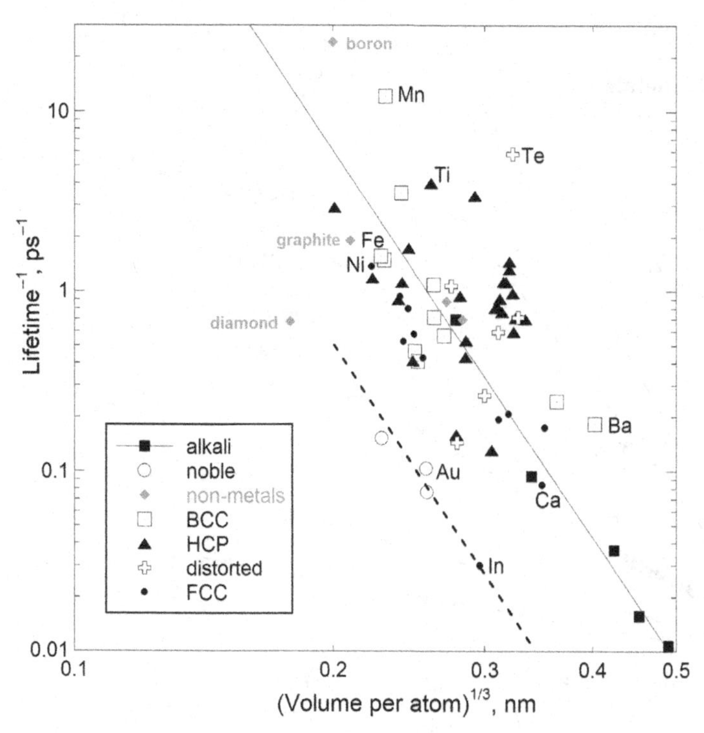

FIGURE 9.13 Comparison of inverse lifetimes with average interatomic distance. Symbols described in the inset. *Source: Data from Table 9.1. After Fig. 23 in Criss, E.M., Hofmeister, A.M., 2017. Isolating lattice from electronic contributions in thermal transport measurements of metals and alloys and a new model. Int. J. Mod. Phys. B 31, No. 175020. doi:10.1142/S0217979217502058.*

coefficient: From Fig. 9.13, $1/\tau$ which decreases with interatomic spacing. The decrease in damping is consistent with longer bonds being weaker.

9.4.1.2 Thermal Diffusivity of Melts

Several of our samples melted during the experimental runs (Fig. 9.11). The temperatures were lower than expected due to the graphite coating serving as a flux. Measurement of D from a suspended disk is possible because time is required for flow. The rate is faster than for silicate glasses (Chapter 10) due to metals having very low viscosity. Thus, we can only capture D_{melt} for a brief interval, about 1 s, before the sample flows, contracts, and falls. In all cases (Si and Ge are shown in Chapter 7; a variety of silicate glasses are shown in Chapter 10), thermal diffusivity associated with melt is lower, regardless of whether the material was crystalline or glassy beforehand. Hence, structural disorder incurred upon melting lowers thermal diffusivity.

The presence of small amounts of C in our flux melts reduces the temperature of the melt. Yet, the values lie on the trends established for pure metals (Fig. 9.14). The very small amount of C does not appear to have an effect, which is consistent with the metal having other impurities and melt being disordered. Linear fits are shown in Fig. 9.14 and more complicated least squares fits given in Table 9.3.

The magnitude of D is perplexing, but seems to be connected with the structure and properties of the solid phase. The crystalline structures should have some relevance because the local coordination is commonly preserved during melting. The melts with small D have high

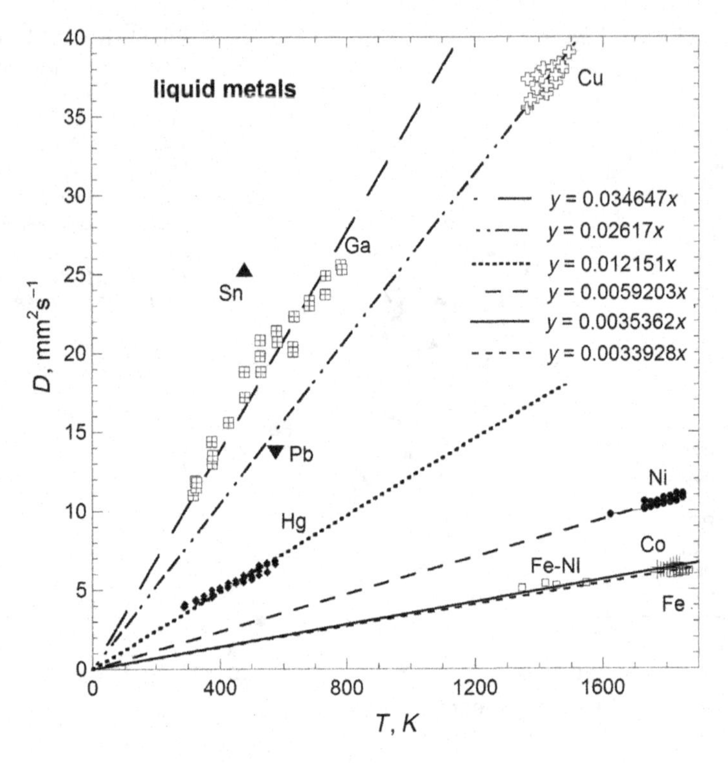

FIGURE 9.14 Thermal diffusivity of liquid metals as a function of temperature from laser-flash analysis. Fits shown (where x is temperature) include the origin. *Data sources and other fits in Table 9.3.*

TABLE 9.3 Thermal Diffusivity Data From LFA on Melts

Name	Fit to D in $mm^2\ s^{-1}$ for T in K	Temp. range, K	Reference	Volume, $cm^3\ mol^{-1}$ [c]
Fe[a]	$2.3005 + 0.0020769T$	1350–1900	[a]	8.04
Co	$-5.0369 + 0.006329T$	1770–1850	Nishi et al. (2003)	7.81
Ni	$1.1394 + 0.0052803T$	1600–1850	[a]	7.80
Cu[b]	$12.731 + 0.01724T$	1360–1550	[b]	8.06
Ga	$0 + 0.041476T - 1.112 \times 10^{-5}T^2$	310–775	Schriempf (1973)	11.47
Sn	25.3	477	This work	18.75
Hg	$0 + 0.015385T - 0.67587 \times 10^{-5}T^2$	270–590	Schriempf (1972)[d]	14.8
Pb	13.7	576	This work	20.17

[a] *Represents data on flux melts (this work) and of Fe from Monaghan and Quested (2001) and/or Nishi et al. (2003). The polynomial fit is* $D = 0 + 0.0049321T - 8.7181 \times 10^{-7}T^2$.
[b] *https://www.netzsch-thermal-analysis.com/us/materials-applications/thermoelectric/pure-copper-thermal-diffusivity/.*
[c] *Near melting. http://www.knowledgedoor.com/2/elements_handbook/iron.html*
[d] *Similar values were obtained by Stankus and Savchenko (2009).*
Notes: Fits in Fig. 9.14 are constrained by the origin.

sound speeds, small cations, and are close packed in the crystalline phase. The melts with large D involve large cations, and generally have more open structures in the solid state. For example, solid crystalline Ga has an orthorhombic structure; solid Hg is trigonal; solid Sn is tetragonal, and solid Cu and Pb are FCC. Solid Ga and Sn having higher sound speeds than Pb and Hg apparently contributes to their higher D values.

9.4.2 Thermal Diffusivity of Alloys

Minor amounts of alloying affect thermal diffusivity values strongly near the end members, but these alloying elements only weakly affect the derivative with temperature (Fig. 9.15). Values at 298K are reduced as the impurity concentration increases, as shown in previous data on K (Fig. 9.4). The initial change in D with T is negative for low amounts of impurities. However, near the middle of the Fe–Ni binary and for large amounts of various cations, D_{lat} of the alloys increases with T. This behavior is observed for both Fe and Cu alloys (parts A and B in Fig. 9.15, respectively). Inconnel behaves similarly to the iron alloys, even though this alloy is mostly Ni. Regarding core materials, these findings show that the type of substitutions in the core should be unimportant if the amounts are moderate to substantial. This situation seems likely given the variety of Ni contents of meteorites (e.g., Buchwald, 1975), coupled with the high temperatures and pressures in the Earth's interior.

Thermal diffusivity of alloys are mostly described by linear fits with temperature (Table 9.2). When ferromagnetism is absent, as occurs for the liquids, the trends point towards the origin. Strong ferromagetism elevates thermal diffusivity of alloys, as seen for the AlNiCo magnet.

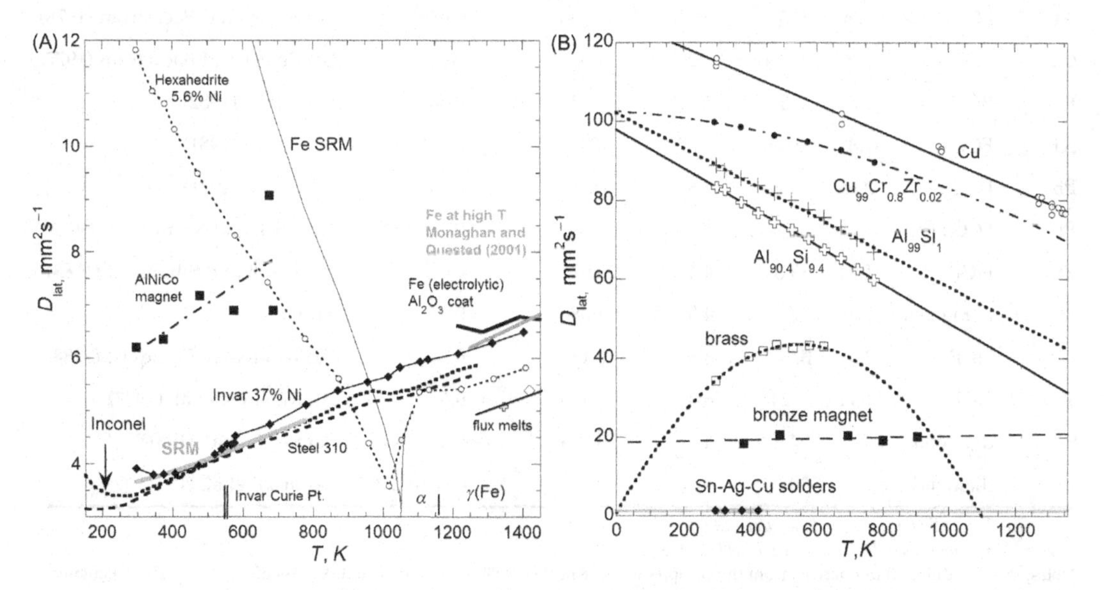

FIGURE 9.15 Thermal diffusivity of alloys from LFA. Data from Criss and Hofmeister (2017) unless listed otherwise. (A) Iron and related alloys. Diamonds: SRM 1460 from Henderson et al. (1998a); dashed line: steel 310 from Blumm et al. (2007); heavy dotted line: inconel from Blumm et al. (2003). (B) Copper alloys. Gray and diamonds: Sn-rich solders from Kim et al. (2009). Dots: slightly impure copper from Rohde et al. (2013). For brass, a least squares polynomical fit is shown that is not constrained to go through the origin.

9.4.3 Thermal Conductivity as a Function of Pressure

Pressure determinations for metals involve thermal conductivity measurements. Reliable data are obtained near ambient pressure (see Chapter 4).

For lattice conduction, dimensional analysis yields (Hofmeister, 2007):

$$\frac{\partial \ln(K)}{\partial P} = \frac{\partial \ln(B_T)}{\partial P} + \frac{\partial \ln(C_V)}{\partial P} + \frac{\partial \ln(\tau)}{\partial P} \tag{9.11}$$

where B is the bulk modulus. The pressure dependence of the heat capacity should be small because the thermodynamic identity:

$$\frac{\partial C_P}{\partial P} = -TV\left(\alpha^2 + \frac{\partial \alpha}{\partial T}\right) \tag{9.12}$$

includes the square of the thermal expansivity, which is $\sim 10^{-10}\,\mathrm{K}^{-2}$ (e.g., Touloukian et al., 1975). This is particularly true for metals, which have high density ($1/V$) and high

TABLE 9.4 Pressure Dependence of Thermal Conductivity Near Ambient Conditions

Phase	Structure	γ_{th}	B_T GPa	$\partial B/\partial P$	$\partial \ln K/\partial P_{calc}$ GPa^{-1}	$\partial \ln K/\partial P_{meas}$ GPa^{-1}	Reference
Al	FCC	2.22	72	5.8[a]	0.081	≈0.065	Sundqvist and Bäckstöm (1977b)
Ag	FCC	2.34	103	5.87[a]	0.057	0.040[b]	Sundqvist and Bäckström (1978)
Au	FCC	2.90	163	5.9[a]	0.036	0.039[b]	Sundqvist and Bäckström (1978)
Cu	FCC	2.02	137.4	5.29[a]	0.039	0.031[b]	Sundqvist and Bäckström (1977a)
Fe	BCC	1.7	163	5.29[a]	0.032	0.035	Sundqvist (1982)
Ni	FCC	1.8	190.5	6.07[a]	0.032	0.041	Sundqvist (1981)
Pb	FCC	2.65	43.2	5.8	0.13	0.18	Bridgman (1921)
Pb	FCC					0.27 at 8K	Itskevick and Kraidenov (1975)
Zn	HCP	1.94	59.8	4.4	0.074	0.087	Jacobsson and Sundqvist (1988)
Sn	tetragonal	2.25	57	4.9	0.086	≈0.13	Bridgman (1921)
Gd	HCP	2.4	37	4.8	0.13	0.22	Jacobsson and Sundqvist (1988)
Te	HCP	1.44	~21	8.4	0.4	≈0.6	Amirkhanov et al. (1973)
Si	diamond		97.9	4.7	0.048	≈0 at 85K	Baranskii et al. (1981)
Ge	diamond		80	≈4	≈0.05	0.04 at 85K	Baranskii et al. (1981)

[a] *Ultrasonic determination; also see Raju et al. (1997).*
[b] *Starr (1938) obtained similar values for the noble metals.*
Notes: Bulk moduli and derivatives from the compilation of Knittle (1995), unless indicated. Values of $\gamma_{th} = \alpha B_T/C_V\rho$ calculated using data in Knittle (1995), Touloukian et al. (1970, 1975), and Touloukian and Buyco (1970).

heat capacity. From the dependence of τ on volume for metals at ambient conditions (Fig. 9.13) and the definition of the bulk modulus:

$$\frac{\partial \ln \tau}{\partial P} = \frac{1}{3B_T} \tag{9.13}$$

Thus, the pressure dependence for lifetime involving the lattice is negligibly small, and $\partial \ln(K)/\partial P = \partial \ln(B_T)/\partial P$. The pressure dependence of heat transport is controlled by the compression of the bonds.

For metals, an electronic mechanism would involve Fermi velocities and electronic heat capacities which are negligibly affected by pressure. Eq. (9.2) yields:

$$\frac{\partial \ln(\tau_{ele})}{\partial P} = -\frac{1}{3B_T} + \frac{\partial \ln(\sigma)}{\partial P} \tag{9.14}$$

Electrical conductivity decreases as pressure increases as demonstrated experimentally (e.g., Eiling and Schilling, 1981). Thus, if electrons conducted heat in metals, the pressure dependence of the thermal conductivity would be negative.

Table 9.4 shows that thermal conductivity increases with pressure, which refutes the electronic mechanism. Measured derivatives are in good agreement with the predicted lattice value and are similar to those of non-metallic elements, as shown.

9.5 SUMMARY AND FUTURE WORK

Virtually all available data on metals are inconsistent with the carriers being electrons. The only support for an electronic mechanism is reasonable agreement at a single temperature for some metals, but this agreement rests on a factor of 3 error in the Lorenz number, traceable to historic models depicting steady-state conditions in negligible thermal gradients (Chapter 5). The underlying problem is that the hypothesis of electron carriers was not tested until Criss and Hofmeister (2017) made transient measurements over short times whereby fast electrons could be resolved from the slower lattice conduction.

The basic misunderstanding is that describing the process of equilibration is possible through many different techniques. In contrast, only transient methods are capable of resolving rapid transport and/or small scale behavior: steady-state or periodic techniques cannot. It is also not widely recognized that transient measurements mostly depict near-equilibrium conditions.

This misunderstanding permeates thermodynamics, which is largely based on equilibrium and reversible processes. The latter is impossible (Chapters 1 and 5) whereas quasi-equilibrium is possible with sufficient time. The latter behavior can be treated with an adiabatic approach. However, transient events involve disequilibrium. Modeling this behavior of electrons in metals is possible because of independent behavior, and requires considering effective temperatures of electrons being greater than ambient. Spectroscopic experiments show that the electrons can be quite hot.

This chapter, which is based on our recent study (Criss and Hofmeister, 2017) shows that the lattice behavior regulates heat transport in metals. The data in this chapter can be used to estimate transport properties for the core, but our understanding is not complete because certain data are missing. For example, optical spectra of the metals are germane, but these data are at 298K, and cover few compositions, mostly on elements. Heat transport data depend strongly on impurity levels, which suggests that the available data underestimates D for pure metals since most measurements are from ~1970s, and purification technique have improved since then. We have confirmed results on Pb and Ag, and added new data on Al, Cu, and W. Our study was focused on surveying electron behavior and so did not explore high temperatures. A faster detection system is needed for accurate measurements of the noble metals and others. Again, transient measurements are needed on a variety of lengths to understand the small scale behavior of metals.

References

Aeschlimann, M., Bauer, M., Pawlik, S., Knorren, R., Bouzerar, G., Bennemann, K.H., 2000. Transport and dynamics of optically excited electrons in metals. Appl. Phys. A 71, 485–491.

Amirkhanov, Kh. I., Magomedov, Ya. B., Émirov, S.N., 1973. Influence of hydrostatic pressure on the thermal conductivity of Te. Sov. Phys. Solid State 15, 1015–1016.

Ashcroft, N.W., Mermin, N.D., 1976. Solid State Physics. Holt, Rinehart and Winston, New York, NY.

Aven, M.H., Craig, R.S., Waite, T.R., Wallace, W.E., 1956. Heat capacity of titanium between 4°K and 15°K. Phys. Rev. 102, 1263–1264.

Baranskii, P.I., Kogutyuk, P.P., Savyak, V.V., 1981. Thermal conductivity of n-type germanium and silicon under strong uniaxial elastic deformation. Sov. Phys. Semicond. 15, 1061–1062.

Bauer, M., Marienfeld, A., Aeschlimann, M., 2015. Hot electron lifetimes in metals probed by time-resolved two-photon photoemission. Prog. Surface Sci. 90, 319–376.

Beaurepaire, E., Merle, J.-C., Daunois, A., Bigot, J.-Y., 1996. Ultra fast spin dynamics in ferromagnetic nickel. Phys. Rev. Lett. 76, 4250–4253.

Bessemer, H., 1905. Sir Henry Bessemer, F.R.S. An autobiography. Offices of Engineering, London, U.K.

Bigot, J.-Y., Vomir, M., Beaurepaire, E., 2009. Coherent ultrafast magnetism induced by femtosecond laser pulses. Nature Phys. 5, 515–520.

Blumm, J., Lindemann, A., Niedrig, B., 2003. Measurement of the thermophysical properties of an npl thermal conductivity standard inconel 600. Proc. 17th European Conference on Thermophysical Properties 621–626.

Blumm, J., Lindemann, A., Niedrig, B., Campbell, R., 2007. Measurement of selected thermophysical properties of the NPL certified reference material stainless steel 310. Intl. J. Thermophys. 28, 674–682.

Bridgman, P.W., 1921. Collected Experimental Papers, vol. 3. Harvard University Press, Cambridge, MA, pp. 1471–1521.

Bron, W.E., 2012. Ultrasonic Processes in Condensed Matter. Springer, New York, UK.

Brorson, S.D., Kazeroonian, A., Moodera, J.S., Face, D.W., Cheng, T.K., Ippen, E.P., et al., 1990. Femtosecond room-temperature measurement of the electron-phonon coupling constant gamma in metallic superconductors. Phys. Rev. Lett. 64, 2172–2175.

Buchwald, V.F., 1975. Handbook of Iron Meteorites, 3 volumes. University of California Press, Berkeley, CA.

Bungardt, K., Spyra, W., 1965. Thermal conductivity of alloyed and plain steels and alloys at temperatures between 20 and 700 degrees. Arch. Eisenbuetten 36, 257-–2267.

Burns, G., 1990. Solid State Physics. Academic Press, San Diego, CA.

Chai, M., Brown, J.M., Slutsky, L.J., 1996. Thermal diffusivity of mantle minerals. Phys. Chem. Minerals 23, 470–475.

Corkum, P.B., Brunel, F., Sherman, N.K., Srinivasan-Rao, T., 1988. Thermal response of metals to ultrashort-pulse laser excitation. Phys. Rev. Lett. 61, 2886–2889.

Criss, E.M., Hofmeister, A.M., 2017. Isolating lattice from electronic contributions in thermal transport measurements of metals and alloys and a new model. Int. J. Mod. Phys. B 31, Available from: https://doi.org/10.1142/S0217979217502058No. 175020.

Drude, P., 1900. Zur elektronentheorie der metalle. Ann. Phys 306, 369–402.

Eiling, A., Schilling, J.S., 1981. Pressure and temperature dependence of electrical resistivity of Pb and Sn from 1-300K and 0-10 GPa-use as continuous resistive pressure monitor accurate over wide temperature range; superconductivity under pressure in Pb, Sn, and In. J. Phys. F: Metal Phys 11, 623–639.

Elsayed-Ali, H.E., Norris, T.B., Pessot, M.A., Mourou, G.A., 1987. Time-resolved observation of electron-phonon relaxation in copper. Phys. Rev. Lett. 58, 1212–1215.

Ewing, C.T., Walker, B.E., Grand, J.A., Miller, R.R., 1957. Thermal conductivity of metals. Liquid metals technology-part 1. Chem. Eng. Progress Symp. Series 53, 19–24.

Fourier, J.B.J., 1822. In: Freeman, A. (Ed.), Théorie Analytique de la Chaleur (translated in 1955 as The Analytic Theory of Heat). Chez Firmin Didot, Paris. Dover Publications Inc, New York.

Gale, W.F., Totemeier, T.C., 2004. Smithells Metals Reference Book, Eighth Edition Elsevier, Oxford, UK.

Gopal, E.S.R., 1966. Specific Heats at Low Temperatures. Plenum Press, New York.

Gorbatov, V.I., Polev, V.F., Korshunov, I.G., Taluts, S.G., 2012. Thermal diffusivity of iron at high temperatures. High Temp. 50, 292–294.

Henderson, J.B., Giblin, F., Blumm, J., Hagemann, L., 1998a. SRM 1460 series as a thermal diffusivity standard for laser flash instruments. Int. J. Thermophys. 19, 1647–1656.

Henderson, J.B., Hagemann, L., Blumm, J., 1998b. Development of SRM 8420 series electrolytic iron as a thermal diffusivity standard. Netzsch Applications Laboratory Thermophysical Properties Section Report No. I-9E .

Hertz, H., 1887. Ueber den Einfluss des ultravioletten Lichtes auf die electrische Entladung" [On an effect of ultra-violet light upon the electrical discharge]. Annalen der Physik. 267, 983–1000. Available from: https://doi.org/10.1002/andp.18872670827.

Ho, C.Y., Ackerman, M.W., Wu, K.Y., Oh, S.G., Havill, T.N., 1978. Thermal conductivity of ten selected binary alloy systems. J. Phys. Chem. Ref. Data 7, 959–1177.

Ho, C.Y., Ackerman, M.W., Wu, K.Y., Havill, T.N., Bogaard, R.H., Matula, R.A., et al., 1983. Electrical resistivity of ten selected binary alloy systems. J. Phys. Chem. Ref. Data 12, 183–322.

Hofmeister, A.M., 2007. Pressure dependence of thermal transport properties. Proc. Natl Acad. Sci. USA 104, 9192–9197.

Hofmeister, A.M., 2010. Scale aspects of heat transport in the diamond anvil cell, in spectroscopic modeling, and in Earth's mantle: implications for secular cooling. Phys. Earth Planet. Inter. 180, 138–147.

Hofmeister, A.M., Dong, J.J., Branlund, J.M., 2014. Thermal diffusivity of electrical insulators at high temperatures: evidence for diffusion of phonon-polaritons at infrared frequencies augmenting phonon heat conduction. J. Appl. Phys. 115 (No. 163517). Available from: https://doi.org/10.1063/1.4873295.

Hust, J.G., Lankford, A.B., 1984. Update of thermal conductivity and electrical resistivity of electrolytic iron, tungsten, and stainless steel. National Bureau of Standards Spec. Pub 260–290. 1-71.

Hust, J.G., Sparks, L.L., 1973. Lorenz ratios of technically important metals and alloys. NBS Technical Note 634 (U.S. Government Printing Office, Washington, DC, 1973).

Ingersoll, L.R., 1920. Some physical properties of nickel-iron alloys. Phys. Rev. 16, 126–132.

Itskevich, E.S., Kraidenov, V.F., 1975. Low-temperature thermal conductivity of lead under pressure. High Temp. High Press. 7, 654–655.

Jacobsson, P., Sundqvist, B., 1988. Thermal conductivity and Lorenz function of zinc under pressure. Int. J. Thermophys. 9, 577–585.

Kim, S.K., Lee, J., Jeon, B.M., Jung, E., Lee, S.H., Kang, K.K., et al., 2009. Thermophysical properties of Sn-Ag-Cu based Pb-Free solders. Int. J. Thermophys 30, 1234–1241.

Kim, Y.M., 2016. The behavior of thermal diffusivity change according to the heat treatment in Al-Si binary system. J. Alloys Compd. 687, 54–558.

Kittel, C., 1971. Introduction to Solid State Physics, Fourth ed. John Wiley and Sons, New York, NY, pp. 224–226, 239-265.

Klemens, P.G., Williams, R.K., 1986. Thermal conductivity of metals and alloys. Int. Metals Rev 31, 197–215.

Knittle, E., 1995. Static compression measurement of equation of state. In: Ahrens, T.J. (Ed.), Mineral Physics and Crystallography. A Handbook of Physical Constants, American Geophysical Union, Washington D.C, pp. 98–142.

Kruglyak, V.V., Hicken, R.J., Ali, M., Hickey, B.J., Pym, A.T.G., Tanner, B.K., 2005. Measurement of hot electron momentum relaxation times in metals by femtosecond ellipsometry. Phys. Rev. B 71, 233104.

Ledbetter, H.M., Reed, R.P., 1973. Elastic properties of metals and alloys, I. iron, nickel, and iron-nickel alloys. J. Phys. Chem. Ref. Data 2, 531–618.

Lee, S., Hippalgaonkar, K., Yang, F., Hong, J., Ko, C., Suh, J., et al., 2017. Anomalously low electronic thermal conductivity in metallic vanadium dioxide. Science 355, 371–374. Available from: https://doi.org/10.1126/science.aag0410.

Lorenz, L., 1872. Determination of heat temperature in absolute units, Ann. Phys. Chem. (Leipzig) 147, 4429–4451.

Lynch, D.W., Hunter, W.R., 1991. An introduction to the data for several metals. In: Palik, E.D. (Ed.), Handbook of Optical Constants of Solids II. Academic Press, Orlando, FL, pp. 341–377.

Meissner, W., 1915. Thermische und elektrische Leitfähigkeit einiger Metalle zwischen 20° und 373° absolut. Ann. der Physik 47, 1001–1058.

Monaghan, B.J., Quested, P.N., 2001. Thermal diffusivity of iron at high temperature in both the liquid and solid states. ISIJ Int. 41, 1524–1528.

Moore, J.P., Kollie, T.G., Graves, R.S., McElroy, D.L., 1971. Thermal transport properties of ordered and disordered Ni3Fe. J. Appl. Phys. 42, 3114–3120.

Nishi, T., Shibata, H., Ohta, H., Waseda, Y., 2003. Thermal conductivities of molten iron, cobalt, and nickel by laser flash method. Metal. Mater. Trans. A 34, 2801–2807.

Ordal, M.A., Bell, R.J., Alexander, R.W., Long, L.L., Querry, M.R., 1985. Optical properties of fourteen metals in the infrared and far infrared: Al, Co, Cu, Au, Fe, Pb, Mo, Ni, Pd, Pt, Ag, Ti, V, and W. Appl. Opt 24, 4493–4499.

Parker, W.J., Jenkins, R.J., Butler, C.P., Abbot, G.L., 1961. Flash method of determining thermal diffusivity, heat capacity, and thermal conductivity. J. Appl. Phys. 32, 1679.

Patz, A., Li, T., Ran, S., Fernandes, R.M., Schmalian, J., Bud'ko, S.L., et al., 2013. Ultrafast observation of critical nematic fluctuations and giant magnetoelastic coupling in iron pnictides. Nat. Commun 5, No. 3229.

Raju, S., Mohandas, E., Raghunathan, V.S., 1997. The pressure derivative of bulk modulus of transition metals: an estimation using the method of model potentials and a study of the systematics. J. Phys. Chem Solids 58, 1367–1373. I997.

Rohde, M., Hemberger, F., Bauer, T., Blumm, J., Fend, T., Häusler, T., et al., 2013. Intercomparison of thermal diffusivity measurements on CuCrZr and PMMA. High Temp.-High Press 42, 469–472.

Rudkin, R.L., Jenkins, R.J., Parker, W.J., 1962. Thermal diffusivity measurements on metals at high temperatures. Rev. Sci. Instrum. 33, 21–24.

Schoenlein, R.W., Lin, W.Z., Fujimoto, J.G., Eesley, G.L., 1987. Femtosecond studies of nonequilibrium electronic processes in metals. Phys. Rev. Lett. 58, 1680–1683.

Schriempf, J.T., 1972. A laser-flash technique for determining thermal diffusivity of liquid metals at elevated remperatures. Rev. Scien. Instrum. 43, 781–786.

Schriempf, J.T., 1973. Thermal diffusivity of liquid gallium. Solid State Commun. 13, 651–653.

Silverman, L., 1953. Thermal conductivity data presented for various alloys and metals up to 900 degrees. J. Metals. 5, 631–632.

Smith, C.S., Palmer, E.W., 1935. Thermal and electric conductivities of copper alloys. Trans. AIME 117, 225–243.

Sommerfeld, A., 1928. Zur Elektronentheorie der Metalle auf Grund der Fermischen Statistik. Zeit. Physik 47, 1–32.

Stankus, S.V., Savchenko, I.V., 2009. Laser flash method for measurement of liquid metals heat transfer coefficients. Thermophys. Aeromech. 16, 589–592.

Starr, C., 1938. The pressure coefficient of thermal conductivity of metals. Phys. Rev. 54, 210–246.

Sundqvist, B., 1981. Thermal conductivity and Lorentz number of nickel under pressure. Solid State Commun. 37, 289–291.

Sundqvist, B., 1982. Transport properties of iron and nickel under pressure. In: Backman C.M., Johannison, T., Tegnér, T. (Eds), High Pressure in Research and Industry, 8th AIRAPT Conference, 19th EHPRG Conference, vol. 1, pp. 432–433.

Sundqvist, B., Bäckström, G., 1977a. Thermal conductivity of copper under high pressure. High Temp.- High Press 9, 41–48.

Sundqvist, B., Bäckström, G., 1977b. Pressure dependence of the thermal conductivity of aluminum. Solid State Commun. 23, 773–775.

Sundqvist, B., Bäckström, G., 1978. Thermal conductivity of gold and silver at high pressures. J. Phys. Chem. Solids 39, 1133–1137.

Touloukian, Y.S., Buyco, E.H., 1970. Specific Heat: Metallic Elements and Alloys. IFI/Plenum, New York, NY.
Touloukian, Y.S., Powell, R.W., Ho, C.Y., Klemens, P.G., 1970. Thermal Conductivity: Non-Metallic Solids. IFI/Plenum, New York, NY.
Touloukian, Y.S., Powell, R.W., Ho, C.Y., Nicolaou, M.S., 1973. Thermal Diffusivity. IFI/Plenum, New York, NY.
Touloukian, Y.S., Kirby, R.K., Taylor, R.E., Desai, P.D., 1975. Thermal Expansion: Metallic Elements and Alloys. IFI/Plenum, New York, NY.
Transtrum, M.K., Machta, B.B., Brown, K.S., Daniels, B.C., Myers, C.R., Sethna, J.P., 2015. Perspective: sloppiness and emergent theories in physics, biology, and beyond. J. Chem. Phys. 143. Available from: https://doi.org/10.1063/1.4923066.
Wakeham, N., Bangura, A.F., Xu, X., Mercure, J.-F., Greenblatt, M., Hussey, N.E., 2011. Gross violation of the Wiedemann–Franz law in a quasi-one-dimensional conductor. Nature Commun. 2 (No. 396).
Watson, T.W., Robinson, H.E., 1961. Thermal conductivity of some commercial iron-nickel alloys. J. Heat Transfer. 83, 403–407.
Wiedemann, G., Franz, R., 1853. Ueber die Wärme-Leitungsfähigkeit der Metalle. Annalen der Physik 165, 497–531.
Zheng, X., Cahill, D.G., Krasnochtchekov, P., Averback, R.S., Zhao, J.C., 2007. High-throughput thermal conductivity measurements of nickel solid solutions and the applicability of the Wiedemann-Franz law. Acta Mater. 55, 5177–5185.
Zheng, X., Cahill, D.G., Zhao, J., 2010. Effect of MeV ion irradiation on the coefficient of thermal expansion of Fe–Ni Invar alloys: A combinatorial study. Acta Mater. 58, 1236–1241.
Ziman, J.M., 1962. Electrons and Phonons: The Theory of Transport Phenomena in Solid. Clarendon Press, Oxford.

Websites

Data on properties of elements exist on many websites. We used: http://www.periodictable.com/Properties/A/ThermalConductivity.v.html

ASTM standard metals. https://www.astm.org/Standards/metal-standards.html

Data on electrical resistivity and other properties of many alloys http://eddy-current.com/conductivity-of-metals-sorted-by-resistivity/, http://www.matweb.com/

Also useful are manufacturer's data http://www.advancedmagnetsource.com/Alnico.pdf.

Another source of data on elements is http://www.knowledgedoor.com/2/elements_handbook/debye_temperature.html

Data on copper https://www.netzsch-thermal-analysis.com/us/materials-applications/thermoelectric/pure-copper-thermal-diffusivity/

Data from the Washington University laboratory are available at http://epsc.wustl.edu/~hofmeist/thermal_data/

CHAPTER

10

Heat and Mass Transfer in Glassy and Molten Silicates

*Alan G. Whittington**

Department of Geological Sciences, University of Missouri, Columbia, MO, United States

OUTLINE

Heat and mass transfer in amorphous materials, such as silicate glasses and melts, is important in industries that process and work with ceramics, metals, and glasses. Silicate melts are also of great importance to geoscience, as a primary means of differentiating the Earth and other silicate-rich planetary bodies. Furthermore, silicate melts have the potential to affect millions of people through volcanic activity, and understanding the properties

* This work was supported by the National Science Foundation, EAR- 1220051 and EAR-0911116 in particular.

DOI: https://doi.org/10.1016/B978-0-12-809981-0.00010-3

of silicate melts over the wide range of composition and temperature relevant to volcanism is an urgent priority (National Academies of Sciences, Engineering, and Medicine, 2017).

Due to the wide-ranging importance of silicate glass and melt properties, many reviews exist; we mention only a handful that are recommended as introductions to the topic. Mills (1995) reviewed the viscosity of slag melts, and Mills and Susa (1995) reviewed their thermal conductivity. The structure and properties of silicate melts and glasses are thoroughly reviewed in books by Stebbins et al. (1995) and Mysen and Richet (2005), primarily from a geological perspective. The review paper by Ni et al. (2015) includes more recent developments in transport properties of silicate melts, including electrical conductivity, viscosity, chemical diffusivity, and heat transfer. While viscosity and mass diffusion can be related in some melts, as can electrical conductivity and mass diffusion, the much more rapid rate of heat transport indicates that it is not closely related to the others. As reviewed in previous chapters in this book, heat is transferred by fundamentally different mechanisms than chemical diffusion or viscous flow.

To aid in assessing the effect of composition on thermal, thermodynamic, and rheological properties, we will focus on three subsets of silicate glasses and melts. Their degree of polymerization can be approximated by the theoretical ratio of nonbridging oxygens to tetrahedral cations (NBO/T; see Mysen et al., 1982). One subset is materials of alkali feldspar-composition $XAlSi_3O_8$ (where X = Li, Na, K), all of which are nominally highly polymerized (NBO/T = 0), indicating that multiple rings and other frameworks are the dominant structural units. The second subset is materials of pyroxene composition $(Ca,Mg)SiO_3$, illustrating the effect of mixing between two end-members with different divalent cations, at a constant nominal NBO/T of 2 (indicating that individual rings and chains of tetrahedra are the dominant structural units). The third is a group of geologically relevant melts that span a wide range of NBO/T, from 0 (silica-rich rhyolites) to >1 (silica-poor basalts).

10.1 THERMAL DIFFUSIVITY AND CONDUCTIVITY OF GLASSES

10.1.1 Structure of Glasses

Glasses are dense solids that possess the structure of a liquid, but lack the ability to flow. The chief difference between glass and crystal is the lack of long-range order in glasses. Both glasses and crystals have atoms that obey the fundamental rules of charge balance, so from the perspective of the nearest-neighbor and next-nearest neighbor environments, their structures look very similar. Bond lengths and angles in crystals are tightly defined and show almost no variation except in the vicinity of defects and substituted ions, while bond lengths and angles in the glass are distributed around a similar mean value to those in the crystal. At distances of three neighbor cations and further, the radial distribution functions of many silicate glasses diverge strongly from those of the crystals, and longer-range order is generally lacking, so that glasses are isotropic.

In detail, the structure of glasses is rather more complex than this simple description. Silicate glasses are built primarily of SiO_4 tetrahedra, which in many cases share their oxygen atoms with adjacent SiO_4 tetrahedra, in which case those are bridging oxygens. In other cases, the negative charge of the corner oxygen is balanced by a metal cation, in which case that is an NBO. If the metal cation is divalent, for example Mg^{2+} or Ca^{2+}, it can bond with two NBO simultaneously. Trivalent cations such as Al^{3+} can also occupy a tetrahedral site, as long as they are also charge-balanced by an additional metal cation, although Al–O bonds are weaker than Si–O bonds so that Al–O–Al bonds are largely avoided (the "aluminum-avoidance principle;" Loewenstein, 1954). Aluminosilicate melts such as alkali feldspars typically exhibit ring structures of between three and eight tetrahedral members, with few NBO, so they are said to be highly polymerized. Melts richer in divalent metal cations, such as the CaO–MgO–Al_2O_3–SiO_2 system commonly used to approximate slag compositions, contain a larger number of NBO, and are less polymerized. They contain fewer framework- or ring-structures and more chain-like structures. It must be emphasized that the NBO/T parameter is only a rough indicator of the actual degree of polymerization, which can be better quantified using NMR spectroscopy (Stebbins, 1995). Additional spectroscopic techniques, combined with molecular dynamics simulations, have led to the modern view of glass and melt structure, reviewed by Henderson (2005). The modified random network model (Greaves, 1985) combines a random continuous covalent SiO_2 network with an interpenetrating lattice of network-modifying metal oxide (e.g., Na_2O, when considering the binary Na_2O–SiO_2 system). Local ordering of network-modifying cations around NBO leads to the existence of percolation channels where bonding is primarily ionic. To date, few spectroscopic or theoretical studies have tackled complex compositions such as are of interest to geologists.

10.1.2 Thermal Diffusivity of Glasses

The simplest silicate glass, at least in terms of chemical formula, is pure silica (SiO_2). This glass has many industrial applications and is available in many specifications from different synthesis techniques, each resulting in particular concentrations of different elemental impurities. The properties of amorphous SiO_2 are very sensitive to the amounts of these impurities, with the glass transition temperature (T_g) and dynamic viscosity (η) dropping abruptly, with the addition of just a few thousand ppm of H_2O or Na_2O, for example (Leko et al., 1977). The thermal diffusivity is also sensitive to impurities, with room-temperature thermal diffusivity values $\sim 0.85 \pm 0.05$ mm^2 s^{-1}, and to strain, with annealed (unstrained) glasses having higher values of D (Hofmeister and Whittington, 2012).

More chemically complex glasses show a wider range in room-temperature thermal diffusivity, but none have a higher D than pure SiO_2 (Fig. 10.1). In general, more polymerized glasses such as feldspars and rhyolites have a higher room-temperature thermal diffusivity ($\sim$0.6–0.7 mm^2 s^{-1}) than less polymerized glasses such as pyroxenes, soda–lime silicates and basalts ($\sim$0.45–0.65 mm^2 s^{-1}). A graph of D_{298} versus NBO/T, as a proxy for polymerization, reveals the trend but is not suited for making any predictions (Fig. 10.1A). When the melt fragility is calculated from viscosity data (see Section 10.4), a tighter trend emerges (Fig. 10.1B). There is still scatter due to chemical substitutions such as Mg for Ca in (Ca,Mg)SiO_3 glasses, with $MgSiO_3$ having a very high D (0.62 mm^2 s^{-1}) compared to $CaSiO_3$ (0.46 mm^2 s^{-1}). The intermediate diopside composition ($CaMgSi_2O_6$) is much closer to the Ca end-member than the Mg end-member. Another substitution series is the (Li,Na,K)$AlSi_3O_8$ glasses, where the lightest cation (Li) again has the highest D, and the heaviest

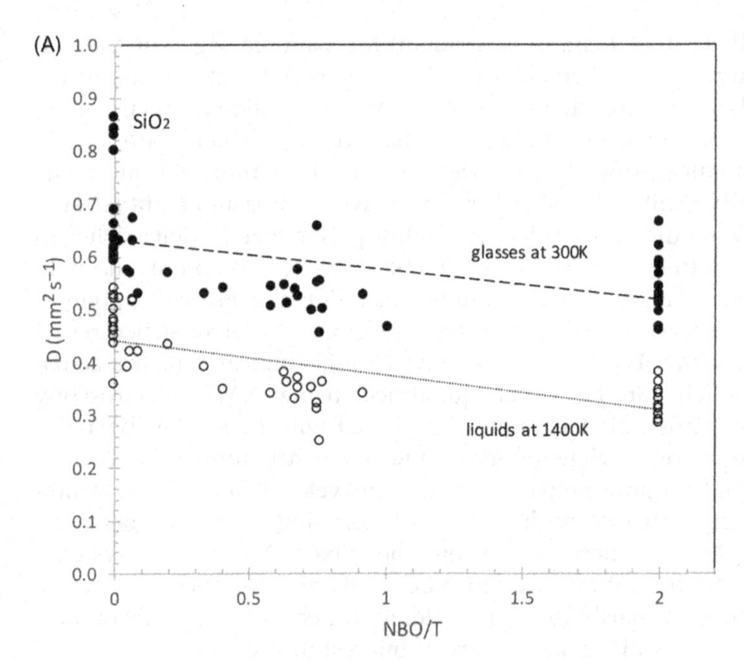

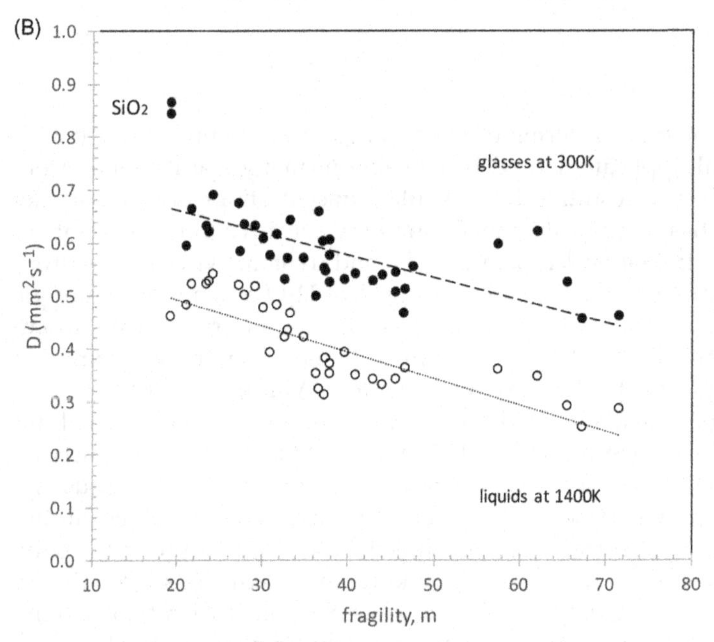

FIGURE 10.1 D_{298} and D_{liq} (A) versus NBO/T and (B) versus fragility, defined as $m = d(\log_{10}\eta)/d(T_g/T)$ for $T = T_g$.

cation (K) has the lowest *D*, but the effect is smaller and within this series *D* does not correlate inversely with fragility.

On heating above room temperature, thermal diffusivity of glasses decreases steeply at first, then more gradually, in many cases approaching a plateau (Fig. 10.2). This pattern

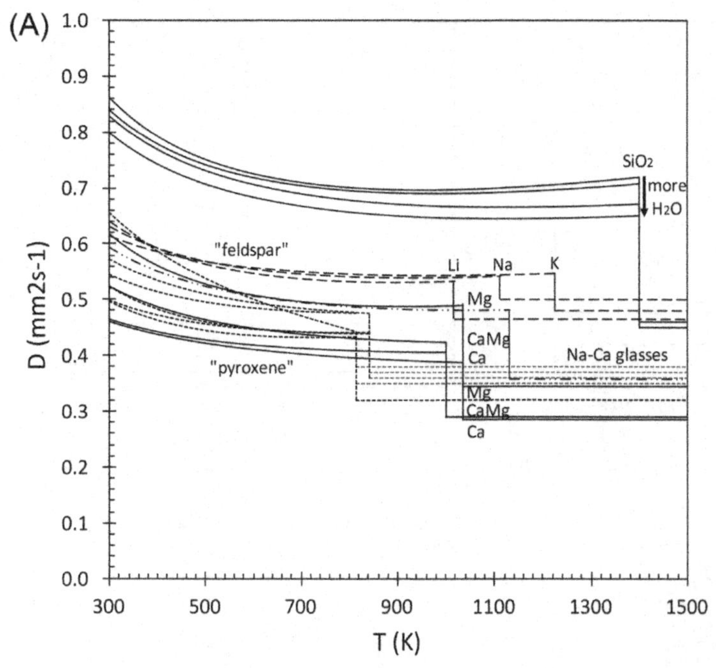

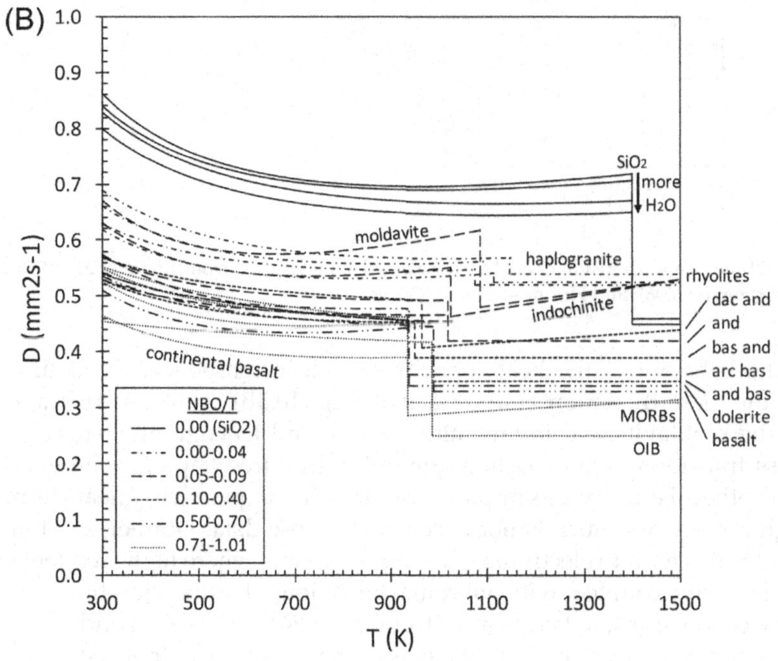

FIGURE 10.2 Thermal diffusivity versus temperature for (A) simple compositions and soda–lime silicates; (B) geologically relevant melts. Fit parameters in Table 10.1.

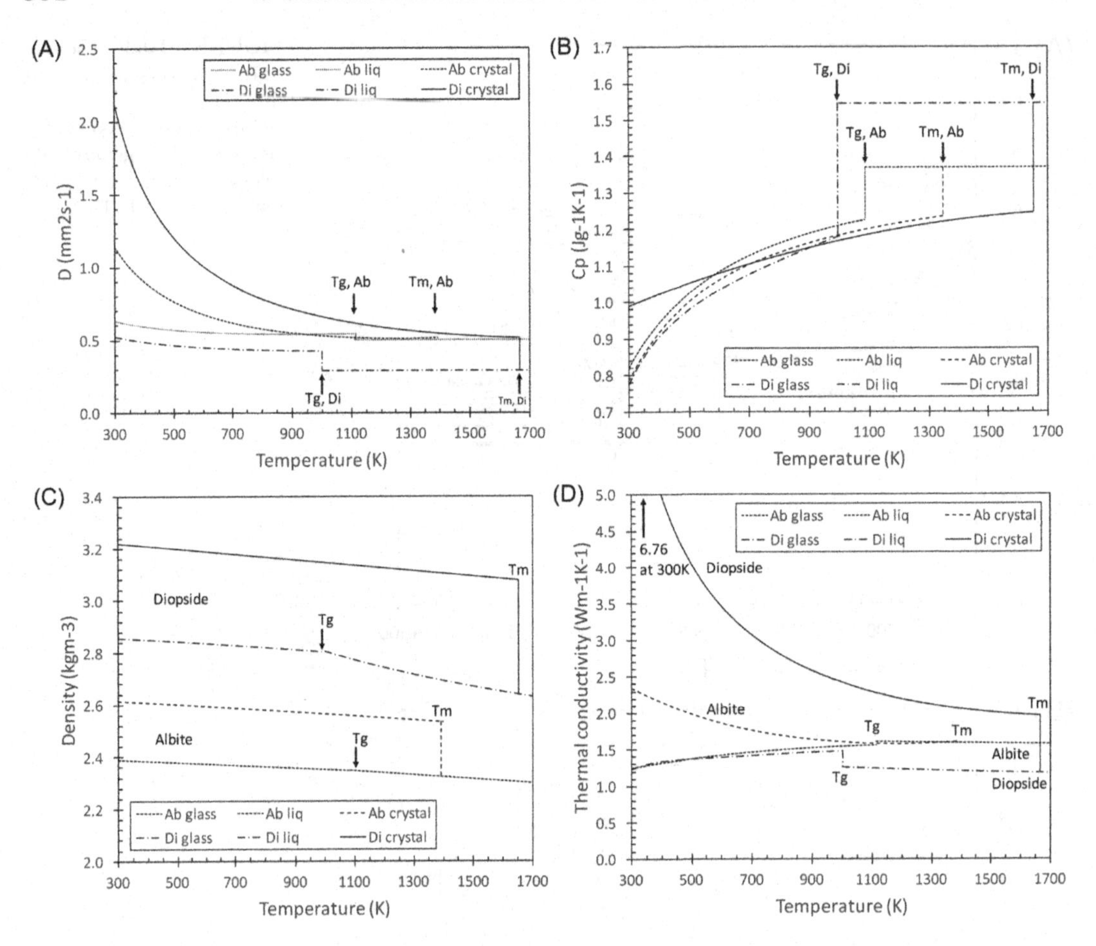

FIGURE 10.3 *D*, C_P, and *k* of albite and diopside crystals, glasses, and liquids. Data from Robie et al. (1978); Fei (1995); Hofmeister and Pertermann (2008); and Hofmeister et al. (2009).

approximately mirrors the behavior of the heat capacity, which increases steeply at first, then more gradually as the glass transition is approached (Fig. 10.3B). Three Al-rich and SiO_2-rich compositions, the moldavite and indochinite tektites, and leucogranite, show an increase in *D* as the glass transition is approached, while the iron-free haplogranite glass does not. Fits to *D* for other Fe-bearing samples also require a positive linear term (Table 10.1), even though a positive upturn is not observed in those data. Hofmeister et al. (2014) attributed the need for this term to electronic–vibronic coupling, where high-frequency electronic transitions of Fe^{2+} are coupled with vibronic transitions. Lattice vibrations are involved with intervalence charge transfer, because electrons cross Fe^{2+}–O–Fe^{3+} bonds.

Compared to their chemically equivalent crystals, glasses have very low thermal diffusivity at room temperature, being only 50% of the crystalline equivalent for $NaAlSi_3O_8$, and ~25% for $CaMgSi_2O_6$ (Fig. 10.3A). This difference becomes much less pronounced at higher temperature, as thermal diffusivity of crystals drops much more steeply than that

TABLE 10.1 Thermal Diffusivity of Glasses, Fit to $D = FT^G + HT$, in $mm^2\ s^{-1}$ for $T < T_{melt}$- ~50K

Name/Composition	L (mm)	D_{298} ($mm^2\ s^{-1}$)	F	G	H	χ^2	T_{melt} (K)	D_{melt} ($mm^2\ s^{-1}$)	Ref.
SiO_2 Infrasil#2 run 4	0.99	0.86	7.9085	−0.40643	0.00020917	0.0004		n.d.	a
Dry Infrasil#2 Run 6	0.90	0.87	8.6602	−0.41814	0.00021457	0.0001	1940	0.46	a
Low H_2O GE124	0.8−1.6	0.85	6.5439	−0.37236	0.00019038	0.0002	1833	0.45	a
Medium H_2O OC7890	1.036	0.84	5.7063	−0.34826	0.0001525	0.0004	1710	0.435	a
High H_2O herasil	1.005	0.800	5.2903	−0.34128	0.00014575	0.0003	1750	crystallized	a
$LiAlSi_3O_8$	0.645	0.650	6.0368	−0.40807	0.000172	0.0002	1200−1320	0.465	b
$NaAlSi_3O_8$	0.789	0.637	3.8003	−0.32703	0.00014244	0.00009	1300−1500	0.50	b
$NaAlSi_3O_8$ remelt	0.793	0.59	4.0808	−0.36002	0.00019629	0.00005		n.d.	b
$KAlSi_3O_8$ W, run 3, dry	0.86	0.620	2.2851	−0.2397	0.00010671	0.0026	1400−1550	0.48	c
$KAlSi_3O_8$ S	0.71	0.596	1.8492	−0.20629	0.000076773	0.0002	1450−1605	0.481	c
$KAlSi_3O_8$ remelt	0.60	0.60	4.8796	−0.38373	0.00019464	0.0006	1400−1660	0.475	c
$CaAl_2Si_2O_8$	0.8−0.9	0.58	4.3031	−0.35805	0.00011838	0.00006	~1300	0.358	b
$LiAlSi_2O_6$ remelt Mad	0.634	0.608	5.4443	−0.39913	0.00013967	0.0001	1150	<0.42[a]	b
$LiAlSi_2O_6$ remelt MG	0.775	0.620	13.776	−0.5676	0.00020521	0.0008	1200	<0.39	b
En = $MgSiO_3$	1.35	0.62	9.098	−0.487	0.000174	<0.0005[b]	1100−1200	0.345	d
$CaMgSi_2O_6$	0.844	0.53	3.305	−0.33255	0.00009162	0.0001	1100−1200	0.29	b
$CaMgSi_2O_6$ remelt	0.878	0.469	1.7703	−0.24343	0.000076518	0.00002	~1100	<0.297	b
Wo = $CaSiO_3$	0.895	0.463	1.559	−0.219	0.000045		1173	0.285	e
Fs_7Wo_1 remelt	0.54	0.598	6.273	−0.429	0.000158		1100−1200	0.36	d
Fs_9Wo_1 remelt	0.41	0.536	4.784	−0.4	0.000131		1100−1200	0.33	d
Fs_{10}	1.12	0.49	10.014	−0.546	0.000213		~1100	0.315	d
$Fs_{11}Wo_1$ remelt	0.78	0.579	8.8	−0.499	0.000185		~1100	0.38	d

(Continued)

TABLE 10.1 (Continued)

Name/Composition	L	D_{298}	F	G	H	χ^2	T_{melt}	D_{melt}	Ref.
	mm	$mm^2\ s^{-1}$					K	$mm^2\ s^{-1}$	
Fs_{21}	0.82	0.514	2.359	−0.278	0.0000995		~1100	0.31	d
$Fs_{22}Wo_4$ remelt[c]	0.44	0.544	3	−0.312	0.000255		1100–1200	0.34	d
Fs_{30}[c]	0.70	0.643	2.739	−0.686	0.000111		~1050	0.38	d
Fs_{40}	0.84	0.678	17.87	−0.587	0.000122		~1000	0.34	d
Fs_{53} has crystals	0.7	0.523	5.376	−0.425	0.000152		~950	<0.39[a]	d
Fs_{54} has crystals	0.98	0.547	11.46	−0.559	0.000228		~1000	0.33	d
Fs_{57} has crystals	0.89	0.517	5.49	−0.437	0.000209		850	<0.37[a]	d
$Fs_{15}Wo_{40}$	0.69	0.5	21.32	−0.688	0.000246		1050	0.30	d
Na-Ca 1895	0.55	0.55	2.9996	−0.31224	0.00013011	0.00008	1000–1200	0.38	f
Na-Ca 1960	0.91	0.66	5.5443	−0.37317	−0.00001388	0.0001	950–1200	0.32	f
Na-Ca Vase	0.576	0.50	5.0046	−0.42463	0.00017506	0.00007	950–1250	0.35	f
Na-Ca Modern	1.124	0.50	2.5057	−0.29559	0.00011506	0.00003	950–1300	0.36	f
Na-Ca 1926 #3	0.848	0.53	6.3079	−0.455	0.0001748	0.00007	940–1170	0.35	f
Na-Ca 1926 #4	0.74	0.57	3.5989	−0.3327	0.00011051	0.00005	1000–1260	0.37	f
Moldavite remelt	0.84	0.68	14.829	−0.56935	0.00031425	0.0006	1250–1420	$0.35 + 1.2 \times 10^{-4}T$	f
Indochinite remelt	0.86	0.54	9.3773	−0.49462	0.00024154	0.0004	1150–1310	$0.32 + 1.4 \times 10^{-4}T$	f
Haplogranite[d]	0.6–0.8	0.70	4.8446	−0.35354	0.00014614	0.0006	1390–1650	0.54	f
Leucogranite remelt	0.77	0.67	5.1196	−0.37323	0.0001706	<0.0005	1370–1600	0.52	f
Rhyolite NCr (extra Al)	0.962	0.626	4.7608	−0.37347	0.00017491	0.0002	1430–1560	0.525	g
Rhyolite remelt NCAR	0.467	0.64	2.9587	−0.28256	0.00012647	<0.0005	1330	0.52	g
Rhyodacite[d]	0.70	0.59	4.4338	−0.36867	0.00010776	<0.0005[b]	1200–1400	0.39	h
Dacite remelt	0.389	0.535	2.2637	−0.2655	0.00010482		1300–1500	0.42	h

Dacite-andesite remelt	0.38	0.57	1.7065	−0.19834	0.000058235	1180–1450	$0.35 + 0.6 \times 10^{-4}T$	h
Andesite[d]	0.886	0.567	2.682	−0.28213	0.0001105	1175–1430	0.42	h
Andesite-basalt[d]	0.875	0.54	1.5661	−0.18983	0.000029777	1150	0.348	h
Basalt-andesite remelt	0.527	0.57	2.9178	−0.31099	0.00011241	1120	0.39	h
Dolerite#3 remelt	0.885	0.543	2.7491	−0.29749	0.00012639	1100	<0.37[a]	h
Dolerite#4 remelt	0.895	0.512	2.5599	−0.29707	0.00011546	1100	0.34	h
Basalt[d]	0.976	0.542	2.5776	−0.28395	0.000083044	1100	0.33	h
Arc basalt remelt	0.71	0.54	8.5283	−0.51659	0.00020721	1080	0.36	h
OIB1 remelt	0.935	0.46	0.67838	−0.0704		1150	~0.25	h
OIB2#1 remelt	1.045	0.53	7.5374	−0.488	0.00019784	1070	<0.36[a]	h
OIB2#2 remelt	1.088	0.49				1400	0.34	h
MORB#1 remelt	0.71	0.56	3.5728	−0.34001	0.00012108	1050–1500[e]	$0.24 + 0.5 \times 10^{-4}T$	h
MORB#2 remelt	0.663					1470	0.31	h
P-MORB remelt	0.846	0.56	1.4342	−0.16726		1500	>0.2[f]	h
Continental remelt	0.748	0.47	4.4707	−0.41267	0.00013284	1030	<0.34[a]	h

[a]*Upper limit provided because the sample crystallized subsequent to melting.*
[b]*All fits for this study were better than $\chi^2 = 0.0005$.*
[c]*Values of fit parameter G for $Fs_{22}Wo_4$ remelt and Wo_{30} were transposed in the original publication and are correct here.*
[d]*Iron-free analog to a natural lava that would have significant iron.*
[e]*Combines melts of wet and dry samples. With the low viscosity, water is not retained upon melting.*
[f]*Lower limit provided because the sample flowed subsequent to melting.*
References: a, Hofmeister and Whittington (2012); b, Hofmeister et al. (2009); c, Pertermann et al. (2008); d, Hofmeister et al.(2014a); e, Hofmeister and Whittington, unpublished data; f, Hofmeister et al. (2014b); g, Romine et al. (2012); h, Hofmeister et al. (2016)
Notes: T_{melt} is about 50K higher than the limit to the glass data. All fits used starting parameters of $F = 100$, $G = 1$, and $H = 0.001$. The values reported for F, G, and H incorporate the digits needed to reproduce the data using Eq. (7.1).

of glasses. Crystals can also have very different thermal diffusivity values along different axes (e.g., Hofmeister and Pertermann, 2008; Hofmeister et al., 2009), while glasses are isotropic.

10.1.3 Thermal Conductivity of Glasses

We calculate the thermal conductivity from the product of thermal diffusivity, density and heat capacity:

$$K = \rho c_P D \tag{10.1}$$

Thermal diffusivity of glasses was discussed in the previous section. Density can be measured to high precision ($\sim$0.1%) at room temperature using the Archimedean method, by weighing in air and again while fully immersed in a liquid of known density. The thermal expansivity of glass is very low ($\alpha \sim 10^{-5}\,K^{-1}$) and has a negligible effect on density (<1%) over the temperature range from ambient to the glass transition (Fig. 10.3). Heat capacity of glasses can be calculated based on the chemical composition since it depends only on the number and type of atoms present, but not on their configuration. Simple linear mixing models predict temperature-dependent heat capacity of glasses very well, typically reproducing data to within experimental uncertainties, that are commonly of the order of $\sim$2% (Richet, 1987). The total uncertainty on the calculated thermal conductivity is therefore expected to be better than $\sim$3%.

At room temperature, thermal conductivity of glass is between about 0.95 and $1.5\,W\,m^{-1}\,K^{-1}$, with many glasses being close to $1.2\,W\,m^{-1}\,K^{-1}$ at 300K. The highest values at 300K are $\sim 1.4\,W\,m^{-1}\,K^{-1}$ for SiO_2, and $\sim 1.5\,W\,m^{-1}\,K^{-1}$ for MORB and dacite-andesite. The lowest values at 300K are $\sim 1\,W\,m^{-1}\,K^{-1}$ for soda–lime silicates, and $\sim 0.95\,W\,m^{-1}\,K^{-1}$ for leucogranite and basaltic–andesite. Clearly, there is no simple relationship between thermal conductivity at room temperature and glass composition.

Thermal conductivity of glasses increases with increasing temperature, because the relative increase in heat capacity outweighs the decrease in thermal diffusivity as temperature increases (Fig. 10.3). At high temperatures, some glasses approach a constant value of conductivity, while others are still increasing as the glass transition is reached. Conversely, thermal conductivity of crystals always decreases with increasing temperature, because the decrease in thermal diffusivity far outweighs the increase in heat capacity (Fig. 10.3). The difference in thermal conductivity between crystals of different composition is much greater than the range for similar glasses.

10.2 CHANGES IN THERMAL PROPERTIES ACROSS THE GLASS TRANSITION

10.2.1 The Glass Transition

The glass transition is a second-order thermodynamic transition, meaning that there is no discontinuous change in the enthalpy or volume on heating from glass to melt, but there is a jump in their derivatives, the heat capacity and thermal expansivity (Fig. 10.3).

The glass transition is also a time-dependent phenomenon. Silicate melts are modeled as viscoelastic materials, which respond to changes in two different ways: an instantaneous elastic response and a slower viscous response that occurs with a characteristic relaxation timescale. Below the glass transition, the material cannot flow and the viscous response is unavailable. Above the glass transition, both elastic and viscous responses happen – and once substantially above the transition, where the melt is relaxed, the viscous response may also appear nearly instantaneous. In the glass transition range, the perceived presence or absence of the viscous response depends on the timescale of observation, and complex time-dependent behavior is normal.

Practically, we can consider the deformation timescale and the relaxation timescale, the latter given by the Maxwell relation:

$$\tau_{\text{relax}} = \frac{\eta}{G_\infty} \tag{10.2}$$

where G_∞ is the shear modulus at infinite frequency. This has a value of approximately 10 GPa for silicates (Dingwell and Webb, 1989). Brillouin spectroscopy measurements compiled by Bansal and Doremus (1986) show that G_∞ ranges from 5 to 42 GPa over a compositional range from 5 to 99 mole% SiO_2, while data for a suite of geologically relevant silicate glasses show that G_∞ varies from ~27–31 GPa for polymerized albite and granite glass, to ~34–37 GPa for highly depolymerized glasses (Whittington et al., 2012).

The ratio of relaxation timescale to characteristic deformation timescale is known as the Deborah number, and as long as this ratio is much less than 1, relaxed (liquid-like) behavior will be observed in viscosity experiments. There are other techniques available for studying the glass transition, for example, differential scanning calorimetry can be used to probe the relaxation behavior of enthalpy. Experiments using different techniques on the same samples have shown that the shear, volume, and enthalpic relaxation timescales are equivalent (Webb and Knoche, 1996; Sipp and Richet, 1998). Relaxation and the glass transition in silicate melts are reviewed in greater detail by Moynihan (1995) and Dingwell (1995a), among others.

10.2.2 Thermal Diffusivity Across the Glass Transition

As glasses are heated towards the glass transition, they typically experience a sharp decrease in thermal diffusivity, followed by a steady value that characterizes the melt. Practically, there are several potential difficulties involved in measuring thermal diffusivity through the glass transition, especially with the laser-flash method. The first is the possibility of relaxation of the sample towards a new fictive temperature (the temperature at which a liquid in equilibrium would have the same structure as the glass), which may occur on a similar timescale to that of the measurement technique, such that the measurement reflects neither a true glass nor a relaxed liquid. The second is the possibility of crystallization, although this is typically only a problem in depolymerized melts whose glass transition temperatures are hundreds of degrees below their solidus, and whose viscosity drops rapidly with heating above the glass transition (Section 10.4). The third is the possibility of volatile loss, which can occur faster in melts than in glasses as bubbles can nucleate and grow in the former. This is of course only a problem in volatile-bearing materials, such as some volcanic glasses. The fourth is the possibility of flow in the sample, resulting in thickness variations that need to be accounted for in calculating thermal diffusivity or conductivity from measurements of time. Examples of several of these behaviors are shown in Hofmeister et al. (2009).

Interpretation of changes in thermal diffusivity across the glass transition is further hindered by its timescale-dependent nature. The glass transition temperature is traditionally taken to be T_{12}, the temperature at which the viscosity is 10^{12} Pa s and relaxation occurs on the order of tens of seconds. However, comparison of laser-flash analysis (LFA) data and viscosity data on the same suite of samples shows that the glass transition is detected by LFA when the viscosity is approximately 2×10^8 Pa s (Fig. 10.4). This $T_{g,\mathrm{LFA}}$ is consistent with the viscosity at which the sample behaves like relaxed melt on the millisecond timescale of the laser pulse. Therefore depolymerized melts, which are prone to crystallization, may undergo physical changes above their viscosimetric glass transition (T_{12}), while still displaying glass-like thermal diffusivity values below their LFA glass transition ($T_{g,\mathrm{LFA}}$). Nevertheless, most melts are strong enough to allow thermal diffusivity measurements at several temperatures prior to irreversible changes. Although the sample is supercooled, well below its solidus temperature, it is metastable, and thermal diffusivity measured above $T_{g,\mathrm{LFA}}$ represents that of structurally relaxed melt up to and including superliquidus conditions.

Every sample shown in Fig. 10.2 shows a decrease in D across the glass transition, consistent with a decrease in order, as the melt can undergo configurational changes faster than the timescale over which the thermal pulse traverses the sample. Decreases are generally larger for more depolymerized melts, such as pyroxenes and soda–lime silicates, than for more polymerized melts, such as feldspars. The exception is that SiO_2 also shows a large decrease in D across the glass transition. Geological melts show the same pattern, with polymerized rhyolites, haplogranite and tektite compositions showing smaller decreases in D than depolymerized melts such as basalts.

10.2.3 Thermal Conductivity Across the Glass Transition

As the glass transition is crossed, the heat capacity increases and the thermal diffusivity decreases, both changes arising from the onset of configurational rearrangements that are possible in the liquid but not in the glass. The decrease in D is often proportionally similar to the increase in C_P, so the thermal conductivity can increase, decrease or stay the same across the transition from glass to liquid (Fig. 10.5). Silica and $CaMgSi_2O_6$ both exhibit large decreases, while feldspars show modest decreases ($LiAlSi_3O_8$, $NaAlSi_3O_8$), no change ($KAlSi_3O_8$), or a small increase ($CaAl_2Si_2O_8$). Geological melts typically exhibit very small changes in thermal conductivity across the transition, with only silica-rich haplogranite and moldavite exhibiting increases. Soda–lime silicates show a mixture of increases and decreases across the glass transition. Therefore, despite the general increase in the configurational heat capacity with decreasing melt polymerization, there is no compositionally consistent pattern to changes in thermal conductivity across the glass transition.

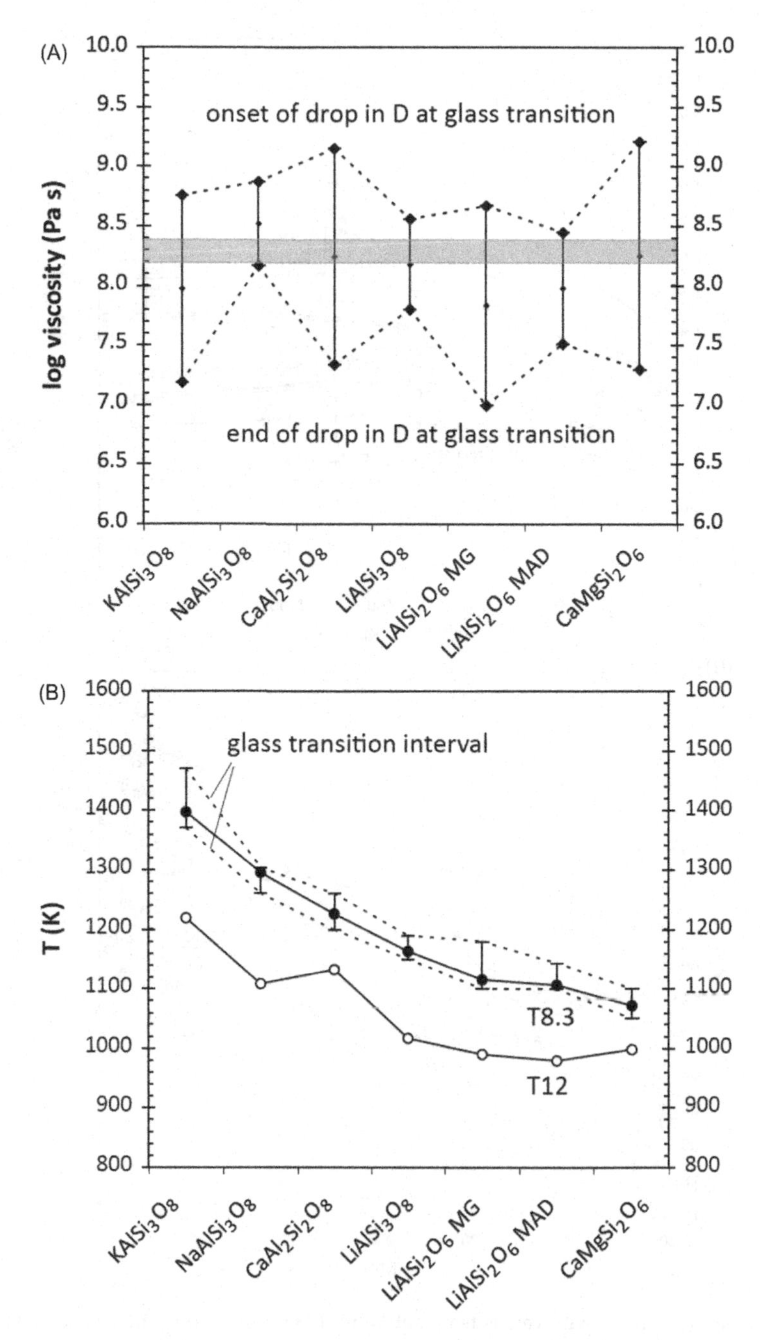

FIGURE 10.4 Correlation of *D* and viscosity for glasses of feldspar, spodumene and diopside composition. Viscosity (left *y*-axis) at the onset and end of the drop in *D* over the glass transition interval (thick lines). Temperature (right *y*-axis) of the glass transition interval observed in LFA experiments (thin dotted lines), of the 2×10^8 Pa s isokom (squares), and of the 10^{12} Pa s isokom (circles). Source: *From Hofmeister, A.M., Whittington, A. G., Pertermann, M., 2009. Transport properties of high albite crystals and near-endmember feldspar and pyroxene glasses and melts to high temperature. Contributions Mineral. Petrol. 158, 381–400, with permission. © Springer-Verlag 2009.*

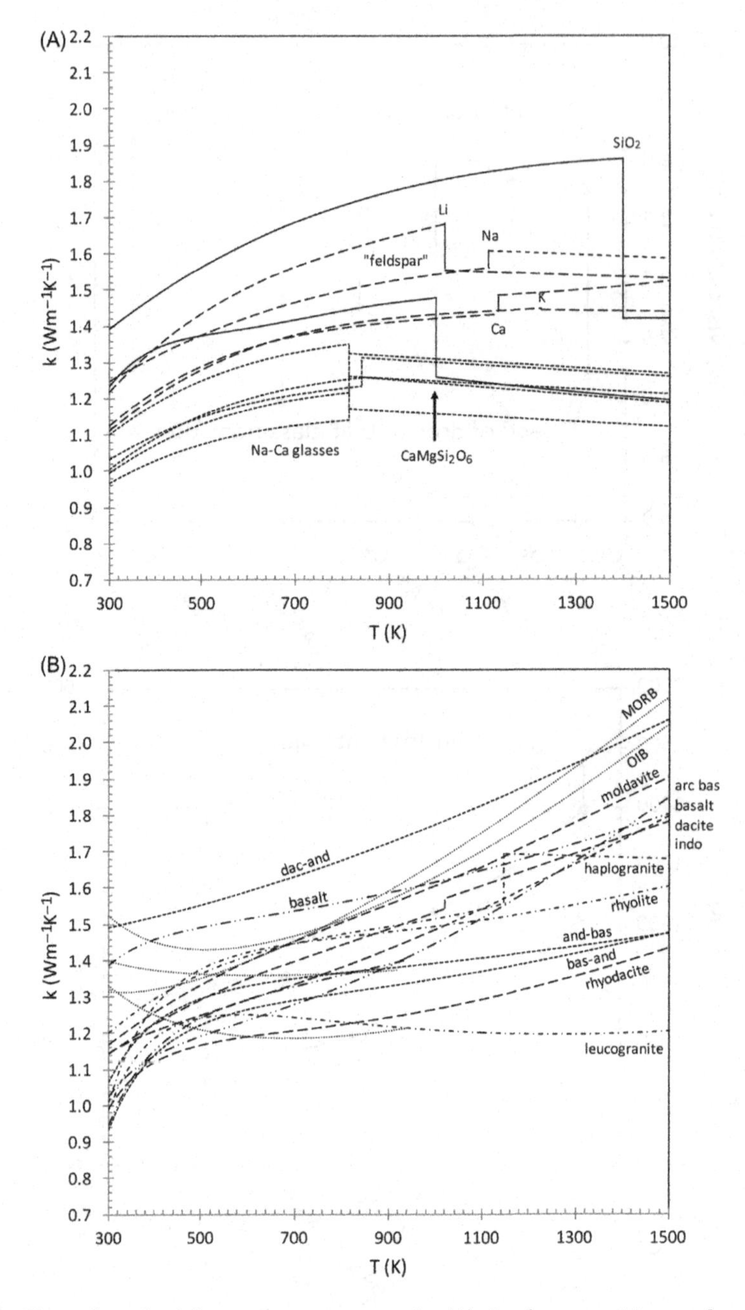

FIGURE 10.5 Thermal conductivity versus temperature for (A) simple compositions and soda–lime silicates; (B) geologically relevant melts. Fit parameters in Table 10.2.

TABLE 10.2 Thermal Conductivity Equations

SiO$_2$ and Mineral Composition Glasses and Melts: $k_{glass} = A + BT + CT^2 + DT^3 + ET^4$ in W m^{-1} K^{-1}									$k_{melt} = A + BT + CT^2 + DT^3$					
Name/Composition	A	$B \times 10^3$	$C \times 10^6$	$D \times 10^{10}$	$E \times 10^{14}$	k_{298}	$k_{gl,Tg}$	T_g	A	B	C	D	$k_{melt,Tg}$	$k_{melt,1500}$
						W m^{-1} K^{-1}	W m^{-1} K^{-1}	K					W m^{-1} K^{-1}	W m^{-1} K^{-1}
SiO_2	1.0331	1.4282	−0.82169	1.5960	0	1.39	1.86	1400.9	1.42				1.42	1.42
$LiAlSi_3O_8$	0.6156	2.7755	−3.00439	16.118	−3.2336	1.22	2.03	1016.8	1.60	−0.043403			1.55	1.53
$NaAlSi_3O_8$	0.9742	1.1016	−0.67129	1.3828	0	1.25	1.56	1108.6	1.67	−0.05508			1.61	1.59
$KAlSi_3O_8$	0.6736	1.9023	−1.62030	4.7596	0	1.11	1.45	1218.7	1.47	−0.01795			1.44	1.44
$CaAl_2Si_2O_8$	0.6161	2.4114	−2.80790	15.029	−3.0239	1.13	1.93	1132.2	1.57	−0.2178	0.12352	−0.22672	1.48	1.52
$CaMgSi_2O_6$	−0.7498	14.966	−42.1960	589.0	−401.05	1.21	1.48	998.5	1.47	−0.265	0.055073		1.26	1.20
Na-Ca 1895	0.669	0.00196	−1.95	7.0		1.27	4.65	806.1	1.39	−0.808			1.32	1.27
Na-Ca 1960	0.647	0.00147	−1.51	5.5		1.10	3.76	808.5	1.23	−0.720			1.17	1.12
Na-Ca Vase	0.601	0.00180	−1.82	6.6		1.15	4.44	815.5	1.32	−0.721			1.26	1.21
Na-Ca Modern	0.572	0.00197	−1.97	7.1		1.17	5.47	870.0	1.38	−0.800			1.31	1.26
Na-Ca 1926	0.695	0.00153	−1.52	5.5		1.16	4.19	842.4	1.35	−1.080			1.26	1.19
Moldavite	0.795	0.00158	−1.26	4.9		1.28	7.29	1085.1	0.98	6.150			1.65	1.90
Indochinite	0.792	0.00151	−1.28	5.1		1.26	6.47	1024.0	1.12	4.410			1.57	1.78
Haplogranite[a]	0.734	0.00213	−2.15	8.0		1.39	12.32	1143.4	1.74	−0.416			1.69	1.68

Rock Remelts: $k = A + BT + CT^{-2} + DT^{1/2}$ and Selected Liquid k (in $W\,m^{-1}\,K^{-1}$) Against Temperature (in K)

Name/Composition	A	$B \times 10^4$	C	$D \times 10^2$	k_{298}	$k_{gl,Tg}$	T_g	$k_{melt,Tg}$	$k_{melt,1500}$
					$W\,m^{-1}\,K^{-1}$	$W\,m^{-1}\,K^{-1}$	K	$W\,m^{-1}\,K^{-1}$	$W\,m^{-1}\,K^{-1}$
Leucogranite	2.8683	11.609	−64532	−8.7197	0.98	1.2	1093.8	1.81	1.80
Rhyolite NCr							1113.1		
Rhyolite NCAR	2.7578	12.292	−71174	−7.6603	1	1.5	1070.2	1.55	1.55
Rhyodacite[a]	2.3706	14.002	−40254	−7.7977	0.99	1.28	1035.8	1.27	1.3
Dacite remelt	1.7444	13.104	−12497	−4.926	1.14	1.51	1019.2	1.48	1.49
Dacite-andesite	1.7973	10.577	−2953.7	−3.4101	1.41	1.76	965.3	1.64	1.73
Andesite-basalt[a]	1.9808	6.6136	−40598	−3.8249	1.06	1.39	988.3	1.4	1.29
Basalt-andesite	2.2837	10.735	−54011	−6.1831	0.93	1.34	948.0	1.31	1.36
Dolerite#3	2.2931	12.6933	−52463	−6.607	0.94	1.41	936.5	1.36	1.36
Basalt[a]	2.2099	11.064	−21139	−5.3125	1.21	1.44	982.9	1.41	1.28
Arc basalt	2.5728	22.7959	−34545	−10.661	1.02	1.41	932.5	1.52	1.44
OIB2#1	2.3872	21.658	8296.5	−9.273	1.33	1.5	919.0	1.52	1.46
MORB#1	1.8148	17.759	1768.9	−6.0827	1.28	1.63	940.3	1.29	1.34
P-MORB	1.5682	4.046	3189.7	−1.8849	1.37	1.38	917.2	1.46	1.22
Continental	1.9863	12.735	7263.5	−6.4536	1.13	1.2	939.4	1.3–1.5	~1.4

[a]*Indicates synthetic iron-free rock composition.*

10.3 LIQUID VISCOSITY MEASUREMENT

10.3.1 Techniques for Measuring Silicate Melt Viscosity

The viscosity of silicate melts typically varies from ~1 Pa s for depolymerized silica-poor melts at ~1500°C to ~10^{12} Pa s at the glass transition (~600–800°C, depending on composition). The most viscous melt composition is pure SiO_2, which still has a viscosity of 10^6 Pa s at 1700°C, while volatile-rich melts that form pegmatitic veins can cross the glass transition at temperatures below 300°C (Sirbescu et al., 2017). Reviews of viscosimetric and rheometric techniques that are tailored to specific applications include Dingwell (1995b), focusing primarily on geological applications; Zanotto and Migliore (1999) focusing primarily on the glass industry; Viswanath et al. (2006), focusing primarily on engineering applications; and Rao (2014) focusing primarily on the food industry. The primary difference between rheological studies of foods and silicate melts is not the range of viscosity, nor even the rheological behaviors they exhibit, but rather the temperature at which measurements are conducted.

In some rheometers, stress is fixed and strain-rate is measured, for example, the parallel-plate or creep viscometer (Gent, 1960), which is commonly used for measurements in the range 10^8–10^{12} Pa s (Fig. 10.6). Here, the linear strain-rate is measured and, assuming a "perfect slip" condition and preservation of cylindrical geometry, the viscosity is simply calculated as:

$$\eta = \frac{Fl^2}{3V(\partial l/\partial t)} \tag{10.3}$$

where F is the applied force, l is the sample height, and V is the sample volume. Other equations should be used if the "perfect slip" condition is violated, for example, in the limiting "no-slip" case which produces a barrel-shape as deformation proceeds:

$$\eta = \frac{2\pi Fl^5}{3V(\partial l/\partial t)(2\pi l^3 + V)} \tag{10.4}$$

Some parallel-plate viscometers operate at constant load, while others can be programmed to operate at constant strain-rate. These are usually modified rock deformation rigs, and typically exert higher stresses (e.g., Hess et al., 2007). Micropenetration techniques are another method used for this high-viscosity range. For a spherical indenter,

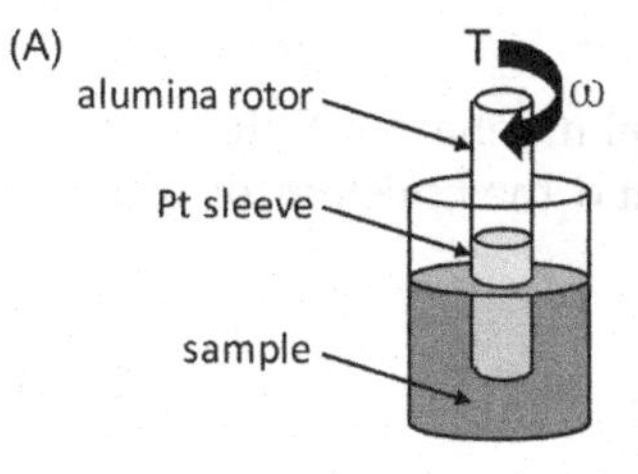

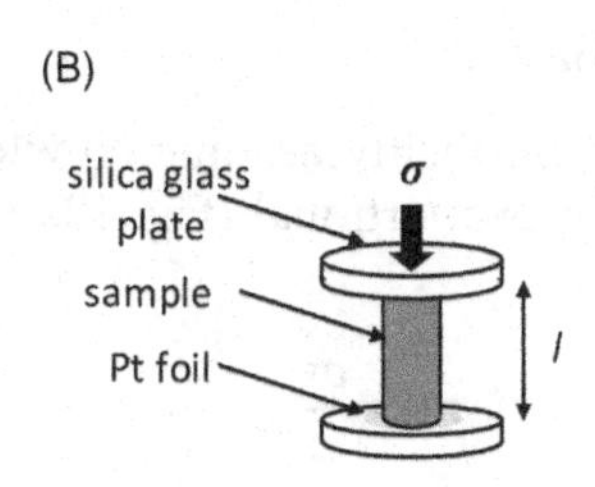

FIGURE 10.6 Cartoon illustrating the principle behind (A) the concentric-cylinder method, and (B) the parallel-plate method for viscosity measurement.

$$\eta = \frac{0.1875Ft}{s^{0.5}l^{1.5}} \tag{10.5}$$

where t is the indentation time, s is the radius of the sphere, and l is the indentation distance.

In other viscometers, the strain-rate is fixed and the stress is measured, for example, in most concentric-cylinder (or "cup and bob" viscometers). In the Couette configuration, the bob is stationary and the cup rotates. In the Searle configuration, which is more commonly used in commercial rheometers, the cup is stationary and the spindle or bob rotates. The viscosity is again the ratio of stress (calculated from the measured torque ξ) to strain-rate (calculated from the measured angular velocity ω), and for a Newtonian fluid the viscosity is:

$$\eta = \frac{\xi(R_c^2 - R_b^2)}{4\pi L\omega R_b^2 R_c^2} \tag{10.6}$$

where R_c is the internal diameter of the crucible, R_b is the bob diameter, and L is the bob length or immersion depth (Fig. 10.6). Different formulae should be used for materials with a power-law or Bingham rheology. Another type of rotational rheometer is the cone and plate viscometer, where the sample material is contained between a downward-pointing cone and a flat plate. As long as the angle between the two surfaces is small, the strain-rate is approximately constant throughout the sample.

Other techniques include the falling-sphere method, commonly employed to determine the viscosity of silicate melts at high pressure, because it can be used in welded capsules. X-ray imaging of platinum spheres can yield precise distance measurements without needing to open the capsule, which can simply be inverted for a repeat measurement. The viscosity is related to the terminal velocity of the falling sphere, of radius s, by Stokes' Law:

$$\eta = \frac{2gs^2\left(\rho_{\text{sphere}} - \rho_{\text{liquid}}\right)}{9\pi u} \tag{10.7}$$

The terminal velocity (u) is determined through multiple runs of different duration, because it takes a finite amount of time for the sphere to accelerate. Multiple spheres of different sizes can be used to calculate viscosity from a single run (e.g., Holtz et al., 1999). The longer the capsule, the lower the relative uncertainty on the falling distance. The Faxen correction needs to be applied to account for the finite capsule diameter and melt volume, although this is usually of similar order to the overall uncertainty of the measurements (Shaw, 1963).

10.3.2 Types of Rheological Behavior

By using the term "viscosity", we implicitly assume Newtonian behavior, where stress (σ) and strain-rate ($\partial\varepsilon/\partial t$) are directly proportional (Fig. 10.7), and the constant of proportionality is the viscosity:

$$\sigma = \eta\frac{\partial\varepsilon}{\partial t} \tag{10.8}$$

In practice, a rheological measurement is carried out at a specific stress–strain-rate condition, and the ratio at any individual measurement condition should be referred to as the

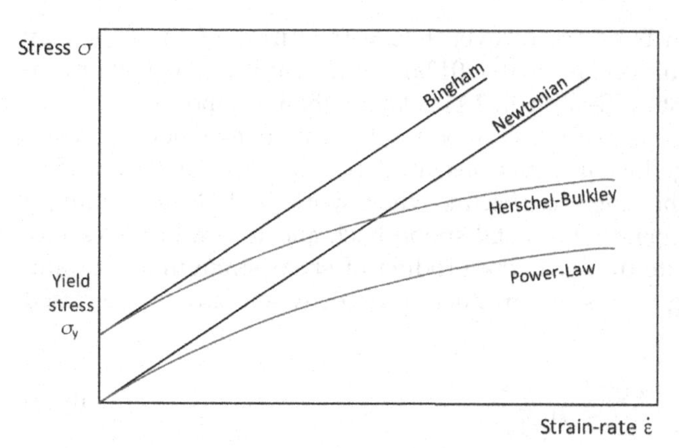

FIGURE 10.7 Stress versus strain-rate diagram showing different types of rheological behavior.

"apparent viscosity," which does not imply any particular rheological law. When measurements are possible over a range of conditions, one can extrapolate (with caution) towards zero strain-rate to assess whether there is a positive intercept on the stress axis or not. If there is, then the rheological behavior is named after Bingham, and the intercept is called the yield strength, σ_y:

$$\sigma = \sigma_y + \eta \frac{\partial \varepsilon}{\partial t} \tag{10.9}$$

Some other rheological behaviors will produce an apparent viscosity that depends on the strain-rate at which the stress is measured (or vice-versa). For example, the relation for power-law rheology includes the consistency κ (equal to the apparent viscosity at a strain-rate of 1 s^{-1}), and the flow index ς:

$$\sigma = \kappa \left(\frac{\partial \varepsilon}{\partial t} \right)^{\varsigma} \tag{10.10}$$

When ς is less than 1, the apparent viscosity decreases as the strain-rate increases, and the flow is termed pseudoplastic. When ς is greater than 1, the apparent viscosity increases as the strain-rate increases, and the flow is termed dilatant. A more generalized behavior is the Herschel-Bulkley type, which allows for both power-law behavior and a yield stress:

$$\sigma = \sigma_y + \kappa \left(\frac{\partial \varepsilon}{\partial t} \right)^{\varsigma} \tag{10.11}$$

10.3.3 Limits to Newtonian Behavior in Silicate Melts

From Section 10.2.1, glass exhibits a purely elastic response to an applied stress, and relaxed liquids will produce an instantaneous elastic response followed by a steady viscous response, but rheological measurements in the glass-transition range will produce complex time-dependent behavior, evolving towards that of a relaxed liquid. If very high strain-rates are imposed, such that the deformation timescale approaches the viscoelastic relaxation timescale, non-Newtonian behavior or even brittle fracturing of the melt may occur (Webb and Dingwell, 1990). Cordonnier et al. (2012a) found that brittle behavior typically occurs when the Deborah number (Section 10.2.1) is larger than around 0.01.

At slightly lower strain-rates, viscous heating may occur. This happens when a fraction of the energy dissipated in driving flow in a viscous fluid is converted to heat, and is somewhat analogous to the shear heating that occurs when work is done deforming a material with a shear strength (e.g., bending a metal spoon back and forth). In the case of the spoon, the volumetric heating rate dQ/dt is the product of shear strength and strain-rate. In the case of the fluid, heating rate is the product of viscosity and strain-rate raised to the second power:

$$\frac{\partial Q}{\partial t} = \eta \left(\frac{\partial \varepsilon}{\partial t} \right)^2 \tag{10.12}$$

Given that densities of silicates are of the order of 2000–4000 kg m^{-3}, and that heat capacities of silicate melts are on the order of 1 to 1.5 J $g^{-1}K^{-1}$, viscous heating is negligible in most experiments, especially those already at high temperature where the furnace can counteract viscous heating by putting out less power. Even a power output of tens of kW m^{-3} will result in heating at a rate of $\sim$0.01 K s^{-1}, and this requires a combination of very high strain-rates and very high viscosity. Once it begins, the decrease in viscosity resulting from heating will tend to focus deformation in that hotter region, increasing strain-rate and potentially initiating a positive feedback cycle. While viscous heating is generally negligible in the lab, except in experiments designed to produce it (e.g., Hess et al., 2007; Cordonnier et al., 2012b), it may have relevance to lavas of any composition as long as strain-rates are sufficiently high. The potential applications include basalt in lava tubes (Dragoni et al., 2002; Costa and Macedonio, 2005), and more silicic magmas in conduits (Hale et al., 2007) or as lava flows (Nelson, 1981; Avard and Whittington, 2012) or even in pyroclastic flows (Robert et al., 2013).

10.4 TEMPERATURE AND COMPOSITIONAL DEPENDENCE OF VISCOSITY

Several theoretical models have been proposed to explain the temperature-dependence of silicate melt viscosity. The simplest model is Arrhenian behavior, whereby the mechanism of viscous flow is interpreted to have a temperature-independent activation energy, E_a:

$$\ln \eta = A + \frac{E_a}{2.303 R_{gc} T} \tag{10.13}$$

where A is the viscosity intercept at infinite temperature, and R_{gc} is the gas constant. Several composition-dependent viscosity models based on Arrhenian behavior exist, and have been applied to both slags (Riboud et al., 1981; Urbain et al., 1981; Urbain, 1987), and geological compositions (Bottinga and Weill, 1972; Shaw, 1972).

Viscosity of a range of silicate melts is plotted on an Arrhenian diagram in Fig. 10.8. It has been known for about a century that silicate melts display non-Arrhenian behavior

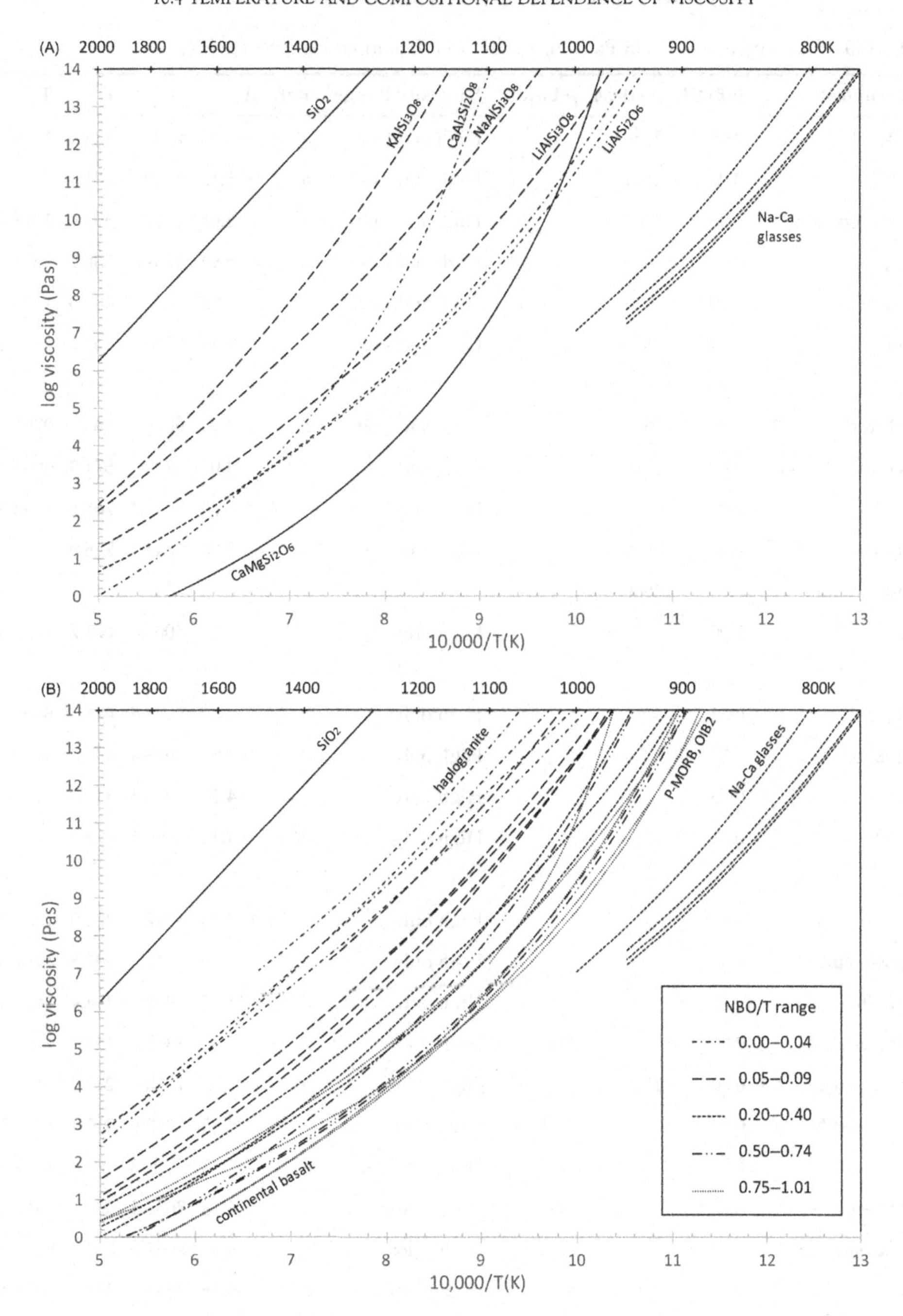

FIGURE 10.8 Viscosity of liquid silicates: (A) simple compositions and soda–lime silicates; (B) geologically relevant melts. Fit parameters in Table 10.3.

TABLE 10.3 Viscosity of Glasses in Pa s, Fit to the VTF Equation, $\log \eta = A + B/(T-C)$

Name/Composition	NBO/T	ρ Glass	ρ Liquid	Viscosity Range[a]	Ref.	A	B	C	T_{12}	m^b
$LiAlSi_3O_8$	0.00	2336		High and low	a	−4.05	7839	528.4	1016.8	33.4
$NaAlSi_3O_8$	0.00	2361		High and low	a	−5.12	11600	431	1108.6	28.0
$NaAlSi_3O_8$ remelt	0.00	2360		High and low	a	−5.65	12621	399.6	1114.7	27.5
$KAlSi_3O_8$	0.00			High and low	b	−5.98	12371	530.7	1218.7	31.8
$KAlSi_3O_8$ remelt	0.00			High and low	b	−6.03	12883	483.3	1197.8	30.2
$CaAl_2Si_2O_8$	0.00	2681		High and low	a	−4.47	5336	808.2	1132.2	57.6
$LiAlSi_2O_6$ remelt Mad	0.00	2379		High and low	a	−4.04	6789	556.3	979.6	37.1
$LiAlSi_2O_6$ remelt MG	0.00	2379		High and low	a	−4.07	6736	571.3	990.5	38.0
CaSiO3	2.00			High only	c	−6.60	5017.1	768.8	1038.5	71.6
$CaMgSi_2O_6$	2.00	2853		High and low	a	−3.90	3842	756.9	998.5	65.7
$CaMgSi_2O_6$ remelt	2.00	2860		None	a					
$MgSiO_3$	2.00			High only	c	−9.90	8006	671.7	1037.3	62.1
Na-Ca 1895	0.63			High only	d	−4.69	5998.5	446.7	806.1	37.4
Na-Ca 1960	0.75			High only	d	−4.51	5998.4	445.2	808.5	36.7
Na-Ca Vase	0.73			High only	d	−4.37	5999.5	449.0	815.5	36.4
Na-Ca 1926 #3	0.68			High only	d	−4.44	5999.5	477.5	842.4	38.0
Moldavite remelt	0.07			High only	d	−3.08	8997	488.5	1085.1	27.4
Indochinite remelt	0.07			High only	d	−3.16	8042	493.5	1024.0	29.3
Haplogranite[c]	0.00			High only	d	−1.87	9001	494.4	1143.4	24.4
Leucogranite remelt	0.00		2335	High + low	d	−7.06	18159	141.1	1093.8	21.9
Rhyolite NCr (extra Al)	0.00	2350		High + low	d	−7.30	17388	212.2	1113.1	23.8
Rhyolite remelt NCAR	0.02		2338	High + low	d	−4.65	12600	313.4	1070.2	23.5
Rhyodacite[c]	0.05		2402	High + low	e	−4.43	9029.8	486.2	1035.8	31.0
Dacite remelt	0.09		2487	High + low	e	−4.68	8655.5	500.1	1019.2	32.7
Dacite-andesite remelt	0.20		2472	High + low	e	−4.50	7967.9	482.4	965.3	33.0
Andesite[c]	0.06		2479	High + low	c	−4.45	7885	539.3	1018.6	35.0
Andesite-basalt[c]	0.40		2538	High + low	e	−4.81	6826.7	582.2	988.3	40.9
Basalt-andesite remelt	0.33		2592	High + low	e	−3.96	6077.6	567.2	948.0	39.7
Dolerite#3 remelt	0.58		2646	High + low	e	−4.09	5293.6	607.5	936.5	45.8

(Continued)

TABLE 10.3 (Continued)

Name/Composition	NBO/T	ρ Glass	ρ Liquid	Viscosity Range[a]	Ref.	A	B	C	T_{12}	m[b]
Basalt[c]	0.67		2591	High + low	e	−5.20	6597.5	599.3	982.9	44.1
Arc basalt remelt	0.64		2687	High + low	e	−3.89	5018.7	616.7	932.5	46.9
OIB1 remelt	0.76		2678	High only	c	0.00	2106.4	809.9	985.4	67.4
OIB2#1 remelt	0.92		2711	High + low	e	−5.10	6230.3	554.7	919.0	43.1
MORB#1 remelt	0.75	2563	2689	High + low	e	−3.94	6411.9	538.0	940.3	37.3
P-MORB remelt	0.76		2722	High + low	e	−2.54	4054.9	638.3	917.2	47.8
Continental remelt	1.01		2720	High + low	e	−4.81	5685.1	601.2	939.4	46.7

[a]*High ~ 10^9–10^{12} Pa s and low ~ 10–1000 Pa s. Viscosity should not be calculated outside the measured range.*
[b]*Fragility, m = $B/[T_{12})1-C/T_{12})^2]$, with parameters B, C, and T_{12} obtained from the VTF fit.*
[c]*Iron-free analog to a natural lava that would have significant iron.*
References: a, Hofmeister et al. (2009); b, Pertermann et al. (2008); c, unpublished; d, Hofmeister et al. (2014b); e, Hofmeister et al. (2016).

over wide temperature ranges, and that a satisfactory fit could be obtained by adding in a third fitting parameter, resulting in the well-known Vogel–Fulcher–Tammann (VFT) equation (e.g., Vogel, 1921):

$$\ln \eta = A + \left(\frac{B}{T - C}\right) \tag{10.14}$$

This equation is the foundation for several empirical models of silicate melt viscosity for geological applications, where viscosities may need to be calculated from the liquidus to the glass transition temperature. In these models, A is typically assumed to be constant (Russell et al., 2003), and B and C are parameterized in terms of chemical composition, typically focusing on the relative amounts of network-forming, charge-balancing, and network-modifying cations. The widest-ranging model taking this approach is by Giordano et al. (2008). Hui and Zhang (2007) formulated an empirical model of silicate melt viscosity using an alternative modification to account for non-Arrhenian behavior:

$$\ln \eta = A + \left(\frac{B}{T}\right) + \exp\left(C + \frac{D}{T}\right) \tag{10.15}$$

Thermodynamically based models hold out the prospect of being able to more confidently predict the viscosity of unmeasured compositions, and to extract thermodynamic data from the existing experimental dataset. Richet (1984) combined the Adam–Gibbs–DiMarzio (AGD) theory of relaxation in silicate melts (Adam and Gibbs, 1965), with the Maxwell relationship (Eq. (10.2)): the resulting equation explicitly relates viscosity to configurational entropy (S^{conf}):

$$\ln \eta = A + \frac{B}{TS^{conf}} \tag{10.16}$$

The variation of S^{conf} with temperature can be obtained from heat capacity measurements:

$$S^{conf}(T) = S^{conf}(T_g) + \int_{Tg}^{T} \frac{C_P^{conf}}{T} dT \tag{10.17}$$

where $C_P{}^{conf}$ is the configurational heat capacity, which is the difference in heat capacity between the liquid and the glass at T_g. For liquids that form by congruent melting of pure crystalline solids, $S^{conf}(T_g)$ can be quantified through a thermodynamic cycle from crystal to liquid to glass, because it is equal to the residual entropy of the glass at zero Kelvin, which is typically of the order of 1.5–3 J $mol^{-1}K^{-1}$ (for a mole of atoms; see Table 3.2 of Mysen and Richet (2003), and references therein).

The degree of non-Arrhenian behavior can be directly related to the rate of increase of S^{conf} with increasing temperature, which in turn relates directly to the magnitude of $C_P{}^{conf}$. In general, more depolymerized melts have higher $C_P{}^{conf}$ and are the most fragile, where increasing "fragility" denotes increasing deviation from Arrhenian behavior (Angell, 1995). Fragility can be quantified as the steepness index, originally defined as:

$$\frac{d(\ln \tau_{relax})}{d(T_g/T)} \text{ for } T = T_g \tag{10.18}$$

(Plazek and Ngai, 1991).

In practice it is often simpler to substitute η instead of τ_{relax}, because the two are directly proportional (Eq. (10.2)), and the fragility m is defined as:

$$m = \frac{d(\log \eta)}{d(T_g/T)} \text{ for } T = T_g \tag{10.19}$$

The "strongest" melt known is pure SiO_2, which has only an 11% increase in C_P on heating across the glass transition (Richet et al., 1982), although $NaAlSi_3O_8$ and $KAlSi_3O_8$ have even smaller relative increases (Richet and Bottinga, 1984). Entropy of mixing contributes to $S^{conf}(T)$, so that viscosity of intermediate $(Ca,Mg)SiO_3$ melts is lower than for end-member $CaSiO_3$ or $MgSiO_3$ (Neuville and Richet, 1991).

The Avramov–Milchev equation is another thermodynamically based approach (Avramov and Milchev, 1988) that relates the non-Arrhenian temperature-dependence of viscosity to $C_P{}^{conf}$ through the fragility:

$$\ln \eta = A + \left(\frac{B}{T}\right)^m \tag{10.20}$$

Once again, B is effectively an activation energy term. Despite both models originating in thermodynamic treatments, the AGD and AM equations extrapolate somewhat differently at both high and low temperatures. Mauro et al. (2009) have proposed an alternative formulation of the AGD model that avoids the divergence of S^{conf} in the AM model at high temperatures, and the vanishing of S^{conf} in the VFT model at low temperatures:

$$\ln \eta = A + \left(\frac{B}{T}\right) \exp\left(\frac{C}{T}\right) \tag{10.21}$$

10.5 THERMAL DIFFUSIVITY AND CONDUCTIVITY OF SILICATE MELTS

Measurements of thermal transport in silicate melts are difficult, because they need to be contained and high temperatures are required. There are two approaches when using the LFA technique. One is to use a three-layer technique, with the outer layers forming the container and the melt filling the sandwich. The other is to use just a single layer of melt, without support. This restricts measurements to temperatures at which the melt viscosity remains high, so it does not flow appreciably on the timescale of the experiment. As can be seen from the viscosity data, this approach restricts depolymerized melts such as basalts and slags to measurements just a few tens of degrees above their glass transition temperature. Such melts are typically prone to both flow and crystallization in the temperature range just above the glass transition, and complex behavior may result (see Hofmeister et al. (2009) for examples of such data and interpretation).

Just above the glass transition, thermal diffusivity of all melts shows a clear correlation with melt polymerization, with polymerized silica-rich melts having higher D and depolymerized silica-poor melts having lower D (Fig. 10.1). Eriksson et al. (2003) found a similar result for slags, using a three-layer technique in the laser-flash apparatus that allowed measurements to 1825K. For simple compositions, SiO_2 and feldspar melts have higher D ($\sim$0.44–0.50 $mm^2\,s^{-1}$) than soda–lime silicate and pyroxene melts ($\sim$0.29–0.38 $mm^2\,s^{-1}$). Rock melts show the trend even more clearly, with D varying from $\sim$0.52 $mm^2\,s^{-1}$ for rhyolite and leucogranite to $\sim$0.25 $mm^2\,s^{-1}$ for ocean island basalt. Within the feldspar melts, the highest D is for Na, then K, then Li, which is different to the order of glass diffusivities. As before, $CaAlSi_2O_8$ resembles the more depolymerized melts much more closely than it does the alkali feldspar melts. Within the pyroxene melts, the Ca end-member once again has a lower D than the Mg end-member, consistent with its greater mass, and the mixed (Ca,Mg) composition is closer to the Ca-endmember, consistent with greater disorder arising from mixing.

The tektite samples (moldavite and indochinite) show D increasing with temperature (Hofmeister et al., 2014), but there is insufficient data to determine dD/dT for most other melts, including pure SiO_2, so the horizontal lines on Fig. 10.2 are speculative. Data for slags published by Eriksson et al. (2003) suggests that dD/dT for slags is about $10^{-9}\,mm^2\,s^{-1}K^{-1}$, but those data explicitly include the effects of ballistic radiative heat transport as well as conduction.

The relative rates of kinematic viscosity ($\nu = \eta/\rho$) and thermal diffusivity are expressed by the dimensionless Prandtl number, Pr:

$$\mathrm{Pr} = \frac{\nu}{D} = \frac{\eta}{\rho D} = \frac{C_P \eta}{K} \tag{10.22}$$

At 20°C, Pr is $\sim$0.015 for mercury, $\sim$0.7 for air, and $\sim$7 for water. From the ranges of viscosity and thermal diffusivity discussed earlier, the Prandtl number is clearly much greater than 1 for all silicate melts under most conditions, and in fact even fully molten basaltic lava will have a Pr of 10,000 or more (based on $\eta \sim 10$ Pa s, $\rho \sim 3000\,kg\,m^{-3}$, and $\kappa \sim 3 \times 10^{-7}\,m^2\,s^{-1}$). The very high Pr of lava allows the equations of motion to be simplified when calculating convection in magma chambers (Clark et al., 1987).

The relative rates of chemical and thermal diffusivity are expressed by the dimensionless Lewis number, Le:

$$\mathrm{Le} = \frac{D}{D_{\mathrm{mass}}} = \frac{\mathrm{Pr}}{\mathrm{Sc}} \tag{10.23}$$

where D_{mass} is the mass self-diffusivity and Sc is the Schmidt number, which is the mass transfer analog of the Prandtl number ($\mathrm{Sc} = \eta/\rho D_{\mathrm{mass}}$). Mass self-diffusivity is not generally measured in silicate glasses and melts. Instead, diffusion of various species (e.g., cations or anions) are generally measured, where the results vary widely, depending on the temperature, the bulk composition, and the diffusing species (Zhang et al., 2010). For example, the diffusivity of alkalis in albitic melt at 1000°C varies from $\sim 10^{-10}$ to $\sim 10^{-17}\ \mathrm{m^2\,s^{-1}}$ depending on cation size; the diffusivity of SiO_2 at 1300°C varies from $\sim 10^{-15}\ \mathrm{m^2\,s^{-1}}$ in rhyolitic melt to $\sim 10^{-11}\ \mathrm{m^2\,s^{-1}}$ in basaltic melt; and the diffusivity of O in soda–lime silicate decreases from $\sim 10^{-10}\ \mathrm{m^2\,s^{-1}}$ in the liquid at ~1400°C to $\sim 10^{-20}\ \mathrm{m^2\,s^{-1}}$ in the glass at ~500°C (see Figs. 2, 4 and 5 in Ni et al. (2015) for sources). In all cases, the thermal diffusivity is orders of magnitude greater than the chemical diffusivity (i.e., $\mathrm{Le} << 1$), meaning that thermal gradients should reach steady-state long before chemical gradients do so in magmatic environments.

Previous experimental investigations of the thermal conductivity of silicate melts have encountered difficulties with radiative transfer (Murase and McBirney, 1970; Murase and McBirney, 1973, critically discussed by Snyder et al., 1994), or with ionic conduction through the melt, interfering with conductivity measurements via a transient hot-wire technique (Snyder et al., 1994, critically discussed by Shore, 1995). This has produced wildly divergent suggestions as to the magnitude of the thermal conductivity of silicate melts, and its variation with temperature. Murase and McBirney (1970) show data for various rocks whose conductivity is $\sim 0.4\text{–}3\ \mathrm{W\,m^{-1}\,K^{-1}}$ on melting, but then increases at $\sim 0.004\ \mathrm{W\,m^{-1}\,K^{-2}}$ at higher temperatures. Snyder et al. (1994) suggested that the lattice conductivity of diopside melt was $0.3\ \mathrm{W\,m^{-1}\,K^{-1}}$ at 1400°C, decreasing to $0.04\ \mathrm{W\,m^{-1}\,K^{-1}}$ at 1600°C, values that are apparently a factor of 4–30 too low (Fig. 10.3D). Büttner et al. (2000) used a viscometric technique to estimate thermal conductivity of Hawaiian basalt (similar to our OIB), and found that it varied from $\sim 1.0\ \mathrm{W\,m^{-1}\,K^{-1}}$ at 1200°C to $\sim 0.6\ \mathrm{W\,m^{-1}\,K^{-1}}$ at 1350°C, although this change is more likely due to the completion of melting than reflecting the temperature-dependence of melt thermal conductivity. Difficulties with various experimental techniques are reviewed by Hofmeister and Branlund (2015) and in Chapter 4.

Here, we calculated the thermal conductivity of various melts from thermal diffusivity and heat capacity measurements, and using volumes calculated from the model of Lange (1997). Heat capacity models for aluminosilicate liquids can be calculated using the model by Stebbins et al. (1984), where experimental data are unavailable. The calculated thermal conductivity values have been fitted using polynomial equations, and the fit parameters are given in Table 10.2. As discussed above, the thermal diffusivity data are usually limited to a few tens of degrees above the relevant glass transition temperature, and extrapolation of fits to diffusivity or conductivity outside this range is not advisable.

Thermal conductivity of molten silica is $\sim 1.42\ \mathrm{W\,m^{-1}\,K^{-1}}$ and independent of temperature (Hofmeister and Whittington, 2012). Simple feldspar compositions have similar values, highest for Na, then Li, then K. Depolymerized melts such as $CaMgSi_2O_6$ and industrial soda–lime silicates have thermal conductivity values in the range 1.1–1.3 W m^{-1} K^{-1}. Thermal conductivity of silica and alkali feldspar-composition melts remains constant or decreases slightly with increasing temperature, because D and C_P are thought to be approximately constant while thermal expansion continually decreases density. The thermal conductivity of $CaAl_2Si_2O_8$ increases at T_g and then increases with temperature, resembling some of the rock melts but none of the other simple mineral melt compositions. It should be emphasized once again that measurements of both D and C_P for silicate melts are rare, and extrapolations to higher temperatures are fraught with uncertainty. Lattice thermal conductivity of depolymerized melts decreases noticeably as temperature increases. Mills and Susa (1995) reviewed thermal conductivity of slags and showed that many have a maximum in thermal conductivity around the glass transition. They attributed this to an increase in thermal expansivity at T_g, which results in a negative dK/dT for the liquid.

The geological melts span a range in K_{melt} from 1.2 to 2.1 W m^{-1} K^{-1} at 1200°C. In contrast to thermal diffusivity, which was clearly lower for less polymerized melts, there is no clear correlation between K_{melt} and degree of polymerization. The geological melts all show increasing thermal conductivity with increasing temperature, except for haplogranite (iron-free) and leucogranite (almost iron-free). However, the rhyodacite and basalt also show increasing thermal conductivity with increasing temperature, but they are synthetic iron-free melts, so electronic–vibronic coupling cannot be the only explanation. Hofmeister et al. (2014) propose an additional mechanism of heat transfer that is responsible for the "$+HT$" term in the fitted equations (Table 10.2).

10.6 FUTURE DIRECTIONS

There remains much work to be done before the transport properties of silicate melts and glasses can be predicted reliably as a function of composition and temperature. Viscosity models are available that do quite well for limited ranges of composition, yet even the best models typically have uncertainties of at least a factor of 2 (e.g., Giordano et al., 2008, which was calibrated against data for a range of geologically relevant melts). Models that focus on narrower compositional ranges can do better, but uncertainties are still of the order of 0.2 log units, or ~50% (e.g., Sehlke and Whittington, 2015, which was calibrated against data for dry tholeiitic melts, a relatively narrow range of lavas that occur on Earth and other moons and planets). Thermodynamically based models hold the best hope for extension to an all-compassing predictive model, and have been shown to work exceptionally well in simple cases like (Ca, Mg)SiO_3 liquids (Neuville and Richet, 1991). A compositionally dependent model based on the equation of Mauro et al. (2009) may be the most fruitful future direction.

It should be possible to develop models for the thermal diffusivity of glasses and melts as a function of composition and temperature, especially for complex melts such as the geological suite shown in Fig. 10.2B. The pattern of high D for strong, polymerized compositions and low D for fragile, depolymerized compositions holds for both glasses and melts (Fig. 10.2). However, there is a better correlation between D and fragility than between D and NBO/T, which in turn is better than the correlation between D and SiO_2 content for example. Consequently, a model relating heat and mass transport may actually be easier to derive than one relating thermal conductivity to composition. It is unlikely that any single model for K as a function of composition could encompass the range in behaviors shown in Fig. 10.4, such as positive or negative dK/dT, and maxima or minima in K near the glass transition. Consequently, thermal conductivity may be best modeled via its constituent terms, as reliable models already exist for volume and heat capacity of glasses and melts (Stebbins et al., 1984; Richet, 1987; Lange, 1997). More high-temperature measurements of heat capacity and thermal diffusivity are needed for such models to be applied with confidence. This is especially important for melt compositions that show more complex behavior, such as the alkali titanosilicates, whose liquid heat capacity decreases at higher temperature (Bouhifd et al., 1999).

References

Adam, G., Gibbs, J.H., 1965. On the temperature dependence of cooperative relaxation properties in glass-forming liquids. J. Chem. Phys. 43, 139–146.

Angell, C.A., 1995. Formation of glasses from liquids and biopolymers. Science 267, 1924–1935.

Avard, G., Whittington, A.G., 2012. Rheology of arc dacite lavas: experimental determination at low strain rates. Bull. Volcanol. 74, 1039–1056. Available from: https://doi.org/10.1007/s00445-012-0584-2.

Avramov, I., Milchev, A., 1988. Effect of disorder on diffusion and viscosity in condensed systems. J. Non-Cryst. Solids 104, 253–260.

Bansal, N.P., Doremus, R.H., 1986. Handbook of Glass Properties. Academic Press, New York, NY.

Bottinga, Y., Weill, D.F., 1972. The viscosity of magmatic silicate liquids: a model calculation. Am. J. Sci. 272, 438–475. Available from: https://doi.org/10.2475/ajs.272.5.438.

Bouhifd, M.A., Sipp, A., Richet, P., 1999. Heat capacity, viscosity, and configurational entropy of alkali titanosilicate melts. Geochim. Cosmochim. Acta 63, 2429–2437.

Büttner, R., Zimanowski, B., Lenk, C., Koopmann, A., Lorenz, V., 2000. Determination of thermal conductivity of natural silicate melts. Appl. Phys. Lett. 77, 1810–1812.

Clark, S., Spera, F.J., Yuen, D.A., 1987. Steady state double-diffusive convection in magma chambers heated from below. In: Mysen, B.O. (Ed), Magmatic Processes: Physicochemical Principles, Geochemical Society, Special Publication 1, Geochemical Society, University Park, PA, pp. 289–305.

Cordonnier, B., Caricchi, L., Pistone, M., Castro, J., Hess, K.-U., Gottschaller, S., et al., 2012a. The viscous-brittle transition of crystal-bearing silicic melt: direct observation of magma rupture and healing. Geology 40, 611–614.

Cordonnier, B., Schmalholz, S.M., Hess, K.-U., Dingwell, D.B., 2012b. Viscous heating in silicate melts: an experimental and numerical comparison. J. Geophys. Res. 117, B02203. Available from: https://doi.org/10.1029/2010JB007982.

Costa, A., Macedonio, G., 2005. Viscous heating effects in fluids with temperature-dependent viscosity: triggering of secondary flows. J. Fluid. Mech. 540, 21–38.

Dingwell, D.B., 1995a. Relaxation in silicate melts: some applications. Rev. Mineral. Geochem. 32, 21–66.

Dingwell, D.B., 1995b. Viscosity and anelasticity of melts. In: Ahrens, T.J. (Ed.), Mineral Physics and Crystallography: A Handbook of Physical Constants, AGU Reference Shelf 2. American Physical Union, Washington, DC, pp. 209–217.

Dingwell, D.B., Webb, S.L., 1989. Structural relaxation in silicate melts and non-Newtonian melt rheology in geologic processes. Phys. Chem. Miner. 16, 508–516.

Dragoni, M., D'Onza, F., Tallarico, A., 2002. Temperature distribution inside and around a lava tube. J. Volcanol. Geotherm. Res. 115, 43–51.

Eriksson, R., Hayashi, M., Seetharaman, S., 2003. Thermal diffusivity measurements of liquid silicate melts. Int. J. Thermophys. 24, 785–797.

Fei, Y., 1995. Thermal expansion. In: Ahrens, T.J. (Ed.), Mineral Physics and Crystallography: A Handbook of Physical Constants, AGU Reference Shelf 2. American Geophysical Union, Washington, DC, pp. 29–44.
Gent, A.N., 1960. Theory of the parallel plate viscometer. Br. J. Appl. Phys. 11, 85–87.
Giordano, D., Russell, J.K., Dingwell, D.B., 2008. Viscosity of magmatic liquids: a model. Earth. Planet. Sci. Lett. 271, 123–134.
Greaves, G.N., 1985. EXAFS and the structure of glass. J. Non-Cryst. Solids 71, 203–217.
Hale, A.J., Wadge, G., Mühlhaus, H.B., 2007. The influence of viscous and latent heating on crystal-rich magma flow in a conduit. Geophys. J. Intl 171, 1406–1429.
Henderson, G.S., 2005. The structure of silicate melts: a glass perspective. Can. Mineal. 43, 1921–1958.
Hess, K.U., Cordonnier, B., Lavallée, Y., Dingwell, D.B., 2007. High-load, high-temperature deformation apparatus for synthetic and natural silicate melts. Rev. Sci. Instrum. 78, 075102. Available from: https://doi.org/10.1063/1.2751398.
Hofmeister, A.M., Branlund, J.M., 2015. Thermal conductivity of the Earth. In: Schubert, G., Price, G.D. (Eds.), Treatise in Geophysics, V. 2 Mineral Physics. Elsevier, Amsterdam, The Netherlands, pp. 584–608.
Hofmeister, A.M., Pertermann, M., 2008. Thermal diffusivity of clinopyroxenes at elevated temperature. Eur. J. Mineral. 20, 537–549.
Hofmeister, A.M., Whittington, A.G., 2012. Thermal diffusivity of fused quartz, fused silica, and molten SiO_2 at high temperature from laser flash analysis: effects of hydration and annealing. J. Non-Cryst. Solids 358, 1072–1082. Available from: https://doi.org/10.1016/j.jnoncrysol.2012.02.012.
Hofmeister, A.M., Whittington, A.G., Pertermann, M., 2009. Transport properties of high albite crystals and near-endmember feldspar and pyroxene glasses and melts to high temperature. Contrib. Mineral. Petrol. 158, 381–400.
Hofmeister, A.M., Dong, J.J., Branlund, J.M., 2014. Thermal diffusivity of electrical insulators at high temperatures: evidence for diffusion of phonon-polaritons at infrared frequencies augmenting phonon heat conduction. J. Appl. Phys. 115, 163517. Available from: https://doi.org/10.1063/1.4873295.
Hofmeister, A.M., Whittington, A.G., Goldsand, J., Criss, R.G., 2014a. Effects of chemical composition and temperature on transport properties of silica-rich glasses and melts. Am. Mineral. 99, 564–577. Available from: https://doi.org/10.2138/am.2014.4683.
Hofmeister, A.M., Sehlke, A., Avard, G., Bollasina, A.J., Robert, G., Whittington, A.G., 2016. Transport properties of glassy and molten lavas as a function of temperature and composition. J. Volc. Geothermal. Res. 327, 330–348. Available from: https://doi.org/10.1002/2015JE004792.
Holtz, F., Roux, J., Ohlhorst, S., Behrens, H., Schulze, F., 1999. The effects of silica and water on the viscosity of hydrous quartzofeldspathic melts. Am. Mineral. 84, 27–36. Available from: https://doi.org/10.2138/am-1999-1-203.
Hui, H., Zhang, Y., 2007. Toward a general viscosity equation for natural anhydrous and hydrous silicate melts. Geochim. Cosmochim. Acta 71, 403–416.
Lange, R.A., 1997. A revised model for the density and thermal expansivity of K_2O-Na_2O- CaO-MgO-Al_2O_3-SiO_2 liquids from 700 to 1900 K: extension to crustal magmatic temperatures. Contrib. Mineral. Petrol. 130, 1–11.
Leko, V.K., Gusakova, N.K., Mescheryakova, E.V., Prokhorova, T.I., 1977. The effect of impurity alkali oxides, hydroxyl groups, Al_2O_3, and Ga_2O_3 on the viscosity of vitreous silica. Sov. J. Glass Phys. Chem. 3, 204–210.
Loewenstein, W., 1954. The distribution of aluminum in the tetrahedra of silicates and aluminates. Am. Mineral. 39, 92–96.
Mauro, J.C., Yue, Y., Ellison, A.J., Gupta, P.K., Allan, D.C., 2009. Viscosity of glass-forming liquids. Proc. Natl Acad. Sci. 106, 19780–19784. Available from: https://doi.org/10.1073/pnas.0911705106.
Mills, K.C., 1995. Viscosities of molten slags. Chapter 9, Slag Atlas, second ed. Verlag Stahleisen mbH, Düsseldorf, Germany, ISBN: 978–3514004573.
Mills, K.C., Susa, M., 1995. Thermal conductivities of slags. Chapter 15, Slag Atlas, second ed. Verlag Stahleisen mbH, Düsseldorf, Germany, ISBN: 978–3514004573.
Moynihan, C.T., 1995. Structural relaxation and the glass transition. Rev. Mineral Geochem. 32, 1–19.
Murase, T., McBirney, A.R., 1970. Thermal conductivity of lunar and terrestrial igneous rocks in their melting range. Science 170, 165–167.
Murase, T., McBirney, A.R., 1973. Properties of some common rocks and their melts at high temperatures. Geol. Soc. Am. Bull. 84, 3563–3592.
Mysen, B.O., Richet, P., 2005. Structure and Properties of Silicate Melts. Elsevier Science, Elsevier, Amsterdam, The Netherlands.

Mysen, B.O., Virgo, D., Seifert, F.A., 1982. The structure of silicate melts: implications for chemical and physical properties of natural magma. Rev. Geophys. Space Phys. 20, 353–383.

National Academies of Sciences, Engineering, and Medicine, 2017. Volcanic Eruptions and Their Repose, Unrest, Precursors, and Timing. The National Academies Press, Washington, DC. Available from: https://doi.org/10.17226/24650.

Nelson, S.A., 1981. The possible role of thermal feedback in the eruption of siliceous magmas. J. Volcanol. Geotherm. Res. 11, 127–137.

Neuville, D.R., Richet, P., 1991. Viscosity and mixing in molten (Ca, Mg) pyroxenes and garnets. Geochim. Cosmochim. Acta 55, 1011–1019.

Ni, H., Hui, H., Steinle-Neumann, G., 2015. Transport properties of silicate melts. Rev. Geophys. 53, 715–744. Available from: https://doi.org/10.1002/2015RG000485.

Pertermann, M., Whittington, A.G., Spera, F., Zayac, J., Hofmeister, A.M., 2008. Thermal diffusivity of orthoclase glasses and single-crystals at high temperature. Contribs. Mineral. Petrol. 155, 689–702. Available from: https://doi.org/10.1007/s00410-007-0265-x.

Plazek, D.J., Ngai, K.L., 1991. Correlation of polymer segmental chain dynamics with temperature dependent time-scale shifts. Macromolecules 24, 1222–1224.

Rao, M.A., 2014. Rheological properties of fluid foods. In: Rao, M.A., Rizvi, S.S.H., Datta, A.K., Ahmed, J. (Eds.), Engineering Properties of Foods, fourth ed. CRC Press, Taylor and Francis, Boca Raton, FL.

Riboud, P.V., Roux, Y., Lucas, D., Gaye, H., 1981. Improvement of continuous casting powders. Fachber. Hüttenprax. Metallweiterverarb. 19, 859–869.

Richet, P., 1984. Viscosity and configurational entropy of silicate melts. Geochim. Cosmochim. Acta 48, 471–483.

Richet, P., 1987. Heat capacity of silicate glasses. Chem. Geol. 62, 111–124.

Richet, P., Bottinga, Y., 1984. Glass transitions and thermodynamic properties of amorphrous SiO_2, $NaAlSi_nO_{2n+2}$ and $KAlSi_3O_8$. Geochim. Cosmochim. Acta 48, 453–470.

Richet, P., Bottinga, Y., Deniélou, L., Petitet, J.P., Téqui, C., 1982. Thermodynamic properties of quartz, cristobalite and amorphous SiO_2: drop-calorimetry measurements between 1000 and 1800K and a review from 0 to 2000 K. Geochim. Cosmochim. Acta 46, 2639–2658.

Robert, G., Andrews, G.D.M., Ye, J., Whittington, A.G., 2013. Rheological controls on the emplacement of extremely high-grade ignimbrites. Geology 41, 1031–1034. Available from: https://doi.org/10.1130/G34519.1.

Robie, R.A., Hemingway, B.S., Fisher, J.R., 1978. Thermodynamic properties of minerals and related substances at 298.15K and 1 bar (105 Pascals) pressure and at higher temperatures. US Geological Survey Bulletin 1452.

Romine, W.L., Whittington, A.G., Nabelek, P.I., Hofmeister, A.M., 2012. Thermal diffusivity of Mono Crater obsidian. Bull. Volc. 74, 2273–2287. Available from: https://doi.org/10.1007/s00445-012-0661-6.

Russell, J.K., Giordano, D., Dingwell, D.B., 2003. High-temperature limits on viscosity of non-Arrhenian silicate melts. Am. Mineral. 88, 1390–1394.

Sehlke, A., Whittington, A.G., 2015. The viscosity of planetary tholeiitic melts: a configurational entropy model. Geochim. Cosmochim. Acta 191, 277–299.

Shaw, H.R., 1963. Obsidian-H_2O viscosities at 1000 and 2000 bars in the temperature range 700° to 900°C. J. Geophys. Res. 68, 6337–6343.

Shaw, H.R., 1972. Viscosities of magmatic silicate liquids: an empirical method of prediction. Am. J. Sci. 272, 870–895.

Shore, M., 1995. Comment on "Experimental determination of the thermal conductivity of molten CaMgSi2O6 and the transport of heat through magmas" by Don Snyder, Elizabeth Gier, and Ian Carmichael. J. Geophys. Res. 100, 22401–22402.

Sipp, A., Richet, P., 1998. Equivalence of volume, enthalpy and viscosity relaxation kinetics in glass-forming silicate liquids. J. Non-Cryst. Solids 298, 202–212.

Sirbescu, M.-L., Schmidt, C., Veksler, I.V., Whittington, A.G., Wilke, M., 2017. Experimental crystallization of undercooled pegmatitic liquids. J. Petrol. 58, 539–568. Available from: https://doi.org/10.1093/petrology/egx027.

Snyder, D., Gier, E., Carmichael, I., 1994. Experimental determination of the thermal conductivity of molten $CaMgSi_2O_6$ and the transport of heat through magmas. J. Geophys. Res. 99, 15503–15516.

Stebbins, J.F., 1995. Dynamics and structure of silicate and oxide melts: nuclear magnetic resonance studies. Rev. Mineral. 32, 191–246.

Stebbins, J.F., Carmichael, I.S.E., Moret, L.K., 1984. Heat capacities and entropies of silicate liquids and glasses. Contrib. Mineral. Petrol. 86, 131–148. Available from: https://doi.org/10.1007/BF00381840.

Stebbins J.F., Richet P., Dingwell D.B., (Eds.), 1995. Structure, Dynamics & Properties of Silicate Melts. Rev. Mineral. 32.

Urbain, G., 1987. Viscosity estimation of slags. Steel Res. Intl. 58, 111–115.

Urbain, G., Cambier, F., Deletter, M., Anseau, M.R., 1981. Viscosity of silicate melts. Trans. Br. Ceram. Soc. 80, 139.

Viswanath, D.S., Ghosh, T.K., Prasad, D.H.L., Dutt, N.V.K., Rani, K.Y., 2006. Viscosity of Liquids: Theory, Estimation, Experiment, and Data. Springer, Dordrecht, The Netherlands.

Vogel, H., 1921. The law of the relation between the viscosity of liquids and the temperature. Physikalische Zeitschrift 22, 645–646.

Webb, S.L., Dingwell, D.B., 1990. Non-Newtonian rheology of igneous melts at high stresses and strain rates: experimental results for rhyolite, andesite, basalt, and nephelinite. J. Geophys. Res. 95, 15695–15701.

Webb, S.L., Knoche, R., 1996. The glass transition, structural relaxation and shear viscosity of silicate melts. Chem. Geol. 128, 165–183.

Whittington, A.G., Richet, P., Polian, A., 2012. Water and the compressibility of silicate glasses: a Brillouin spectroscopic study. Am. Mineral. 97, 455–467.

Zanotto, E.D., Migliore, A.R., 1999. Viscous flow of glass forming liquids: experimental techniques for the high viscosity range. In: Ricon, J.M., Romero, M. (Eds.), Characterization Techniques of Glasses and Ceramics. Springer-Verlag, Berlin, Heidelberg, pp. 138–150.

Zhang, Y., Ni, H., Chen, Y., 2010. Diffusion data in silicate melts. Rev. Mineral. Geochem 72, 311–408.

Websites

Thermopedia is an online encyclopedia of thermodynamics, heat and mass transfer, and fluids engineering, from Begell House, Inc., http://www.thermopedia.com/

Viscopedia is an online encyclopedia of viscometry, from Anton Paar, Inc. https://wiki.anton-paar.com/en/basics-of-rheology/

Further Reading

Hofmeister, A.M., Whittington, A.G., 2018. Thermal diffusivity and conductivity of glasses and melts. In: Richet, P. (Ed.), Encyclopedia of Glass Science, Technology, History and Culture. John Wiley and Sons, Hoboken, NJ, US, in press. ISBN 978–1118799420.

Stebbins, J.F., 2016. Glass structure, melt structure, and dynamics: some concepts for petrology. Am. Mineral. 101, 753–768.

[illegible]

Websites

[illegible]

Further reading

[illegible]

CHAPTER

11

Modeling Diffusion of Heat in Solids

OUTLINE

If we ever found the conservation law for energy appearing to fail, we would recover it by discovering a new form of energy. – ***attributed to Henri Poincaré by Bridgman (1941) and others.***

Models of heat conduction in solids assume that the kinetic theory of gas applies to scattering of phonons, after Debye (1914). This approach neglects the equivalence of heat and light, the required inelastic interactions and optically thick conditions, and uses summations that violate energy conservation. Diffusive radiative transfer is currently viewed as a distinct phenomenon from conduction (e.g., Brewster, 1992; Siegel and Howell, 1972), although not by Fourier (Chapter 3). Because the ever-present flow of radiation into and out of a material can explain several thermodynamic laws (Text Box 5.1), the present chapter

Measurements, Mechanisms, and Models of Heat Transport
DOI: https://doi.org/10.1016/B978-0-12-809981-0.00011-5

focusses on radiative diffusion at infrared frequencies that are associated with heat. The presentation is guided by experiments and theory presented in previous chapters, with the goal of elucidating heat transfer processes in solids, particularly electrical insulators.

Interestingly, formulae for the effective radiative thermal conductivity k_{rad} can be obtained from the mean free gas model (Hofmeister, 2014). Other findings need to be incorporated: Criss and Hofmeister (2017) showed that the factor of ⅓ originating in the kinetic theory of gas does not apply to heat flow, and provided new sum rules based on the adiabatic approximation (Chapter 3 and Chapter 9). Lifetimes inversely summing means that brief interaction intervals control transport properties, rather than the time between collisions (Chapter 5). These interactions are the source of inelastic losses. A numerical error of 4 in the blackbody intensity function (Chapter 8) needs to be addressed. The dependence of thermal diffusivity on length (Chapter 7) and differences between transient and long-time behavior (Chapter 9) are important. The universal formula for electrical insulators (Hofmeister et al., 2014) needs an explanation.

A macroscopic model is appropriate, for several reasons. Foremost, diffusion requires optically thick conditions, and so averaging is appropriate. Second, infrared light interacts with many millions of atoms as it enters the solid (Wooten, 1972). Infrared light penetrates only $\sim 10^{-9}$ m into metals, but $\sim 10^{-6}$ m into electrical insulators at IR frequencies where strongly absorbed (Chapter 2). Third, a microscopic model needs information on the frequency dependence of thermal diffusivity and conductivity, but such does not exist and only indirect experiments may be possible.

Yet, dialog on the interactions of light with matter is steeped in quantum physics. Some perplexing questions exist regarding how mid-IR light with a wavelength near 10,000 nm can interact with vibrations occurring on ~nm scales. Explaining recent data on the macroscopic process of heat diffusion should provide some clues regarding microscopic behavior. The obvious needs stating, namely that the mostly elastic collisions of unrestricted motions of molecules in a gas, which carry both heat and mass, are not a suitable analog for behavior of solids. Contrastingly, vibrational collisions in solids are localized, substantially inelastic, and wholly disconnected with mass diffusion.

Section 11.1 describes models used in physics and applied sciences. Section 11.2 amends existing formulae in accord with recent theoretical findings. Section 11.3 devises a macroscopic model for radiative diffusion of low frequency heat. The evaluation is semi-quantitative because additional data on spectra and heat transfer properties are needed, as discussed in Section 11.4.

11.1 HISTORICAL PERSPECTIVE

11.1.1 Models of Conductive Heat Transfer

Existing models of heat transport in electrical insulators are based on Debye's (1914) analogy of the scattering of phonons to collisions of molecules in a gas. Peierls (1929) and Klemens (1958) derived summations similar to:

$$K_{lat} = \frac{1}{3}\frac{\rho}{ZM}\sum_{j=1}^{3}\sum_{i=1}^{3NZ} C_{ij}u_{ij}^{2}\tau_{i} \tag{11.1}$$

where is M is the molar formula weight, Z is the number of formula units in the primitive unit cell, u_{ij} is the group velocity ($=\mathrm{d}\omega_i/\mathrm{d}q_j$), q_j is the wavevector, $\omega_i = 2\pi\nu_i$ is the circular frequency of a given mode, ν_i is frequency, τ_i is the mean free lifetime, $i = 1$ to $3NZ$ sums the normal modes of a crystal with N atoms in the formula unit, and $j = 1$ to 3 sums the three orthogonal directions. The Einstein heat capacity for the ith vibrational mode is calculated at constant volume on a per atom basis:

$$C_i = k_B \left(\frac{h\nu_i}{k_B T}\right)^2 \exp\left(\frac{h\nu_i}{k_B T}\right) \Big/ \left[1 - \exp\left(\frac{h\nu_i}{k_B T}\right)\right]^2 \tag{11.2}$$

The difficulty in computing K_{lat} lies in determining lifetimes, which has prompted a variety of approximate formulae. Early publications assume lifetimes go as $1/T$ above $\sim$298K (e.g., Ziman, 1962) or as T^3 below (Callaway, 1959). Very different forms were needed to explain why K_{lat} generally increases with T at low temperature but decreases with T at high T. Fig. 11.1 shows typical behavior: note that end-member nonmetallic crystals behave similarly to the metallic elements whereas insulating glasses and highly disordered crystals behave similarly to alloys (e.g., Cahill et al., 1992; Section 11.3 gives more examples).

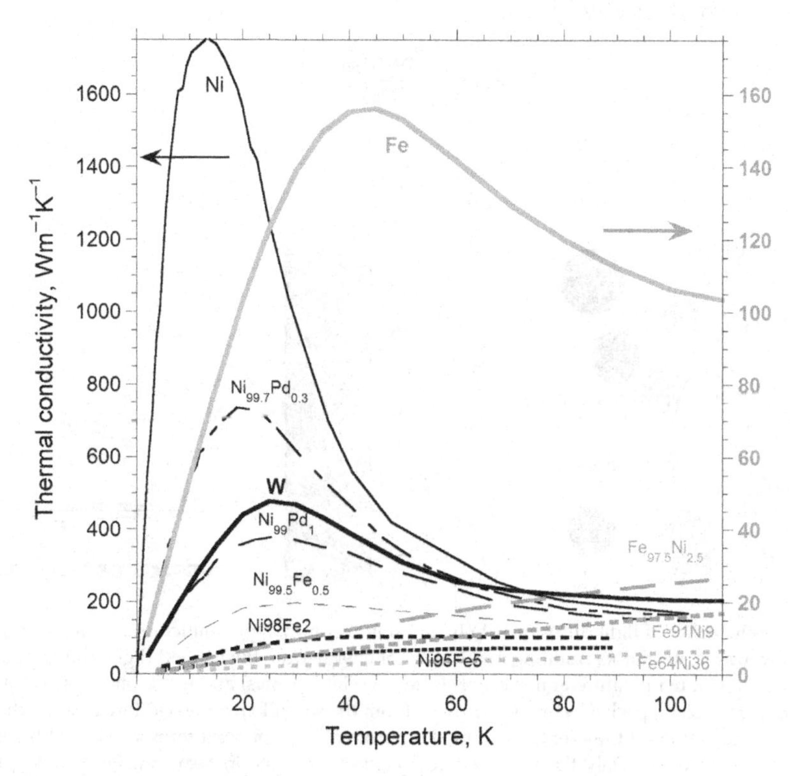

FIGURE 11.1 Dependence of measured thermal conductivity of selected metals and alloys on temperature. Left axis: Ni and Ni alloy data from Farrell and Greig (1969); W from Hust and Lankford (1984). Right axis: Fe from Hust and Lankford (1984): invar ($Fe_{64}Ni_{36}$) and Fe alloys from Bradley and Radebaugh (2013).

Lifetimes have also been quantified via the damped harmonic oscillator model of Lorentz and IR spectral data (e.g., Hofmeister et al., 2007). Because IR modes are mixed with each other and with acoustic modes, extracting the damping parameters from quantitative analysis of IR reflectivity spectra provides the average lifetimes of all modes. More recently, the focus has been on computational studies of the various vibrations. This approach is not covered here, because computational methods contain the flaws listed above and detailed in Section 11.2.

11.1.2 The Effective Thermal Conductivity of Radiative Diffusion

Diffusive radiative transfer is currently modeled as progressive absorption and re-emission of light from grain to grain down a temperature gradient which is sufficiently gradual to maintain local radiative thermal equilibrium (Fig. 11.2A; Hottel and Sarofim, 1967). This process has been considered as important in the deep mantle due to high temperatures (e.g., Lubimova, 1958), and long mean free paths in the associated near-IR regions.

The formula commonly used in geophysics (e.g., Keppler et al., 2008) for an effective radiative thermal conductivity is:

$$K_{\text{rad}} = \frac{16n^2\sigma_{\text{SB}}}{3A_R}T^3 \tag{11.3}$$

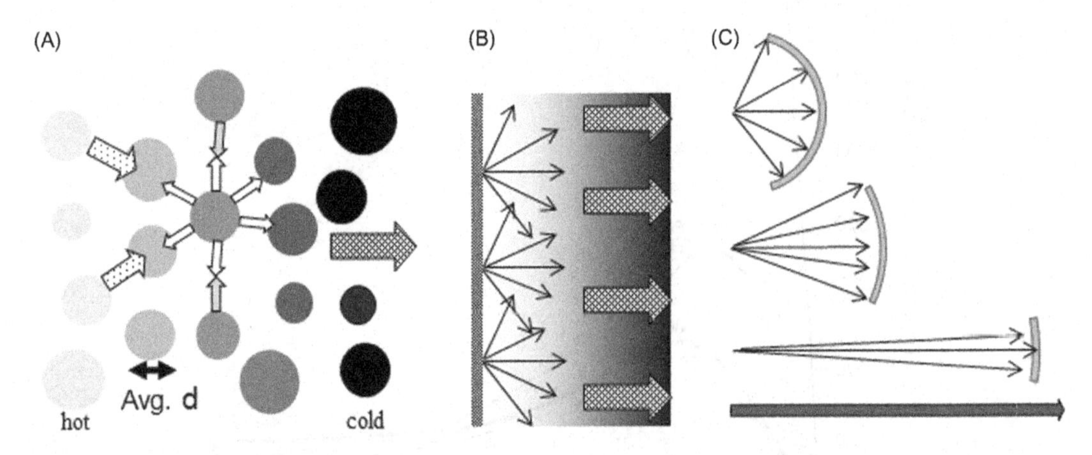

FIGURE 11.2 Schematics of light diffusion. (A) In a grainy medium, light emitted from any given grain of a certain temperature warms its neighbors resulting in diffusion of energy from the hot end (light gray) to the cold (black). Emissions from equivalent temperature grains cancel (gray vs white arrows) and emissions from hotter grains overwhelm emissions from cooler (speckled vs white arrows). From the second law, the net flow is along the temperature gradient (stippled arrow). (B) Light rays entering a medium from any given point form a cone, but the perpendicular components balance each other. Only the parallel components participate in heat transfer, yielding forward flow down the temperature gradient (stippled arrow), so diffraction plays no role in radiative transfer. (C) Comparison of radial flow from a point to parallel rays flowing from a plane source. For the latter, flux is constant, irrespective of distance, so the solid angle is irrelevant.

where n is the index of refraction of the solid medium, σ_{SB} is the Stephan-Boltzmann constant, A_R is the Rosseland (1936) mean absorption coefficient:

$$\frac{1}{A_R} = \frac{\int_0^\infty \frac{1}{A(\nu,T)} \frac{\partial I_{BB}(\nu,T)}{\partial T} d\nu}{\int_0^\infty \frac{\partial I_{BB}(\nu,T)}{\partial T} d\nu}, \tag{11.4}$$

and I_{BB} is the Planck blackbody function (Eq. (2.3); Eq. (8.1)). Frequency (ν) is in Hertz, see also Chapter 8. Shankland et al. (1979) provided a more general formulation:

$$K_{rad}(T) = \frac{4\pi}{3} \int_0^\infty \Lambda(\nu, T) \frac{\partial\{[n(\nu,T)]^2 I_{BB}(\nu,T)\}}{\partial T} d\nu \tag{11.5}$$

where the mean free path, after Lee and Kingery (1960), addresses absorption and scattering, using an average grain size (d):

$$\frac{1}{\Lambda(\nu,T)} = \frac{1}{d} + A(\nu,T) \tag{11.6}$$

Eq. (11.4) is obviously problematic at the near-IR frequencies, where the measured absorption coefficient (A) can be too small to be resolved for typical sample sizes of mm to cm (see Chapter 2; Palik, 1998). Including scattering reduces the impact of A being zero within uncertainty, but this formula assumes the grains are not well annealed and have mismatched n. The integral of Eq. (11.4) is only valid for certain frequency regions: for this reason, books on radiative transfer tabulate frequency intervals for integrals of I_{BB} and its temperature derivative. Lastly, Hofmeister (2010a) argued that the high speeds in transparent spectral regions restrict applicability to transient events.

The factor of n^2 in both Eqs. (11.3) and (11.5) is apparently limited to the geophysics literature: engineering and astrophysics textbooks such as Siegel and Howell (1972) and Rybicki and Lightman (2004) do not include this erroneous factor, see below.

The factor of 4π in Eq. (11.5) stems from I_{BB} being alledgedly provided as per steradian (cf. Chapter 8). This factor is present in Eq. (11.3) as part of $\sigma_{SB} = 2\pi^5 k_B{}^4/(15c^2h^3)$. Radiative diffusion models utilizing these formulae assume spherical symmetry and diffusion along the radius, although many situations involve Cartesian geometry, see below.

11.2 CORRECTIONS TO EXISTING HEAT TRANSPORT FORMULATIONS

11.2.1 Amendments for Directionality

One geometrical error stems from not accounting for the one-dimensional nature of heat flow down a temperature gradient. The numerical factor of ⅓ in summations describes fluctuations in gas under isothermal conditions (Chapter 5). Fluctuations in the transverse directions to the thermal gradient are not important to heat flow and re-equilibration. The sum rules of Criss and Hofmeister (2017) derived in an adiabatic approximation, independently show that the factor of ⅓ from the kinetic theory of gas is in error (Chapter 3 and Chapter 9).

A second geometrical error occurs in formulae used in geophysics for the effective radiative conductivity. The factor of n^2 in Eqs. (11.3) and (11.5) was obtained by Kellet (1952) and Clark (1957), who multiplied the blackbody function by this factor to describe a single cone of radiation entering the medium (see Hottel and Sarofim, 1967). The bending of light at an interface of materials with different n-values is shown in Fig. 2.4B,C. However, diffraction is not relevant to propagation of heat for two reasons (Hofmeister, 2014; Fig. 11.2). First, many such cones of refraction exist because entry is over a plane, not at a single point. Second, only components of rays parallel to the thermal gradient contribute based on thermodynamic law. The same arguments hold for a section of a sphere.

Hofmeister (2014) also showed that Eq. (11.5) can be obtained from Eq. (11.1). She converted the summation to an integral and used $u = c/n$, Planck's (1914) definition of the energy density (e) of a blackbody on a per volume basis at any given frequency:

$$e(\nu) = 4\pi \frac{n}{c} I_{BB} \tag{11.7}$$

plus a thermodynamic definition:

$$C_V = \left.\frac{\partial e}{\partial T}\right|_V \tag{11.8}$$

Note that the heat capacity from Eqs. (11.7) and (11.8) involves energy density of a moving front, so this differs from a material property.

From the above equivalence, the erroneous factor of ⅓ in the summation for conduction affects all equations for K_{rad}. This factor presumes that diffusion occurs in the two angular directions as well as radial.

11.2.2 Implications of the Revised Blackbody Function

For Planck's blackbody function to agree with experimental determinations of σ_{SB} requires:

$$I^*_{BB}(\nu, T) = \frac{h\nu^3}{2c^2} \frac{1}{\exp\left(\frac{h\nu}{k_B T}\right) - 1} = \frac{I_{BB}(\nu, T)}{4} \tag{11.9}$$

where frequency is in Hertz. The origin of the two extra factors of 2 in the historical formulae (I_{BB} without the asterisk) are discussed in Chapter 8. Both functions provide flux per area, per steradian, and per frequency yielding units of $\mathrm{W\,m^{-2}\,sr^{-1}\,Hz^{-1}}$ ($=\mathrm{J\,m^{-2}\,sr^{-1}}$).

This correction requires that Eq. (11.5) be divided by 4. Eq. (11.3) can be obtained from (11.5) using (11.6) with $d = 0$ and A as a numerical constant. Shankland et al. (1979) used the correct formula:

$$\sigma_{SB} = \frac{2\pi^5 k_B^4}{15 c^2 h^3} \tag{11.10}$$

when making this connection. Consequently, one factor of 4 in Eq. (11.3) is also superfluous. In addition, Eq. (11.3) also assumes radiative diffusion in spherical symmetry because $\sigma_{SB}T^4$ describes the total flux over a spherical surface. Another correction is provided in Section 11.2.3.

Clark (1957) and Shankland et al. (1979) used the correct equation for the Stefan-Boltzmann constant based on numerical agreement. Neither mention the integrations performed by Fikhtengol'ts (1948) or Gradshteyn and Ryzhik (1965), and so did not notice the numerical inconsistency of I_{BB}.

Taking the temperature derivative of I_{BB}^* to represent the heat capacity of light per area at some radius under steady-state conditions, as in Stefan's experiment (Fig. 8.7A) gives:

$$\frac{\partial I^*{}_{BB}(\nu, T)}{\partial T} = \frac{\nu^2}{2c^2} C_E(\nu, T) \tag{11.11}$$

Einstein's heat capacity is defined in Eq. (11.2). The connection of BB emissions to the heat capacity of an oscillator does not show that light is produced by quantum oscillations, because Planck's function does not satisfy $E_{tot} = jh\nu$, where j is an integer (Section 11.1 of Chapter 8). Instead, Eq. (11.11) indicates that the heat capacity of a material is related to the ever-present flux of light via inelastic collisions inside the material, and the ability of the material to store this heat. Importantly, I_{BB} was derived by assuming that frequency and temperature are independent variables (Chapter 8). Thus, Eq. (11.11) further shows that Einstein's heat capacity function requires that frequency does not depend on T. Real materials have vibrational frequencies that actually depend on T. Such anharmonic behavior is strong for certain materials, and leads to their measured heat capacity differing from summing the Einstein heat capacities of the modes.

The factor of ½ in (11.11) stems from I_{BB} combining energy with a distribution function, as discussed in Chapter 8. The factor ν^2/c^2 is related to speed of diffusion, after integrating: see Section 11.3.3.2. Fig. 8.7 compares the constant density of light itself, with the decrease in energy density in space as light propagates from a point source.

11.2.3 Point Versus Planar Sources

Because laboratory and astronomical measurements of emissions typically involve small spherical angles and point-like sources, describing emissions as radial suits most applications. For clarity, the factor of 4π should be removed from the various formulations, and it should be stated that K_{rad} is provided per steradian.

In contrast, planar heat sources are used to measure D and K. A visualization (Fig. 11.2C) relates planar to point sources. As the radius grows, the heat front form a point source becomes quite large, and, sections with equal area become progressively flatter with distance. Eventually, rays from a point become effectively parallel. A plane source is composed of many point sources, so the flux is constant with distance, and the steradian no longer applies. To covert radial to Cartesian flow, the factor of 4π that is explicit in Eq. (11.5), and implicit in Eq. (11.3), should be deleted. For constant A and diffusion in a Cartesian system, the correct formula is $K_{rad} = \sigma_{SB}T^3/(\pi A_{const})$ and is not per steradian.

11.2.4 Summation Rules

By conserving energy during heat flow under parallel conditions, Criss and Hofmeister (2017) derived Eq. (3.17) for K, which involves heat capacity on a per volume basis ($C = \rho c$). Solving Eq. (3.17) requires making additional assumptions. Cases relevant to heat transfer in solids are summarized here.

For N independent mechanisms, $C = \Sigma C_i$, yielding:

$$K = N\frac{\sum C_i K_i}{\sum C_i} \text{ for } C_i \approx C_j \text{ and } D_i \approx D_j \tag{11.12}$$

Dimensionally, $D_i = u_i^2 \tau_i$. When the C_is and D_is are each similar, the change in the temperature of the sample with time represents the average behavior of the various carriers. For this averaging to occur requires either accidental similarity of the physical properties or that the processes exchange energy (i.e., are *not* independent), thereby creating similar properties. A single mechanism is discussed below.

Criss and Hofmeister (2017) showed that the case of dissimilar Cs, but similar Ds, leads to a more complex summation:

$$K \cong \sum K_i C_i \frac{\sum C_i}{\sum C_i^2} \tag{11.13}$$

Eq. (11.13) is relevant to independent mechanisms (e.g., electrons vs lattice vibrations in metals).

If the specific heats of each mechanism differ greatly, the result is simple. If $C_1 << C_2 \approx C$, and $K_1 >$ or $\sim K_2$, then $D_1 >> D_2$ is required. For either case:

$$K = K_2 \text{ for } C_1 << C_2 \tag{11.14}$$

Eq. (11.14) describes metals at ordinary temperatures, where the lattice dominates, so fast electronic transport is limited to small amounts of heat, transiently.

If instead $K_1 < K_2$, while $C_1 < C_2$, then Eq. (11.14) still holds. For this case, $D_1 < D_2$ and both the amount of heat and how fast it is transported by mechanism 1 are negligibly small. Unsurprisingly, mechanism 1 contributes negligibly to heat transfer.

For a single mechanism with a total thermal conductivity of k_{single}, then $C_{single} = \Sigma C_i / N$. For a single mechanism, the carriers interact and a single temperature field is a good approximation. Eq. (3.17) simplifies to:

$$K_{single} = \frac{\sum C_i K_i}{\sum C_i} \tag{11.15}$$

If the heat capacities are quite similar, then Eq. (11.15) reduces to:

$$K_{single} = \frac{\sum K_i}{N} \equiv \langle K_i \rangle \text{ for } C_i \approx C_j \tag{11.16}$$

Eq. (11.16) proves that the factor of ⅓ in Eq. (11.1) from the kinetic theory of gas is erroneous. Furthermore, Eq. (11.1) assumes similar D_i. Thus, the terms $u_i^2 \tau_i$ must be quite similar. Lifetimes are discussed next.

None of the previous formulations adhere to energy conservation, as all models for thermal conductivity of solids are based on Eq. (11.1), derived from the kinetic theory of gas, which does not weigh the contributions.

11.2.5 Importance of Lifetimes Adding Inversely

Lifetimes sum *inversely* Eq. (5.10), because these are connected with inverse probabilities (Reif, 1965). If the same speed pertains to both processes, then mean free paths also sum *inversely*. Eq. (11.1) sums the lifetimes while Eq. (11.5) integrates mean free paths, which is equivalent to summing: both formulations are therefore inconsistent with underlying statistical premises. Eq. (11.6) does not fix this problem, even though it contains an inverse summation. For emphasis, we note that absorption coefficients add (i.e., peaks superimpose and are superimposed on the continuum: Fig. 11.3; also see Chapter 2), but the formulation sums A's inversely when Eqs. (11.5) and (11.6) are combined.

Brief interactions dominate lifetimes in correct, inverse summations. Thus, existing models overestimate lifetimes substantially. The weighting factors (heat capacity) show that mechanisms with low heat carrying ability contribute negligibly, whether fast or slow. The difference in speeds is also important, because high speed mechanisms are associated with transient behavior, whereas slow speed mechanisms govern near-equilibrium conditions. Virtually none of these aspects are incorporated in the previous models. Three factors multiplying in summations permits obtaining reasonable agreement of models with measurements, even with numerical and theoretical errors. This behavior is the hallmark of a "sloppy" model (Transtrum et al., 2015).

11.2.6 Inelastic Interactions and the Internal Blackbody Emissions

Although blackbody radiation permeates all matter, it does not stay put. Thus, heat sources are required even for isothermal conditions, and reversibility does not truly exist.

Interestingly, chemists who practice spectroscopy were well aware of the existence of an internal blackbody gas, and devised formulae to describe how its departure from a medium is affected by material properties and thermal gradients (McMahon, 1950; Gardon, 1956; Bates, 1978). Perhaps due to overspecialization, this knowledge has not transferred into models of heat conduction.

It is essential to incorporate inelasticity in heat transport models, which means that the motions of both the internal and applied BB radiation must be considered. It is also essential to distinguish solids from gases. Simply put, heat in gas is mostly transported by long range motions of the molecules, but is modified by their interactions during collisions. In solids, heat transport is entirely based on interactions, since the atoms are fixed in space.

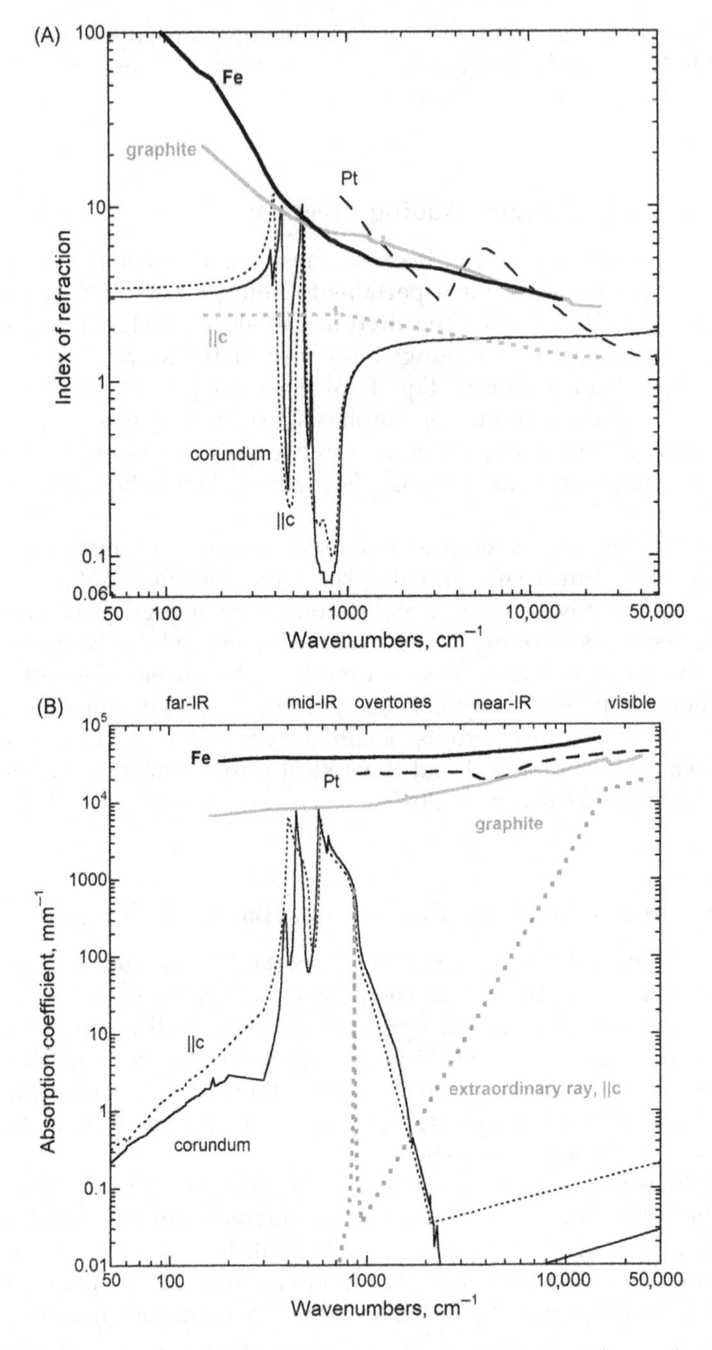

FIGURE 11.3 Optical spectra of materials with diverse properties, determined from reflectivity data shown in Chapter 2 and Chapter 8. Corundum represents electrical insulators. Graphite and metals are nearly featureless, and so best represent the continuum. The compilation of Palik (1998) provides similar values for other electrical insulators and metals. (A) Indices of refraction. (B) Absorption coefficients calculated from the optic index. These values are not affected by back reflections. The very low numerical values in the near-IR are not constrained. Upper axis shows approximate positions for the various spectral regions.

11.3 A REVISED FORMULATION FOR DIFFUSION OF HEAT IN SOLIDS

The numerical corrections and incorporating the heat capacity as a weighting factor, as described in Section 11.2, are straightforward to implement. How to sum lifetimes inversely while accounting for inelasticity is not obvious, requiring us to revisit behavior of gas.

11.3.1 Transport Properties in Gas

For monatomic gases, two processes occur: collisions and interactions during collisions (Fig. 11.4A). Due to inverse lifetimes adding, the governing lifetime is that connected with the brief interval over which the gas molecules or atoms interact. Consequently, the translational speed of the molecules is independent of their interaction lifetimes, and so the mean free path associated with the collisions is irrelevant to gas transport properties. Data on gas support these findings (Chapter 5). To describe the data, Eq. (5.29) was developed, which is based on averages. However, averaging neglects the sum rules (Sect. 11.2.4). For averaging to uphold the first law requires as a minimum that the lifetime of the interactions are similar for all collisions. The following summation is a possibility:

$$K_{\text{monatomic}} \cong \sum u_i^2 C_i^2 \frac{\sum C_i}{\sum C_i^2} \frac{N}{\sum 1/\tau_{\text{interact},i}} \quad \text{for } \tau_i \cong \tau_j \tag{11.17}$$

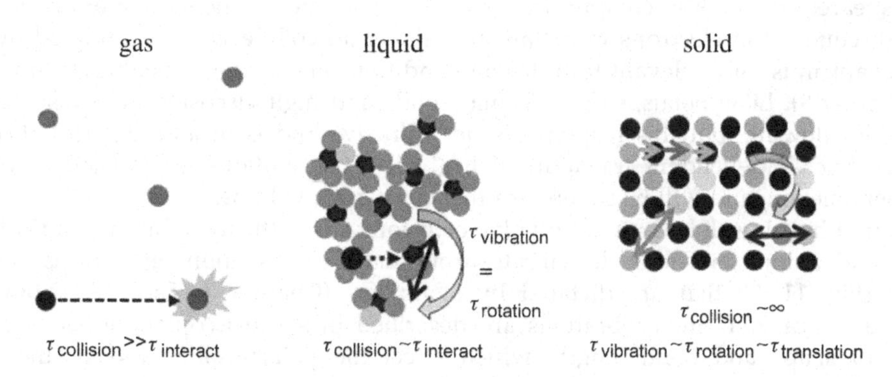

FIGURE 11.4 Schematic of heat transfer in gas, liquids, and solids, showing the different processes and lifetimes. Dots: atoms. Arrows indicate motions. Left = monatomic gas. Speed, mean free path and collisional lifetimes are interconnected. The much shorter lifetime of the interaction determines *D*. Middle = liquid. The collisional process involves "jostling," giving a short lifetime, which is distinct from internal motions of the atoms in the molecules, and from the interaction time. Right = solid. The array of atoms is fixed, so collisions are unlike behavior in gases. Rather, the atoms continually bump into each other in various ways, with similar lifetimes due to interchange of energy among the various reservoirs. Gray arrows at top indicate an acoustic (translational) mode. Black double arrows indicate an optic (stretching) mode. Curved arrow indicates an optic (rocking or rotational) mode.

If lifetimes vary substantially, as is possible for a mixture of very different gases, then this summation is invalid. Monatomic gas is specified because other gases have additional heat reservoirs of vibrations and rotations. If the interaction times of collisions are similar despite multiple reservoirs existing, then Eq. (11.17) pertains to molecular gases as well. This relationship is testable, but is deferred to focus on solids in this chapter.

11.3.2 Transport Properties of Pourable Liquids

Viscosity, mass, and thermal diffusivity of molecular liquids behave in different ways (Chapter 6). The differences become larger as the size of the molecules increase. Viscosity is better described as drag for large molecules. This process hinders mass diffusion, and limits the motions of the molecules (Fig. 11.4B). Rather than collisions, the behavior is better described as jostling, where energy is transferred via both the weak collisions and the vibrational and rotational motions. A much different model than that for gas is needed because pourable liquids transport heat more like solids than gases, and are further complicated by mass diffusion varying with molecule size. Models of heat transport for solids need improving before the complex behavior of liquids can be addressed. However, glasses and molten silicates (Chapter 10) can be treated as condensed matter due to low mass diffusivity and high kinematic viscosity.

11.3.3 Transport Properties of Solids

Strong ionic-covalent bonding within the three-dimensional network of nonmetallic solids severely limits mobility (Fig. 11.4C). Metals and alloys behave similarly to insulators if not sheared or highly compressed, which cause these ductile materials to flow. Although conduction electrons in metals and alloys do collide when stimulated by heat, this mechanism is only relevant to transient conditions and moves a negligible amount of heat (Chapter 9). Like metals, soft molecular solids and high viscosity melts also respond to stress by flowing, but this response is not diffusive and is in addition to lattice heat transport. Such independent transport of heat via mass motions (advection) is a larger scale phenomenon, and will be covered in the companion volume.

Motion of heat in solids does not involve collisions and diffusion of atomic constituents. Cations and anions in solids do vibrate around a fixed position, in various types of motions (Fig. 11.4C) that are dictated by symmetry (Chapter 2; Table 7.1), but these motions are localized. Such vibrations are described in spectroscopic studies as cyclical changes of shape and bond length, which affect the polarizability and/or the dipole moment. The various motions exchange energy even under isothermal conditions: for this reason heat capacities of vibrational modes are averaged in models, after Debye's and Einstein's formulations. That the vibrational modes interact with each other and that those involving changes in the dipole moment furthermore directly uptake and re-emit infrared light points to radiative diffusion as being important to lattice conduction.

Before we present a radiative diffusion formulation, problems in the phonon scattering model need elaborating and characteristic behaviors of solids need to be summarized. This chapter focuses on Al_2O_3, graphite, and iron metal, which represent end-member bond types (insulating, semi-conducting, and metallic), have distinct, well-studied spectra (Figs. 11.3, 2.10, 2.12, 2.14, and 8.6), and have well-constrained transport properties (Figs. 11.1, 7.9, 7.10, 7.14, and 9.11). Metals are included because their spectra are quite similar to that of graphite (Fig. 11.3) and their thermal conductivity behaves much the same with temperature (Fig. 11.5). Importantly, the shape of K_{meas} versus T for metallic elements bears a strong resemblance to the blackbody curve of I_{BB} versus frequency (Fig. 11.1).

11.3.3.1 *Inconsistencies of the Phonon Scattering model*

A single mechanism with many carriers describes both thermal conductivity of the lattice (Eq. (11.15)) and its heat capacity because the various vibrational modes exchange energy. However, mode heat capacities (Eq. (11.2)) are not identical because the frequencies vary. Thus, weighting K for the modes by the heat capacities is needed if a summation is to be used. Summing over different directions is not needed, nor is the factor of ⅓. Hence, Eq. (11.1) should be modified to:

$$K_{\text{lat}} = \frac{\rho}{ZM} \frac{\tau \sum_{i=1}^{3NZ} C_i^2 u_i^2}{\sum Ci} \quad \text{for } \tau_i = \tau_j = \tau \tag{11.18}$$

where $C_{\text{lat}}=\Sigma C_i/N$. For identical lifetimes, an inverse sum is not needed. It is likely that the lifetimes are quite similar, in which case separately averaging these is reasonable, that is, $\tau = N/\Sigma\tau_i^{-1}$, as suggested for gas (Eq. (11.17)). To use this summation, the nature of the heat carrier must be understood.

Acoustic modes with large velocities ($\sim$10 km s^{-1}) are generally considered to dominate heat transfer (e.g., Tang and Dong, 2009, 2010) because the optic modes have low, near zero velocities from dispersion across the Brillouin zone. Crucially, dispersion is measured under isothermal conditions, and thus data on dispersion velocities represent vibrational modes that are not experiencing the disequilibrium conditions inherent to heat transfer. The velocity of optic modes under nonequilibrium conditions cannot be zero, even at zone center, because infrared light (applied heat) of any given frequency is transmitted when samples are sufficiently thin. Otherwise, optical spectra could not be measured. Furthermore, acoustic modes are frequently observed in IR spectra, either as overtones or resonances with IR fundamentals. Substances which lack IR fundamentals (e.g., diamond or Si) still have weak IR bands at modest frequencies, such as overtones associated with acoustic modes and impurity bands (see figures in Chapter 2 and Chapter 8). Exchange of energy between acoustic and optic modes is unavoidable, so acoustic modes cannot dominate heat transport.

Yet, reasonable results are obtained from computational models using sound speeds. Existence of trade-offs among multiple unconstrained parameters underlies the agreement: (1) speeds and lifetimes multiply in all models, so trade-offs exist, i.e. the sum is a sloppy model. (2) Shear and compressional modes are both considered to be effective transmitters in these models, but shear modes do not change the lattice volume (Hofmeister and Mao, 2002) and thus do not participate in heat transfer (Criss and Hofmeister, 2017). Including the two nonparticipating shear modes with the compressional mode compensates for the incorrect factor of ⅓. (3) The terms in the sum were not weighted for heat capacity, which

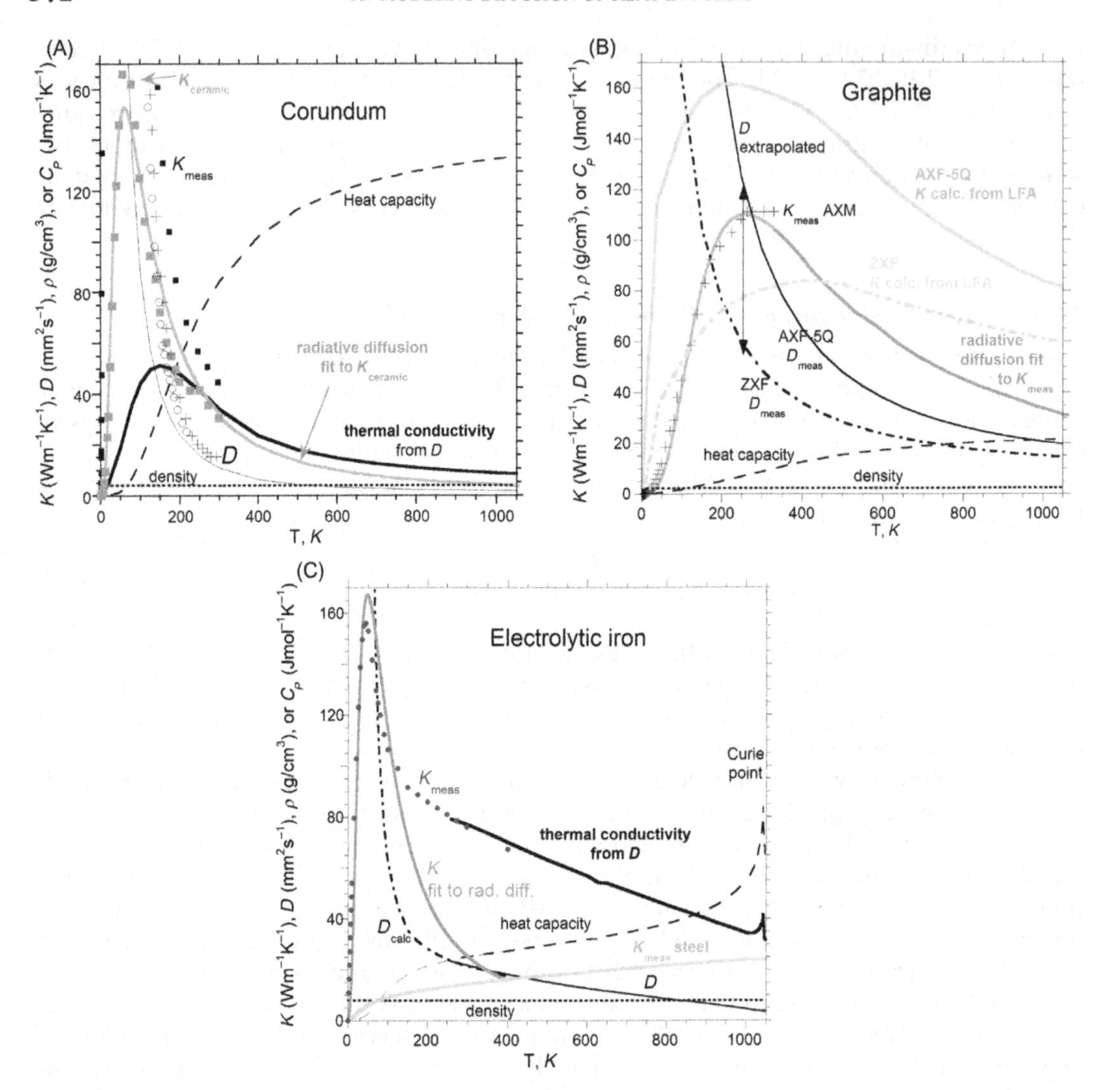

FIGURE 11.5 Transport and thermodynamic properties of a partially transparent insulator, a semiconductor, and a metal. The *X*- and *Y*-axes have the same scale in all panels. The radiative diffusion model (discussed below) uses two free parameters to fit the data. (A) Insulating sapphire, pure Al_2O_3, comparing single-crystal sapphire to ceramics. Dotted line: density (Fiquet et al., 1999). Dashed line: heat capacity (Desai, 1986). Circle: *D* for sapphire; +: *D* for Al23 ceramic, both are LFA data from Burghartz and Schulz (1994), which include ballistic radiation. Fine line: *D* for sapphire (Hofmeister, 2014). Squares: measured *K* (Lake Shore Cryogenics, 2018), which almost certainly are data on crystals from Berman (1951) and on ceramic from Berman et al. (1960), see Childs et al. (1973). Slack (1962) measured gemmy samples and recommended even higher *K* values than Berman did. (B) Graphite, grainy and nearly pure. Thermal diffusivity, density and high *T* heat capacity from the Poco graphite website. Thermal conductivity at low *T* from Woodcraft and Gray (2009); C_P at low *T* on pyrolytic graphite from Pérez-Castañeda et al. (2013). Double arrow indicates that LFA data on *D* are extrapolated below 298K, using the universal equation. (C) Polycrystalline electrolytic iron, Fe, which is NIST standard SRM-8421. Dark gray curve: measured *K* from Hust and Lankford (1984), which agrees with LFA measurements above 298K (fine line). Dashed line: heat capacity from Desai (1986). Because the Curie point is sharp, this region is not amenable to calculations. Light gray curve is K for stainless steel, SRM-1641 from Hust and Lankford (1984). Thermal expansion from Liu et al. (2004).

violates conservation of energy. Optic modes carry most of the heat in a solid, and thus will dominate a properly weighted sum. (4) Importantly, acoustic waves exist whether or not a heat is applied, and so describe a phenomenon other than heat transfer: specifically, acoustic modes are a response to sound or pressure disturbances. (5) Crucially, acoustic modes are not strongly attenuated: for example, seismic waves travel thousands of km, which implies a ballistic response. In contrast, heat is extremely damped: rocks are used as hot plates, because such materials inhibit transmittal of heat beyond ~cm distances.

Another questionable notion of the phonon scattering model is that different scattering mechanisms operate at high and low temperatures (Figs. 11.1 and 11.5). Instead, one mechanism for heat transport is expected because one mechanism provides the lattice heat capacity.

In more detail, the two scattering mechanisms are described as umklapp and normal processes, which refer to the directions of the phonon scattering vectors (Peierls, 1929). Energy flow is reversed by umklapp events, which are considered to dominate at ordinary temperatures (e.g., Burns, 1990). However, flow reversal is inconsistent with thermodynamic law, which mandates that heat flows from higher to lower temperatures. As pointed out by Hofmeister (2006; Fig. 7 therein), the peak in K is clearly due to the contrasting responses of C_P and D with temperature, which multiply to provide K. In detail, the maximum in K results from C_P strongly increasing with T above cryogenic temperatures and then plateauing, whereas D decreases as T increases, first strongly and then weakly. The materials considered in this section were chosen because these exemplify the wide range of behaviors for solids. Fig. 11.5 concerns typically used crystal sizes of a few mm, where thermal diffusivity is independent of thickness, and measured D decreases continuously with T as T increases (Chapter 7), noting that D is rarely measured at cryogenic temperatures.

Some additional materials warrant mention. Continuous behavior is shown for LFA data on D for $YAlO_3$ from ~100K (Aggarwal et al., 2005) to 1600K (Hofmeister, 2010b). Cryogenic contact measurements show that D for $LaAlO_3$ perovskite also decreases continuously from ~10K upwards (Schnelle et al., 2001) and merges with elevated temperature LFA data, if contact losses are accounted for. Both of these perovskites are shown in Fig. 5 of Hofmeister (2010b). Twinning in $LaAlO_3$ inhibits radiative transfer, thereby permitting comparison of LFA data with contact studies of certain orientations. Alkali halide data only roughly agree (Yu and Hofmeister, 2011) because strong radiative transfer exists in these highly transparent single-crystals at low temperatures, as noted in earlier studies. Regarding glasses and highly disordered solids, such as solid-solution feldspars (e.g., Cahill et al., 1992), Fig 9 in Pertermann et al. (2008) showed that K being a product of D and C_P also explains the lack of a peak in K. For disordered materials, D very weakly decreases with temperature, so heat capacity increasing with T dominates thermal conductivity, which then slowly increases with T. This behavior is clearly demonstrated for silica, for which many studies of thermal transport properties exist from ~0.05 to 2000K (see Figs. 7 and 8 in Hofmeister and Whittington, 2012). Similarly, the lack of a peak in K for alloys (Fig. 11.1) is due to their thermal diffusivity being lower and flatter than in metallic elements and generally increasing with T above 298K (Figs. 9.9, 9.11, and 9.15).

Inconsistencies in the phonon scattering models arise because the focus has been on thermal conductivity data, which combine the rate of transport with the amount of heat transported, and because models were based on simple summations, apparently derived by analogy to electrical currents (Chapter 3). Therefore, we have focused on thermal diffusivity, which describes behavior near and far from equilibrium.

11.3.3.2 *Why Radiative Diffusion is the Mechanism of Lattice Conduction*

Phonons are pseudoparticles that are convenient for representing vibrations in a solid. Phonon properties, being measured under isothermal conditions, are disconnected from heat transfer, which is a response to disequilibrium conditions. Phonon scattering models are elastic, where momentum conservation involves momentum of the crystal lattice. However, for temperature to evolve, inelastic losses must exist. In a solid, interactions of vibrating atoms have an inelastic component (damping) and measurably interact with light (Chapter 2).

In contrast to phonons, heat is a real entity, regardless of whether it is best represented as a particle or a wave. Because heat enters and exits solids as light, it must also cross the solid. Observations besides the existence of emissions at virtually any temperature (Chapter 2, Chapter 5, and Chapter 8) support a radiative mechanism for lattice conduction. (1) Any amount of heat may be transmitted. Continuum behavior is required because heat is linked to temperature, which is the parameter actually measured in experiments. (2) Light applied to a solid is absorbed and re-emitted, as shown by diverse experiments, and so a simple mechanism exists for heat transfer. However, heat transfer via radiative diffusion is necessarily complicated, because it depends on temperature, frequency, and the material characteristics. (3) Light crossing partially transparent insulators ballistically is a well-known artifact in heat transfer experiments (Chapter 7). Whether conditions are optically thin or optically thick depends on sample size, and gradations exist between these end-members, so radiative diffusion must also exist during these experiments. (4) Thickness not only affects thermal diffusivity, but the specific response to thickness depends on the IR spectra of the material (Chapter 7). Thus, the process of heat transfer is tied to absorption strengths in the material. (5) Thermal diffusivity increases with temperature at high T, which shows that some type of radiative process is involved (Hofmeister et al., 2014). The increase actually exists at all T, but is only revealed at high temperature due to the small size of the coefficient describing the increase, and because the trend at low temperature involves a strong decrease of D with T. For thin samples, an additional increase in D with T exists at low temperatures, which is also a radiative process. (6) Metals have similar transport properties to insulators, although not identical, and under all but highly transient conditions conduct heat via the lattice vibrations (Chapter 9; Criss and Hofmeister, 2017). Metals, like insulators, emit, transmit, absorb, and reflect light in finite amounts (Fig. 11.3).

All observations point to heat transfer involving light. Hence, one model should describe all heat transport phenomena. Notably, converting the sum of Eq. (11.1) to an integral provides a form like Eq. (11.5), used previously for an effective radiative thermal conductivity (Hofmeister, 2014). Furthermore, a discrete summation with a large number of modes, which was previously used to represent phonon scattering, is not mathematically distinct from an integral. Lastly, the Einstein heat capacity is closely related to the blackbody heat capacity (Eq. (11.11)).

11.3.3.3 Basic Formulation for Radiative Diffusion Inside Solids by a Single Mechanism

We begin with a basic model for an ideal case, where all radiative transfer is diffusive, reflections are neglected, and blackbody radiation describes the entity that is diffusing. This basic model requires that the material is optically thick at all frequencies. It is appropriate for single-crystals or a glassy samples that do not involve significant internal scattering. Although weighting by heat capacity is required to conserve energy, for simplicity we begin by neglecting the weighting factors: this approach is robust if one mechanism dominates. Multiple mechanisms are considered afterwards.

For radiative transfer, the mean free path is germane because this is related to the inverse of the absorption coefficient. Actually, A is not a mean free path, but instead describes attenuation, because A is defined as $-I^{-1}\partial I/\partial z$ (Chapter 2). Hence, Eq. (11.5) describes a single mechanism with multiple carriers: specifically, this mechanism is absorption and the carriers are the various frequencies. The resulting formula for diffusion of blackbody radiation in a Cartesian direction (or per steradian if radial) is:

$$K_{\text{rad}}(T) = \int_0^\infty \frac{1}{A(\nu,T)} \frac{\partial I_{\text{BB}}{}^*(\nu,T)}{\partial T} \text{d}\nu \tag{11.19}$$

It is straightforward to extend the basic macroscopic formulation to condensed matter. Eq. (11.10) relates the derivative in Eq. (11.19) to a radiative heat capacity (C_E), but is on a per atom basis. The prefactor of $\rho/(MZ)$ in Eq. (11.1) that is needed to describe any given material is relevant, since the radiative equation was derived from a simplified summation (Hofmeister, 2014). Combining these equations gives:

$$K_{\text{lat,rad}}(T) = \frac{V_{\text{unit cell}}\rho N_a}{MZ} \int_0^{\text{cut-off}} \frac{1}{A(\nu,T)} \frac{\nu^2}{2c^2} C_E(\nu,T)\text{d}\nu \tag{11.20}$$

A finite upper limit (the cut-off frequency) is used instead of infinity for reasons discussed below.

The factor of the unit cell volume is needed to represent the speed, as follows: Thermal conductivity is the product of diffusion length × speed × heat capacity, where the identity of the diffusion length ($1/A$) and the heat capacity (C_E) are obvious. The factor of ½ exists because the blackbody formulation combines energy with a distribution function. (Chapter 8 showed that considering casting the integral in terms of energy involves increments of energy equal to ½$h\nu$). The quantity $f = V_{\text{unitcell}}\nu^2 d\nu/c^2$ has the correct units of length/time. That f is equivalent to a speed (u) can be confirmed by considering weakly absorbing regions and one particular frequency (i.e., $d\nu$ is approximated as ν). Dispersion relations under weak absorption provide $u = c/n = 2\pi\nu/\kappa$, where κ is the wavevector and equals $2\pi/V^{1/3}$ for a cubic lattice (Burns, 1990: note that wavevectors represent geometry of the Brillouin zone and are used in scattering theory). Thus, $f = c/n^3 = u/n^2$. The extra factor of n^2 arises because optically thin spectral regions were considered whereas optically thick conditions describe diffusion.

More precisely, the heat exchanged at any given frequency is inseparable from the speed. From the definition of the absorption coefficient:

$$A = \frac{\text{power dissipation per unit volume}}{\text{energy density} \times \text{energy velocity}} \tag{11.21}$$

(Wooten, 1972). Inverting Eq. (11.21), and recognizing that power is energy per time, shows that the mean free path is a characteristic velocity times some lifetime. This is consistent with dimensional analysis of diffusion (Chapter 3).

The convolutions of key parameters in both Eqs. (11.20) and (11.21) serves to emphasize that considerable differences exist between radiative diffusion and the collisional mechanism of gas. The sole carrier of heat in solids is not a particle, but is energy itself. Therefore, transport is slow at certain frequencies and temperatures, where the solid is strongly absorbing (interacting). In spectral regions where the solid is not strongly absorbing, interactions still exist with some characteristic faster speed, low absorption coefficient, and longer lifetimes. Eq. (11.20) represents one process, radiative diffusion, and for that reason the basic model does not sum the mean free paths inversely and does not weigh by heat capacity (not yet, at least). From another point of view, blackbody intensity was constructed by assuming isothermal diffusion of a heat front, and so the phenomenon meets thermodynamic requirements for a single mechanism.

Eq. (11.20) does not specify that the unit cell volume and density depend on T because their temperature responses cancel. Likewise, the responses of $V_{\text{unit cell}}$ and ρ to pressure cancel, and so the pressure dependence of K rests on that of A (Section 11.3.5).

11.3.3.4 Scattering in Grainy Solids

To address the additional mechanism of scattering in a grainy medium, Eq. (11.6) inversely summed mean free paths. This equation is incorrect for three reasons. (1) This approach assumes that the speeds of the two mechanisms are the same, which is generally untrue since scattering should occur at the speed c/n whereas the speed associated with absorption depends on the frequency. (2) Scattering depends on frequency if the particles are small. (3) Heat is not transferred during a scattering event.

One possible realistic approach would be to use A that is measured for the medium which has scattering. Testing is needed, such as of well-characterized ceramics. For the present, phenomenological equations based on measurements (Chapter 7) can be used to address mixtures of phases and the effect of porosity in solids.

11.3.3.5 Both Transport Properties Near the 0K Limit

Attenuation is convoluted with speed in Eq. (11.21) and furthermore with frequency and heat capacity inside the integrals of Eq. (11.20). Hence, thermal diffusivity is not simpler to model than thermal conductivity. This finding seems surprising, considering dimensional analysis of the heat equation (Chapter 3). However, the heat equation is governed solely by D only when the temperature changes are small: otherwise heat capacity and density are also involved. Hence, in actuality, thermal conductivity describes the simplest situation, that of near-equilibrium conditions, whereas thermal diffusivity describes thermal evolution for idealized, simple situations. One region where heat transfer should be simple is near the limit of absolute zero; another simplifying circumstance is constant A. These special cases are considered here:

If the absorption coefficient is independent of frequency, then this can be extracted from the integral of Eq. (11.20). Such behavior does not well describe solids (Fig. 11.3). Moreover, extracting A still leaves the speed and heat capacity convoluted, so not much was gained in considering A as independent of ν. Nonetheless, one temperature region can be explored, that near the limit of 0K, where only low frequencies pertain. Eq. (11.20) becomes:

$$K_{\text{lat,rad}}(T) = \frac{V_{\text{unitcell}}\rho N_a}{MZ}\frac{\sigma_{\text{SB}}}{\pi A(T)}T^3 \text{ for } T \sim 0; A \neq f(\nu) \tag{11.22}$$

Heat capacity goes as T^3 near 0K from the Debye model and measurements. Using Eq. (11.10) and Debye's model, specifically the formula of Burns (1990) which provides heat capacity per mole, gives:

$$D_{\text{lat,rad}}(T) = \frac{V_{\text{unitcell}}(T)}{Z}\frac{c}{18}\frac{\Theta^3_{\text{Debye}}}{A(T)}\left(\frac{k_B}{ch}\right)^3 \text{ for } T \sim 0; A \neq f(\nu) \tag{11.23}$$

Volume decreases with temperature, but only slightly. Near the 0K limit, A should increase with T as more vibrational states become populated. Thus, D should decrease as T increases at low T, which is consistent with the data, noting that measurements involve much higher temperatures of $\sim$100K. As $T \rightarrow 0$ both $\nu \rightarrow 0$ and $A \rightarrow 0$, so the size of D could be very large near 0K. Measurements of thermal diffusivity do not exist near this limit, and low temperature measurements of A are rare. Thus, Eq. (11.23) can only be used qualitatively. For sapphire, $A \sim 0.1\ \text{mm}^{-1}$ at low ν (Fig. 11.3); $V = 296\ \text{Å}^3$ contains six formula units; and the Debye temperature is near 600K. These values give $D = 10{,}656\ \text{mm}^2\ \text{s}^{-1}$ at very low T. This estimate agrees with the projected LFA data at 8.5K (Fig. 11.5A).

The above calculation could also be compared to D inferred from measured K of sapphire. However, cryogenic measurements of K for insulators involve frequencies where single-crystals are transparent, which permits unwanted ballistic radiative transfer. Ballistic effects in cryogenic measurements were demonstrated for single-crystal perovskites and alkali halides, and fused silica glass, (e.g., Hofmeister, 2010b; Yu and Hofmeister, 2011; Hofmeister and Whittington, 2012). The difference between K of sapphire and ceramic Al_2O_3 also suggests ballistic transport at low T (Fig. 11.5A).

Despite uncertainties in heat transfer data and the shortage of relevant spectra, Eq. (11.22) can be evaluated qualitatively. First, K should be weaker than T^3. Measurements from $\sim$0.1 to a few Kelvins show that K follows a power law with T (Table 11.1). The power is largest for end-member crystals ($\sim$2.8), slightly less for ceramics and solid-solution crystals ($\sim$2); moderate for glasses, plastics, and brass alloys ($\sim$1.4), and low for metals and most alloys (<1.3; mostly $\sim$1). Platinum has a broad peak at low T which affects the fit. All other elemental metals are fit linearly.

Second, initial $\partial K/\partial T$ should relate to spectral changes at low ν. For the ionic crystals MgO and LiF, the calculated absorption index (k), and therefore A, are low and weakly increase with on both frequency and T (Jasperse et al., 1966). For metals, A increases strongly with frequency (Fig. 11.3). Data on molten Fe (Fig. 2.13) suggest that the temperature response of A for metals is weak. At low frequency, we can extract A out of the integral by using the approximate relationship of $k_BT \sim h\nu$. This approximation provides an effective temperature increase of A that is linear with T. Thus, thermal conductivity of metals should be increase much more weakly with T than does K of electrical insulators, as observed (Table 11.1). Graphite has spectra intermediate to metals and oxides (Fig. 11.3) which is compatible with power law for K at low T also being intermediate (Table 11.1).

TABLE 11.1 Initial Temperature Dependence of Thermal Conductivity as a Power Law

Substance	Power	T Range	Power	T Range	T_{peak}	T_{Debye}[a]
		K		K	K	K
$KBr_{0.53}I_{0.47}$[b]	3	0.05–0.3			4.6	180
Li_3N[b]	2.9	0.08 – 0.8			12	
Sapphire[c]	2.80	1–2			35	990
Al_2O_3 ceramic[b,c]	2.70	1–2	2.5	2–8	50	990
La_2CuO_4[b]	2.50	0.1 – 4			?	
KBr[b]	2.4	0.3–1.3			5	192
TiO_2[b]	2.0	2 – 4.6			16	450
$La_{1.9}Sr_{0.1}CuO_4$[b]	1.9	0.07 – 1.1	1.5	1.1–7.5	?	
BeO ceramic[c]	1.89	1–2.5			120	1280
Macor ceramic[d]	1.86	0.05–1			100	
Teflon PTFE[c]	1.65	1–2			30 weak	
Graphite AXM-5Q[b,d]	1.50	0.05–1	2.37	1–80	300	413
Fused quartz glass[c]	1.45	1–3			5 weak	~250
$Cu_{94.8}Sn_5P_{0.2}$[c]	1.40	1–9			?	~300
$Cu_{84}Mn_{12}Ni_4$[c]	1.35	1–7			?	~340
Pyrex borosilicate glass[c]	1.31	1–4			?	
Platinum[c]	1.30	1–2	1.0	2–7	8	225
Cu[e]	0.7	1–8			22	310
Cu, OFHC[c]	1.03	1–10			32	310
Brass ~$Cu_{68}Zn_{32}$[c]	1.10	1–20			?	~300
Al 6063 (>98% Al)[c]	1.02	1–17			40	390
Electrolytic iron[f]	1.00	1–20			44	373
Stainless steel[c,f]	1.00	1–20	1.24	20–30	?	~350
Tungsten[f]	1.00	1–20			25	312
Gold[c]	0.95	1–10			15	178
Al 1100[e]	0.98	1–20			40	390
Be[e]	1.0	1–20			30	1031
Mo[e]	1.00	1–20			30	377
Pb[e]	n.a.	>4			<3	87
Varnish 7031[c]	0.24	1–12			?	

[a]*Debye temperatures are at room temperature, from various sources but mostly Burns (1990) and http://www.knowledgedoor.com/2/elements_handbook/debye_temperature.html. For mixtures and alloys,* T_{Debye} *was estimated by averaging end-members.*
[b]*From Ventura and Perfetti (2014), who provide power-law fits to various crystalline materials and compile data on metals, alloys, and superconductors.*
[c]*Taken from a graph of thermal conductivity from 1 to 300K of 16 different materials used in cryogenic studies (Lake Shore Cryogenics, 2018).*
[d]*From Woodcraft and Gray (2009), who provide data from 0.05 to 300K on several materials.*
[e]*From Bradley and Radebaugh's (2013) compilation of older data.*
[f]*Hust and Lankford (1984).*
Notes: High temperature powers are provided only if different from the low temperature trend. For many samples, particularly alloys, a peak was not observed in the temperature range explored. For additional data see Bradley and Radebaugh (2013), the references cited below and the Website list.

Qualitatively, the basic model is consistent with low temperature behavior. Additional measurements of D or K where ballistic transport is avoided, and coordinated measurements of low frequency spectra at temperatures below ~70K are needed for a quantitative test.

11.3.3.6 Lattice Conduction at Laboratory Temperatures and Truncation of the Integral

An analytical formula is easier to understand than an integral. Relevant spectral data are quite limited, so generic spectra are considered here. Our analytic approach addresses the need to weight contributions from different mechanisms by their heat capacities, and utilizes that measurements of K at temperatures typically accessed in the lab represent interactions of heat with vibrating atoms.

For temperatures considerably below several thousand Kelvins, lattice heat capacity is large whereas electronic heat capacities (representing free electrons in metals, or d–d transitions in colored insulators) are small. Thus, only vibrational interactions need be considered to evaluate lab data. Consequently, the integral should not extend to infinity. Specifically, for electrical insulators and semi-conductors, the fundamental vibrational spectra are limited to the mid-IR, whereas the overtones and light cation fundamental (e.g., O–H stretching) occur at higher frequencies, but below ~3600 cm^{-1}. Lattice vibrations do not produce light at high frequencies (e.g., the visible) for two reasons: Inelastic exchanges can provide no more than 100% of the energy associated with any given vibration, and energies beyond the third overtone in solids are not observed, even when directly stimulated.

The concept of a cut-off frequency underlies Debye's model for heat capacity which reasonably describes diverse bonding types (e.g., metallic or ionic) and crystalline or amorphous structures (e.g., Burns, 1990). High accuracy is achieved for complex structures with minor modification (Kieffer, 1979). Debye's model does not require infrared activity or discrete transitions: it simply states that the heat content of a solid is represented by a single frequency or a single temperature. Other properties of elements correlate with T_{Debye} (Grimvall and Sjödin, 1974), including the peak in thermal conductivity (Table 11.1; Fig. 11.6).

Distribution of heat energy to infinity is not compatible with experiments describing interactions of light with matter. Different responses of matter to the various frequencies of light (Fig. 2.20) stem from energy density of light varying with ν (Chapter 8). High energy gamma rays can cause nuclear events. Absorption of X-rays probes electron levels and electronic configurations of cations, but mainly X-rays are elastically scattered by arrays of atoms, permitting crystal structures to be determined. Visible light is capable of displacing electrons, producing the photoelectric effect in metals (Chapter 8) and color in insulators (Chapter 2). None of these interactions are important to the slow diffusion at low energies. At low temperatures, some other source than heat is needed to stimulate the above mentioned, high energy interactions. Infrared light is not only thermally produced at low temperature, but it interacts with many atoms in a solid, which process is compatible with diffusion and thus with a macroscopic model. If electronic transitions are coupled to vibrations, then some heat transfer can be conducted electronically: this coupled process

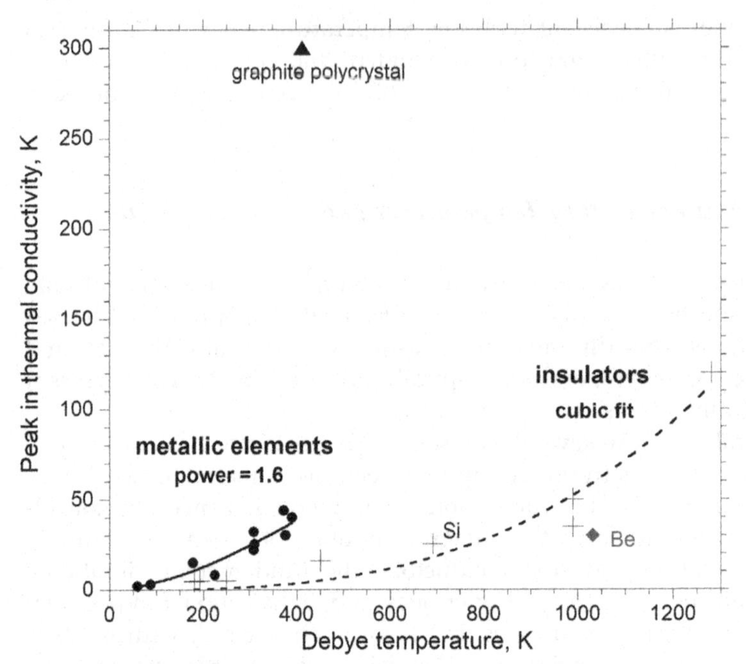

FIGURE 11.6 Connection of the peak in thermal conductivity with the Debye temperature. Dots: metals. Plus: electrical insulators, including Si, which is a single-crystal. Triangle: polycrystalline graphite. Diamond: Be metal, which has Raman activity (Feldman et al., 1968) and a Debye temperature more than double that of any other metallic element. Data in Table 11.1. Fits are power laws, as indicated.

was proposed to operate in glasses with transition elements, in order to explain the diverse responses of D with T at high temperature in colored silicate glasses (Chapter 10).

Importantly, radiative diffusion of heat requires optically thick conditions. Many insulators are extremely transparent in the near-IR (i.e., minerals with negligible to low amounts of Fe^{2+} in large sites), and so a ballistic model is needed for this region and into the visible. The requirement of optically thick conditions explains the dependence of measured D on sample thickness (Chapter 7). For thin colorless samples, only the mid-IR fundamentals can diffuse light (Fig. 11.3), whereas for thick samples of the same composition, overtone/combination bands in the near-IR also participate. Absorption coefficients of overtones being <0.01 times those of fundamentals, explains large D for ~mm sized samples, whose IR overtones can diffuse. The gradual increase of D with thickness up to ~1 mm is due to IR fundamentals having tails which overlap with the overtones, rather than being square shapes. Although these tails extend to infinity in an ideal case, experiments show that mid-IR tails are negligible by ~3000 cm^{-1} for ~cm-sized insulators (Chapter 2).

Therefore, a truncated integral is needed to represent lab data on solids. Because the limit will depend on the spectra of the material, we need to evaluate the indefinite version of Eq. (11.20) for a wide range of functions describing A, or more usefully, its inverse. However, we did not find *any* form for A^{-1} for which the indefinite integral that could be solved, using either Mathematica or an online calculator. In contrast, many analytical solutions exist when a simple exponential is integrated. This simplification of Eq. (11.20) is related to Wein's approximation to the blackbody curve, which is reasonable except at very low frequency (Chapter 8). To evaluate whether a radiative heat capacity after Wein's approach provides a reasonable estimate for K, we compare results from integrating out to the limit of infinity:

Eq. (11.20) can be evaluated analytically from $\nu = 0$ to ∞ when $A_0/A = \nu^j$ for certain integer values of j. Note that A_0 may be a function of T. Using the definitions:

$$b = \frac{h}{k_B T} \quad \text{and} \quad \aleph = \frac{V_{\text{unitcell}}\rho N_a}{MZA_0} \tag{11.24}$$

gives:

$$K_{\text{lat,rad}}(T) = \aleph \frac{b^2}{2c^2} \int_0^\infty \nu^{4+j} \frac{e^{b\nu}}{\left(e^{b\nu}-1\right)^2} d\nu = \aleph \frac{b^2}{2c^2} \int_0^\infty \nu^{4+j} \frac{e^{-b\nu}}{\left(1-e^{-b\nu}\right)^2} d\nu = \frac{\aleph}{2c^2} X \tag{11.25}$$

The less commonly used form for I_{BB} on the RHS shows how Planck's and Wein's formula are linked, and that these historic formulae are equivalent at high frequency. The integral can therefore be approximated as:

$$K_{\text{lat,rad}}(T) = \aleph \frac{b^2}{2c^2} \int_0^\infty \nu^{4+j} e^{-b\nu} d\nu = \frac{\aleph}{2c^2} Y \tag{11.26}$$

Solutions for either Eqs. (11.25) or (11.26) can be summed to explore more complex functions, such as $1/A = (1 + \text{constant} \times \nu)/A_0$. The possibility of summation is why the inverse of A is used in our analysis. Table 11.2 lists results. Note that noninteger values of j can be investigated using Eq. (11.26).

Numerical values of Wein's function are smaller than Planck's such that the discrepancy grows with temperature (cf Eq. (11.25) to (11.26)). Yet, integrating reduces the discrepancy (Fig. 11.7). The ratio of the exact and approximate integrals cannot depend on frequency with the limits chosen, and surprisingly does not depend on temperature. The fit shown in Fig. 11.7 suggests that $K_{\text{rad,diff}}$ can be reasonably estimated using Eq. (11.26). For strong decreases in A with frequency, this scaling scarcely perturbs integration to infinity. We postulate that results of Table 11.2 can be similarly scaled, which is certainly reasonable for a qualitative evaluation of Eq. (11.20).

Regarding which solutions in Table 11.2 usefully represent real materials, the spectra in Fig. 11.3 were fit in various ways. These fits (Table 11.3) oversimplify the data, in order to focus on their essential character as well as the strongly changing spectral regions.

A very useful approximation to modeling radiative diffusion is the "boxcar" shape, where constant A (or A^{-1}) is assigned to any given spectral region, based on the connection of each region with a particular process (Fig. 2.2). Furthermore, the boxcar approximation is instructive because this idealization, with an upper limit of infinity, led to the T^3 law (Eq. (11.3)).

Fig. 11.8 visually compares a boxcar spectrum to approximations for A and A^{-1} depending linearly on ν, and $A \sim \nu^2$, and also shows the results of integrating these three functions. The calculated peak in thermal conductivity depends largely on the cut-off frequency, and is little affected by the power law used for the absorption coefficient (nearly vertical line in Fig. 11.8B). The findings are summarized in the gray boxes. Also shown is that the peak in K depends on T^3 for the boxcar cut-off. Thus, the cut-off frequency depends on the formula used for A. This finding is related to the area under an absorption

TABLE 11.2 Exact (X) and Approximate (Y) Solutions for Radiative Transfer Integral, Utilizing Power-Law Descriptions of the Inverse Absorption Coefficient

j	X	Y	Indefinite Solution for Eq. (11.26)
−4	Nonconverging	b	$-be^{-b\nu}$
−3	Nonconverging[a]	1	$-e^{-b\nu}(1+b\nu)$
−2	$\frac{\pi^2}{3b}=\frac{3.28}{b}$	$\frac{2}{b}$	$-\frac{e^{-b\nu}}{b}(2+2b\nu+b^2\nu^2)$
−3/2	Nonconverging	$\frac{15\sqrt{\pi}}{8b^{3/2}}$	$\frac{15\sqrt{\pi}\mathrm{erf}(\sqrt{b\nu})}{8b^{3/2}}-\frac{\sqrt{\nu}e^{-b\nu}(15+10b\nu+4b^2\nu^2)}{4b}$
−1	$\frac{6\zeta(3)}{b^2}=\frac{7.21}{b^2}$	$\frac{6}{b^2}$	$-\frac{e^{-b\nu}}{b^2}(6+6b\nu+3b^2\nu^2+b^3\nu^3)$
−1/2	Nonconverging	$\frac{105\sqrt{\pi}}{16b^{5/2}}$	$\frac{105\mathrm{erf}(\sqrt{b\nu})}{16b^{5/2}}-\frac{\sqrt{\nu}e^{-b\nu}(105+70b\nu+28b^2\nu^2+8b^3\nu^3)}{8b^2}$
0[b]	$\frac{4\pi^4}{15b^3}=\frac{26}{b^3}$	$\frac{24}{b^3}$	$-\frac{e^{-b\nu}}{b^3}(24+24b\nu+12b^2\nu^2+4b^3\nu^3+b^4\nu^4)$
1/2	Nonconverging	$\frac{945\sqrt{\pi}}{32b^{7/2}}$	$\frac{945\mathrm{erf}(\sqrt{b\nu})}{32b^{5/2}}-\frac{\sqrt{\nu}e^{-b\nu}(16b^4\nu^4+72b^3\nu^3+252b^2\nu^2+630b\nu+945)}{16b^2}$
1	$\frac{120\zeta(5)}{b^4}=\frac{124}{b^4}$	$\frac{120}{b^4}$	$-\frac{e^{-b\nu}}{b^4}(120+120b\nu+60b^2\nu^2+20b^3\nu^3+5b^4\nu^4+b^5\nu^5)$
2	$\frac{720\pi^6}{945b^5}=\frac{732}{b^5}$	$\frac{720}{b^5}$	$-\frac{e^{-b\nu}}{b^5}(720+720b\nu+360b^2\nu^2+120b^3\nu^3+30b^4\nu^4+6b^5\nu^5+b^6\nu^6)$
j[c]	$\frac{(4+j)!\zeta(4+j)}{b^{3+j}}$	$\frac{(4+j)!}{b^{1+j}}$	$-\frac{e^{-b\nu}}{b^{5+j}}\left[\begin{matrix}(4+j)!+(4+j)!b\nu+\frac{(4+j)!}{2!}b^2\nu^2+\frac{(4+j)!}{3!}b^3\nu^3\\+\dots+\frac{(4+j)!}{(3+j)!}b^{3+j}\nu^{3+j}+b^{4+j}\nu^{4+j}\end{matrix}\right]$

[a] *The recursion relation suggests that* X (j = −3) = ζ(1), *which asymptotes to positive infinity from above. Specifically, for large* y, ζ (1 + 1/y) = y.

[b] *This value for* j *(constant* A) *returns the* T^3 *result prevalent in radiative transfer models (Section 11.1).*

[c] *Integers must be* ≥ −2 *for the Planck function, but can be smaller,* ≥ −4, *for the Wein function.*

Notes: The column for the definite integral Y should be added the indefinite integral to provide results from $\nu = 0$ to $\nu = \nu_{max}$. At the limit of $\nu = \infty$, Y = 0. ζ is Riemann's zeta function.

peak being proportional to the number of atoms producing the peak. Lastly, Fig. 11.8B shows that the $\partial K/\partial T$ above the peak depends on the parameters and the formula, but weakly.

11.3.3.7 Accounting for the Effect of Temperature on Absorption

If A can be separated into a function of frequency that is multiplied by another function of temperature, then the latter can be extracted from the integral. Although the absorption coefficient is expected to be a complex convolution of T and ν, using some average dependence of A on T (or P) is reasonable because integrating averages. Truncating the integral provides an average for the mechanism associated with that particular frequency range.

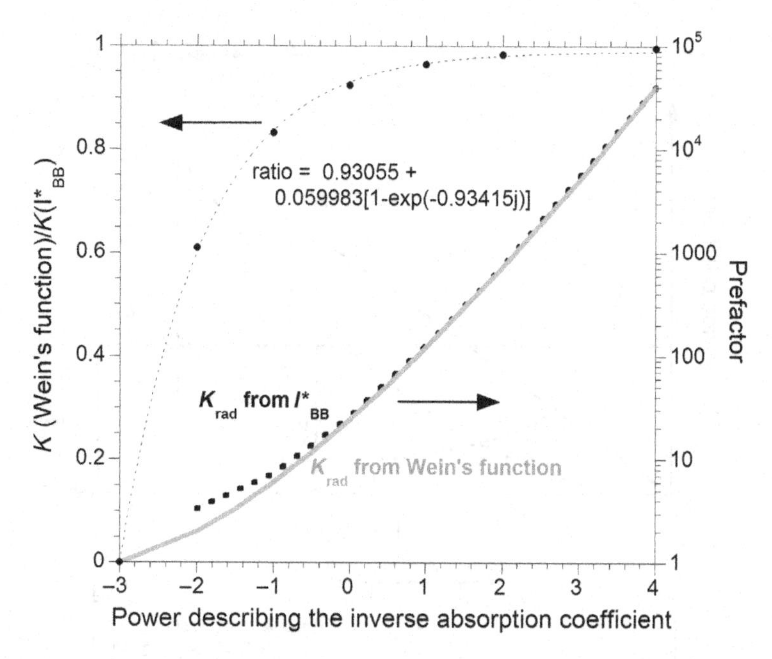

FIGURE 11.7 Results from Table 11.2 as function of j, the power of A^{-1}. Dotted line with symbols: ratio of $K_{rad,diff}$ from Eqs. (11.26) to (11.25). Heavy dots: the numerical prefactor in X. Gray line: the numerical prefactor in Y. Note that both X and Y depend on the same power of b and thus of T, and both K's are multiplied by $B/(2c^2)$.

TABLE 11.3 Fits to Spectral Data on Sapphire, Graphite, and Iron in Figs. 11.3 and 2.13

Substance	Range of ν cm^{-1}	Fit to A^{-1} mm	Residual
Sapphire E‖a[a]	~0–600	$9877.2\nu^{-2}$	0.946
Sapphire E‖c	~0–600	$9218.1\nu^{-2}$	0.960
Sapphire E‖a	~4000–2500	$3.6309 \times 10^{-39}\nu^{12}$	0.973
Sapphire E‖c	~400–2500	$1.595 \times 10^{-32}\nu^{10}$	0.999
Graphite E‖a	~100–24,000	Ninth-order polynomial	0.995
Graphite E‖a	~800–24,000	$0.0023783\nu^{-0.44699}$	0.996
Iron	~0–200	$0.0005608\nu^{-1}$	0.99
Iron[b]	~0–2000	$[38372(1-e^{-0.012297\nu})]^{-1}$	0.998
Iron	~200–22,000	$2.5826 \times 10^{-5} - 6.9549 \times 10^{-10}\nu$	0.99
Iron	~200–52,000	$2.8442 \times 10^{-5} - 1.1584 \times 10^{-9}\nu + 1.7662 \times 10^{-12}\nu^2$	0.99

[a]*The spectroscopic orientation of E‖a corresponds to heat flow along the c-axis.*
[b]*This fit was made to A, but is listed as inverse of A.*

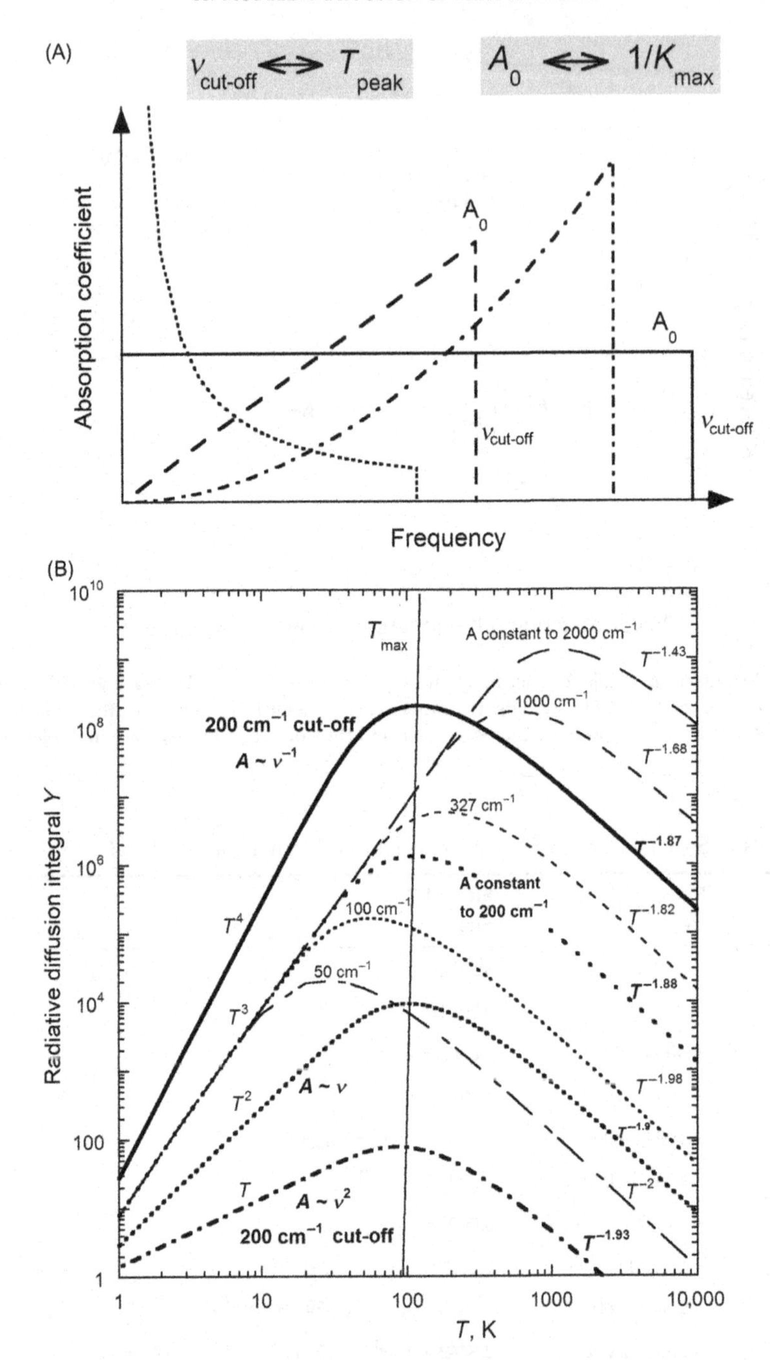

FIGURE 11.8 Idealized spectra for integrals and results. (A) Cases of the absorption coefficient decreasing inversely with frequency (dotted curve), being constant (solid) and increasing linearly (dashes) or by ν^2 (dot-dashes). Maximum A and cut-off frequencies are indicated for the linear and boxcar idealizations. Gray boxes show how these parameters link to thermal conductivity at low temperatures. (B) Results for the approximate radiative diffusion integral of Eq. (11.26). Heavy lines all use a cut-off frequency of 200 cm^{-1}, while comparing the various synthetic spectra of part A, as labeled. Dashed lines all depict the boxcar with constant A but different cut-off frequencies, as labeled. The slopes to either side of the peak follow power laws in T, as indicated. As shown, T_{max} is weakly affected by the spectral form, but strongly by the cut-off frequency.

11.3.3.8 Why One Model Represents All Solids

Inelastic interactions of atoms exist during any vibrational motion, and this production of energy provides the idealization known as the blackbody curve. All solids can emit as a blackbody under certain conditions, but these conditions are difficult to meet for insulators, due to surface gradients, strong peaks, and reflections at interfaces. The smooth, nearly featureless spectra of graphite and metals (e.g., Fig. 11.3) are why these materials are used in blackbody experiments.

However, spectra record the response of the solid to stimulation by applied light, and do not indicate which frequencies are present at any given temperature and can participate in thermal conduction. This difference is the origin of the cut-off frequency, which correlates with the Debye temperature (Fig. 11.6) and represents the heat being carried (C_P). For metals, the Debye frequency clearly correlates with acoustic modes, although the calorimetric value differs from the elastic value. Why is obvious: acoustic modes are the only fundamental vibrations in most metals. Triangularly shaped spectra, as in Fig. 11.8A, could represent heat conduction in metals. Importantly, the dispersion relations of acoustic modes are irrelevant to heat transfer since the triangle shape is a function of *frequency*, rather than *inverse lattice constant*.

For solids with IR active modes (most minerals), either a boxcar or a triangular spectrum can be used to model heat capacity, but accuracy for many atom structures is improved with combinations thereof plus delta functions at certain frequencies (Kieffer, 1979). As in the model for C_P, accuracy of the model for K will depend on the complexity of the lattice structure. Because the boxcar shape better reproduces C_P of insulators than the triangular shape, the former is used as the first step to model K.

11.3.3.9 Extension of the Basic Model to Multiple Mechanisms

To allow for additional mechanisms that are associated with slightly higher frequencies, we sum the integral for the fundamentals with other integrals representing additional processes. Each integral represents one mechanism and involves different absorption coefficients and different frequency limits: the boxcar approximation with both cut-on and cut-off frequencies is appropriate for mechanisms other than fundamental vibrations. For the overtones only, integration can begin with $\nu=0$ because some overtones occur in the range of the IR fundamentals, or overlap with the fundamentals, and therefore exchange energy. However, the extreme difference in A values (about 1000-fold) requires that different integrals be used for overtones and fundamentals. The need to weight the contributions is accounted for when summing the various integrals.

The importance of overtones to heat conduction of insulators in experimental measurements is suggested by thermal diffusivity and overtone absorbance depending similarly on thickness (Figs. 7.9 and 2.10). The cut-off frequency for the overtones increases with thickness because the overtones become progressively weaker with order and thus with frequency. The overtone region terminates near 3000 cm^{-1}, such that the next spectral region, the near-IR, is highly transparent and transmits light ballistically, not diffusively, at ~mm thicknesses used for heat transfer experiments. On the basis of Fig. 11.8, the gradual increase in the cut-off frequency will gradually increase K at any given temperature and then plateau, as observed for transparent insulators (Figs. 7.9, 7.10).

Certain materials have Raman, but not IR, fundamental modes: examples are the diamond structure and hcp metals. Their overtones are IR active. For Si, overtones are present from ~500 to 2000 cm^{-1} (Fig. 8.2). Finite but low absorption coefficients at high frequency lead to the high K of the diamond structure.

Few species are active in the near-IR spectral region. Stretches of O–H occupy a narrow frequency range and have moderate absorption coefficients, so these contributions are likely overshadowed by lattice conduction. Of greater relevance to planets is the d-d electronic transitions ranging from the near-IR to the UV. It is unexpected that electronic processes contribute to heat transport under lab conditions of low temperatures and small size. Relevance of this mechanism to planets (which have high temperatures and large size) will be covered elsewhere.

11.3.4 Comparison of Thermal Conductivity Data to the Radiative Transfer Model

The radiative diffusion model can be utilized without fitting parameters if IR spectra are available at low frequency and appropriate temperature. Spectral data exist on MgO and LiF to cryogenic temperatures (Jasperse et al., 1966), but the low temperature K values of MgO differ among studies and only one dataset on LiF exists (Ventura and Perfetti, 2014). Rather than focusing on two compounds with the same structure, we investigate a variety of solids by fitting the data on thermal conductivity using the analytical formulae of Table 11.2.

The basic model for a single mechanism involves one parameter (the cut-off frequency) and a choice of equations in Table 11.2. The prefactor is unimportant to the evaluation because K can be nondimensionalized. Importantly, the choice of formulae is dictated by the initial derivative of either A with ν or of K with T, which are unequivocally linked, as discussed above. Power laws for $K(T)$ in Table 11.1 show that insulators have constant A at low frequency, whereas metallic elements have A increasing proportional to ν^2. The deduced behavior is consistent with available spectral data near 298K (Fig. 11.3). Other solids (glasses, ceramics, semi-conductors, alloys) show intermediate behavior (Table 11.1; Ventura and Perfetti, 2014). The equations used in the fitting are restricted by spectra and/or very low T data on K, and moreover the formulae represent broad classes of solids, rather than specific substances. Therefore, one parameter is needed in our radiative transfer model if only one absorption mechanism exists.

11.3.4.1 *Expected Behavior of* K *with* T *from the Basic Model*

From Fig. 11.8, the peak temperature for K should increase linearly with the cut-off frequency. This deduction is consistent with the data in Fig. 11.6 that demonstrate a simple dependence of the position of the peak in K on T_{Debye}. Metals and insulators occupying differ trends in Fig. 11.6 is consistent with different spectral shapes being needed to accurately model each of K for these two types, as is known to be the case for C_V (cf. Burns, 1990 to Kieffer, 1979).

Fig. 11.8B indicates that for K associated with fundamental vibrations should steeply decrease immediately above the peak, and then become flat at high T. This general behavior exists for all types of solids at moderate temperature (Fig. 11.5). Departure from the basic model increases as temperatures climb, which is expected because only one process is described, namely diffusion of heat by the low frequency fundamental vibrations. This result is due to the nature of the integral (Eq. (11.25)), which begins at ~0K. The model provides the essential behavior of K at low T, which is common to all types of solids.

11.3.4.2 Fits Below ~300K Using Simple Frequency Functions for A

Fits were made to representative data for different types of solids. Absolute values of K are used, because these are important in applications, so the prefactor is included in the fitting. However, this parameter is inconsequential to the physics.

For the insulator Al_2O_3, the boxcar model was used. Data on the ceramic sample in Fig. 11.5A were fit to:

$$K(T) = \text{const.}\left[\frac{24}{b^3} - \frac{e^{-b\nu}}{b^3}\left(24 + 24b\nu + 12b^2\nu^2 + 4b^3\nu^3 + b^4\nu^4\right)\right]; \quad b = \frac{1.44}{T} \tag{11.27}$$

This yielded a cut-off frequency of 108.6 cm^{-1} and a prefactor of 0.000713 W m^{-1} K^{-1} with a residual of 0.986. The gem-quality sample could not be fit, which is attributed to ballistic augmentation. Projecting the above fit to higher temperatures parallels the trend for sapphire at high T, which is based on LFA measurements from which ballistic effects were removed. That the model (based on ceramic data) provides slightly lower thermal conductivity than single-crystal data is expected, and is consistent with LFA measurements on the same type of ceramic (Al23) and sapphire (Fig. 7.9).

Agreement of the radiative model with data on graphite is excellent (Fig. 11.5B). The fit is based on A increasing linearly with ν which gives K increasing as T^2, as observed. An initial increase in A is required (Table 11.3) because graphite absorption is high in the mid-IR, but fairly flat (Fig. 11.3). The fitting equation:

$$K(T) = \text{const.}\left[\frac{6}{b^2} - \frac{e^{-b\nu}}{b^2}\left(6 + 6b\nu + 3b^2\nu^2 + b^3\nu^3\right)\right]; \quad b = \frac{1.44}{T} \tag{11.28}$$

gave a cut-off frequency of 293.1 cm^{-1} and a constant prefactor of 3.7589×10^{-5} W m^{-1} K^{-1} with a residual of 0.999. The projection of the fit to 1100K is compatible with available data on other Poco graphites with similar porosity and grain size. As temperature climbs further, measured thermal conductivity next becomes flat and subsequently increases linearly with T, which involves another mechanism, that of the overtones (Section 11.3.4.3).

Measured thermal conductivity of sapphire and graphite are similar to results for other nonmetallic crystals and ceramics (see websites) as well as glasses (Chapter 10). However, measurements below and above 298K are generally made in different labs using different techniques on nonidentical samples. Spurious ballistic transfer is a common problem. For these reasons, and because the radiative diffusion model describes the general behavior of insulators and semi-conductors, and fits corundum and graphite well to ~1100K, additional examples of these types are not explored.

Metallic iron is less well-represented by the model (Fig. 11.5C). The fitting equation:

$$K(T) = \text{const.}\left[\frac{2}{b} - \frac{e^{-b\nu}}{b}\left(2 + 2b\nu + b^2\nu^2\right)\right]; \quad b = \frac{1.44}{T} \tag{11.29}$$

gave a cut-off frequency of 94 cm^{-1} and a prefactor constant of 0.0797 W m^{-1} K^{-1} with a residual of 0.8. Iron has a magnetic transition at 1000K, which affects both its D and C_P. Although the effect on K is small at the Curie point, $\partial K/\partial T$ is impacted. Because ballistic transport is not a problem for metals, additional elements and alloys are investigated:

For pure and nearly pure metals, K reaches a peak below ~100K (Fig. 11.9). The best fits were obtained for $A \sim \nu^2$, which is consistent with K increasing linearly with T above ~0K, as is observed. The basic model describes metals at low T, including the peak in K, but temperatures near ambient and above are not well-represented. How well the peak is fit depends on several factors, one being the range selected for the fitting. Another factor is that the height of the peak varies considerably: for example, different studies of high purity tungsten provided a peak K-value that varies from 400 to 1000 W m^{-1} K^{-1} (Hust and Lankford, 1984). Impurities have a great effect on near end-members, metals or insulators.

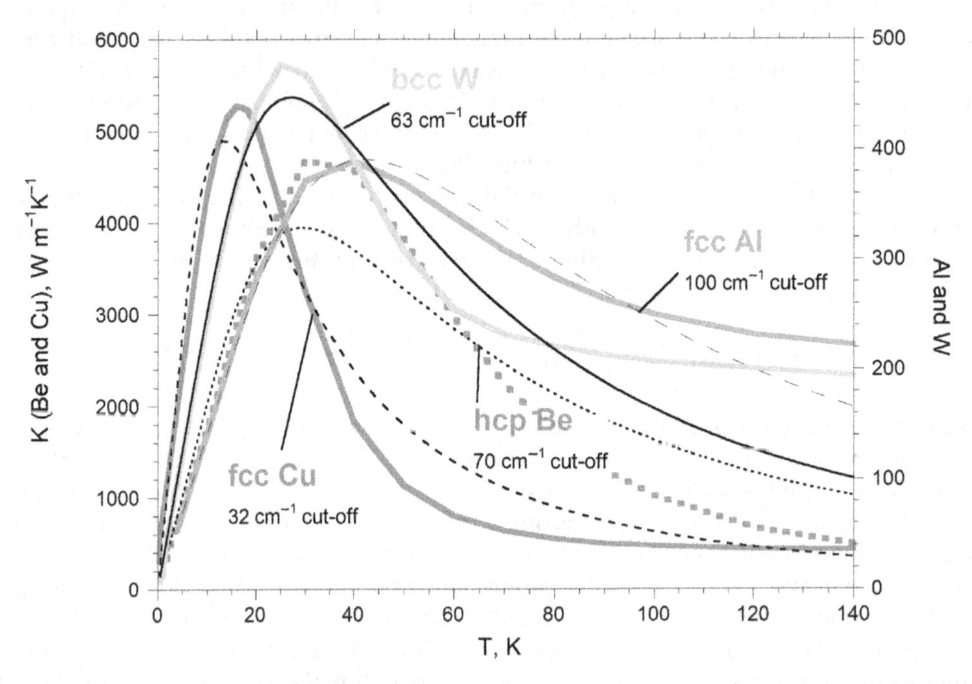

FIGURE 11.9 Fits of thermal conductivity data on metals to the basic radiative diffusion model assuming the absorption coefficient increases as ν^2. Data (gray curves) are fit up to 140K. Sources are Hust and Lankford (1984) and the compilation of Bradley and Radebaugh (2013). Al-1100 is an alloy with about 1% impurities, usually Si or Fe. Copper is oxygen free, but the processing has greatly improved with times, and so these data represent a metal with low impurities.

Both types of mismatch show that a second mechanism exists, which involves higher frequencies than the fundamentals, and becomes increasingly important as T increases, but is present at low temperature as well. In support of this deduction, the spectra in Fig. 11.3 show that A is flat in the mid- to near-IR and then increases in the visible: Thus, the initial trend in A from the far-IR that governs cryogenic behavior of metals is not representative of A that governs higher T behavior. Above the Curie transition, $K(T)$ for Fe is fairly flat. The shape of $K(T)$ for Be differs, which is consistent with the hcp structure having Raman activity and distinct overtones. For all metals, multiple mechanisms for radiative transfer exist, and need to be included in fitting to correctly portray high T behavior.

11.3.4.3 Accuracy of the Approximate Integral

From Figs. 11.5 and 11.9, better fits are obtained for materials that have peaks at higher temperature, and therefore cut-off frequencies that occur at higher temperature. High cut-off frequencies are associated with better fits because the approximation of Eq. (11.26) becomes closer to the exact integral Eq. (11.25) as ν increases.

11.3.4.4 Thermal Conductivity at Higher Temperature Associated With Vibrational Overtones

For multiple processes, the contributions of the various integrals must be weighted by their heat capacities. Because dimensionless data could be fit, the prefactor associated with integrating over the fundamental vibrations is inconsequential. However, the ratio of the prefactors for the overtone to the fundamental integral is important to the fitting. This ratio combines the ratio of in absorption coefficients with the ratio of weighting factors, and also compensates for differences between the exact and approximate integrals. Thus, accounting for two mechanisms involves three parameters. We focus on metals because these lack ballistic transfer.

The metals and Al-1100 alloy were reasonably fit by using the boxcar approximation or the absorption constant increasing with frequency for the overtones (Fig. 11.10). However, the alloy steel required $A \sim \nu$ for the overtones to provide the nearly linear increase in K over the range of measurements. Thermal conductivity of steel could not be fit with one integral. Likewise, the elements all need two formulae, representing each of fundamentals and overtones, to represent thermal conductivity over a wide T range. The fit for iron is affected by the Curie transition, causing one of the prefactors to be negative. Fits for the other metals involve a much smaller prefactor for the contribution of the overtones, as expected.

With two integrals (i.e., two formulae), fitting provides two peaks in $K(T)$ in all materials (Fig. 11.10). In the model, each peak in K is associated with a cut-off frequency, such that $T_{max} = \sim {}^1\!/_2\nu_{cut\text{-}off}$. Peaks at higher T are broader because the FWHM is related to ${}^1\!/_2\nu_{cut\text{-}off}$ (i.e., $\sim {}^1\!/_2T_{max}$). Indeed, two peaks in measured K with T are often observed experimentally. Aluminum has well-resolved peaks in $K(T)$ at ~100 and 400K. Tungsten has a strong peak at 30K and a weak peak at 220K. Iron has a strong peak at 150K and seems to have a weaker, overlapping peak at 180K. Steel has two weak peaks at about 175 and 1200K, which blend. The measurements also provide FWHM of about twice the peak position. The model best fits metals and alloys where the measurements provide well separated peaks in $K(T)$.

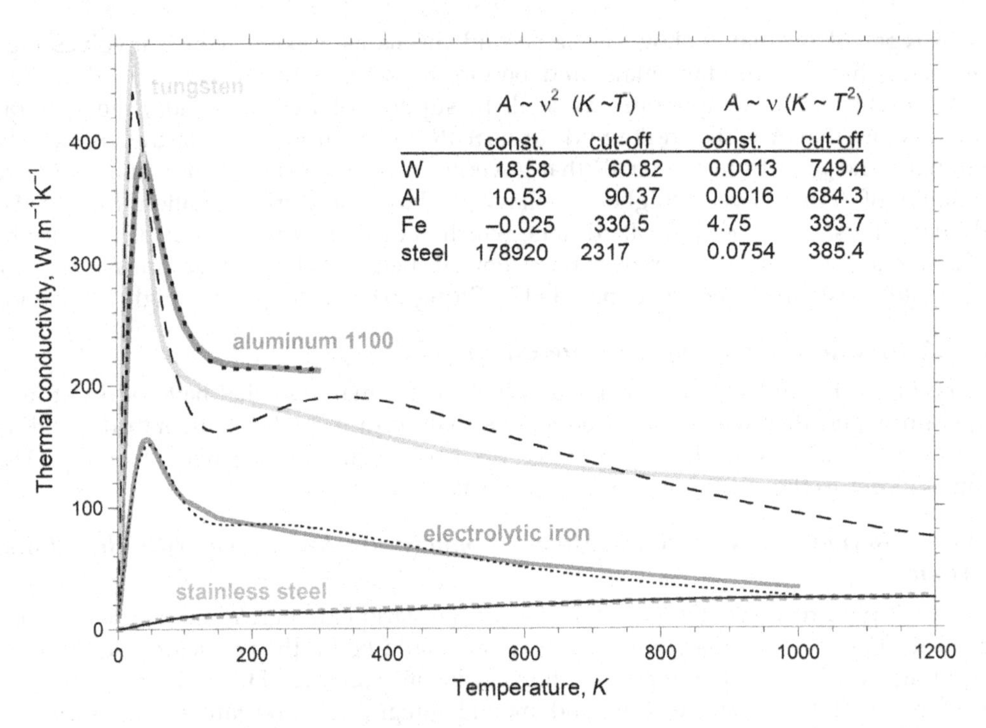

	A ~ ν² (K ~T)		A ~ ν (K ~ T²)	
	const.	cut-off	const.	cut-off
W	18.58	60.82	0.0013	749.4
Al	10.53	90.37	0.0016	684.3
Fe	−0.025	330.5	4.75	393.7
steel	178920	2317	0.0754	385.4

FIGURE 11.10 Fits to various metals using the radiative diffusion model with two mechanisms. We assumed $A \sim \nu^2$ for the fundamentals as above and $A \sim \nu$ for the overtones. Parameters are listed in the inserted table. If K had been nondimensionalized, only three parameters would be needed. Data are in gray; fits are in black. Various colors and patterns are for different metals and alloys, as labeled. Aluminum 1100 has 1% other elements. Recommended values for pure Al from Touloukian and Sarksis (1970). Thermal conductivity increasing with T for nonmagnetic steel (SRM-1461) is typical of alloys with substantial substitution, see Chapter 9. Iron has <0.1 wt % impurities. Tungsten is sintered, with a density of 19.23 g cm^{-3} and thus minor porosity exists.

Similarly, two narrow and well-resolved peaks are seen for $K(T)$ of ceramic corundum at ~100 and 150K (Fig. 11.5A). A high-T peak was not seen in K for the single-crystal, which is attributed to the peak lying above the T range of the measurements.

Graphite has one broad peak in $K(T)$ and a shallow rise in K above the temperatures shown in Fig. 11.5B. However, the thermal diffusivity data at high temperature do not extrapolate well below 250K and thermal conductivity data of one material do not extend from ~0 to high temperatures, and so fitting one dataset is not possible over a broad T range. The existing data show a linear increase in D and therefore of K with T at high T, per the mathematics of the formulae and results of fitting steel. This increase is consistent with very high frequencies for the overtones of graphite, as expected since the fundamental IR modes reach ~1600 cm^{-1}.

At high T where C_P is nearly constant, LFA data on over 100 insulators clearly show that D increases with T (Chapter 7) and thus K must also increase with T at high T for these same phases, which are silicate minerals and complex oxides. Both species have high frequency overtones. In contrast, alkali halides and simple substances like MgO have D

that decreases with T in a power law at high temperatures. For these materials, the overtones are at low frequencies, and so the integral formula for the overtones would likewise produce a minor peak in K at low temperature and a decrease in K with T at high T. Corundum spectra and thermal conductivity are both intermediate to these behaviors, which lead to a unique fit for $D(T)$ for single-crystals (Table 7.2; Hofmeister et al., 2014).

Thus, the model for radiative diffusion is compatible with the universal formula for $D(T)$ for insulators and semi-conductors. The model for two radiative processes also describes thermal conductivity of metals and alloys reasonably well. Because the fitting formulae are based on integrals, it is not necessary to consider the dependence of IR spectra on temperature. However, the simple approximations used for the spectral data limit the accuracy of the fitting. For metals in particular, a third formula with a very high cut-off temperature and another prefactor is needed to describe transport properties at very high T. This need exists for metals because they diffuse near-IR light in lab experiments on heat transport, unlike insulators for which heat transport at near-IR frequencies is generally ballistic. Black minerals with low frequency d–d transitions would require an integral representing the near-IR, as is suggested by variations in the D-data on basaltic glasses (Chapter 10).

11.3.4.5 Thermal Conductivity of Gas as a Function of Temperature

For gas, thermal conductivity depends on a power law at all temperatures, not just immediately above absolute zero. From Fig. 11.8, very high cut-off frequencies extend the positive power-law increase in K to very high T. Gases should have very high cut-offs because vibrational overtones exists into the UV (Fig. 2.7). This result is consistent with thermal conductivity of gas increasing with T overall temperatures accessed, sometimes reaching ~2000K (Fig. 5.10).

The power laws for measured K cover a narrow range of $T^{1.69}$–$T^{1.84}$ to very high T. The power laws for K with T shown in Fig. 11.8B assume that A does not depend on T. Importantly, $K(T)$ is controlled by how the absorption coefficient depends on both frequency and temperature. For gas, the absorption coefficient decreasing with frequency is offset by the bands broadening with temperature, so that the overall absorption increases with T. The combination provides trends in K for gas that follow various power laws, which are weaker than T^3 expected for A constant out to infinity.

Heat transfer in gas is thus also described by our radiative diffusion model. Motions of the molecules may contribute, but to ascertain the proportions of the two mechanisms requires modeling kinematic viscosity and mass diffusivity in addition to heat transport properties. Such future studies are needed to better understand the interiors of the gas giants.

11.3.5 Effect of Pressure on Heat Conduction

The theoretical dependence of K on pressure is far less complicated than of K on temperature, because the absorption coefficient, but not C_E, depends on P. Derivatives near room pressure are addressed here, due to problems in the high P experiments (see Chapter 4) which are benchmarked against ambient P measurements on samples with much larger size and thus much larger thermal diffusivity (Ch. 7).

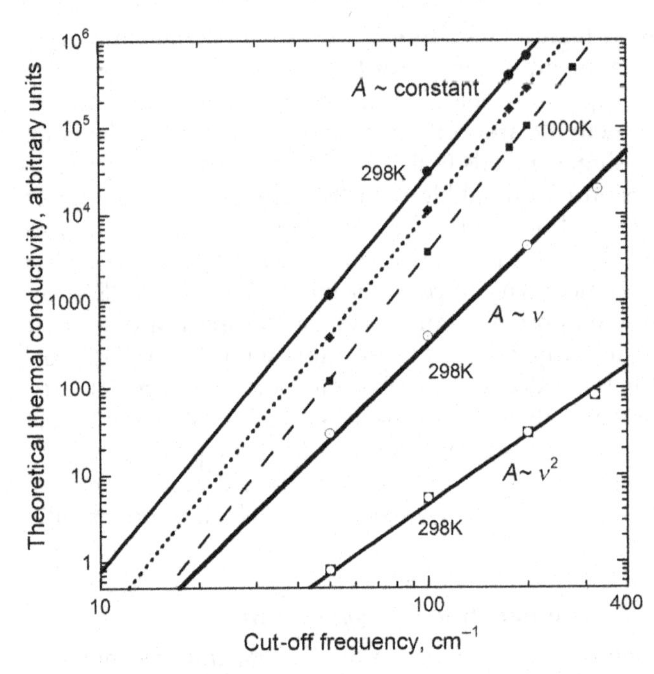

FIGURE 11.11 Isothermal values of thermal conductivity plotted as a function of the frequency cut-off in Eq. (11.25). Most of the values are shown in Fig. 11.8. Power-law fits were made. The specific power depends on the cut-off, but Figs. 11.5, 11.10, and 11.11 show that the range is narrow for the fundamentals. At room temperature, the overtones have a small effect.

Pressure generally shifts absorption bands to higher frequency. Shifting the cut-off frequency to higher ν should increase K at constant T, when temperature is moderate to high. A small magnitude for the derivative is expected because frequency shifts with P are small. To quantify the derivative, thermal conductive for various isotherms was plotted against the cut-off frequency for idealized spectra (Fig. 11.11). For cut-off frequencies ranging from 32 cm^{-1} as in copper to 293 cm^{-1} as in graphite, K for the boxcar approximation (insulators) is proportional to $\nu^{4.6}$ and K for the triangular approximation (metals) is proportional to $\nu^{2.65}$. Consequently, the pressure derivative near ambient conditions for a single mechanism is predicted to be:

$$\left.\frac{\partial \ln(K_{\text{lat,rad}})}{\partial P}\right|_T \cong \text{const.}\frac{\partial \ln(\nu_{\text{cut-off}})}{\partial P} \cong \text{const.}\frac{\gamma_{th}}{B_T} \tag{11.30}$$

where the numerical constant is 4.6 for insulators, but 2.65 for metals. The thermodynamic relationship on the RHS is based on the thermal Grüneisen parameter (γ_{th}) being proportional to the average of the mode Grüneisen parameters, which are defined as $\gamma_i = B_T^{-1}[\partial \ln (\nu_i)/\partial P]$, and also that the cut-off frequency is related to the upper frequency of the fundamental vibrations. The proportionality constants involved in these connections all cancel, due to taking a logarithmic derivative. Predictions of Eq. (11.30) agree with the most reliable, low pressure experimental determinations available (Table 11.4). Both insulators and metals (Table 9.4) are well-represented, as shown in Fig. 11.12, which supports radiative diffusion as the mechanism for heat conduction at laboratory conditions in all solids. Although previous equations are similar to Eq. (11.30), no other simple formula predicts derivatives

TABLE 11.4 Pressure Derivatives of Thermal Conductivity Near Ambient Conditions From Mostly Contact Methods for Single-Crystal and Glassy Electrical Insulators and Semi-Conducting Elements

Sample	P_{max} GPa	$\partial \ln(K_{meas})/\partial P$ % GPa^{-1}	Ref.	γ_{th}	B_T GPa	Eq. (11.30) % GPa^{-1}
MgO	1.2	5	a	1.54	160	4.4
	5	2	b			
$Mg_{1.8}Fe_{0.2}SiO_4$	4.8	4.8[a]	c	1.31	128	4.7
	5.6	~5[b]	d			
	8.3	3.2–3.8	e			
NaCl	5.6	18	b	1.58	23.8	30.5
	5	17[b]	d			
	4	31[b]	f			
	1.7	27[a]	g			
SiO_2 glass	9	− 3.7	h	0.036	36.5	0.5
	1	− 4	i			
	1.0	6	j			
Quartz ⊥c	1.2	50	j	0.667	37.5	8.2
Quartz, mixed	1.2	29	j			
$Py_{25}Al_{74}Gr_1$	8.3	4.6	e	~1.1	177	
~$NaAlSi_2O_6$	3	4.6	e	1.06	125	3.9
CaF_2	1.0	11	k	1.83	86.3	9.8
LiF	1	12	k	1.60	65	11.3
Si	~1	~0 (85 K)	l	0.46	98	2[c]
Ge	~1	4 (85 K)	l	0.80	80	5[c]

[a]*Optical technique that should not be impacted by contact resistance changing with pressure.*
[b]*Pressures are not well constrained in these early studies.*
[c]*For 298K. Pressure derivatives should be lower at cryogenic temperatures, depending on the cut-off frequency.*
Notes: Measurements involving polycrystals and cracked samples were excluded due to compression of pore space and sliding along fractures artificially enhancing the increase with pressure. Bulk moduli and the thermal Gruneisen parameter are compiled by Hofmeister and Mao (2003).
References: a, Andersson and Bäckström (1986); b, MacPherson and Schloessin (1982); c, Chai et al. (1996); d, Beck et al. (1978); e, Osako et al. (2004); f, Yukatake and Shimada (1978); g, Pangilinan et al. (2000); h, Katsura (1993); i, Andersson and Dzhavadov (1992); j, Horai and Sasaki (1989); k, Andersson and Bäckström (1987); l, Baranskii et al. (1981).

for both insulators and metals with this level of agreement (see the reviews of Ross et al., 1984 and Hofmeister et al., 2007). All formulae provide similar derivatives because compression of the lattice is described by the bulk modulus always, and because some type of Grüneisen parameter is used in all models.

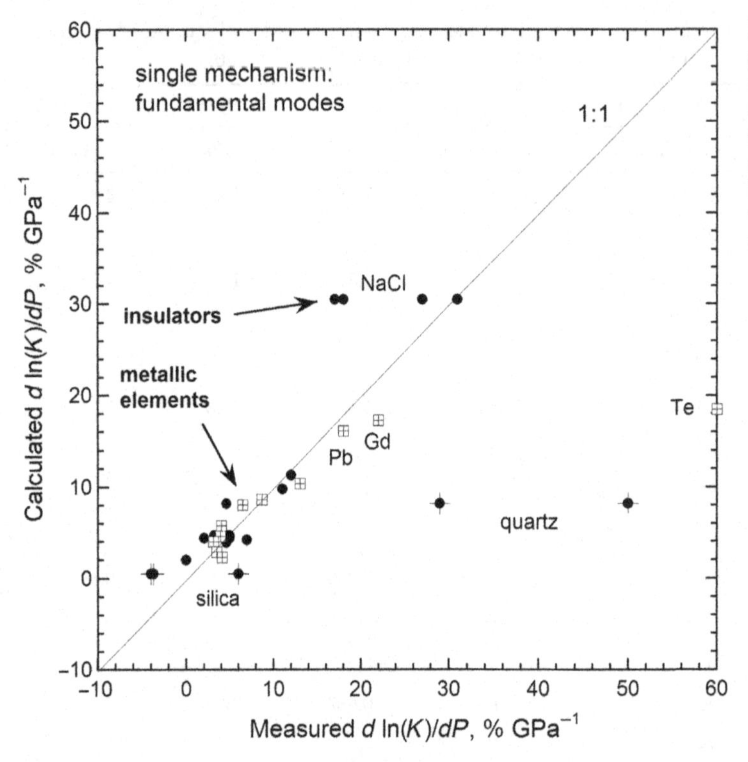

FIGURE 11.12 Comparison of pressure derivatives calculated for insulators (dots) and metals (squares) from Eq. (11.30) to measurements listed in Tables 11.4 and 9.4. Data on SiO_2 are suspect because cracking occurred whereas data on Te are approximate.

11.4 CONCLUSIONS AND FUTURE WORK

The radiative diffusion model connects absorption spectra with heat conduction and describes all types of solids with few parameters. The number of parameters depends on the number of conduction mechanisms, and their relative heat capacities. The following trends in thermal conductivity and thermal diffusivity, and the factors which common to insulators, semi-conductors, and metals, are explained by the model:

- Thermal conductivity profiles from ~10K to melting of highly pure metallic elements and end-member insulating crystals with simple structures are extremely similar in appearance.
- The exponent j describing $K \sim T^j$ below ~10K is controlled by the dependence of A on ν at very low ν, which differ for metals and insulators, in accord with their spectral characteristics.
- The peak in K at cryogenic temperatures is controlled by a cut-off frequency associated with the lowest lying fundamental vibrations, which set an upper limit to light produced via inelastic interactions of these vibrations.
- The same spectral profiles used to compute heat capacity for any given mechanism also describe thermal conductivity associated with that mechanism.
- Changes in K with T are continuous and are linked to ranges in optical spectra where absorption exists, but not to specific peaks.

- The temperature dependence of K for all matter differs from the T^3 law.
- The decrease in K and D with T up to very high temperatures is associated with the fundamental vibrations and overtones being limited to low frequencies.
- As temperature increases, overtones with higher frequencies increasingly participate, yielding a second, higher temperature peak in K, which depends on overtone positions.
- The universal law for $D(T)$ for insulators above ~250K has two terms because IR interactions differ at high and low frequencies.
- Strong involvement of overtones and a high cut-off frequency produces a linear increase in K and in D with T, which is resolved at elevated temperatures.
- Disorder has a similar effect on $K(T)$ for electrical insulators and alloys.
- Thermal diffusivity increases with sample length much like overtone spectra, both yielding a flat trend for lengths typically used in heat transport experiments.
- The height of the cryogenic peak is controlled by impurity concentration and disorder.
- Pressure derivatives of thermal conductivity are governed by the ratio of the thermal Grüneisen parameter to the bulk modulus.

The radiative diffusion model for conduction of heat at laboratory temperatures explains virtually all facets of heat conduction while lacking the problems associated with phonon scattering models, which explain rather few of the features and behaviors in heat transport properties, and moreover violate the first and second laws of thermodynamics. Few parameters are needed to fit $K(T)$ over wide ranges of temperature, making the radiative model useful for extrapolation. A predictive formula is provided for the pressure dependence of thermal conductivity, and was verified against the most reliable data available.

Further tests require complete IR spectral characterization and measurements of thermal conductivity and diffusivity data of the same material over wide temperature ranges and as a function of length. Length has been reported in few studies, and is probably a major cause of discrepancies between laboratories.

References

Aggarwal, R.L., Ripin, D.J., Ochoa, J.R., Fan, T.Y., 2005. Measurement of thermo-optic properties of $Y_3Al_5O_{12}$, $Lu_3Al_5O_{12}$, $YAlO_3$, $LiYF_4$, $LiLuF_4$, BaY_2F_8, $KGd(WO_4)_2$, and $KY(WO_4)_2$ laser crystals in the 80-300 K temperature range. J. Appl. Phys. 98, 103514.

Andersson, S., Bäckström, G., 1986. Techniques for determining thermal conductivity and heat capacity under hydrostatic pressure. Rev. Sci. Instrum. 57, 1633–1639.

Andersson, S., Bäckström, G., 1987. Thermal conductivity and heat capacity of single-crystal LiF and CaF_2 under hydrostatic pressure. J. Phys. C. Solid State Phys. 20, 5951–5962.

Andersson, S., Dzhavadov, L., 1992. Thermal conductivity and heat capacity of amorphous SiO_2: pressure and volume dependence. J. Phys. Condensed Matter. 4, 6209–6216.

Baranskii, P.I., Kogutyuk, P.P., Savyak, V.V., 1981. Thermal conductivity of n-type germanium and silicon under strong uniaxial elastic deformation. Sov. Phys. Semicond. 15, 1061–1062.

Bates, J.B., 1978. Infrared emission spectroscopy. Fourier Transform IR Spect 1, 99–142.

Beck, A.E., Darba, D.M., Schloessin, H.H., 1978. Lattice conductivities of single-crystal and polycrystalline materials at mantle pressures and temperatures. Phys. Earth Planet. Int. 17, 35–53.

Berman, R., 1951. The thermal conductivities of some dielectric solids at low temperatures. Proc. R. Soc. Lond. A 208, 90–108.

Berman, R., Foster, E.L., Schneidmesser, B., Tirmizi, S.M.A., 1960. Effects of irradiation on the thermal conductivity of synthetic sapphire. J. Appl. Phys. 31, 2156–2159.

Bradley, P.E., Radebaugh, R., 2013. Properties of selected materials at cryogenic temperatures. NIST Publication 680, 1–14. Available from: https://ws680.nist.gov/publication/get_pdf.cfm?pub_id = 913059 (accessed 14.04.18.).

Brewster, M.Q., 1992. Thermal Radiative Transfer and Properties. John Wiley & Sons, New York, NY.

Bridgman, P.W., 1941. The Nature of Thermodynamics. Harvard University Press, Cambridge, MA.

Burghartz, S., Schulz, B., 1994. Thermophysical properties of sapphire, AlN and MgAl2O4 down to 70 K. J. Nucl. Mater. 212-215, 2065–2068.

Burns, G., 1990. Solid State Physics. Academic Press, San Diego.

Cahill, D.G., Watson, S.K., Pohl, R.O., 1992. Lower limit of thermal conductivity of disordered solids. Phys. Rev. B. 46, 6131–6140.

Callaway, J., 1959. Model for lattice thermal conductivity at low temperatures. Phys. Rev. 113, 1046–1051.

Chai, M., Brown, J.M., Slutsky, L.J., 1996. Thermal diffusivity of mantle minerals. Phys. Chem. Minerals 23, 470–475.

Childs, G.E., Ericks, L.J., Powell, R.L., 1973. Thermal Conductivity of Solids at Room Temperature and Below: A Review and Compilation of the Literature. National Bureau of Standards, Washington, DC.

Clark, S.P., 1957. Radiative transfer in the earth's mantle. Trans. Amer. Geophys. Union 38, 931–938.

Criss, E.M., Hofmeister, A.M., 2017. Isolating lattice from electronic contributions in thermal transport measurements of metals and alloys and a new model. Int. J. Mod. Phys. B 31, No. 175020. Available from: https://doi.org/10.1142/S0217979217502058.

Debye, P., 1914. Vortrage über die kinetische Theorie der Materie und der Electrizität. B.G. Teuber, Berlin.

Desai, P.D., 1986. Thermodynamic properties of iron and silicon. J. Phys. Chem. Ref. Data 15, 967–983.

Farrell, T., Greig, D., 1969. The thermal conductivity of nickel and its alloys. J. Phys. C: Solid State Phys. 2, 1465–1473.

Feldman, D.W., Parker Jr., J.H., Ashkin, M., 1968. Raman scattering by optical modes of metals. Phys. Rev. Lett. 21, 607–609.

Fikhtengol'ts, G.M., 1948. Kurs Differentsialnogo I Integral'nogo Ischisleniya (Course in Differential and Integral Calculus), Vol. 2. Gostekhizdat, Moscow.

Fiquet, G., Richet, P., Montagnac, G., 1999. High-temperature thermal expansion of lime, periclase, corundum and spinel. Phys. Chem. Miner. 27, 103–111.

Gardon, R., 1956. The emissivity of transparent materials. J. Am. Ceram. Soc. 39, 278–287.

Gradshteyn, I.S., Ryzhik, I.M., 1965. Table of Integrals, Series, and Products. Academic Press, New York, NY.

Grimvall, G., Sjödin, S., 1974. Correlation of properties of materials to Debye and melting temperatures. Phys. Scr. 10, 340–352.

Hofmeister, A.M., 2006. Thermal diffusivity of garnets at high temperature. Phys. Chem. Minerals 33, 45–62.

Hofmeister, A.M., 2010a. Scale aspects of heat transport in the diamond anvil cell, in spectroscopic modeling, and in Earth's mantle. Phys. Earth Planet. Inter. 180, 138–147. Available from: https://doi.org/10.1016/j.pepi.2009.12.006.

Hofmeister, A.M., 2010b. Thermal diffusivity of perovskite-type compounds at elevated temperature. J. Appl. Phys. 107, No. 103532.

Hofmeister, A.M., 2014. Thermodynamic and optical thickness corrections to diffusive radiative transfer formulations with application to planetary interiors. Geophys. Res. Lett. 41, 3074–3080.

Hofmeister, A.M., Mao, H.K., 2002. Redefinition of the mode Gruneisen parameter for polyatomic substances and thermodynamic implications. Proc. Natl Acad. Sci. 99, 559–564.

Hofmeister, A.M., Mao, H.K., 2003. Pressure derivatives of shear and bulk moduli from the thermal Gruneisen parameter and volume-pressure data. Geochem. Cosmo. Acta 66, 1207–1227.

Hofmeister, A.M., Whittington, A.G., 2012. Thermal diffusivity of fused quartz, fused silica, and molten SiO2 at high temperature from laser flash analysis: effects of hydration and annealing. J. Non-Cryst. Solids 358, 1072–1082. Available from: https://doi.org/10.1016/j.jnoncrysol.2012.02.012.

Hofmeister, A.M., Dong, J.J., Branlund, J.M., 2014. Thermal diffusivity of electrical insulators at high temperatures: evidence for diffusion of phonon-polaritons at infrared frequencies augmenting phonon heat conduction. J. Appl. Phys. 115, No. 163517. Available from: https://doi.org/10.1063/1.4873295.

Hofmeister, A.M., Pertermann, M., Branlund, J.M., 2007. Thermal conductivity of the Earth. In: Schubert, G., Price, G.D. (Eds.), Treatise in Geophysics, V. 2 Mineral Physics. Elsevier, Amsterdam, The Netherlands, pp. 543–578.
Horai, K., Sasaki, J., 1989. The effect of pressure on the thermal conductivity of silicate rocks up to 12 kbar. Phys. Earth Planet. Int 55, 292–305.
Hottel, H.C., Sarofim, A.F., 1967. Radiative Transfer. McGraw-Hill Book Company, St. Louis, MI.
Hust, J.G., Lankford, A.B., 1984. Update of thermal conductivity and electrical resistivity of electrolytic iron, tungsten, and stainless steel. Natl Bureau Stand. Spec. Pub 260-290, 1–71.
Jasperse, J.R., Kahan, A., Plendl, J.N., Mitra, S.S., 1966. Temperature dependence of infrared dispersion in ionic crystals LiF and MgO. Phys. Rev. 146, 526–542.
Katsura, T., 1993. Thermal diffusivity of silica glass at pressures up to 9 GPa. Phys. Chem. Minerals 20, 201–208.
Kellett, B.S., 1952. Transmission of radiation through glass in tank furnaces. J. Soc. Glass Tech. 36, 115–123.
Keppler, H., Dubrovinsky, L.S., Narygina, O., Kantor, I., 2008. Optical absorption and radiative thermal conductivity of silicate perovskite to 125 gigapascals. Science 322, 1529–1532.
Kieffer, S.W., 1979. Thermodynamics and lattice vibrations of minerals: 3. Lattice dynamics and an approximation for minerals with application to simple substances and framework silicates. Rev. Geophys. Space Phys 17, 20–34.
Klemens, P.G., 1958. Thermal conductivity and lattice vibrational modes. Solid State Phys 7, 1–98.
Lake Shore Cryogenics, 2018. Appendix I: Cryogenic reference tables—Lake Shore Cryotronics, Inc. <https://www.lakeshore.com/Documents/LSTC_appendixI_l.pdf> (accessed 15.04.18.).
Lee, D.W., Kingery, W.D., 1960. Radiation energy transfer and thermal conductivity of ceramic oxides. J. Am. Ceram. Soc. 43, 594–607.
Liu, Y.C., Sommer, F., Mittemeijer, E.J., 2004. Calibration of the differential dilatometric measurement signal upon heating and cooling; thermal expansion of pure iron. Thermochim. Acta 413, 215–225.
Lubimova, H., 1958. Thermal history of the earth with consideration of the variable thermal conductivity of the mantle. Geophys. J. R. Ast. Soc. 1, 115–134.
MacPherson, W.R., Schloessin, H.H., 1982. Lattice and radiative thermal conductivity variations through high P, T polymorphic structure transitions and melting points. Phys. Earth Planet. Inter. 29, 58–68.
McMahon, H.O., 1950. Thermal radiation from partially transparent reflecting bodies. J. Opt. Soc. Am. 40, 376–380.
Osako, M., Ito, E., Yoneda, A., 2004. Simultaneous measurements of thermal conductivity and thermal diffusivity for garnet and olivine under high pressure. Phys. Earth Planet. Inter. 143σ144, 311–320.
Palik, E., 1998. Handbook of Optical Constants of Solids. Academic Press, San Diego, CA.
Pangilinan, G.I., Ladouceur, H.D., Russell, T.P., 2000. All-optical technique for measuring thermal properties of materials at static high pressure. Rev. Sci. Instru 71, 3846–3852.
Peierls, R.E., 1929. Zur kinetische theorie der warmeleitung in kristallen. Ann. Phys. Leipzig 3, 1055–1101.
Pérez-Castañeda, T., Azpeitia, J., Hanko, J., Fente, A., Suderow, H., Ramos, M.A., 2013. Low-temperature specific heat of graphite and CeSb2: Validation of a quasi-adiabatic continuous method. J. Low Temp. Phys. 173, 4–20.
Pertermann, M., Whittington, A.G., Hofmeister, A.M., Spera, F.J., Zayak, J., 2008. Thermal diffusivity of low-sanidine single-crystals, glasses and melts at high temperatures. Contrib. Mineral. Petrol. 155, 689–702. Available from: https://doi.org/10.1007/s00410-007-0265-x.
Planck, M., 1914. In: Masius, M.P. (Ed.), The Theory of Heat Radiation. P. Blakiston's Son and Co., Philadelphia, PA.
Reif, F., 1965. Fundamentals of Statistical and Thermal Physics. McGraw-Hill, New York, NY.
Ross, R.G., Andersson, P., Sundqvist, B., Bäckström, G., 1984. Thermal conductivity of solids and liquids under pressure. Rep. Prog. Phys. 47, 1347–1402.
Rosseland, S., 1936. Theoretical Astrophysics: Atomic Theory and the Analysis of Stellar Atmospheres and Envelopes. Clarendon, Oxford.
Rybicki, G.B., Lightman, A.P., 2004. Radiative Processes in Astrophysics. Wiley-VCH, Weinheim.
Schnelle, W., Fischer, R., Gmelin, E., 2001. Specific heat capacity and thermal conductivity of NdGaO3 and LaAlO3 single crystals at low temperatures. J. Phys. D 34, 846–851.
Shankland, T.J., Nitsan, U., Duba, A.G., 1979. Optical absorption and radiative heat transport in olivine at high temperature. J. Geophys. Res. 84, 1603–1610.
Siegel, R., Howell, J.R., 1972. Thermal Radiation Heat Transfer. McGraw-Hill, New York, United States.

Slack, G., 1962. Thermal conductivity of MgO, Al2O3, MgAl2O4, and Fe3O4 crystals from 3° to 300° K. Phys. Rev. 126, 427–441.

Tang, X., Dong, J.J., 2009. Pressure dependence of harmonic and anharmonic lattice dynamics in MgO: a first-principles calculation and implications for lattice thermal conductivity. Phys. E. Plan. Int 174, 33–38.

Tang, X., Dong, J.J., 2010. Lattice thermal conductivity of MgO at conditions of Earth's interior. Proc. Natl Acad. Sci. 107, 4539–4543. Available from: https://doi.org/10.1073/pnas.0907194107.

Touloukian, Y.S., Sarksis, Y., 1970. Thermal Conductivity: Metallic Elements and Alloys. IFI/Plenum, New York, NY.

Transtrum, M.K., Machta, B.B., Brown, K.S., Daniels, B.C., Myers, C.R., Sethna, J.P., 2015. Perspective: sloppiness and emergent theories in physics, biology, and beyond. J. Chem. Phys. 143, 010901. Available from: https://doi.org/10.1063/1.4923066.

Ventura, G., Perfetti, M., 2014. Thermal Properties of Solids at Room and Cryogenic Temperatures. Springer, Heidelburg, Germany.

Woodcraft, A.L., Gray, A., 2009. A low temperature thermal conductivity database. AIP Confer. Proc. 1185, 681–684.

Wooten, F., 1972. Optical Properties of Solids. Academic Press, Inc, San Diego, CA.

Yu, X., Hofmeister, A.M., 2011. Thermal diffusivity of alkali and silver halides. J. Appl. Phys. 109, 033516. Available from: https://doi.org/10.1063/1.3544444.

Yukatake, H., Shimada, M., 1978. Thermal conductivity of NaCl, MgO, coesite and stishovite up to 40 kbar. Phys. Earth Planet. Int 17, 193–200.

Ziman, J.M., 1962. Electrons and Phonons: The Theory of Transport Phenomena in Solids. Clarendon Press, Oxford.

Websites

Online integrator (accessed April 9, 2014)
www.wolframalpha.com/calculators/integral-calculator/
Poco graphite properties (accessed Feb. 2, 2018)
http://poco.com/Portals/0/Literature/Semiconductor/IND-109441-0115.pdf
http://poco.com/MaterialsandServices/Graphite.aspx
Compiled historic cryogenic data on K (accessed April 14, 2018)
https://ws680.nist.gov/publication/get_pdf.cfm?pub_id = 913059
https://www.nist.gov/mml/acmd/nist-cryogenic-materials-property-database-index
Properties of elements (accessed April 20, 2018). This site includes many references.
http://www.knowledgedoor.com/

CHAPTER

12

Conclusions and Future Work

OUTLINE

Accurate and minute measurement seems to the non-scientific imagination, a less lofty and dignified work than looking for something new. But nearly all the grandest discoveries of science have been but the rewards of accurate measurement and patient long-continued labour in the minute sifting of numerical results. (Lord Kelvin, Presidential inaugural address, to the British Association for the Advancement of Science, 1871)

You tell me whar a man gits his corn pone, en I'll tell you what his 'pinions is.' (quote from a former slave by Mark Twain, 1901)

This book sifts through copious data on heat transfer for the three states of matter, and takes a critical look at ideas in classical physics which pertain to heat, especially those that underlie geophysical or planetary studies, which are the subject of our forthcoming companion volume. Many fundamental errors have been revealed by our recent studies in heat transfer, solid mechanics, and gravitation of oblate bodies. This book uncovers several additional long-standing misconceptions, and tries to improve current understanding of heat, its motion, and interaction with matter. Five historical errors (Fig. 12.1, top row), which were never amended to address more recent results (second row), have had a domino effect on studies of heat transfer in large bodies, which are more complex than lab studies (bottom three rows). Several of our new findings involve fundamental physical principles.

The skeptical reader may think "How is this possible?" The short answer is that many historical ideas were based on insufficient information and developed out-of-synch, so they inevitably embody some misconceptions. Unfortunately, given their long-standing

Measurements, Mechanisms, and Models of Heat
DOI: https://doi.org/10.1016/B978-0-12-809981-0.00017-6

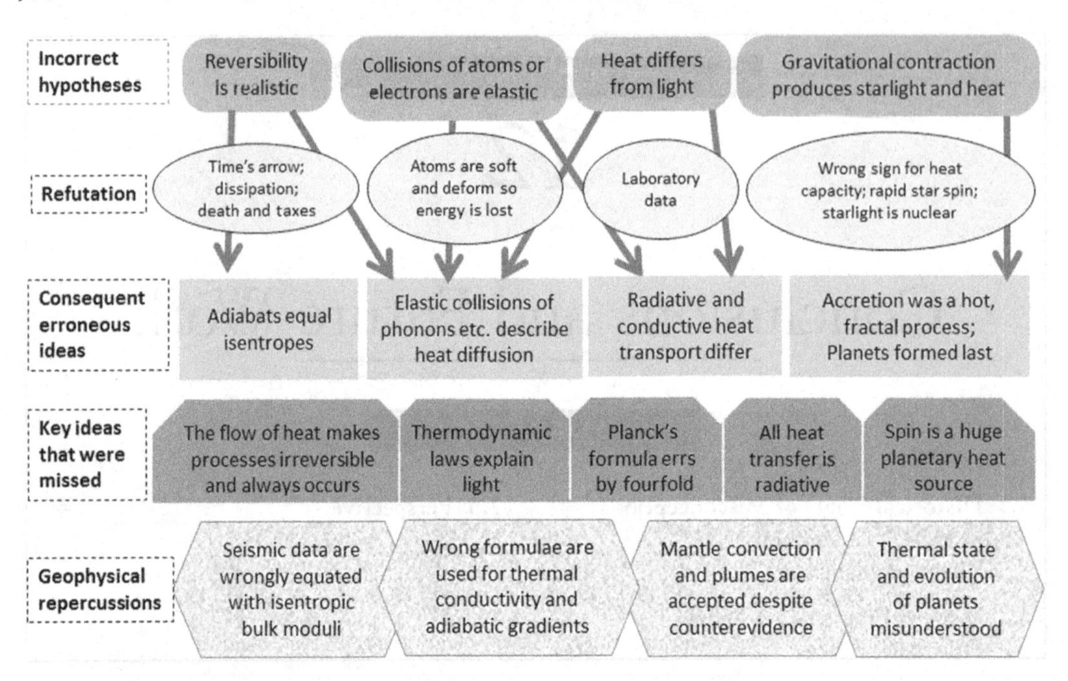

FIGURE 12.1 Schematic illustrating how certain antiquated ideas concerning heat and heat transfer impede modern geophysics, even though counter evidence is compelling. The left column summarizes each row. From top to bottom: Rounded rectangles: historical hypotheses originating in the 1800s, where the gray arrows connect these to slightly more current ideas. Ovals: counter evidence appearing from 1850 to the late 1900s. Rectangles: key wrong idea from the 1900s stemming from the earlier misconceptions. Barn-shaped boxes: ideas developed recently and in this book by recognizing that key information was missing in the 1800s and 1900s when the fundamental hypotheses were developed. Hexagons: problematic approaches and ideas in geophysics stemming from the incorrect hypotheses of the 1800s. Many cross-connections exist of the bottom row with those above, but are not shown. The right column and mantle convection is addressed in our recent papers (Hofmeister and Criss, 2012, 2015, 2016, 2018) and are a focus of our companion volume.

nature, some of these misconceptions have seriously impacted subsequent work in heat transfer, thermodynamics, and their applications. The discussion here emphasizes backlash in geophysics and planetary science. Section 12.1 covers the historical development of these problematical ideas, with hope of learning from the past. Section 12.2 discusses the driving forces that maintain clearly incorrect notions as scientific truth, and points to unrecognized self-defeating approaches in modern science. Section 12.3 makes some suggests for improving how science is done.

12.1 HISTORICAL ROOTS OF MISCONCEPTIONS IN HEAT TRANSFER AND THERMODYNAMICS

Four incorrect historical hypotheses are the basis for inferring the thermal state and thermal evolution of planets (Fig. 12.1, top row). Some of the links of these antiquated notions to modern planetary studies are obvious, as shown, but several connections are buried. The long accepted misconceptions, along with their cross-links and hidden assumptions, have made it very difficult to fix ensuing problems. That self-gravitation affects the thermodynamics of large bodies has complicated matters greatly, especially since classical thermodynamic analyses neglect the effect of gravity. However, the microscopic picture of heat transfer can be corrected through experimental studies, which are covered in considerable detail in many different chapters. Here, we attempt to summarize and synthesize.

Key ideas of what heat and light are and how they move and interact with matter were developed prior to any knowledge of atomic structure, and before heat and light were accepted as being the same phenomenon (Chapter 1). This error stems from ignoring the experimental work of Melloni, despite the support of Ampere. Another factor is the change of the focus in physics to quantum phenomena, under the mistaken belief that classical physics was complete and correct.

Although the equivalence of light and heat is now accepted, the motion of heat (infrared wavelengths) in matter is treated differently than the motion of light (visible wavelengths) in matter. The criterion for diffusive (as opposed to ballistic) behavior is not the wavelength, but optical thickness. Although spectroscopic research is based on the proper division (Chapter 2), the necessary concepts were not utilized in heat transfer studies until the late 1990s and only in laser-flash analysis (Chapter 4). Optical thickness does not enter into heat transfer theory (Chapter 3). Another obvious problem is the focus on steady-state experiments, which do not reveal how re-equilibration occurs, which is largely transient. Consequently, that thermal diffusivity depends on sample thickness, and that light, not electrons, moves heat in metals have heretofore gone unnoticed (Chapter 7 and Chapter 9). Equivalence of heat and light, which was bitterly contested by the scientific community for 50 years, also has repercussions on our understanding of light itself. Errors include a numerical factor in Planck's formula for blackbody emissions and several mistakes in formulations for radiative transfer (Chapter 8 and Chapter 11). Lastly, equivalence of conduction with diffusion of light, primarily infrared, was completely missed in previous work (Chapter 11) due to the compartmentalization of spectroscopic and heat transfer studies, and to overemphasis of Debye's acoustic model for heat capacity, which lead to acceptance of his acoustic model for thermal conductivity.

A central problem in probing the behavior of light and heat is that experiments always involve interactions with matter. However, because the nature of atoms was not well-understood circa 1900, some experimental results were misinterpreted, and observed behavior was attributed to light when matter was mostly the control, for example, the photoelectric effect (Chapter 8). Specifically, not realizing that high energy light is physically small and dense prevented recognizing that its interactions with electrons are largely governed by the exchange of momentum. Conversely, not knowing that infrared light is large and dilute prevented recognizing that its interactions with condensed matter are macroscopic and diffusive, even as atomic scales are approached (Chapter 11). Energy has been a primary focus of physical science for quite some time, although early studies in physics emphasized forces and the conservation of momentum. The directional nature of the latter quantities makes them more difficult to understand than energy, which is a scalar. Although heat is also a scalar, its flow is not, resulting in mathematical errors since heat flows only in one direction, down the thermal gradient. Additional errors result from overuse on analogies to electrical currents, for which conservation laws are simpler. Regarding planets, the most important type of intrinsic momentum is spin, which is a type of angular momentum that is much more difficult to envision than linear momentum. Thus, the approaches commonly taken in science have contributed to misunderstandings the gross behavior of planets, that is, how axial spin originated and how it is being dissipated (Hofmeister and Criss, 2012, 2016, 2017), and to misunderstanding the nature of conduction, which is simply diffusion of thermal radiation.

The concept of reversibility, although known to be absurd from common experience, is a pillar of thermodynamics, and continues to influence research. For example, the hypothetically reversible Carnot cycle is the subject of a recent issue in the journal *Entropy*. Three non-sequiturs important to geophysics stem from taking the idealization of reversibility seriously: (1) The isentropic bulk modulus (B_S) is identical to the adiabatic bulk modulus; (2) seismically determined velocities are adiabatic; and (3) the adiabatic gradient can be computed from thermodynamic properties with a formula valid only for gas. None of these is correct, see Chapter 1. The mistake of treating solids by analogy to gases traces back to the mid-1800s focus on gas experiments and the requirement that such experiments be nearly isothermal to avoid convective motions. One severe consequence of accepting this misguided analogy, plus the one involving electrical currents, is mantle convection models, which are based on equations for bottom heated, low viscosity liquids which will flow under any applied stress, unlike rocks (Hofmeister and Criss, 2018). The companion volume focusses on convection, utilizing findings in the present book.

Reversibility supports the peculiar concept that atomic and molecular collisions are elastic, yet can transfer heat. This combination is preposterous, as reversible systems cannot evolve or even cool. For gas, adjusting a few parameters is sufficient to make the elastic kinetic theory of gas agree reasonably well with measurements. This agreement stems from mass and heat being transferred inseparably in gas, so that in this case the two diffusivities are nearly equal. Due to the similarity of both diffusivities to the kinematic viscosity, existence of two types of collisions went unnoticed, even though Newton had solved the problem for glancing collisions (drag), see Chapter 5. Drag overtakes diffusion in controlling viscosity as the molecular weight of liquids increases (Chapter 6), whereas molten silicates have viscosities that originate almost entirely in drag (Chapter 10). For solids, mass and heat transfer are entirely separate, and so there is no reason that the kinetic theory of gas in any form should describe solids. Using wrong formulae for thermal diffusivity or conductivity of solids has contributed to misunderstanding the thermal state of planetary interiors.

Collisions of atoms are not elastic, as revealed by subsequent studies of atomic behavior. Atoms deform, which converts kinetic energy of the collisions into heat (light). Because this deduction was missed historically, the root of thermodynamic law in the behavior of light went unrecognized, and some of the laws remain incomplete. This missed deduction has made entropy difficult to understand and lead to incorrect descriptions of ideal gas expansion (Chapter 1) and of contraction of nebulae during accretion (as discussed by Hofmeister and Criss, 2012). The former error supported reversibility as being a reasonable approximation. The tie of inelastic collisions to a gas of light was missed, and the very basic idea that heat is always evolved and is invariably associated with irreversibility was missed. These ideas underlie thermal evolution of large bodies, which have been completely misunderstood, due to the above omissions and to a serious error by an important scientist. Details are provided below, as an introduction to the companion volume, and to illustrate how slowly wrong ideas fade.

To explain production of starlight, Kelvin (and to a lesser extent, Helmholtz) proposed that the quantity "gravitational energy" existed, and is −1 times the gravitational potential, rather than using a potential difference of appropriate sign, which is permissible. Why Newton's discovery was ignored shall remain a mystery, although gravitation (action at a distance) is difficult to understand. The conversion of a positive energy source to heat is thermodynamically robust, although this process does not produce starlight; nuclear reactions do. Rather than relegate Kelvin's mistake to history, the correct sign for gravitational potential was used in the 1970s to argue that nebulae are "special" by having a *negative* heat capacity. Dilute hydrogen gas is about the most ideal gas in existence. Plus, a nebula with negative heat capacity would cool upon receipt of light, so illumination could possibly cause nebula temperatures to fall below absolute zero! Yet this rationalization was accepted, and Kelvin's wrong hypothesis morphed into gravitational contraction providing internally retained heat, which then allegedly leads to star ignition and is a key feature of accretion. Instead, potential energy is transformed into kinetic energy, and then heat is generated, as is presented in many examples in elementary physics books. But because high rates of axial spin for large young astronomical bodies is a recent discovery, and not part of the historic literature, the consequences of high spin state and the gradual dissipation of this large amount of energy were only recently introduced into astronomical and planetary literature, see Fig. 12.1.

12.2 MODERN PRACTICES AND THEIR DRAWBACKS

Currently, considerable research in Earth and planetary science involves computer models rather than on gathering and interpreting data. The problem with computational approaches to extremely complex natural phenomena is that the solutions cannot be validated or verified, which is quite different than in engineering, where heat flow in multi-component devices is modeled during the design period, but once the optimal design suggested by the modeling is settled on, the prototype is extensively tested. Measurements provide the final say regarding industrial production.

In addition, numerical models are complex and cannot easily be analyzed, so the source of errors is not easily traced. Numerical models rest on assumptions, just as analytical models do, but problems and assumptions are much more evident in the latter. The compactness and simplicity of analytical results also renders them more amenable to testing and making definitive predictions, which involve experiments. Here, we point to Kelvin's quote at the beginning of the chapter. When measurements are used to evaluate existing ideas, this leads to new ideas, but mustering data with the purpose of supporting an existing idea is a philosophically sterile practice.

As is well-known, most great scientific advances have been made by individuals or very small research groups, and not by agencies or bureaucracies. The Voyager missions are an exception, being the first of their kind. Yet, the current trend is towards "mission science" involving huge research groups and collaborative efforts. Although many hands make light work, the work being done is agreed upon in advance. Once consensus is reached, fundamental discovery stops, whether or not any problem has been solved. It is extremely harmful to science to promote group mentality at the expense of individual efforts. This, however, has been the trend for 30 years.

The trend towards large groups is partially driven by funding. More funds means more people supported, and more clout for the head of the group. Large groups appear to be important, even if their work only provides small modifications of some well-worn theme. Partially, this occurs because we evaluate the influence of papers through the H index, which is rather a marker of their popularity, and one skewed to the advantage of papers with large numbers of authors.

The destructive importance of status has long been a blight on the sciences. Kelvin's early discoveries were monumental, but he became an obstructionist in his later years. Innovation will always play a secondary role when maintenance of the status quo is essential for securing support, funding, and equipment. This explains why the factor of 4 in Planck's blackbody curve has not been discussed. To say something which differs from majority opinion is to risk disapproval, which impacts a scientist's livelihood. For a graduate student, dissent is even more risky, as the approval of the advisor is essential to obtain a doctoral degree. Mark Twain's quote at the chapter head summarizes these problems.

12.3 PERSPECTIVE

Never have we had more access to data and publications, more equipment, more computational power, and more manpower, yet the physical sciences seem to be running in circles. We believe in absurdities far beyond the several in geophysics. Examples are non-baryonic dark matter, dark energy, and black holes. Billions of dollars have been squandered on the search for dark matter over the past 30 years, with no results, yet the physics community still follows this fruitless path. Similarly, the geophysics community continues to pursue mantle convection as it searches for plumes via poorly constrained modeling, even as it molds data to support popular beliefs. The Vesta mission exemplifies the latter.

Many of the problems rest in how we teach science. Facts and assumptions are muddled in efforts to present a coherent picture, when much remains unknown. The current generation grew up deluged by TV commercials, so lies and distortions are the new norm, not the exception. The quest for scientific truth now pales in comparison with the scramble for funding, so the number of fundamental discoveries has faded. Hyperbole and "awards" can cover up the fact that running in circles is common place in the physical sciences, but this will not last forever.

To fix the problems requires some draconian and some trivial measures.

- The H index needs to be revised to dilute "credit" for papers according to the number of authors. Effecting the latter change would be trivial, but its benefit would be far-reaching.

- Funding should not be directed to the same groups, over and over.
- Funding should not include overhead, or nominal amounts should be provided.
- Funding should not cover salary for the overseer of a project, because a component of professorial duty is directing the efforts of graduate students.
- The review system needs overhauled. Anonymous reviews are the rule because providing a signed review carries risk, but this system is prone to subjectivity and unfairness. This risk is tied to funding, as is the explosion of modern literature and second-class journals. As a consequence, it is not scholarly articles that are being read, but distillations and "cameos".

Fixing the problems with the H index and with funding, particularly the overhead, would permit researchers to focus on producing fewer publications of higher quality, and would reduce the motivation to impede competitors via destructive reviews.

Now, the mundane: what data are needed for future advances in heat transfer? Spectroscopic measurements are most essential. The second largest need is for flux measurements under transient conditions. These will confirm or refute the ideas presented in this book, and will generate new ideas.

True scientific advances are not predictable. History has repeatedly shown that paradigms fall and their supporters are forgotten. Paradigms exist for two reasons: one is that academia is hierarchical and modeled after medieval churches, and the other is that a basic tenet of science is infrequently followed, which is to entertain multiple hypotheses, while weighing ideas against evidence.

References

Hofmeister, A.M., Criss, R.E., 2012. A thermodynamic model for formation of the solar system via 3-dimesional collapse of the Dusty Nebula. Planet. Space Sci. 62, 111–131.

Hofmeister, A.M., Criss, R.E., 2015. Evaluation of the heat, entropy, and rotational changes produced by gravitational segregation during core formation. J. Earth Sci. 26, 124–133.

Hofmeister, A.M., Criss, R.E., 2016. Spatial and symmetry constraints as the basis of the virial theorem and astrophysical implications. Can. J. Phys. 94, 380–388.

Hofmeister, A.M., Criss, R.E., 2017. Implications of geometry and the theorem of gauss on newtonian gravitational systems and a caveat regarding poisson's equation. Galaxies 5 89–100. Available from: http://www.mdpi.com/2075-4434/5/4/89.

Hofmeister, A.M., Criss, E.M., 2018. How properties that distinguish solids from fluids and constraints of spherical geometry suppress lower mantle convection. J. Earth Sci. 29, 1–20. Available from: https://doi.org/10.1007/s12583-017-0819-4.

APPENDIX

A

Conventions, Abbreviations, and Variables Used

CONVENTIONS USED

d	differential or total derivative
∂	partial derivative
δ	very small change; delta function
Δ	large change
e	mathematical function; the number 2.718281828...
π	geometrical constant; the number 3.141592654...
$\zeta(3)$	Riemann Zeta function of 3 (Apéry's constant), 1.202056903159594...
f, f^{+}	some unspecified function, when not a subscript
i	root of -1, when not a subscript
f, i	when used as subscripts, refer to the initial and final states
j, k, n	used in summations, refers to an integer
0	initial or reference value (commonly subscripted)
∞	limit of infinity

ABBREVIATIONS

d	dimensions (not italicized)
EOS	equation of state
LHS	left hand side (of an equation)
RHS	right hand side (of an equation)
STP	standard temperature (0°C) and pressure (1 atm)
NTP	normal temperature (20°C) and pressure (1 atm)
BB	blackbody
EM	electromagnetic
IR	infrared
UV	ultraviolet
NMR	nuclear magnetic resonance
vis	visible

refl reflected
trns transmitted
abs absorbed
emit emitted
rfrt refracted
ln natural logarithm
log common logarithm
exp exponential function
erf error function
TO transverse optic
LO longitudinal optic
TA transverse acoustic
LA longitiudinal acoustic
fwhm full width at half maximum
mfp mean free path
red reduced mass
LFA laser-flash analysis
3ω a transient technique
Gr dimensionless Grashof number
Nu dimensionless Nusselt number
Pr dimensionless Prandtl number
Ra dimensionless Rayleigh number
Le dimensionless Lewis number
Sc dimensionless Schmidt number
V.T. Virial Theorem of Clausius
SHO simple harmonic oscillator
DHO damped harmonic oscillator
RE rotational energy
KE kinetic energy
PE potential energy

FITTING CONSTANTS

A, B, C, D Fits in Chapter 10
F, G, H Fits in Chapter 7 and Chapter 11
b Fitting constant in many equations
R_s Correlation coefficient in fitting

PHYSICAL CONSTANTS AND THEIR VALUES

c speed of light = 2.99792458×10^8 m s^{-1}
h Planck's constant = $6.62609923 \times 10^{-34}$ J s
k_B Boltzmann's constant = 1.380658×10^{-23} J atom^{-1} K^{-1}
N_a Avogadro's number = 6.0221×10^{23} atoms mol^{-1}
R_{gc} gas constant = $N_a k_B = 8.314$ J mol^{-1} K^{-1}
σ_{SB} Stephan–Boltzmann constant = 5.670×10^{-8} W m^{-2} K^{-4}
G Gravitational constant = 6.67×10^{-11} m^3 kg^{-1} s^{-2}

USEFUL CONVERSION FACTORS

1 atm = 101325 J m^{-3}
$hc/k_B = 1.43877$ (for use with wavenumbers)

VARIABLES, WHERE *DENOTES INTENSIVE

a absorptivity (fitting parameter in Chapter 5)
A absorption coefficient
∀ absorbance
Å area
b an unknown constant; fitting parameter; attenuation
B bulk modulus
C heat capacity (per mole and per volume are apparent from units in the equations)
c_m specific heat* (heat capacity per mass)
d grain size or scattering length when used as a variable
D thermal diffusivity, sometimes D_{heat} for clarity
D_∞ thermal diffusivity, long length limit
D_m mass diffusion coefficient, sometime subscripted with mass
$\Im$ flux in energy (Joules per time per area)
$\Im_m$ flux in mass (mass per times per area)
¢ effusivity
Ǝ Electric field intensity, a vector
E_{emit} Energy of a specific type, as indicated by the subscript
E Internal energy
E_a Activation energy
F Helmholtz free energy; also fitting parameter for $D_{heat}(T)$
F_{drag} Force of some type, as indicated by the subscript
G Gibbs function (free enthalpy); also fitting parameter for $D_{heat}(T)$
G_∞ Shear modulus at infinite frequency (Chapter 10)
h hardness
H enthalpy; also fitting parameter for $D_{heat}(T)$
Ħ Magnetic field intensity
I intensity
I.I. integrated intensity or peak area
j number of atoms in a molecule
J heat current or electric current
k absorption index or force constant when subscripted
K thermal conductivity (used by Fourier)
l number of rotational modes, indentation distance
L length or pathlength
M, m mass (sometimes per mole)
n (real) index of refraction*
$\tilde{n}$ complex index of refraction
N number of atoms, molecules, or moles
O oscillator strength
p_x momentum; subscripts indicate the direction
P pressure
q heat, surface oscillation
Q applied external heat
R reflectivity, or radius when subscripted (Chapter 10)
Ʀ electrical resistivity
$\Re$ radiant energy emitted per area and per unit time
r radius in cylindrical or spherical coordinates, as indicated
S entropy
t time
T temperature*
T_g glass transition temperature

u speed or velocity; dummy variable in integrals
v voltage
V volume
w work or dummy frequency in integrals
x, y, z Cartesian, orthogonal directions
X, Y variables or functions, as needed (Chapter 11)
Z number of formula units in the unit cell
ℵ prefactor in radiative diffusion equation of solids (Chapter 11)
ι, ς fitting constants or parameters
э dielectric function with real and imaginary parts
$\in$ permittivity
μ permeability (or micro = 10^{-6}, when combined with a dimension)
κ consistency (Chapter 10) or wavevector (Chapter 11)
ε emissivity or strain rate (Chapter 10)
ρ density*
α thermal expansivity
β compressibility
ξ concentration (in mass diffusion) or torque (Chapter 10)
η dynamic viscosity (sometimes called absolute viscosity)
υ kinematic viscosity
χ fraction or ratio
χ^2 chi squared, indicates goodness of fits
σ electrical conductivity (Chapter 9)
σ_{xy} stress, with directions
γ_{th} Grüneisen parameter
γ_i mode Grüneisen parameter
γ ratio of C_P/C_V; constant for ideal gas
Υ transmittivity
τ time constant or lifetime or relaxation time (note subscripts)
Λ mean free path or attenuation (Chapter 4)
λ wavelength
ν frequency
ω circular frequency
ϖ pore fraction
θ, ϕ angle variables
φ shape function
Θ phase angle in complex analysis
ζ thermalization time
Π power
Φ phase lag (Chapter 4) or thermal transmissivity through a surface
Γ irreducible representation of fundamental vibrations

APPENDIX

B

Guide to Electronic Deposit of Thermal Diffusivity Data

Data acquired at Washington University are available at http://epsc.wustl.edu/~hofmeist/thermal_data/. This site meets NSF requirements to provide public access to data obtained under federal grant support. Most of the data have been published, with details. Some of the data were summarized in published papers, but details were not provided. A few measurements were presented in the book. Some data are unpublished and papers are in preparation. For the last three categories, essential details are provided.

The website ilists temperature calibrated LFA measurements (i.e., thermal diffusivity vs temperature), notes on the measurement such as sample thickness, examples of temperature–time curves, microprobe analyses, spectra (mostly IR and visible, i.e., of hydrous impurities and d–d transition elements), and rarely X-ray diffraction data. The data can be read by the Microsoft Excell (CVS format) and many other plotting programs. Sample graphs are included if permissions were not needed. Regarding collaborative studies, only the data acquired at Washington University are archived.

The data are organized as follows: One folder is associated with each paper or study in preparation, which concerned either one crystallographic structure, a mineral family, rocks of a certain provenance, metals, or glasses of a certain composition crystallographic structure. The folder contains a text file with a link to the original publication, if one exists, and a list of the samples studied, with notes. Subfolders contain the types of data listed above. The website is being updated as more data are collected and analyzed.

B

Guide to Electronic Depository of Thermal Diffusivity Data

Data acquired at Washington University are available at http://[illegible] [illegible]. This site meets NSF requirements to [illegible] data obtained under [illegible] grant support. Most of the [illegible] have been published [illegible] details. Some of the data were [illegible] in part, and [illegible] details were [illegible] vided. A few measurements were [illegible] in the book, [illegible] papers are in preparation, [illegible] are provided.

The website [illegible] calibrated [illegible] temperature, notes on the measurement [illegible] examples of [illegible] time curves, [illegible] impurities and [illegible] on [illegible] can be read by [illegible] Excel [illegible] plots [illegible] graphs are [illegible] data [illegible].

The data [illegible] preparation, [illegible] in a certain [illegible] [illegible] [illegible] [illegible].

APPENDIX

C

Summary of the Literature on Heat Capacity and Density (or Thermal Expansivity) as a Function of Temperature

LFA experiments measure D directly. The related property of thermal conductivity, which is $K(T) = \rho C_P D$ can be calculated from $D(T)$ when density $\rho(T)$ and isobaric heat capacity $C_P(T)$, are known. At high temperatures, K of insulators depends linearly on temperatures, partially due to C_P generally increasing with T at high T. However, we do not find any single model that universally fit all $K(T)$ data of the crystals studied over a wide temperature range (Chapter 10 provides examples). This is in part due to the complicated dependence of C_P on T and to opposing trends among ρ, C_P, and D as T increases. For these reasons, this appendix provides a list of articles and websites with the needed data.

There is no shortage of heat capacity data. The items below are a small proportion of what exists, although many of the papers are limited to certain chemical compositions. Heat capacities of minerals can be reasonably computed from sums of oxides. The same holds for glasses (Chapter 10).

NIST has several databases on heat capacity, where the focus is on fluids:
https://srdata.nist.gov/gateway/gateway?property = heat + capacity
NASA has a user friendly database:
https://www.grc.nasa.gov/WWW/CEAWeb/ceaThermoBuild.htm
Many books exist, for example:
Chase, M. W. (1998). NIST - JANAF Thermochemical Tables (Fourth ed.). Journal of Physical and Chemical Reference Data.

Cox JD, Wagman DD, Medvedev VA, editors. 1989. *CODATA Key Values for Thermodynamics*. New York: Hemisphere Publishing Corporation. Available at http://www.worldcat.org/oclc/18559968.
Robie RA, Hemingway BS. 1995. Thermodynamic properties of minerals and related substances at 298.15K and 1 bar (10^5 Pascals) pressure and at higher temperatures. U. S. Geological Survey. (Bulletin 2131). Available at https://pubs.er.usgs.gov/publication/b2131.

For a comparison of databases, see:
Blasco, M., Gimeno, M.J, Auque, L.F., 2017. Comparison of different thermodynamic databases used in a geothermometrical modelling calculation. *Procedia Earth and Planetary Science 17*, 120–123. https://doi.org/10.1016/j.proeps.2016.12.023

For minerals, the following papers are often used:
Berman RG. 1988. Internally-consistent thermodynamic data for minerals in the system $Na_2O-K_2O-CaO-MgO-FeO-Fe_2O_3-Al_2O_3-SiO_2-TiO_2-H_2O-CO_2$. *Journal of Petrology 29*(2): 445–522.
Holland TJB, Powell R. 2011. An improved and extended internally consistent thermodynamic dataset for phases of petrological interest, involving a new equation of state for solids. *Journal of Metamorphic Geology 29*(3): 333–383. https://doi:10.1111/j.1525-1314.2010.00923.x

Low temperature data on minerals are less common, but see:
Gamsjäger, E., Wiessner, M., 2018. Low temperature heat capacities and thermodynamic functions described by Debye–Einstein integrals. *Monatsh Chem. 149*, 357–368. doi:10.1007/s00706-017-2117-3

Typically, densities are measured at ambient conditions and X-ray data are used to determine values at higher temperature. This approach is not accurate due to the small size of the thermal expansivities of minerals, $\sim 10^{-5}\,K^{-1}$. Much higher accuracy is obtained by dilatometry or interferometry on macroscopic samples, but these data are few for geologic substances. Therefore, density is often linearized.

NIST has a database on thermal expansivity of ceramics and technologically important materials:
https://srdata.nist.gov/gateway/gateway?keyword = thermal + expansion

For silicate glasses, see Chapter 10.

For commercial glasses, see the website of Alexander Fluegel (2005, 2007)
http://glassproperties.com/expansion/ExpansionMeasurement.htm
http://glassproperties.com/

Books and reviews are relatively few:
Ho, C.Y., Taylor, R.E, 1998. Thermal expansion of solids. ASM International.
Fei Y 1995 Thermal expansion. In: Ahrens T J (ed.) *Mineral Physics and Crystallography. A Handbook of Physical Constants*, American Geophysical Union, Washington, DC, pp. 29–44

Index

Note: Page numbers followed by "*f*," "*t*," and "*b*" refer to figures, tables, and boxes, respectively.

C

D

T